FIELDS INSTITUTE COMMUNICATIONS

THE FIELDS INSTITUTE FOR RESEARCH IN MATHEMATICAL SCIENCES

Universality and Renormalization

From Stochastic Evolution to Renormalization of Quantum Fields

Ilia Binder
Dirk Kreimer
Editors

American Mathematical Society
Providence, Rhode Island

The Fields Institute for Research in Mathematical Sciences
Toronto, Ontario

FIELDS

The Fields Institute for Research in Mathematical Sciences

The Fields Institute is a center for mathematical research, located in Toronto, Canada. Our mission is to provide a supportive and stimulating environment for mathematics research, innovation and education. The Institute is supported by the Ontario Ministry of Training, Colleges and Universities, the Natural Sciences and Engineering Research Council of Canada, and seven Ontario universities (Carleton, McMaster, Ottawa, Toronto, Waterloo, Western Ontario, and York). In addition there are several affiliated universities and corporate sponsors in both Canada and the United States.

2000 *Mathematics Subject Classification.* Primary 81T15, 81T16, 76F30, 60K35, 30C99, 60J65, 82B41, 82B43.

Library of Congress Cataloging-in-Publication Data

Universality and renormalization / Ilia Binder, Dirk Kreimer, editors.
p. cm. — (Fields Institute Communications, ISSN 1069-5265 ; v. 50)
Includes bibliographical references.
ISBN 978-0-8218-4273-7 (alk. paper)
1. Quantum field theory—Congresses. 2. Mathematical physics—Congresses. I. Binder, Ilia, 1970– II. Kreimer, Dirk, 1960–

QC174.45.A1U55 2007
530.14′3—dc22 2006048035

∞ The paper used in this book is acid-free and falls within the guidelines
established to ensure permanence and durability.
This publication was prepared by the Fields Institute.
`http://www.fields.utoronto.ca`
Visit the AMS home page at `http://www.ams.org/`

10 9 8 7 6 5 4 3 2 1 12 11 10 09 08 07

Contents

Preface

The Fields Institute organized a thematic program entitled "Renormalization and Universality in Mathematics and Mathematical Physics" in the fall of 2005. Part of the program were two workshops, "Percolation, SLE and Related Topics" (September 20-24 2005) and "Renormalization and Universality in Mathematical Physics" (October 18-22 2005). This volume contains the proceedings of the two workshops.

These workshops covered a wide range of topics in the context of universality and renormalization. Self-similarity and the study of scaling behaviour and its anomalous breaking away from a fixpoint of the renormalization group are phenomena which permeate the natural sciences everywhere, as those workshops have convincingly demonstrated, and so, we hope, does this volume.

The sixteen contributions in this volume cover mathematical aspects as well as detailed physical applications. They touch areas of mathematical specialization as diverse as number theory, infinite dimensional Lie and Hopf algebras, noncommutative geometry, algebraic geometry, differential equations, nonlinear dynamics, complex analysis, probability theory.

The physics discussed ranged from the microscopic ranges of high energy physics and quantum fields to the macroscopic areas of matter in its various phases, to the large scales of gravity. Both the history of the subject and current developments were well represented, and we were encouraged to see the underlying concepts of renormalization and universality influencing contemporary scientific thinking as profoundly as ever.

We enjoyed excellent support from the staff at the Fields Institute, and are grateful for the hospitality extended to us. A special thanks goes to Debbie Iscoe, whose help was mandatory in assembling this volume.

October 2006
Ilia Binder and Dirk Kreimer

Fields Institute Communications
Volume **50**, 2007

Manifestly Gauge Invariant Exact Renormalization Groups

Stefano Arnone
Dipartimento di Fisica
Università degli Studi di Roma "La Sapienza"
Roma, Italy
stefano.arnone@roma1.infn.it

Tim R. Morris
School of Physics and Astronomy
University of Southampton
Southampton, UK
T.R.Morris@soton.ac.uk

Oliver J. Rosten
School of Physics and Astronomy
University of Southampton
Southampton, UK
ojr@phys.soton.ac.uk

Abstract. We construct a manifestly gauge invariant Exact Renormalization Group for $SU(N)$ Yang-Mills theory, in a form suitable for calculations without gauge fixing at any order of perturbation theory. The effective cutoff is incorporated via a manifestly realised spontaneously broken $SU(N|N)$ gauge invariance. Diagrammatic methods are developed which allow the calculations to proceed without specifying the precise form of the cutoff structure. We confirm consistency by computing for the first time both the one and two loop beta function coefficients without fixing the gauge or specifying the details of the cutoff. We sketch how to incorporate quarks and thus compute in QCD. Finally we analyse the renormalization group behaviour as the renormalized coupling becomes large, and show that confinement is a consequence if and only if the coupling diverges in the limit that all modes are integrated out. We also investigate an expansion in the inverse square renormalized coupling, and show that under general assumptions it yields a new non-perturbative approximation scheme corresponding to expanding in $1/\Lambda_{QCD}$.

2000 *Mathematics Subject Classification.* Primary 81T13, 81T17; Secondary 81T15, 81T16, 81T18.

1 Adapting Exact RG to gauge invariant systems

1.1 Motivation: obvious advantages. We want to take the Exact RG (Renormalization Group),[1] which encapsulates Wilson and Fisher's ideas [2, 3] about the non-perturbative RG, adapted to the continuum, and apply these ideas to gauge invariant systems. We use the continuum version because we are interested in the quantum field theory description of particle physics. The reason we want to bring gauge invariance and the Exact RG together is because each of these ideas is important and powerful in itself.

Although the Exact RG was conceived in the 1970s, it has only been understood more recently that it is much more than just a formal device. It is a powerful framework for doing computations in quantum field theory . In brief, the reasons for this are as follows. (For reviews, see [3–6]. We will cover some aspects in more detail later.) Firstly, from the way it is constructed, RG invariance is built in from the beginning. This means it is particularly easy to define the main object of interest in quantum field theory, namely the continuum limit: the Exact RG is an equation that describes how the Wilsonian effective action changes as the Wilsonian effective cutoff is changed. To get a continuum limit one simply searches for a self-similar solution (the "functional self-similarity" of Shirkov [7]), *i.e.* an effective action whose only dependence on the cutoff (after writing all quantities as dimensionless ratios using the cutoff) is through a finite set of couplings.[2] Secondly, all the information one could want can be extracted directly from the Wilsonian effective action. This is because the action is effectively the generator of connected Green functions, which in turn gives the S-matrix elements and thus anything that can be asked of the quantum field theory. Finally there is a great deal of freedom in the choice of approximations that can be applied. Of course the equations can be solved in perturbation theory; however there is also a limitless variety of non-perturbative approximations: the exact RG equations can simply be truncated, or more motivated model approximations can be made – adapted to the physics being studied. All these approximations still preserve the property of RG invariance and existence of continuum limits. These are the reasons why this framework has been a popular avenue of exploration in the last decade or so [4].

It would be hard to overemphasise the importance of gauge invariance to particle physics. It underlies the Standard Model and attempts to go beyond the Standard Model, and in fact much more besides than particle physics. Therefore if the ERG framework is to be useful in particle physics and more generally, we do need to understand how best to combine it with gauge invariance.

1.2 The problem. Given that gauge invariance and the non-perturbative RG are so important, this problem would have been solved long ago if it were not for the fact that their combination presents special difficulties. Indeed at first sight the two concepts are incompatible. In the Wilsonian RG the first step is a Kadanoff blocking [8] and in the continuum this means integrating out momentum modes down to some momentum cutoff. But a momentum cutoff breaks gauge invariance.

There are only two ways to proceed. Either the gauge invariance is broken by the cutoff or the Exact RG is generalised so as to incorporate a gauge invariant notion of a cutoff. In the first alternative it is now well understood how to recover

---

[1]The Exact RG was discovered and christened simultaneously by Wegner and Wilson [1, 2]

[2]The term "coupling" stands for masses also.

the gauge invariance at least in principle: the breaking can be quantified by a set of broken Ward identities, which if satisfied exactly at some point on the flow[3] remain satisfied elsewhere on the flow, and furthermore turn into the unbroken Ward identities once all momentum modes are integrated out.

1.3 Hidden advantages: manifest gauge invariance. We will explain how to implement the second alternative. In this case gauge invariance is exactly preserved at all points along the flow, allowing us to exploit its elegance and power. However, as we will see, we will need to introduce some extra regularisation structure.

As well as the obvious benefits of this approach that we have been describing, it turns out there are a couple of surprise benefits. Because we have to generalise what we mean by an Exact RG, we are forced to realise that there are in fact infinitely many different forms of Exact RG [10]. We can use this inherent freedom to help solve the system, in particular in this case to make the flow equation itself manifestly gauge invariant. More suprising however, when we come to solve the flow equation we find that we do not have to fix the gauge. As a result there are no BRST ghost fields [11], there are no Gribov problems [12],[4] there is no wavefunction renormalization for the gauge field, and the expressions are simple and tightly constrained by the exact preservation of gauge invariance. In particular, expressions can be built purely from covariant derivatives. (This should be compared to the much more involved procedure of using BRST invariance, Lee-Zinn-Justin identities and so forth [13, 14].)

1.4 Yang-Mills without gauge fixing. We will concentrate on $SU(N)$ Yang-Mills. It is defined through the covariant derivative $D_\mu = \partial_\mu - iA_\mu$, where $A_\mu \equiv A^a_\mu \tau^a$ is the connection, or gauge field, contracted into the generators of the $SU(N)$ Lie algebra.[5] The field strength $F_{\mu\nu} = i[D_\mu, D_\nu]$ is just a commutator of covariant derivatives and the effective action, being gauge invariant, is just an expansion built out of covariant derivatives:

$$S[A](g) = \frac{1}{4g^2}\,\mathrm{tr}\int F^2_{\mu\nu} + \text{higher dimension operators} + \text{vacuum energy}. \qquad (1.1)$$

This is an example of a continuum limit solution in this framework. The $F^2_{\mu\nu}$ term is already dimension four so we know there is only one coupling g; we require a solution $S[A]$ which is a function of the scale only through this coupling. In the quantum field theory we need to define what we mean by g once we go beyond the classical level. The expansion above acts as this required renormalization condition: our coupling is defined to be the coefficient of the $F^2_{\mu\nu}$ term as in (1.1).

As we have already emphasised we preserve exactly the local invariance

$$\delta A_\mu = [D_\mu, \omega(x)] \qquad (1.2)$$

which implies that there is no wavefunction renormalization, so it really is the case that only g runs. (The higher dimensional operators in (1.1) are all irrelevant and are fixed, computable, functions of g. If wavefunction renormalization were needed we would have to write $A_\mu = Z^{1/2} A^R_\mu$ where A^R_μ is the renormalized field, but in

---

[3]This is the hard part. It can be achieved in perturbation theory with some work [9].

[4]This is an infamous non-perturbative problem arising from an incomplete gauge fixing which has as yet no practical solution.

[5]For simplicity of exposition, we normalise the generators to $\mathrm{tr}\,\tau^a\tau^b = \delta^{ab}$, differing from standard practice and our papers [6, 15–26].

terms of this the invariance becomes $\delta A^R_\mu = Z^{-1/2}\partial_\mu\omega - i[A^R_\mu,\omega]$. In other words, it is preserved only if $Z=1$ and $A_\mu = A^R_\mu$. This simple argument fails in the usual framework only because ω gets replaced by a ghost field, leading to a divergent product of quantum fields, which requires further renormalization.)

1.5 Hidden advantages: diagrammatic approach. A second surprise benefit that is basically forced on us is as follows. As we have already intimated, in order to make everything gauge invariant, we will have to add quite a bit of extra structure. We can make a particular choice for this extra regularisation but it does not seem helpful to do so: when computing Feynman diagrams this corresponds to a particular choice of vertices and propagators and there does not appear to be any choice that makes the integrals easy to do.

It turns out that it is better to specify just a set of schemes satisfying certain general properties (including of course gauge invariance). This then leads to vertices and propagators that are not completely defined but instead satisfy certain properties. The physics in quantum field theory is encoded in universal quantities whose values are independent of a particular scheme. Therefore it ought to be possible to extract these values by manipulating the diagrams, even though its elements are not completely specified.

It turns out that this is indeed the case. With all these elements unknown, there is so little freedom in manipulating these expressions that the procedure is essentially algorithmic. This computational method also furnishes an automatic check of universality since the final result is either unique, *i.e.* independent of the details of the scheme, or it was not universal after all. Inspired by the fact that this procedure is essentially algorithmic, Dr. Rosten went on to find directly the solution to the algorithm [20, 25], so although in this report we will demonstrate how to manipulate these diagrams it is now understood how to jump directly to the answer.

2 Generalised Exact RGs

2.1 Kadanoff blockings. One way to see that there are infinitely many different Exact RGs is to start from the fact that there are infinitely many different Kadanoff blockings. Apart from the fact that we write this in the continuum the standard definition for a Kadanoff blocking is

$$\mathrm{e}^{-S[\varphi]} = \int \mathcal{D}\varphi_0\, \delta\Big[\varphi - b[\varphi_0]\Big]\, \mathrm{e}^{-S_{bare}[\varphi_0]}. \tag{2.1}$$

The integration is performed over the bare (or microscopic) field φ_0 (we take for simplicity a single component real scalar field) weighted by the Boltzmann factor containing S_{bare}, the microscopic action. The effective field, φ, is related to the microscopic field through the blocking functional b_x. A simple linear blocking is

$$b_x[\varphi_0] = \int_y K(x-y)\,\varphi_0(y)$$

for some kernel $K(z)$ which is steeply decaying once $z\Lambda > 1$, Λ being the momentum cutoff. It can also be non-linear, for which there is a huge freedom of choice.

Eqn. (2.1) is the standard definition because, integrating over the effective field, we immediately see that the effective partition function is equal to the microscopic one:

$$\mathcal{Z} = \int \!\mathcal{D}\varphi\, \mathrm{e}^{-S[\varphi]} = \int \mathcal{D}\varphi_0\, \mathrm{e}^{-S_{bare}[\varphi_0]},$$

therefore no information has been lost, it has only been recast. We get an Exact RG by differentiating with respect to Λ:

$$\Lambda\frac{\partial}{\partial\Lambda}\,\mathrm{e}^{-S[\varphi]} = -\int_x \frac{\delta}{\delta\varphi(x)}\int \mathcal{D}\varphi_0\;\delta\Big[\varphi - b[\varphi_0]\Big]\,\Lambda\frac{\partial b_x}{\partial\Lambda}\;\mathrm{e}^{-S_{bare}[\varphi_0]}.$$

We can interpret the integral as, up to normalization, the rate of change, Ψ_x, of the blocking functional in the measure provided by (2.1):

$$\Lambda\frac{\partial}{\partial\Lambda}\,\mathrm{e}^{-S[\varphi]} = \int_x \frac{\delta}{\delta\varphi(x)}\left(\Psi_x\,\mathrm{e}^{-S[\varphi]}\right). \tag{2.2}$$

We see that we have a flow equation for each choice of blocking, or equally we have a flow equation for each choice of Ψ. There are infinitely many such flow equations and by construction they leave the partition function, $\mathcal{Z}$, invariant. This is equally clear from (2.2), since $\partial_\Lambda\mathcal{Z}$ is the integral over φ of a total derivative with respect to φ, which vanishes for sensible S.

2.2 Polchinski's exact RG for a single massless scalar field. Every exact RG can be written in the form (2.2). One of the most popular is Polchinski's [27] for which the rate of change is

$$\Psi_x = \frac{1}{2}\int_y \dot{\Delta}_{xy}\,\frac{\delta\Sigma_1}{\delta\varphi(y)}. \tag{2.3}$$

We have introduced what we will call the exact RG kernel, $\dot{\Delta} \equiv -\Lambda\partial_\Lambda\Delta$. Here and later the over-dot is the differential with respect to 'renormalization group time'. In this case we are differentiating the 'effective propagator'

$$\Delta = \frac{c(p^2/\Lambda^2)}{p^2}. \tag{2.4}$$

Here at last we make explicit the effective cutoff in the theory: the cutoff function c is a smooth function such that $c(0) = 1$ and c vanishes for large momenta sufficiently rapidly that all Feynman diagrams are regularised when Δ is used in place of $1/p^2$. $\Sigma_1 = S - 2\hat{S}$ contains the 'seed action' $\hat{S}$, something that one is free to choose and which for the Polchinski flow equation just corresponds to what we expect for the regularised kinetic term $\hat{S} = \frac{1}{2}\partial_\mu\varphi\cdot c^{-1}\cdot\partial_\mu\varphi$, yielding the inverse of (2.4). We will later use this as our starting point for generalisation. We write, here and later, the momentum kernel contracted into functions in a number of equivalent ways:

$$\begin{aligned} f\!\cdot\! W\!\cdot\! g &= \int_{x,y} f(x)\,W_{xy}\,g(y) = \int_x f(x)W(-\partial^2/\Lambda^2)\,g(x)\\ W_{xy} &= W(-\partial^2/\Lambda^2)\,\delta(x-y) = \int_p W(p^2/\Lambda^2)\,\mathrm{e}^{ip\cdot(x-y)}. \end{aligned}$$

Substituting (2.3) into the general form (2.2), it easy to see that we get the following equation for the flow of the effective action with respect to RG time:

$$\dot{S} = \frac{1}{2}\frac{\delta S}{\delta\varphi}\!\cdot\!\dot{\Delta}\!\cdot\!\frac{\delta\Sigma_1}{\delta\varphi} - \frac{1}{2}\frac{\delta}{\delta\varphi}\!\cdot\!\dot{\Delta}\!\cdot\!\frac{\delta\Sigma_1}{\delta\varphi}. \tag{2.5}$$

The two terms are straightforward to interpret diagrammatically, as shown in fig. 1. On the LHS (left hand side) of the diagram we have a differentiated vertex with

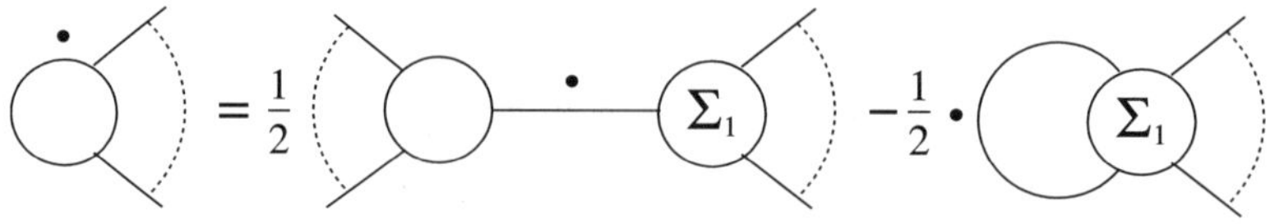

Figure 1 Diagrammatic representation of the flow equation for a single scalar field.

some number of legs. As already intimated, this is constructed from connected (one-particle-reducible) Feynman diagrams, where instead of the usual propagator, an effective propagator is used.[6] Thus the differential with respect to RG time either hits a propagator which divides the diagram in two, giving the first, tree-level type, diagram on the RHS (Right Hand Side), or it hits a propagator carrying loop momenta, in which case the result is the second, one-loop-like, diagram.

Now consider only S_0, the classical part of the effective action. Its flow is governed by just the classical part:

$$\dot{S}_0 = \frac{1}{2}\frac{\delta S_0}{\delta\varphi}\cdot\dot{\Delta}\cdot\frac{\delta\Sigma_1}{\delta\varphi}.$$

If we concentrate just on the two-point vertex, diagrammatically the flow is of the

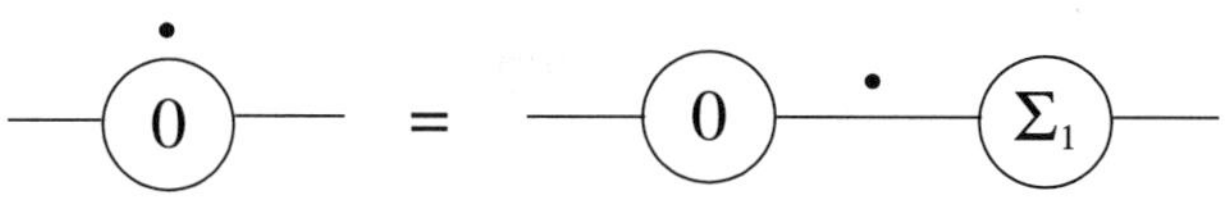

Figure 2 The Polchinski flow equation for the classical two-point vertex

form of fig. 2. It turns out that for the Polchinski equation one can set the classical two-point vertex equal to the seed action two-point vertex. Then the flow equation has a straightforward solution:

$$S_0^{\varphi\varphi}(p) = \hat{S}^{\varphi\varphi}(p) \implies \dot{S}_0^{\varphi\varphi} = -S_0^{\varphi\varphi}\dot{\Delta}S_0^{\varphi\varphi} \implies \Delta = \left(S_0^{\varphi\varphi}\right)^{-1}, \tag{2.6}$$

so Δ is just the inverse of the classical effective action two-point vertex, justifying our naming it an effective propagator.

Later we will turn this analysis around, and insist that the classical and seed action two-point vertices can be identified and then the equation above determines what we use for the exact RG kernel, or equivalently the effective propagator. This step is not really necessary but it does lead to technical simplifications.

2.3 Generalised exact RG for a single scalar field. It is now straightforward to generalise this. For example we can replace the seed action by $\hat{S} = \frac{1}{2}\partial_\mu\varphi\cdot c^{-1}\cdot\partial_\mu\varphi + \cdots$, where the ellipses stands for some higher point vertices, or even an infinite number of higher point vertices. The only constraint, apart from the usual generic requirements, is that these vertices' ultraviolet behaviour should not be so violent as to destroy the regularisation properties of c. In fact since the two-point vertex is the most general two-point vertex consistent with there being only one scale, Λ, and Poincaré invariance, in effect we replace the seed action by

[6] In fact the effective propagator here is the infrared regulated propagator $(1-c)/p^2$, with differential $-\dot{\Delta}$, corresponding to the modes already integrated out [28].

any effective action of our choice, as long as it corresponds to a regularised massless scalar field.

What changes? Hardly anything! The equation (2.5) and its diagrammatic interpretation, fig. 1, are still the same. The only difference is that one must remember that $\hat{S}$ vertices can now have more than two legs. It might seem surprising but the universal answers extracted from this equation are exactly the same as in standard quantum field theory. We have checked this explicitly. For example, for a single scalar field in four dimensions the first two beta function coefficients come out exactly the same, independent of the details of c and the higher point vertices [29]. The underlying reason for this is that these extra terms amount to some reparametrisation of the field as the cutoff is lowered [10], and of course such a change of field variables does not change the physics.

3 Manifestly gauge invariant Exact RG for Yang-Mills

Replacing the scalar field φ by a gauge field A_μ, we can just require that the seed action is any choice of effective action: we do not even have to specify that the field is massless, since gauge invariance will ensure that. However making these simple replacements clearly does not turn (2.5) into a gauge invariant equation, because the gauge invariance is broken by the kernel. We can solve this problem by replacing $\dot{\Delta}(-\partial^2/\Lambda^2)$ by $\dot{\Delta}(-D^2/\Lambda^2)$. This is only one of infinitely many ways to covariantize a kernel. We will let braces stand for some fixed choice, which we will never have to specify precisely. Now the flow equation is gauge invariant:

$$\dot{S} = \frac{1}{2}\frac{\delta S}{\delta A_\mu}\{\dot{\Delta}\}\frac{\delta \Sigma_g}{\delta A_\mu} - \frac{1}{2}\frac{\delta}{\delta A_\mu}\{\dot{\Delta}\}\frac{\delta \Sigma_g}{\delta A_\mu}. \tag{3.1}$$

It corresponds to the gauge covariant rate of change:

$$\Psi = \frac{1}{2}\{\dot{\Delta}\}\frac{\delta \Sigma_g}{\delta A_\mu}.$$

At the diagrammatic level, the only difference is that we have to remember that there are now fields hidden in the kernels, so the kernels themselves have vertices, as illustrated in fig. 3.[7] One last detail: we scale the coupling constant out of the action, as in (1.1), and some thought shows that it ends up in front of the action so that in place of Σ_1 we have $\Sigma_g = g^2 S - 2\hat{S}$.

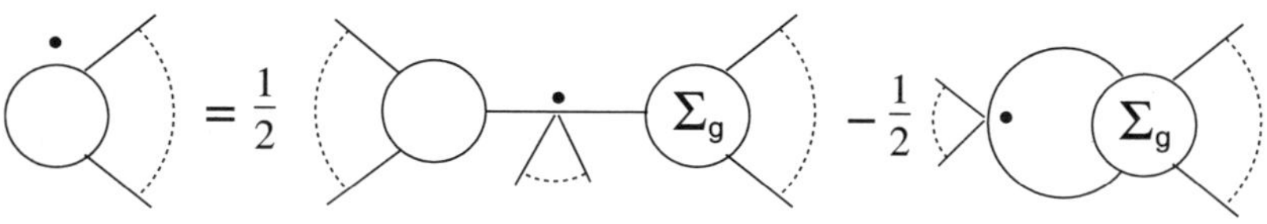

Figure 3 The flow equation in gauge theory.

[7] Generically there are also diagrams where $\delta/\delta\mathcal{A}_\mu$ acts on $\{\dot{\Delta}\}$, which can be excluded by a careful choice of covariantization [6, 19].

Having scaled g outside the action, we expect it to count powers of $\hbar$, and it is easy to check that (3.1) has this form for its weak coupling expansion:[8]

$$\begin{aligned} S &= \frac{1}{g^2} S_0 + S_1 + g^2 S_2 + \cdots \\ \beta &:= \Lambda \partial_\Lambda g = \beta_1 g^3 + \beta_2 g^5 + \cdots , \end{aligned}$$

where S_n and β_n are the n-loop pieces of the effective action and β function respectively. Of course these components have yet to be computed.

3.1 Consequences at the classical level. Classical effective action vertices follow from the $g \to 0$ limit of (3.1), which is easily seen to be

$$\dot{S}_0 = \frac{1}{2} \frac{\delta S_0}{\delta A_\mu} \{\dot{\Delta}\} \frac{\delta \Sigma_0}{\delta A_\mu}, \qquad \text{where} \qquad \Sigma_0 = S_0 - 2\hat{S}.$$

Diagrammatically, this takes the form in fig. 4.

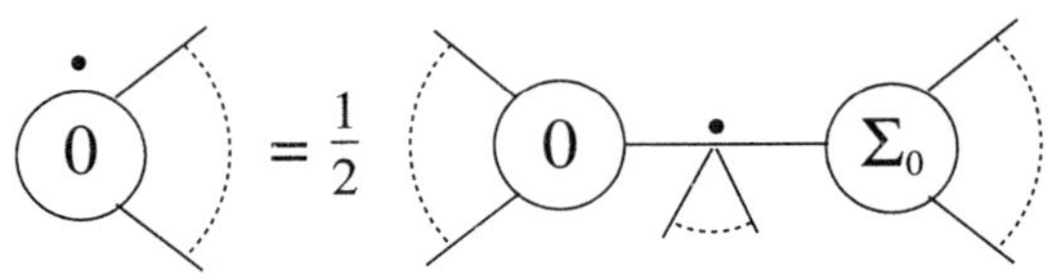

Figure 4 The classical flow equation for the classical part of the effective action.

Recall that we determine the kernel by insisting on the equality of the classical and seed action two-point vertices. The classical vertex has the following general form

$$S_0{}^{AA}_{\mu\nu} = \Box_{\mu\nu}(p)/c(p^2/\Lambda^2), \qquad \text{where} \qquad \Box_{\mu\nu}(p) = \delta_{\mu\nu} p^2 - p_\mu p_\nu .$$

This form is fixed uniquely by masslessness, gauge invariance and Poincaré invariance. The arbitrary function c will again play the rôle of a cutoff function. $\Box_{\mu\nu}(p)$ is the standard transverse kernel in the kinetic term of a gauge field. (From here on we suppress its dependence on p.) Since it has a longitudinal zero-mode: $\Box_{\mu\nu} p_\nu = 0$, it is not invertible and this is the standard obstruction in gauge theory that motivates gauge fixing.

Following our procedure for determining the Exact RG kernel we see, using $\Box_{\mu\nu}\Box_{\nu\lambda} = p^2 \Box_{\mu\lambda}$, that we get precisely the same solution for $\dot{\Delta}$ and thus also Δ, as before:[9]

$$S_0{}^{AA}_{\mu\nu} = \hat{S}^{AA}_{\mu\nu} \implies \dot{S}_0{}^{AA}_{\mu\nu} = -S_0{}^{AA}_{\mu\alpha} \dot{\Delta} S_0{}^{AA}_{\alpha\nu} \implies \Delta = c/p^2 .$$

This satisfies

$$\Delta S_0{}^{AA}_{\mu\nu} = \delta_{\mu\nu} - p_\mu p_\nu / p^2 . \tag{3.2}$$

Although we will keep calling Δ an effective propagator, it is not really an effective propagator any more because the two-point vertex has no inverse. Instead of unity on the RHS above, we get the transverse projector, or equivalently $\delta_{\mu\nu}$ minus the purely longitudinal term $p_\mu p_\nu / p^2$. This latter term when multiplying other elements in a diagram can be simplified by using the Ward identities that follow from invariance under small gauge transformations.

[8] In practice it is convenient to let $\hat{S}$ also have an $\hbar$ expansion [20, 23].

[9] We do not yet see the vertices from covariantizing the kernel because that would require one-point vertices for the gauge field which are not allowed, by Lorentz invariance for example.

Specialising fig. 4 to the three-point vertex we get the diagrams in fig. 5. We now

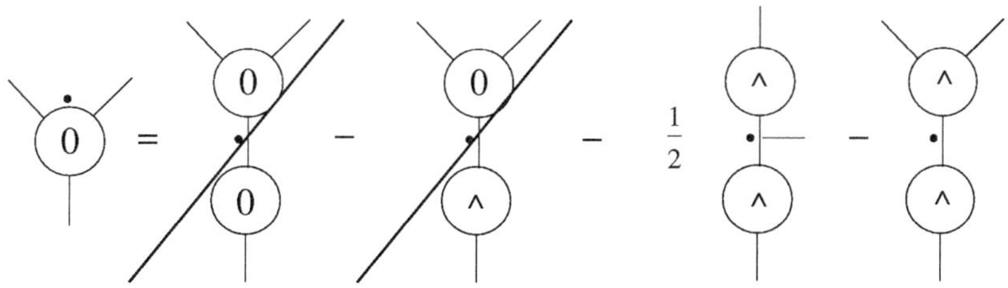

Figure 5 The flow equation for the classical three-point vertex. The seed-action vertices have a caret in the centre.

see the effect of covariantizing the kernel, in the third term on the RHS. The first two terms have the classical three-point vertex on the top, but they cancel because of the equality of classical and seed action two-point vertices. The remaining terms depend only on the covariantized kernel and seed action vertices, *i.e.* on terms that we put in by hand. Thus everything is known on the RHS and it easy to integrate this with respect to RG time to get the full classical three-point vertex. Notice that we have computed this without having to try to invert the (non-invertible) kinetic term.

This cancellation works for all the other vertices at the classical level: the flow is only into terms that have already been determined. For example, the flow of the four-point vertex is shown in fig. 6. On the RHS there are non-vanishing terms

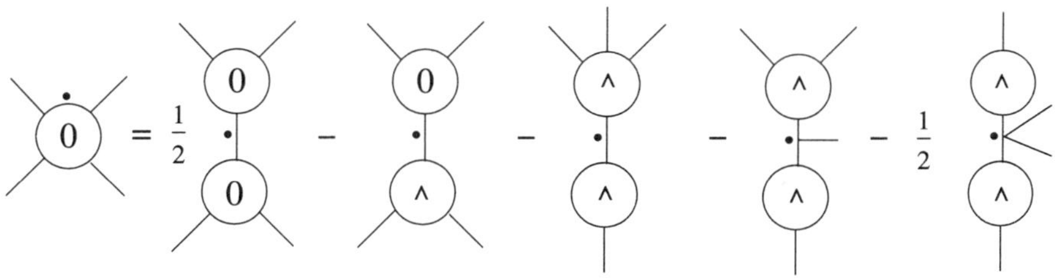

Figure 6 The flow equation for the classical four-point vertex.

with effective action vertices this time, but these are the three-point vertices that we have just determined. Therefore this equation is straightforward to integrate, giving the four-point vertex. In the same way we can determine all the vertices at the classical level.

3.2 One loop. Let us illustrate what happens at one loop by considering just one particular piece – the part that yields the famous asymptotically free β function for the coupling constant g. Recall that it is defined through the renormalization condition (1.1). This tells us that the two-point vertex $S_{\mu\nu}(p) = \Box_{\mu\nu}/g^2 + O(p^4)$, where the higher powers of momentum come exclusively from the higher dimension operators. Therefore when we compute the flow of $S_{\mu\nu}(p)$ at $O(p^2)$, all we pick up is the flow of g, *i.e.* the β function. At one loop, from (3.1), we find

$$-2\beta_1\Box_{\mu\nu}(p) + O(p^4) = \frac{1}{2}\frac{\delta}{\delta A_\alpha}\{\dot{\Delta}\}\frac{\delta\Sigma_0}{\delta A_\alpha}\bigg|^{AA}_{\mu\nu}, \tag{3.3}$$

which, when expanded, gives the diagrams of fig. 7. However, when we compute

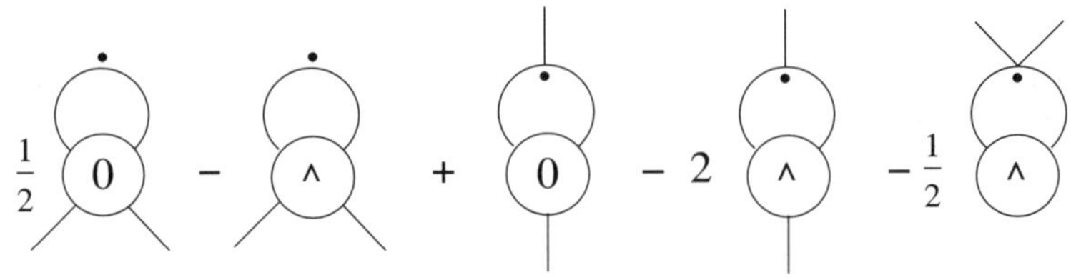

Figure 7 Two-point vertex at one loop and $O(p^2)$.

these diagrams we find that the integrals do not converge. The ultraviolet divergences are not regulated. This is not a surprise. Covariantizing the cutoff function in the way that we have done is equivalent to implementing a covariant higher derivatives regularisation and this is known to fail, precisely at one loop [30].

We have to replace this effective covariant higher derivatives with some other gauge invariant regularisation which can naturally be incorporated in this Exact RG framework, such that it cures this problem. One way to proceed is to add covariant Pauli-Villars fields [31] in such a way to as to cancel the leading divergences. This leads to further divergences which one cancels by introducing new regulator fields and interactions. Continuing in this way one finds eventually that the whole framework closes; the interactions are essentially unique and once in place, all diagrams, including the new ones, are regularised [6].

After staring at the result for a while, one realises that there is a much more elegant way to arrive at the same answer [16–19].

4 $SU(N|N)$ regularisation

4.1 $SU(N|N)$ **Yang-Mills.** Take the $SU(N)$ Yang-Mills and embed $SU(N)$ in the supergroup $SU(N|N)$. Its Lie algebra is realised by supermatrices, which we employ in the defining representation. The gauge field becomes a supergauge field,

$$\mathcal{A}_\mu = \begin{pmatrix} A^1_\mu & B_\mu \\ \bar{B}_\mu & A^2_\mu \end{pmatrix} + \mathcal{A}^0_\mu \mathbb{1},$$

where A^1 and A^2 are traceless Hermitian matrices and the off-diagonal terms B and $\bar{B}$, are Grassmann valued, and thus anticommuting (fermionic). It turns out we also have to add the $\mathcal{A}^0$ term, proportional to the $2N \times 2N$ unit matrix.

A^1 gauges the original $SU_1(N)$ group. We add an index to the group to distinguish it from an extra copy, $SU_2(N)$, gauged by A^2. The bosonic subgroup is therefore $SU_1(N) \otimes SU_2(N) \otimes U(1)$, where the $U(1)$ factor is carried by $\mathcal{A}^0$. If we had been dealing with a compact Lie algebra, as is normally used in gauge theory, we would be able to discard this $U(1)$ factor, but we cannot here because the $SU(N|N)$ superalgebra is an example of an indecomposable superalgebra; $\mathcal{A}^0$ gauges a central term produced by some of the fermionic generators, for example:[10]

$$\left\{ \begin{pmatrix} 0 & \mathbb{1} \\ \mathbb{1} & 0 \end{pmatrix}, \begin{pmatrix} 0 & \mathbb{1} \\ \mathbb{1} & 0 \end{pmatrix} \right\} = 2\mathbb{1}.$$

[10]In a Lie superalgebra, fermionic generators obey anticommutation laws amongst themselves.

This leads to some technicalities which are nevertheless key, so we briefly describe them. To form invariants we must use the supertrace

$$\mathrm{str}\begin{pmatrix} X^{11} & X^{12} \\ X^{21} & X^{22} \end{pmatrix} = \mathrm{tr}_1 X^{11} - \mathrm{tr}_2 X^{22}, \tag{4.1}$$

because it is this that is cyclically invariant for supermatrices and thus this which leads to the construction of $SU(N|N)$ invariants. The action will thus be built from a supercovariant derivative $\nabla_\mu = \partial_\mu - i\mathcal{A}_\mu$ as

$$S \sim \frac{1}{4g^2}\,\mathrm{str}\int \mathcal{F}^2_{\mu\nu}. \tag{4.2}$$

This has a neat effect on the central term because the kinetic term for $\mathcal{A}^0$ will just be a $U(1)$ field strength squared times $\mathrm{str}\mathbb{1}$. But from (4.1), $\mathrm{str}\mathbb{1} = N - N = 0$. Therefore $\mathcal{A}^0$ has no kinetic term. Furthermore, its interactions are formed from commutators, but $\mathbb{1}$ commutes with everything and thus $\mathcal{A}^0$ has no interactions. It therefore does not appear at all in the action displayed above.

We now raise this to a principle and require that the theory be invariant under the local shift symmetry

$$\delta\mathcal{A}^0_\mu = \lambda_\mu(x). \tag{4.3}$$

This "no-$\mathcal{A}^0$ symmetry" simply ensures that nothing in fact depends on $\mathcal{A}^0$. It means that in practice we gauge the coset space $SU(N|N)/U(1)$.

The $SU(N|N)/U(1)$ theory has very nice properties at high momenta. In particular one can show that the quadratic Casimir vanishes, which it turns out is sufficient to ensure that the effective action is finite at one loop. At higher loops, wherever the $SU(N)$ theory yielded $\mathrm{tr}\mathbb{1} = N$, one now obtains $\mathrm{str}\mathbb{1} = 0$, thus in the 't Hooft large N limit this theory has no quantum corrections at all.

Note that there is no sense in which this can be regarded as a physical theory, in particular the B_μ are fermionic integer spin particles which thus violate Pauli's spin statistics theorem [32]. The kinetic term for A^2 has the wrong sign, as follows from (4.2) and (4.1). This leads to negative norm states in the Fock space [17]. However we will be able to ensure that these sicknesses are felt by the physical $SU(N)$ Yang-Mills sector only at the cutoff scale.

4.2 $SU(N|N) \to SU_1(N) \otimes SU_2(N)$ spontaneous breaking. At high energies this more symmetric theory has some very useful ultraviolet properties. At low energies we want to recover just the $SU(N)$ Yang-Mills. In particle physics we know how to go from one regime to the other: we use the Higgs mechanism. We therefore introduce a super-Higgs field, the superscalar

$$\mathcal{C} = \begin{pmatrix} C^1 & D \\ \bar{D} & C^2 \end{pmatrix},$$

the C^i being bosonic and the off-diagonal D being fermionic. We can arrange a potential so that $\mathcal{C}$ picks up the expectation value[11]

$$\langle\mathcal{C}\rangle = \Lambda\begin{pmatrix} \mathbb{1} & 0 \\ 0 & -\mathbb{1} \end{pmatrix}. \tag{4.4}$$

[11]Note that $\mathcal{C}$ is actually a $U(N|N)$ adjoint field, the supertraceful generator being σ_3. This nevertheless transforms under $SU(N|N)$, another peculiarity compared to compact bosonic groups [17].

It is easy to show that this Pauli σ_3-matrix form breaks spontaneously all the fermionic directions, and only the fermionic directions. Thus, via the Higgs mechanism, the Bs eat the Ds and they become massive with masses of order the cutoff, where we will ultimately be able to ignore them. The 'physical' Higgs fields C^i can also be given masses at the cutoff scale, so all that remains at low energies are the two massless gauge fields A^1 and A^2. However in this gauge invariant approach, because the gauge fields are charged under different groups, the lowest dimension interaction between them is

$$\frac{1}{\Lambda^4}\,\mathrm{tr}_1\left(F^1_{\mu\nu}\right)^2\mathrm{tr}_2\left(F^2_{\mu\nu}\right)^2.$$

This is already dimension 8, so by the Symanzik-Appelquist-Carazzone theorem [33] the two sectors decouple into a direct product in the low energy limit.

At finite N, and with this extra $\mathcal{C}$ sector, we need to supply some covariantized cutoff functions in order to make the theory ultraviolet finite. It is then possible to show this to all orders in perturbation theory. For completeness we give the precise statement and sketch the steps necessary to prove it [17].

Theorem 4.1 *Let*

$$S=\mathrm{str}\int\frac{1}{4g^2}\mathcal{F}_{\mu\nu}c^{-1}\left(-\frac{\nabla^2}{\Lambda^2}\right)\mathcal{F}_{\mu\nu}+\frac{1}{2}\nabla_\mu\mathcal{C}\,\tilde{c}^{-1}\left(-\frac{\nabla^2}{\Lambda^2}\right)\nabla_\mu\mathcal{C}+\frac{\lambda}{4}\left(\mathcal{C}^2-\Lambda^2\right)^2,$$

where c^{-1} is a polynomial of rank r and $\tilde{c}^{-1}$ is a polynomial of rank $\tilde{r}$. Shifting $\mathcal{C}\mapsto\mathcal{C}+\Lambda\begin{pmatrix}1\!\!1 & 0\\ 0 & -1\!\!1\end{pmatrix}$, all amplitudes are finite to all orders in perturbation theory providing $r>\tilde{r}+1$ & $\tilde{r}>1$.

We use standard techniques of perturbative field theory for the proof, to avoid confusing the issue of regularisation with those of manifest gauge invariance and the Exact RG description. The first step therefore is to gauge fix to 't Hooft gauge. We introduce a further cutoff function $\hat{c}^{-1}$ (rank $\hat{r}$) for the gauge fixing term and thus also the BRST ghosts. Providing that $\hat{r}\geq r>\tilde{r}-1$ and $\tilde{r}>-1$ the leading contributions at high momentum have the symmetries of the symmetric regime. By power counting, one can show that if $\hat{r}\geq r$, $r-\tilde{r}>1$ and $\tilde{r}>1$, all diagrams are superficially finite except certain "remainder contributions": symmetric phase parts of one-loop graphs with no external $\mathcal{C}$ or ghost lines and up to four external $\mathcal{A}$s. Of these, two-point and three-point diagrams vanish by properties of the superalgebra since one can show that they involve only the following forms: $\mathrm{str}\mathcal{A}\,\mathrm{str}\mathcal{A}$, $\mathrm{str}\mathcal{A}\mathcal{A}\,\mathrm{str}1\!\!1$, $\mathrm{str}\mathcal{A}\mathcal{A}\,\mathrm{str}\mathcal{A}$ and $\mathrm{str}\mathcal{A}^3\,\mathrm{str}1\!\!1$. These all vanish because $\mathrm{str}1\!\!1=0$ and $\mathrm{str}\mathcal{A}=0$. This leaves only the symmetric phase four-point diagram which can be proved to be finite by Lee-Zinn-Justin identities [13]. The result is only conditionally convergent. This can be made well defined by an appropriate limiting procedure. Defining the theory in dimension $D\to 4^-$ is sufficient.

5 Manifestly $SU(N|N)$ gauge invariant Exact RG

We now combine the ideas in the last two sections. Given the field content, this amounts to finding an appropriate choice for Ψ in (2.2).[12]

$$\begin{aligned}\dot{S} \;=\; & \frac{1}{2}\frac{\delta S}{\delta \mathcal{A}_\mu}\{\dot{\Delta}^{\mathcal{A}}\}\frac{\delta \Sigma_g}{\delta \mathcal{A}_\mu} + \frac{1}{2}\frac{\delta S}{\delta \mathcal{C}}\{\dot{\Delta}^{\mathcal{C}}\}\frac{\delta \Sigma_g}{\delta \mathcal{C}} \\ & + \frac{1}{2}[\mathcal{C}, \frac{\delta S}{\delta \mathcal{A}_\mu}]\{\dot{\Delta}^{\mathcal{A}}_m\}[\mathcal{C}, \frac{\delta \Sigma_g}{\delta \mathcal{A}_\mu}] + \frac{1}{2}[\mathcal{C}, \frac{\delta S}{\delta \mathcal{C}}]\{\dot{\Delta}^{\mathcal{C}}_m\}[\mathcal{C}, \frac{\delta \Sigma_g}{\delta \mathcal{C}}] \\ & + \frac{1}{2}\frac{\delta S}{\delta \mathcal{A}_\mu}\{\dot{\Delta}^{\mathcal{A}}_\sigma\}\left(\{\mathcal{C}, \frac{\delta \Sigma_g}{\delta \mathcal{A}_\mu}\}\,\mathrm{str}\,\mathcal{C} - 2\mathcal{C}\,\mathrm{str}\left\{\mathcal{C}\frac{\delta \Sigma_g}{\delta \mathcal{A}_\mu}\right\}\right) + (S \leftrightarrow \Sigma_g) \\ & + \text{ quantum terms.}\end{aligned} \tag{5.1}$$

On the first line we have the term expected already which gives the flow equation for the gauge (now supergauge) field, together with its kernel which we have yet to determine. We also have a piece for the super-Higgs and its corresponding kernel. The whole flow equation must be invariant under

$$\delta \mathcal{C} = -i[\mathcal{C}, \Omega], \quad \delta \mathcal{A}_\mu = [\nabla_\mu, \Omega] + \lambda_\mu 1\!\!1, \quad \delta\left(\frac{\delta}{\delta \mathcal{A}_\mu}\right) = -i\left[\frac{\delta}{\delta \mathcal{A}_\mu}, \Omega\right] + \tilde{\lambda}_\mu 1\!\!1, \tag{5.2}$$

i.e. the combination of $SU(N|N)$ supergauge transformations, and the no-$\mathcal{A}^0$ symmetry that ensures we divide out by the central term.[13] The top line is already invariant under these but it is not enough because we want to keep the technical simplifications that follow from setting the classical two-point vertex equal to the seed action two-point vertex. This means that in the spontaneously broken phase we have to allow the kernels for the fermionic fields to grow a mass term, because the corresponding effective propagators must be inverses of their now-massive two-point vertices. The simplest such terms are displayed on the second line of (5.1): using (4.4), the commutators isolate (at the two-point level) just the fermionic directions as required. We are not quite done yet. We have to remember that really we now have two coupling constants, one for our original gauge field A^1 and one for the unphysical copy A^2. At the quantum level we can no longer identify the two couplings because the wrong sign action for A^2 results in its coupling running with the wrong sign. (It is asymptotically trivial rather than asymptotically free.) In the way we are doing things, this changes the normalisation of the A^2's kinetic term. Thus, if we are to keep the equality of $\hat{S}$ and S_0 two-point vertices we have to allow for different effective propagators for the two gauge fields. The simplest term that allows for this in the flow equation, whilst still obeying the symmetries (5.2), is the multi-supertrace term on the third line of (5.1). Finally the quantum terms are just analogous to the last term in (3.1), as determined by the general form (2.2).

(One can further understand the appearance of these multisupertrace terms as follows. Classically the effective action contains the terms:

$$S_0 = \frac{1}{4g^2}\,\mathrm{str}\int\begin{pmatrix}(F^1_{\mu\nu})^2 & 0 \\ 0 & (F^2_{\mu\nu})^2\end{pmatrix} + \cdots .$$

[12]For technical reasons it is helpful to make $\mathcal{C}$ dimensionless [15, 16, 19]. For simplicity of exposition we do not do this in what follows.

[13]The natural definition of $\delta/\delta\mathcal{A}$ is traceless. This determines $\tilde{\lambda}$ [19].

Since the one-loop correction has opposite sign for A^2, it takes the form

$$-\frac{\beta_1}{g^2}\ln\frac{\Lambda}{\mu}\,\mathrm{str}\begin{pmatrix}1\!\!1 & 0\\ 0 & -1\!\!1\end{pmatrix}\begin{pmatrix}(F^1_{\mu\nu})^2 & 0\\ 0 & (F^2_{\mu\nu})^2\end{pmatrix}.$$

At first sight such a correction is a puzzle since naïve guesses for the full contribution as a functional of $\mathcal{A}$, do not obey the no-$\mathcal{A}^0$ symmetry. In fact, as we confirmed by direct calculation, this term arises from a contribution of the form

$$\mathrm{str}\,\mathcal{C}\,\mathrm{str}\,\mathcal{C}\mathcal{F}^2_{\mu\nu} - (\mathrm{str}\,\mathcal{C}\mathcal{F}_{\mu\nu})^2\,|_{\mathcal{C}\to\mathcal{C}+\Lambda\sigma_3} + \cdots,$$

where the ellipses stands for higher dimension invariant terms. This does obey no-$\mathcal{A}^0$ symmetry as can be seen by using $\delta\mathcal{F}_{\mu\nu} = (\partial_\mu\lambda_\nu - \partial_\nu\lambda_\mu)1\!\!1$. The terms we added in the flow equation have a similar structure and indeed allow contributions such as above to be incorporated at tree-level, as now required.)

5.1 The kernels. The kernels in (5.1) are determined in the same way as before. We first write down the classical two-point vertices in a form as general as required and such that they are consistent with broken supergauge invariance and the renormalization conditions (and Poincaré invariance *etc.*). We then insist that the $\hat{S}$ two-point vertices are equal to these and this determines the kernels. For example, the two-point classical flow equations for the gauge fields take the form

$$\dot{S}_{0\,\mu\nu}^{\,A_1A_1} = -S_{0\,\mu\alpha}^{\,A_1A_1}\dot{\Delta}^1 S_{0\,\alpha\nu}^{\,A_1A_1} \qquad \text{and} \qquad \dot{S}_{0\,\mu\nu}^{\,A_2A_2} = -S_{0\,\mu\alpha}^{\,A_2A_2}\dot{\Delta}^2 S_{0\,\alpha\nu}^{\,A_2A_2},$$

determining Δ^1 and Δ^2, which are the linear combinations $\Delta^1 = \Delta^{\mathcal{A}} + 8N\Delta^{\mathcal{A}}_\sigma$ and $\Delta^2 = \Delta^{\mathcal{A}} - 8N\Delta^{\mathcal{A}}_\sigma$. In the same way, the other two-point classical flow equations determine linear combinations of the original kernels, and inverting we finally find expressions for the original kernels. The superscalar field classical two-point vertex flows in the same way as the scalar field considered earlier:

$$\dot{S}_0^{CC} = -S_0^{CC}\dot{\Delta}^C S_0^{CC}.$$

However, the B and D fields mix which other. This is just the Higgs effect in this manifestly gauge invariant description. We therefore use 2×2 matrices to describe the flow:

$$\begin{pmatrix}\dot{S}_{0\,\mu\nu}^{\,\bar{B}B} & \dot{S}_{0\,\mu}^{\,\bar{B}D}\\ -\dot{S}_{0\,\nu}^{\,\bar{B}D} & -\dot{S}_0^{\bar{D}D}\end{pmatrix} = -\begin{pmatrix}S_{0\,\mu\alpha}^{\,\bar{B}B} & S_{0\,\mu}^{\,\bar{B}D}\\ -S_{0\,\alpha}^{\,\bar{B}D} & -S_0^{\bar{D}D}\end{pmatrix}\begin{pmatrix}\dot{\Delta}^B\delta_{\alpha\beta} & 0\\ 0 & -\dot{\Delta}^D\end{pmatrix}\begin{pmatrix}S_{0\,\beta\nu}^{\,\bar{B}B} & S_{0\,\beta}^{\,\bar{B}D}\\ -S_{0\,\nu}^{\,\bar{B}D} & -S_0^{\bar{D}D}\end{pmatrix}.$$

A much more elegant description involves combining these into five-dimensional fields but for simplicity we will not do so here.

5.2 Effective propagator relations. From the above flow equations we determine the effective propagators and thus get a relation of the same form as (2.6):

$$S_0^{CC}\Delta^C = 1,$$

and relations analogous to (3.2):

$$S_{0\,\mu\,\nu}^{\,A_1A_1}\Delta^1 = \delta_{\mu\nu} - p_\mu p_\nu/p^2 \qquad \text{and} \qquad S_{0\,\mu\,\nu}^{\,A_2A_2}\Delta^2 = \delta_{\mu\nu} - p_\mu p_\nu/p^2.$$

Once again, the effective propagators here are not really propagators: multiplying the two-point vertex they give unity up to terms that simplify via gauge transformations. For the Bs and Ds one finds

$$\begin{pmatrix}S_{0\,\mu\alpha}^{\,\bar{B}B} & S_{0\,\mu}^{\,\bar{B}D}\\ -S_{0\,\alpha}^{\,\bar{B}D} & -S_0^{\bar{D}D}\end{pmatrix}\begin{pmatrix}\Delta^B\delta_{\alpha\nu} & 0\\ 0 & -\Delta^D\end{pmatrix} = \begin{pmatrix}\delta_{\mu\nu} & 0\\ 0 & 1\end{pmatrix} - \begin{pmatrix}f\frac{p_\mu}{\Lambda^2}\\ g\end{pmatrix}(p_\nu\ 2).$$

This is unity up to a term which simplifies using spontaneously broken supergauge transformations: the (p_ν 2) part operating on a vertex can be simplified via broken Ward identities. Its coefficient contains the cutoff dependent terms f_p and g_p, whose precise structure depends on the parameterisation we used for the two-point vertices [19]. We can unify all of these relations diagrammatically as in fig. 8. We write the coefficient parts as a single unfilled arrow *i.e.* as ">" and fill this arrow in when we include also the terms that generate gauge transformations (null, p_ν and (p_ν $-$ 2) for the C, A and B/D sectors respectively). Then in all sectors, the two-point classical action (or seed action) vertex attached to an effective propagator gives 1 minus a term which simplifies on using Ward identities.

Figure 8 Diagrammatic representation of the effective propagator relations.

5.3 'Naïve' Ward identities. To see what these Ward identities look like, it is enough to consider the simple case in pure $SU(N)$ gauge theory of a vertex of the form

$$S = \cdots + \int S_{\mu_1\mu_2\cdots\mu_n}(p_1, p_2, \cdots, p_n)\,\mathrm{tr}\, A_{\mu_1}(p_1)A_{\mu_2}(p_2)\cdots A_{\mu_n}(p_n) \;+\cdots.$$

This is because the diagrammatic representation turns out to be the same for any vertex in all sectors [20–22]. Requiring that the action be invariant under infinitesimal gauge transformations (1.2), *i.e.* $\delta A_\mu = \partial_\mu \omega - i[A_\mu, \omega]$, results in the so-called naïve Ward identities:

$$p_1^{\mu_1} S_{\mu_1\mu_2\cdots\mu_n}(p_1, p_2, \cdots, p_n) =$$
$$S_{\mu_2\cdots\mu_n}(p_1+p_2, p_3, \cdots, p_n) - S_{\mu_2\cdots\mu_n}(p_2, p_3, \cdots, p_n+p_1).$$

They are called "naïve" because in the gauge fixed approach one should really be using BRST invariance which leads to complications and ultimately the Lee-Zinn-Justin identities [13, 14]. Here, because the gauge invariance is preserved we get the above simple form. (The commutator part of the non-Abelian gauge transformation provides the RHS which would just be zero in an Abelian gauge theory.) In the diagrammatic representation we see that the contraction of the term generating gauge transformations results in the momentum of that leg being 'pushed forward' onto the following leg (with a plus sign) or 'pulled back' onto the previous leg (with a minus sign) as illustrated in fig. 9. In fact, since the vertex has been

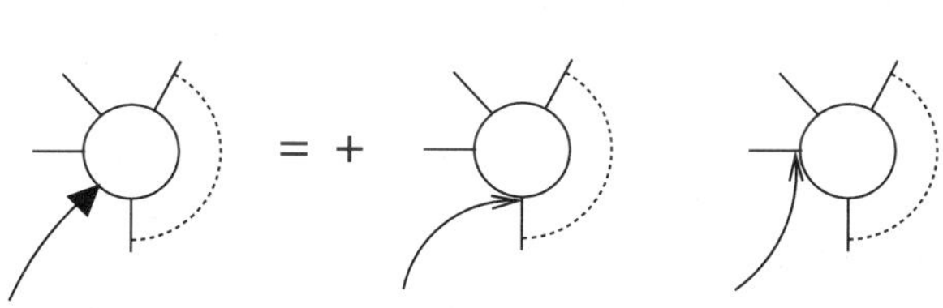

Figure 9 Diagrammatic representation of the Ward identities. Only one push-forward (counterclockwise arrow) and pull-back (clockwise arrow) have been drawn. By symmetry they appear on all legs of the vertex.

symmetrised over permutations (as usual in Feynman diagrams), the pull-backs and push-forwards operate on all legs.

6 Diagrammatic method

We now have all the elements necessary to describe the full diagrammatic method. The flow equation has the same diagrammatic form as before, *viz.* fig. 3. All we have to remember, in fact most of the time we can forget, is that propagating between vertices is now not just the gauge field A^1 but all the other partners (A^2, B, C, D), which are necessary for regularisation. This means only that internal legs (kernels or effective propagators) must be interpreted as summed over these flavours.

As a consequence the one-loop diagrams look the same as before, *viz.* fig. 7, however these diagrams converge, not only in the infrared (as guaranteed by the Wilsonian formalism for smooth cutoffs) but also now in the ultraviolet. Therefore, we can just compute them. However we want to proceed without having to specify the regularisation structure (cutoff functions, covariantization, and seed action) precisely. This leaves us with very little room to maneouver. The only worthwhile operation is to shift the RG-time derivative in the first diagram by integration by parts, resulting in fig. 10. Since the integral in the first diagram is a contribution

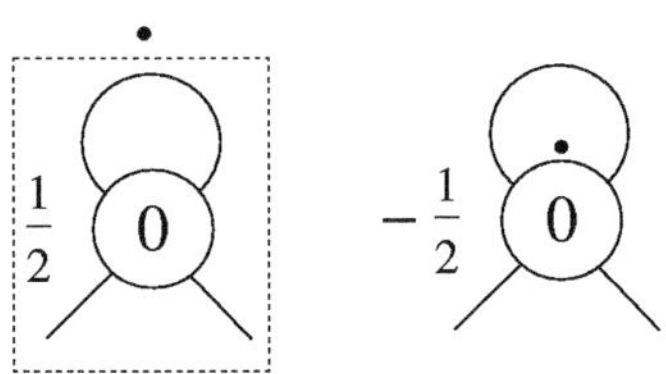

Figure 10 Shifting the RG-time derivative. The differential is now outside the integral in the first expression.

to the β function, it is dimensionless. *A priori* the integral can depend only upon Λ. However, since Λ has dimensions of mass and there is no other dimensionful parameter, the integral is in fact independent of this and is thus a pure number. Differentiating it gives zero.

We have thus effectively shifted the RG-time derivative onto the classical vertex, as in the remaining diagram of fig. 10. We can now expand this flow using fig. 6. This results in a bunch of diagrams, corresponding to all possible ways of tieing up two legs in the diagrams on the RHS of fig. 6. Let us concentrate on just two of these. The first diagram displayed in fig. 11 corresponds to tieing up a top and a bottom leg from the first diagram on the RHS of fig. 6. The second diagram in fig. 11, the result of tieing up the top and bottom legs of the third diagram on the RHS of fig. 6, can be processed further. We use the effective propagator relations, fig. 8. The result is fig. 12. The first diagram in this figure is equal and opposite to the second diagram in the original list, fig. 7, so that now two diagrams have disappeared from this list. The second diagram in fig. 12 can be processed further by using the Ward identities, fig. 9, pushing-forwards and pulling-backwards to give fig. 13. We continue to proceed in the same way, for example the first diagram of fig. 11 can be expressed as a total derivative minus terms where the RG-time derivative hits the classical three-point vertices. These can thus be expanded. Evaluating

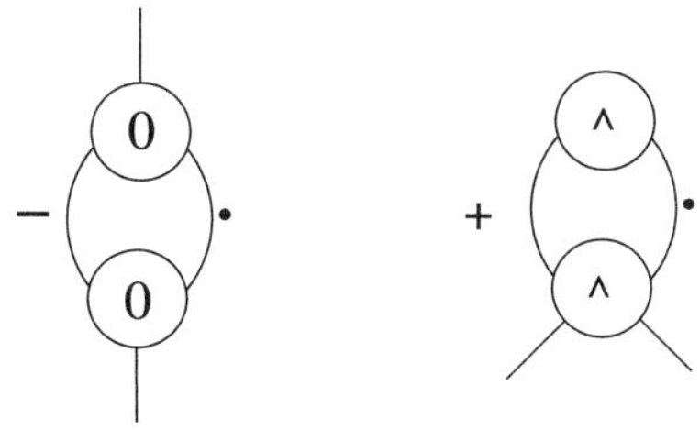

Figure 11 Examples of diagrams obtained from expanding the second diagram in fig. 10, together with their combinatorics.

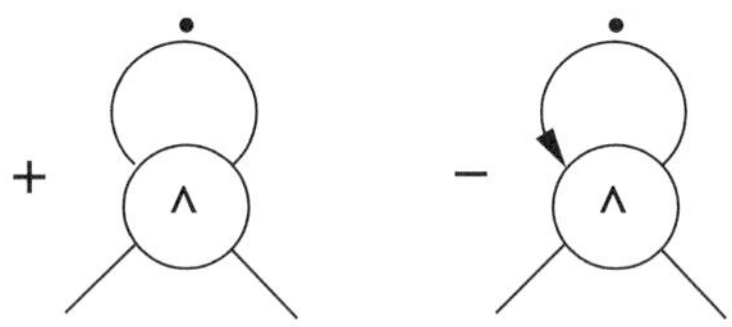

Figure 12 The result of applying effective propagator relations to the second diagram in fig. 11.

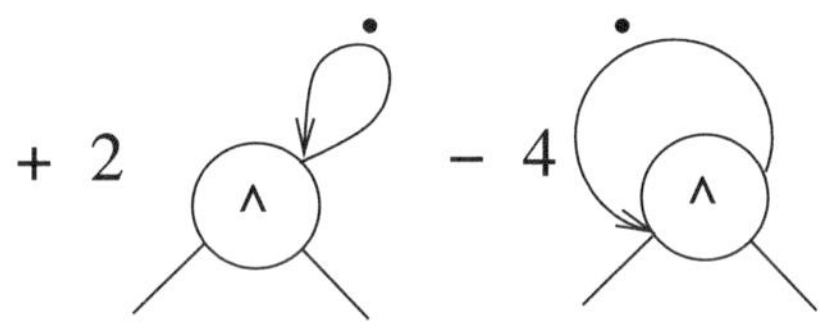

Figure 13 The result of applying the Ward identities to the second diagram in fig. 12.

these and others in the same way, we find that all diagrams cancel, including the originals in fig. 7, except for the set of total derivative diagrams in fig. 14 [20, 21]. (The third diagram is the result of expanding to $O(p^2)$, as required from (3.3),

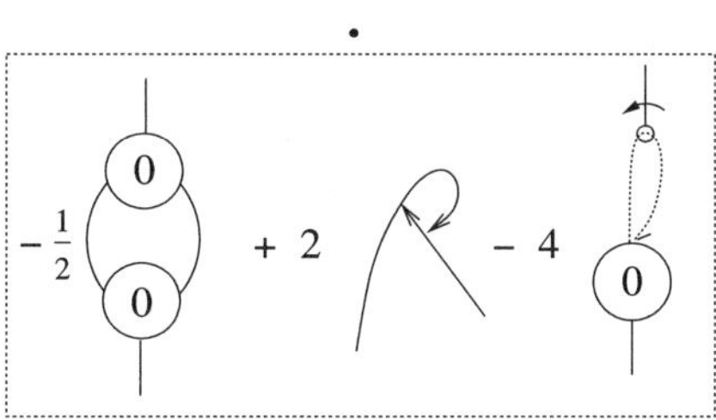

Figure 14 The final diagrams contributing the one-loop β function.

and processing the resulting derivatives with respect to p^λ using differential Ward identities [20, 21].)

We saw in fig. 10 that the total RG-time derivative of a diagram contributing to β is zero because the integral can only give a pure number. The only reason that these diagrams are not also zero is that in the exchange of limits, placing the differential with respect to RG-time outside the momentum integral, we have introduced infrared divergences. In other words, strictly speaking we were not allowed to exchange limits in this way, however doing so has the advantage that we see very clearly that the results for these integrals depend only on the massless fields (*i.e.* the A^i) propagating in the far infrared. For such vanishing momenta, all the elements in the integral are then determined by the renormalization conditions. That is why the result is universal. It is easy to compute the resulting integrals, and thus we obtain [19, 21]:

$$\frac{22N}{3(4\pi)^2}\,\Box_{\mu\nu}(p) + O(p^4).$$

By comparing to our starting point, (3.3), we extract the famous asymptotically free one-loop beta function for Yang-Mills.

The procedure generalises to two loops and beyond. The main new feature is that subtraction diagrams need to be defined in order to isolate the universal parts [20, 23]. In this way, we have confirmed that the universal two-loop beta function is also reproduced by these methods.

7 QCD

Spinors transforming in the fundamental representation can be incorporated in the framework, so we can compute in the physically important theory of Quantum Chromodynamics, without fixing the gauge [34]. These fermions are the quarks; the gauged group is $SU(3)$; A^1 is thus the gluon field.

The obvious thing to try, is to extend the quark spinor ψ to a fundamental representation Ψ of $SU(3|3)$. However, gauge invariance then fixes the interaction to be of the form $\Psi\not{\mathcal{A}}\Psi$ which violates the no-$\mathcal{A}^0$ invariance (4.3). The only way out is to try a different embedding which does not have this problem. In fact we can embed ψ in the off-diagonal part of a 3×3 supermatrix (transforming as a complexified $U(3|3)$ adjoint field):

$$\Psi = \begin{pmatrix} \varphi^1 & \psi \\ \varrho & \varphi^2 \end{pmatrix}.$$

In this way ψ is fermionic by default. It transforms as the fundamental under colour, $SU_1(3)$, and complex conjugate fundamental under $SU_2(3)$. Thus we are forced into a situation where Ψ represents multiples of three flavours of quark. From this point of view it is at least serendipitous that nature decided that as well as three colours, there are also three families. One super-spinor Ψ_u can thus contain the up-type quarks u, c and t, and another super-spinor Ψ_d hold the down-type quarks d, s and b. Of course at this stage the family symmetry $SU_2(3)$ is gauged, moreover by the unphysical field A_2. To give the quarks different masses, we need to break it spontaneously (so that we keep the regularising properties of $SU(3|3)$ at high energies).

(Although it is cute that families come in multiples of the number of colours, there is no restriction to handling different numbers of families since we can always send individual quark masses to infinity at the end of the calculation.)

8 Other applications

We note that these ideas have been generalised, or rather simplified, to the case of QED [35], and the steps necessary for computing general gauge invariant matrix elements have also been fleshed out and incorporated in this general scheme, using the example of the one-loop contribution to the Wilson loop [26]. We were also able to use these ideas to make explicit the cutoff in the AdS/CFT construction [36].

So far we have only discussed explicitly the perturbative expansion of the flow equations, and concentrated on the evaluation of the first two β function coefficients, without gauge fixing. The fact that this all works however, gives us experience and confidence in applying the framework to the non-perturbative regime. As we intimated in the introduction, there is a wide variety of possible approximations that can be applied.

Indeed within this framework, we have already studied the large N approximation to some extent in [6, 15]. The similarities to string theory that this uncovered, together with the gauge (and Poincaré) invariant cutoff definition in this framework, and suggestions of exact RG from the AdS/CFT side [37], provided the initial motivation to search for some connection with the AdS/CFT construction.

We would also like to mention that it is possible to perform a sort of Eguchi-Kawai reduction [38] of the equations, this time in the continuum, casting the equations in an equivalent way in terms of vertices that carry no momentum [39].

Of course one can also simply resort to truncations, a strategy that has been very popular in the exact RG field [4, 5].

We would like to finish by showing that under some very general conditions, confinement is a natural consequence in this framework [39]. We then describe in some detail one further possibility adapted to this observation, namely a strong renormalized coupling expansion, which indeed could be combined with any of the above approaches and could have more general applicability to other exact RGs [39]. As well as being a desirable starting point for further approximation, it means that one can test for confinement within this framework by instead testing whether these general conditions hold true or not.

9 Confinement

From the perturbative β function we know that $g(\Lambda)$ grows as Λ gets smaller. Although we cannot trust perturbation theory in this regime, the natural expectation is that g grows without limit. Indeed in any formulation where the gauge invariance is manifest, including lattice gauge theory, the gluon itself cannot have a mass gap. Intuitively, in this case confinement only follows if the coupling diverges in the infrared. (In the gauge fixed theory, arguments based on the Kugo-Ojima criterion [40] actually come to the opposite conclusion: that g freezes out at some finite value. There is some evidence for this in gauge fixed formulations [41], however these arguments depend by their very nature on the ghost fields and thus have nothing to say about a manifestly gauge invariant approach. In any case, *a priori* both pictures can be correct since they refer to different non-perturbative definitions of g.)

Our first condition then is that $g(\Lambda) \to \infty$ as Λ decreases. In principle, we could find that g diverges at a finite critical value $\Lambda \to \Lambda_{cr} > 0$ but in this case we would have an obstruction to completing the computation in this way, which requires that we integrate out all the modes, *i.e.* that we take the limit $\Lambda \to 0$. It may be that a change of variables will allow us to pass smoothly over Λ_{cr}, *i.e.* that the problem is a kind of 'coordinate singularity'. If this can be done in terms of some $SU(N|N)$ fields $\mathcal{A}$ and $\mathcal{C}$, it amounts to a different choice for Ψ in (2.2), *i.e.* a different choice of exact RG.

Therefore we require that $g(\Lambda) \to \infty$, but only as $\Lambda \to 0$; we require this for at least one of the infinitely many choices that we have for gauge invariant exact RGs, if we are to find confinement in this framework.

Remember that g is defined by the renormalization condition (1.1), which tells us that the effective two-point A^1 vertex has the form

$$S_{\mu\nu}(p) = \Box_{\mu\nu}(p)/g^2 + O(p^4). \tag{9.1}$$

It follows that in this case, in the limit $\Lambda \to 0$ we find $S_{\mu\nu}(p) \sim O(p^4)$. This is a well-known signal of confinement. Indeed the three dimensional Fourier transform of the (gauge fixed) full effective propagator gives the effective potential $V(r)$ between two colour charges. In the perturbative regime one has for large r, the standard Coulomb potential, as follows trivially from dimensional analysis:

$$V(r) \sim \int \frac{d^3p}{p^2}\, \mathrm{e}^{i\mathbf{p}\cdot\mathbf{r}} \sim \frac{1}{r}.$$

In this non-perturbative regime we have instead the famous linearly rising form for large r, corresponding to a constant Yang-Mills string tension and confinement:

$$V(r) \sim \int \frac{d^3p}{p^4}\, \mathrm{e}^{i\mathbf{p}\cdot\mathbf{r}} \sim r,$$

as again follows trivially by dimensional analysis. In what amounts to much the same computation, one can show that changing the propagator from $1/p^2$ to $1/p^4$ changes the lowest order contribution to the expectation value of a Wilson loop from perimeter law to area law (another standard signal of confinement).

We therefore come to the conclusion that in this manifestly gauge invariant exact RG formalism, confinement occurs if and only if $g(\Lambda) \to \infty$ as $\Lambda \to 0$.

10 Renormalized strong coupling expansion

Although this observation is neat, we cannot use it as a basis for computation in this regime because it only makes sense for momentum $p \approx 0$ (which is in fact all that is needed to confirm the form of the potentials above for large r). The reason is as follows. Since we are dealing with a continuum limit the effective action must be in self-similar form (*cf.* the introduction), which means that the only explicit scale in the solution is Λ. This means that the expansion in (9.1) is in p^2/Λ^2, and thus as $\Lambda \to 0$ the approximation $S_{\mu\nu}(p) \sim O(p^4)$ becomes valid only for vanishing momenta.

We should expect that dimensional transmutation takes place so that the scale in the problem is really set by Λ_{YM}.[14] In this way, we ought to be able to address the case of non-vanishing momenta.

We will show that we can do so within a new form of expansion in the renormalized coupling, namely in $1/g^2$. We leave as an open problem whether such an expansion actually exists. Indeed if such an expansion of the effective action is substituted into the exact RG (5.1), the equations (unsurprisingly) do not close, so its existence can probably only be tested within some further approximation.

[14] We restrict the discussion here to the Yang-Mills we have been addressing, although it can be applied to QCD also, where the corresponding scale is Λ_{QCD}.

(Of course strong coupling expansions have a long history [42] but these were in the bare coupling $g_0 \sim g(\Lambda_0)$. They are known to have a finite radius of convergence, which is problematic, since by asymptotic freedom we require $g_0 \to 0$ in the continuum limit.)

By gauge (and Poincaré) invariance and dimensions, the A^1 two-point vertex takes the form $S_{\mu\nu}(p) = s(p,\Lambda)\,\Box_{\mu\nu}$, for some dimensionless function s. By the requirement of self-similarity this function can be written as $s(p/\Lambda, g)$, where $g(\Lambda)$ runs according to $\Lambda\partial_\Lambda g = \beta(g)$.

We now assume that in the low energy regime, S is analytic in $1/g^2$ for large g, and therefore can be written as an expansion in $1/g^2$. For s, this implies

$$s = s_0(p/\Lambda) + \frac{1}{g^2} s_1(p/\Lambda) + \frac{1}{g^4} s_2(p/\Lambda) + \cdots, \tag{10.1}$$

for some functions s_i. This will have the consequence that β itself has an analytic strong coupling expansion:

$$\Lambda\partial_\Lambda 1/g^2 = \tilde\beta(1/g^2) = \tilde\beta_0 + \tilde\beta_1/g^2 + \tilde\beta_2/g^4 + \cdots.$$

The flow $\Lambda\partial_\Lambda 1/g^2 = \tilde\beta$ can now be solved in the regime of large g with the result that if g is to diverge but only as $\Lambda \to 0$, then $\tilde\beta_0 = 0$ and $\tilde\beta_1 > 0$, and

$$\frac{1}{g^2} = \left(\frac{\Lambda}{\Lambda_{YM}}\right)^{\tilde\beta_1} + \frac{\tilde\beta_2}{\tilde\beta_1}\left(\frac{\Lambda}{\Lambda_{YM}}\right)^{2\tilde\beta_1} + \cdots, \tag{10.2}$$

where the unit coefficient for the first RHS term amounts to our definition of Λ_{YM} and the neglected terms are of order $(\Lambda/\Lambda_{YM})^{3\tilde\beta_1}$. (It is also possible that $\tilde\beta_1 = 0$ and $\tilde\beta_2 > 0$, however in this case Λ_{YM} would indicate a maximum Λ above which the renormalized strong coupling expansion breaks down. We will not investigate this interesting possibility further here.)

We assume that S stands only for the part of the effective action that obtains a finite limit as $\Lambda \to 0$. (We will not address how to isolate this part of the effective action from the regularisation structure which of necessity diverges in this limit. For the Polchinski effective action it would be everything except the regularisation in the kinetic term. It is not so simple here, however all expectation values of gauge invariant operators and their correlators, are automatically isolated from these divergent parts. They are introduced into the action by being coupled via source terms [15, 26].)

Since g is a function of Λ/Λ_{YM}, we can write the limit $S|_{\Lambda=0}$ as an RG invariant action S^{inv}. To illustrate with the two-point vertex $s(p/\Lambda, g)$, we write it equivalently as the function $\tilde s(p/\Lambda_{YM}, \Lambda/\Lambda_{YM})$. Its limit, $\tilde s(p/\Lambda_{YM}, 0)$, can instead be written as the Λ independent function $s^{inv}(p/\Lambda, g)$.

If we now assume that S^{inv} also has a strong coupling expansion then, from

$$0 = \Lambda\partial_\Lambda S^{inv} = \Lambda\partial_\Lambda|_g S^{inv} + \tilde\beta\,\partial_{1/g^2} S^{inv},$$

we constrain the form of its expansion coefficients, and in particular for s^{inv} we find

$$s_0^{inv} = \sigma_0, \qquad s_1^{inv} = \sigma_1\left(\frac{p}{\Lambda}\right)^{\tilde\beta_1}, \qquad s_2^{inv} = \sigma_2\left(\frac{p}{\Lambda}\right)^{2\tilde\beta_1} - \sigma_1\frac{\tilde\beta_2}{\tilde\beta_1}\left(\frac{p}{\Lambda}\right)^{\tilde\beta_1},$$

and so on, where the σ_i are numerical coefficients. Combining with the expansion (10.2), we find the expected dimensional transmutation:

$$s^{inv} = \sigma_0 + \sigma_1 \left(\frac{p}{\Lambda_{YM}}\right)^{\tilde{\beta}_1} + \sigma_2 \left(\frac{p}{\Lambda_{YM}}\right)^{2\tilde{\beta}_1} + \cdots. \tag{10.3}$$

We see that the assumption of a renormalized strong coupling expansion and the existence of a limit as $\Lambda \to 0$, implies that this expansion turns into an expansion in small p/Λ_{YM}. Since we have a mass gap, we should find that the resulting expansion is analytic in p (otherwise there would be long-distance structure in position space). In this case we would have to find that $\tilde{\beta}_1$ is a positive even integer (probably $\tilde{\beta}_1 = 2$).

Subtracting S^{inv} from S, leaves a remainder R which thus also has a strong coupling expansion:

$$S = S^{inv} + R_0 + R_1/g^2 + R_2/g^4 + \cdots.$$

Since $S \to S^{inv}$ as $\Lambda \to 0$, and since g can be varied independently of the R_i (by varying Λ_{YM}), we must have that $R_n/g^{2n} \to 0$ as $\Lambda \to 0$ at fixed finite p and Λ_{YM}. From (10.2), this implies $R_n\Lambda^{n\tilde{\beta}_1} \to 0$ as $\Lambda \to 0$. Of course this bounds above the amount by which R_n can diverge as $\Lambda \to 0$. For the two-point vertex, let us call the remainder coefficients $r_n(p/\Lambda)$. Then by dimensions, $r_n/p^{n\tilde{\beta}_1} \to 0$ as $p \to \infty$. On the contrary, from (10.3) and (10.1), we have that the s expansion coefficients have a unique finite limit: $s_n/p^{n\tilde{\beta}_1} \to \sigma_n/\Lambda^{n\tilde{\beta}_1}$ as $p \to \infty$. We therefore have an upper bound on the behaviour of r_n and s_n as $p \to \infty$.

Recall that we are interested in the limit $\Lambda \to 0$ at fixed finite p. We have thus seen that the renormalized strong coupling expansion is restricted to the regime $\Lambda \ll p \ll \Lambda_{YM}$. We therefore cannot use the definition of g in (9.1). If we can extract the coefficients (*e.g.* the r_n and s_n above) from the exact RG equations, then at finite Λ we can expect to find that they have a complicated form, certainly not polynomial in p/Λ. From the above analysis we see that they are bounded above by some power of p/Λ. We therefore expect that these functions can be expanded as a series in Λ/p. The natural generalisation of (9.1) to this case is to define $1/g^2$ to be the momentum independent coefficient[15] in the expansion of $s(p/\Lambda, g)$ in Λ/p. We emphasise however that this is a different definition from the perturbative one in (9.1). There is no reason to expect them to coincide.[16]

Since $\lim_{p/\Lambda\to\infty} s_0(p/\Lambda) = \sigma_0$, we have from (10.1) and this renormalization condition that $\sigma_0 = 0$. Since $s \to s^{inv}$, we recover confinement from (10.3), by the arguments of the previous section, providing indeed that we find that $\tilde{\beta}_1 = 2$.

Acknowledgement

TRM thanks the Fields Institute, and the Royal Society and University of Helsinki for financial support to attend the workshop and conference respectively.

[15]Remember that s is the coefficient of $\Box_{\mu\nu}$ in the effective two-point vertex.

[16]This follows from simple analysis; compare for example the x independent terms in the small x expansion, and small $1/x$ expansion, of $1/(1+x)$.

References

[1] F.J. Wegner and A. Houghton, "Renormalization group equation for critical phenomena," Phys. Rev. A8 (1973) 401.

[2] K. Wilson and J. Kogut, "The Renormalization group and the epsilon expansion," Phys. Rep. 12C (1974) 75.

[3] M. E. Fisher, "Renormalization group theory: Its basis and formulation in statistical physics," Rev. Mod. Phys. 70 (1998) 653.

[4] D. F. Litim and J. M. Pawlowski, in *The exact renormalization group* ed A. Krasnitz *et al*, World Sci (1999) 168, "On gauge invariant Wilsonian flows," [arXiv:hep-th/9901063]. K. Aoki, "Introduction to the nonperturbative renormalization group and its recent applications," Int. J. Mod. Phys. B **14** (2000) 1249. C. Bagnuls and C. Bervillier, "Exact renormalization group equations: An introductory review," Phys. Rept. **348** (2001) 91 [arXiv:hep-th/0002034]. J. Berges, N. Tetradis and C. Wetterich, "Non-perturbative renormalization flow in quantum field theory and statistical physics," Phys. Rept. **363** (2002) 223 [arXiv:hep-ph/0005122]. J. Polonyi, "Lectures on the functional renormalization group method," Central Eur. J. Phys. **1** (2003) 1 [arXiv:hep-th/0110026]. M. Salmhofer and C. Honerkamp, "Fermionic renormalization group flows: Technique and theory," Prog. Theor. Phys. **105** (2001) 1. B. Delamotte, D. Mouhanna and M. Tissier, "Nonperturbative renormalization group approach to frustrated magnets," Phys. Rev. B **69** (2004) 134413 [arXiv:cond-mat/0309101].

[5] T. R. Morris, "Properties of derivative expansion approximations to the renormalization group," Nucl. Phys. Proc. Suppl. **42** (1995) 811 [arXiv:hep-lat/9411053], "New developments in the continuous renormalization group," in *Zakopane 1997, New developments in quantum field theory* (1998) 147 [arXiv:hep-th/9709100], "Elements of the continuous renormalization group," Prog. Theor. Phys. Suppl. **131** (1998) 395 [arXiv:hep-th/9802039].

[6] T. R. Morris, review in *The exact renormalization group* ed A. Krasnitz *et al*, World Sci (1999) 1. The later lectures "A manifestly gauge invariant exact renormalization group," also available at arXiv:hep-th/9810104.

[7] D. V. Shirkov, "Renormalizations In Quantum Field Theory," *Theor. Math. Phys.* **60** (1985) 778.

[8] L.P. Kadanoff, "Scaling laws for Ising models near T(c)," *Physics* **2** (1966) 263.

[9] C. Becchi, "On the construction of renormalized gauge theories using renormalization group techniques," in Seminario Nazionale di Fisica Teorica, Parma (1991) Eds M. Bonini, G. Marchesini and E. Onofri [arXiv:hep-th/9607188]; M. Bonini and E. Tricarico, "Fine-tuning and the Wilson renormalization group," Nucl. Phys. B **585** (2000) 253 [arXiv:hep-th/0006183].

[10] J. I. Latorre and T. R. Morris, "Exact scheme independence," JHEP **0011** (2000) 004 [arXiv:hep-th/0008123], "Scheme independence as an inherent redundancy in quantum field theory," Int. J. Mod. Phys. A **16** (2001) 2071 [arXiv:hep-th/0102037].

[11] C. Becchi, A. Rouet and R. Stora, "Renormalization Of The Abelian Higgs-Kibble Model," *Comm. Math. Phys.* **42** (1975) 127; "Renormalization Of Gauge Theories," in *Renormalisation Theory*, eds. G. Velo and A. S. Wightman (Reidel, Dordrecht, 1976); "Renormalization Of Gauge Theories," *Ann. Phys.* **98** (1976) 287; I. V. Tyutin, "Gauge Invariance In Field Theory And Statistical Physics In Operator Formalism," Lebedev Institute preprint N39 (1975).

[12] V. Gribov, "Instability Of Nonabelian Gauge Theories And Impossibility Of Choice Of Coulomb Gauge," Nucl. Phys. B139 (1978) 1; I. Singer, "Some Remarks On The Gribov Ambiguity," Comm. Math. Phys. 60 (1978) 7.

[13] J. Zinn-Justin, "Renormalization Of Gauge Theories," in *Trends in Elementary Particle Physics (Lecture Notes in Physics 37)* Bonn 1974, eds. H. Rollnik and K. Dietz, (1975) Springer-Verlag, Berlin; B. W. Lee, "Gauge theories. Course 3," in *Methods in Field Theory*, Les Houches 1975, eds. R. Balian and J. Zinn-Justin (1976) North-Holland, Amsterdam.

[14] J. Zinn-Justin, *Quantum Field Theory and Critical Phenomena* (1993) Clarendon Press, Oxford.

[15] T. R. Morris, "A gauge invariant exact renormalization group. I," Nucl. Phys. B **573** (2000) 97 [arXiv:hep-th/9910058].

[16] T. R. Morris, "A gauge invariant exact renormalization group. II," JHEP **0012** (2000) 012 [arXiv:hep-th/0006064].

[17] S. Arnone, Y. A. Kubyshin, T. R. Morris and J. F. Tighe, "Gauge invariant regularisation via SU(N|N)," Int. J. Mod. Phys. A **17** (2002) 2283 [arXiv:hep-th/0106258]; "Gauge invariant regularisation in the ERG approach," in *15th International Workshop on High-Energy Physics and Quantum Field Theory (QFTHEP 2000), Tver, Russia, 14-20 Sep 2000* (2000) 297 [arXiv:hep-th/0102011]; "A gauge invariant regulator for the ERG," Int. J. Mod. Phys. A **16** (2001) 1989 [arXiv:hep-th/0102054].

[18] T. R. Morris, "An exact RG formulation of quantum gauge theory," Int. J. Mod. Phys. A **16** (2001) 1899 [arXiv:hep-th/0102120].

[19] S. Arnone, A. Gatti and T. R. Morris, "Towards a manifestly gauge invariant and universal calculus for Yang-Mills theory," Acta Phys. Slov. **52** (2002) 621 [arXiv:hep-th/0209130]; "A proposal for a manifestly gauge invariant and universal calculus in Yang-Mills theory," Phys. Rev. D **67** (2003) 085003 [arXiv:hep-th/0209162].

[20] O. J. Rosten, "The manifestly gauge invariant exact renormalisation group," *PhD Thesis*, arXiv:hep-th/0506162.

[21] O. J. Rosten, T. R. Morris and S. Arnone, "The gauge invariant ERG," in *13th Int. Seminar on High-Energy Physics: Quarks 2004, Pushkinskie Gory, Russia, 2004*, arXiv:hep-th/0409042; "A generalised manifestly gauge invariant exact renormalisation group for SU(N) Yang-Mills," arXiv:hep-th/0507154.

[22] O. J. Rosten, "A primer for manifestly gauge invariant computations in SU(N) Yang-Mills," arXiv:hep-th/0507166.

[23] T. R. Morris and O. J. Rosten, "A manifestly gauge invariant, continuum calculation of the SU(N) Yang-Mills two-loop beta function," Phys. Rev. D **73** (2006) 065003 [arXiv:hep-th/0508026].

[24] O. J. Rosten, J. Phys. A 39 (2006) 8141 in *Proceedings of Renormalization Group 2005 (RG 2005), Helsinki, Finland, 30 Aug - 3 Sep 2005*, "Scheme independence to all loops," [arXiv:hep-th/0511107].

[25] O. J. Rosten "A manifestly gauge invariant and universal calculus for SU(N) Yang-Mills," arXiv:hep-th/0602229.

[26] O. J. Rosten, "General computations without fixing the gauge," arXiv:hep-th/0604183.

[27] J. Polchinski, "Renormalization And Effective Lagrangians," Nucl. Phys. B **231** (1984) 269.

[28] T. R. Morris, "The Exact renormalization group and approximate solutions," Int. J. Mod. Phys. A **9** (1994) 2411 [arXiv:hep-ph/9308265].

[29] S. Arnone, A. Gatti and T. R. Morris, "Exact scheme independence at one loop," JHEP **0205** (2002) 059 [arXiv:hep-th/0201237], "A demonstration of scheme independence in scalar ERGs," Acta Phys. Slov. **52** (2002) 615 [arXiv:hep-th/0205156]; S. Arnone, A. Gatti, T. R. Morris and O. J. Rosten, "Exact scheme independence at two loops," Phys. Rev. D **69** (2004) 065009 [arXiv:hep-th/0309242].

[30] A.A. Slavnov, "Invariant regularization of gauge theories," Theor. Math. Phys. 13 (1972) 1064; B.W. Lee and J. Zinn-Justin, "Spontaneously Broken Gauge Symmetries. I. Preliminaries," Phys. Rev. D5 (1972) 3121.

[31] T.D. Bakeyev and A.A. Slavnov, "Higher covariant derivative regularization revisited," Mod. Phys. Lett. A11 (1996) 1539; M. Asorey and F. Falceto, "On the consistency of the regularization of gauge theories by high covariant derivatives," Phys. Rev. D54 (1996) 5290, Nucl. Phys. B327 (1989) 427; C.P. Martin and F. Ruiz Ruiz, "Unitarity violation in nonAbelian Pauli-Villars regularization," Nucl. Phys. B436 (1995) 545; B.J. Warr, "Renormalization Of Gauge Theories Using Effective Lagrangians. 1," Ann. Phys. 183 (1988) 1.

[32] N. Burgoyne, "A proof of the spin statistics theorem", Nuovo Cim. **8** (1958) 607.

[33] K. Symanzik, "Infrared Singularities And Small Distance Behavior Analysis," Comm. Math. Phys. 34 (1973) 7; T. Appelquist and J. Carazzone, "Infrared Singularities And Massive Fields," Phys. Rev. D11 (1975) 2856. See also J. Collins, *Renormalization* (CUP, 1984).

[34] T. R. Morris and O. J. Rosten, "Manifestly gauge invariant QCD", SHEP 06-20, hep-th/0606nn.

[35] S. Arnone, T. R. Morris and O. J. Rosten, "Manifestly gauge invariant QED," JHEP **0510** (2005) 115 [arXiv:hep-th/0505169].

[36] N. Evans, T. R. Morris and O. J. Rosten, "Gauge invariant regularization in the AdS/CFT correspondence and ghost D-branes," Phys. Lett. B **635** (2006) 148 [arXiv:hep-th/0601114].

[37] J. de Boer, E. P. Verlinde and H. L. Verlinde, "On the holographic renormalization group," JHEP **0008** (2000) 003 [arXiv:hep-th/9912012]; E. P. Verlinde and H. L. Verlinde, "RG-flow, gravity and the cosmological constant," JHEP **0005** (2000) 034 [arXiv:hep-th/9912018]; S. Hirano, "Exact renormalization group and loop equation," Phys. Rev. D **61** (2000) 125011 [arXiv:hep-th/9910256]; M. Li, "A note on relation between holographic RG equation and Polchinski's RG equation," Nucl. Phys. B **579** (2000) 525 [arXiv:hep-th/0001193].

[38] T. Eguchi and H. Kawai, "Reduction Of Dynamical Degrees Of Freedom In The Large N Gauge Theory," Phys. Rev. Lett. **48** (1982) 1063.

[39] T. R. Morris, unpublished.

[40] T. Kugo and I. Ojima, "Local Covariant Operator Formalism Of Nonabelian Gauge Theories And Quark Confinement Problem," Prog. Theor. Phys. Suppl. **66** (1979) 1.

[41] L. von Smekal, R. Alkofer and A. Hauck, "The infrared behavior of gluon and ghost propagators in Landau gauge QCD," Phys. Rev. Lett. **79** (1997) 3591 [arXiv:hep-ph/9705242], "A solution to coupled Dyson-Schwinger equations for gluons and ghosts in Landau gauge," Annals Phys. **267** (1998) 1 [Erratum-ibid. **269** (1998) 182] [arXiv:hep-ph/9707327]; D. Atkinson and J. C. R. Bloch, "Running coupling in non-perturbative QCD. I: Bare vertices and y-max approximation," Phys. Rev. D **58** (1998) 094036 [arXiv:hep-ph/9712459]; J. M. Pawlowski, D. F. Litim, S. Nedelko and L. von Smekal, "Signatures of confinement in Landau gauge QCD," AIP Conf. Proc. **756** (2005) 278 [arXiv:hep-th/0412326], "Infrared QCD and the renormalisation group," arXiv:hep-th/0410241, "Infrared behaviour and fixed points in Landau gauge QCD," Phys. Rev. Lett. **93** (2004) 152002 [arXiv:hep-th/0312324].

[42] K. Wilson, "Confinement of Quarks," Phys. Rev. **D10** (1974) 2445.

Fields Institute Communications
Volume **50**, 2007

SLE(8/3) and Brownian Excursions in Annuli

Robert O. Bauer
Department of Mathematics
Altgeld Hall
University of Illinois at Urbana-Champaign
1409 West Green Street
Urbana, IL 61801, USA
rbauer@math.uiuc.edu

Abstract. We give a new derivation of the initial/boundary value problem for the probability $F(a,x)$ that a chordal SLE(8/3) in the unit disk $\mathbb{D}$ from e^{ix} to 1 stays in the annulus $A_q \equiv \{q < |z| < 1\}$, where $q = e^a < 1$. We give an exact expression for the probability of this event if SLE(8/3) is replaced by a Brownian excursion and find its asymptotics as $q \nearrow 1$.

1 Introduction

In this paper we give a new derivation of the partial differential equation satisfied by the probability $F(a,x)$ that a chordal SLE(8/3) in the unit disk $\mathbb{D}$ from e^{ix} to 1 stays in the annulus $A_q \equiv \{q < |z| < 1\}$, where $q = e^a < 1$. This partial differential equation was first derived in [1]. We also derive an explicit expression for the probability that a Brownian excursion $\hat{B}$ from e^{ix} to 1 in $\mathbb{D}$ stays in A_q. If q is near 1, then the probability that the Brownian excursion from e^{ix} to 1 stays in A_q is close to the probability that the filling of the Brownian excursion stays in A_q. In fact, it can be shown that the difference of the probabilities is small compared to the probability of either. The filling of a Brownian excursion is a restriction measure with exponent $\alpha = 1$, while SLE(8/3) is a restriction measure of exponent 5/8. Our results show in particular that the asymptotic behavior of $F(a,x)$ established in [1] is the same as the asymptotic behavior of $P(\hat{B} \subset A_q)^{5/8}$ as $q \nearrow 1$.

Chordal restriction measures are studied in [8], and SLE(8/3) in particular is the conjectured scaling limit of the self-avoiding random walk, [9]. SLE in doubly or, more generally, multiply connected domains has been studied in [16], [4], [2].

2 Chordal Loewner equation

2.1 Chordal Loewner equation in the upper half-plane. Suppose γ is a simple curve in the upper half-plane $\mathbb{H}$ with one end on the real axis and the other

---

2000 *Mathematics Subject Classification.* Primary 60K35; Secondary 60H30.

in the interior of $\mathbb{H}$. Then there is a unique conformal map g from $\mathbb{H}\backslash\gamma$ onto $\mathbb{H}$ with

$$g(z) = z + 2a/z + o(1/|z|)$$

at infinity. Here a must be real as g maps reals to reals near infinity. If $\gamma' \subset \gamma$ also has one end on $\mathbb{R}$, and $g(z) = z + 2b/z + o(1/|z|)$ is the unique conformal map from $\mathbb{H}\backslash\gamma'$ onto $\mathbb{H}$, then $a \geq b$, and the inequality is strict if $\gamma\backslash\gamma' \neq \emptyset$, [10]. We may in fact assume that γ is parametrized on $[0, T]$ such that if g_t is the normalized conformal map from $\mathbb{H}\backslash\gamma(0, t]$ onto $\mathbb{H}$, then $g_t(z) = z + 2t/z + o(1/|z|)$. We write $K_t = \gamma(0, t]$. It is well known that there exists a continuous function $t \in [0, T] \mapsto w_t \in \mathbb{R}$, such that for each $z \in \mathbb{H}$

$$\partial_t g_t(z) = \frac{2}{g_t(z) - w_t}, \quad t < t_z, \tag{2.1}$$

where $t_z = \inf\{t : \lim_{s \nearrow t} g_s(z) = w_t\}$. Then $K_t = \{z : t_z \leq t\}$. Equation (2.1) is known as the chordal Loewner equation, [7], [12]. Note that $g_t(\gamma_t) = w_t$.

2.2 Chordal Loewner equation in the unit disk. Let u be the conformal map from the upper half-plane $\mathbb{H}$ onto the unit disk $\mathbb{D}$ given by $u(z) = (z-i)/(z+i)$. In particular,

$$u^{-1}(z) = i\frac{1+z}{1-z}, \quad \text{and} \quad u'(u^{-1}(z)) = -i\frac{(1-z)^2}{2}. \tag{2.2}$$

Set $x_t = u(w_t)$. Then $u \circ g_t \circ u^{-1}$ maps $\mathbb{D}\backslash u(K_t)$ conformally onto $\mathbb{D}$, sending the "tip" of $u(K_t)$ to x_t. We have

$$\partial_t(u \circ g_t \circ u^{-1}(z)) = \Phi(u \circ g_t \circ u^{-1}(z), t),$$

where

$$\Phi(z, t) = \frac{2u'(u^{-1}(z))}{u^{-1}(z) - u^{-1}(x_t)}. \tag{2.3}$$

The residue of $\Phi(\cdot, t)$ at x_t is

$$2u'(w_t)^2 = -\frac{(1-x_t)^4}{2}. \tag{2.4}$$

2.3 Chordal Loewner equation in an annulus. Consider an annulus

$$A_Q = \{Q < |z| < 1\}.$$

Suppose that γ is a simple curve in the annulus with one end on the unit circle, say at $x \neq 1$, and the other in the interior of A_Q. Then there is a unique conformal map $f = f(\cdot, Q)$ from $A_Q\backslash\gamma$ onto an annulus A_{Q^*} $(Q^* > Q)$ such that $f(1) = 1$. In particular, we may assume that γ is parametrized over the interval $[Q, Q^*]$, such that $\gamma(Q) = x$, and for each $q \in [Q, Q^*]$, $A_Q\backslash\gamma[Q, q]$ is conformally equivalent to the annulus A_q. Denote $f(\cdot, q)$ the unique conformal equivalence from $A_Q\backslash\gamma[Q, q]$ onto A_q fixing 1. Then

$$\frac{\partial \ln f(z, q)}{\partial \ln q} = \Xi(f(z, q), q), \tag{2.5}$$

where the vectorfield Ξ is given by

$$\Xi(z, q) = \frac{2\omega_1}{\pi i}\left[\zeta\left(\frac{\omega_1}{\pi}(i\ln z - i\ln y_q)\right) - \frac{\eta_1}{\pi}i\ln z - \zeta\left(-\frac{\omega_1}{\pi}i\ln y_q\right)\right]. \tag{2.6}$$

Here ζ is the Weierstrass-ζ-function with half-periods ω_1 real and ω_2 positive imaginary satisfying

$$q = \exp\left(-\frac{\pi\omega_2}{i\omega_1}\right), \tag{2.7}$$

$\eta_1 = \zeta(\omega_1)$, $\eta_2 = \zeta(\omega_2)$, and $y_q = f(\gamma_q, q)$ is the image of the "tip of the slit," [6]. Note that q only determines the ratio ω_1/ω_2. Ξ is holomorphic in z with a pole at y_q with residue $-2y_q$, and a zero at $1 = f(1,q)$. Because of the parametrization of γ, we have $\Re(\Xi(z,q)) = 1$ on the circle $|z| = q$. This can also be observed directly form (2.6) by noting that ζ maps reals to reals (except for poles), using (2.7), and applying the Legendre relation $\eta_1\omega_2 - \eta_2\omega_1 = \pi i/2$. If $t = t(q)$ is monotone and we set $f_t(z) = f(z,q)$, $y_t = y_q$, then

$$\partial_t f_t(z) = f_t(z)\Xi(f_t(z), q)\frac{\partial \ln q}{\partial t}, \tag{2.8}$$

and the vectorfield $z \mapsto z\ \Xi(z,q)\partial \ln q/\partial t$ has residue $-2y_t^2 \partial \ln q/\partial t$ at $z = y_t$.

2.4 Interior perturbation. Let $0 < q < 1$. If $u(K_t) \subset A_q$ denote f_t the unique conformal equivalence from $A_q \backslash u(K_t)$ onto $A_{q(t)}$. Then f_t satisfies (2.8). Set

$$h_t = f_t \circ u \circ g_t^{-1} \circ u^{-1}.$$

Then h_t is the unique conformal equivalence from $u \circ g_t \circ u^{-1}(\mathbb{D}\backslash u(K_t))$ onto $A_{q(t)}$ fixing 1. Note that $h_t(x_t) = y_t$.

Lemma 2.1 *The vectorfield*

$$z \in u \circ g_t \circ u^{-1}(\mathbb{D}\backslash u(K_t)) \mapsto \partial_t h_t(z)$$

extends continuously to the boundary.

Proof By (2.3), the domains $u \circ g_t \circ u^{-1}(\mathbb{D}\backslash u(K_t))$ change smoothly in t. The map h_t can be written explicitly in terms of domain functionals, namely the harmonic measures and their conjugates. The lemma follows then from Hadamard's formula for the variation of domain functionals under smooth boundary perturbations, [13]. □

Corollary 2.1 *We have*

$$\partial_t h_t(z) = h_t(z)\Xi(h_t(z), q)\partial_t \ln q + \frac{h_t'(z)(1-z)^2}{\frac{1+z}{1-z} - \frac{1+x_t}{1-x_t}},$$

where q is determined by

$$\partial_t \ln q = h_t'(x_t)^2 \frac{(1-x_t)^4}{4h_t(x_t)^2}.$$

Proof By the chain rule

$$\partial_t h_t(z) = \partial_t f_t(u \circ g_t^{-1} \circ u^{-1}(z)) + (f_t \circ u)'(g_t^{-1} \circ u^{-1}(z))\partial_t g_t^{-1}(u^{-1}(z)).$$

Since

$$\partial_t g_t^{-1}(u^{-1}(z)) = -\frac{2(g_t^{-1})'(u^{-1}(z))}{u^{-1}(z) - u^{-1}(x_t)},$$

we get

$$\partial_t h_t(z) = \partial_t f_t(u \circ g_t^{-1} \circ u^{-1}(z)) - \frac{2h_t'(z)u'(u^{-1}(z))}{u^{-1} \circ h_t^{-1}(h_t(z)) - u^{-1} \circ h_t^{-1}(y_t)}.$$

In particular, the second term on the right above has a pole at y_t with residue

$$-2h_t'(x_t)^2 u'(w_t)^2 = h_t'(x_t)^2 \frac{(1-x_t)^4}{2}.$$

The corollary now follows from (2.2) and Lemma 2.1. □

To simplify notation we now set $\omega_1 = \pi$. Then, indicating derivative with respect to time by a dot and suppressing the t subscripts, we have

$$\begin{aligned}\dot{h}(z) = h(z)&\left[\frac{2}{i}\zeta\left(i\ln\frac{h(z)}{h(x)}\right) - \frac{2\eta_1}{\pi}\ln h(z) + \frac{2}{i}\zeta(i\ln h(x))\right]\\ &\times\frac{h'(x)^2(1-x)^4}{4h(x)^2} + \frac{h'(z)(1-z)^2}{\frac{1+z}{1-z}-\frac{1+x}{1-x}}.\end{aligned} \tag{2.9}$$

This implies

$$\begin{aligned}&\dot{h}'(z)\\ &= 2h'(z)\frac{h'(x)^2(1-x)^4}{4h(x)^2}\\ &\times\left[\frac{1}{i}\zeta\left(i\ln\frac{h(z)}{h(x)}\right) - \frac{\eta_1}{\pi}(1+\ln h(z)) + \frac{1}{i}\zeta(i\ln h(x)) - \wp\left(i\ln\frac{h(z)}{h(x)}\right)\right]\\ &+\frac{h''(z)(1-z)^2 - 2h'(z)(1-z)}{\frac{1+z}{1-z}-\frac{1+x}{1-x}} - \frac{2h'(z)}{\left(\frac{1+z}{1-z}-\frac{1+x}{1-x}\right)^2},\end{aligned} \tag{2.10}$$

where $\wp(z) = -\zeta'(z)$ denotes the Weierstrass-$\wp$-function. Using

$$\zeta(z) = \frac{1}{z} + O(|z|^3), \quad \text{and} \quad \wp(z) = \frac{1}{z^2} + O(|z|^2),$$

we find

$$\begin{aligned}\lim_{z\to x}\dot{h}(z) = &\frac{h'(x)^2(1-x)^4}{4h(x)}\left[\frac{2}{i}\zeta(i\ln h(x)) - 3 - \frac{2\eta_1}{\pi}\ln h(x)\right]\\ &+\frac{3}{4}h''(x)(1-x)^4 - \frac{3}{2}h'(x)(1-x)^3,\end{aligned} \tag{2.11}$$

and

$$\begin{aligned}\lim_{z\to x}\dot{h}'(z) = &\frac{h'(x)^3(1-x)^4}{4h(x)^2}\left[\frac{2}{i}\zeta(i\ln h(x)) - \frac{5}{6} - \frac{2\eta_1}{\pi}(1+\ln h(x))\right]\\ &+\frac{3}{2}(1-x)^2h'(x) - \frac{3}{2}(1-x)^3h''(x)\\ &-\frac{h''(x)^2(1-x)^4}{8h'(x)} + \frac{1}{3}h'''(x)(1-x)^4.\end{aligned} \tag{2.12}$$

Equation (2.11) has previously been derived in [4].

3 Chordal stochastic Loewner evolution

Denote B_t a one-dimensional real standard Brownian motion, κ a positive real number, and let $W_t = \sqrt{\kappa}B_t$. For z in the upper half-plane $\mathbb{H}$ consider the ordinary differential equation

$$\partial_t g_t(z) = \frac{2}{g_t(z) - W_t}, \quad g_0(z) = z. \tag{3.1}$$

The solution exists up to a random time $T(z) \in (0,\infty]$ such that $(g_t(z) - W_t) \to 0$ as $t \nearrow T(z)$. Set $K_t = \{z \in \mathbb{H} : T_z \le t\}$. Then g_t is the conformal map from $\mathbb{H}\backslash K_t$ onto $\mathbb{H}$ with $g_t(z) = z + 2t/z + o(1/|z|)$. The construction of the random growing sets $K_t \equiv \{z \in \mathbb{H} : T(z) \le t\}$ by means of the Loewner equation (3.1) goes back to Schramm, [14]. It can be shown that K_t is almost surely generated by a curve

$t \in [0,\infty) \mapsto \gamma_t \in \bar{\mathbb{H}}$ in the sense that K_t is the complement of the unbounded component of $\mathbb{H}\backslash\gamma(0,t]$, [12]. For $\kappa \le 4$, $K_t = \gamma(0,t]$ is known to be a simple curve, while for $4 < \kappa < 8$ the curve γ has double points. For $\kappa > 8$ γ is space filling. The family of conformal maps $\{g_t : t \ge 0\}$ is known as chordal Schramm-Loewner (or stochastic Loewner) evolution.

Set $X_t = u(W_t)$. If $u(K_t) \subset A_q$, define f_t (as before in the deterministic case) as the unique conformal equivalence from $\mathbb{D}\backslash u(K_t)$ onto $A_{q(t)}$. Because K_t and thus also $u(K_t)$ only grow at a single point at a time, f_t still satisfies (2.8), sending the point of growth (the tip of the slit) to a random process Y_t on S^1. Set $h_t = f_t \circ u \circ g_t^{-1} \circ u^{-1}$, as before. Then $h_t(X_t) = Y_t$, and all results from Section 2 continue to hold. In particular, replacing x by X, equations (2.11) and (2.12) also hold in the stochastic case. Thus, from an appropriate version of Itô's formula ([11]), we find for the Itô differentials $dh_t(X_t)$ and $dh_t'(X_t)$,

$$\begin{aligned} dh(X) &= \dot{h}(X)\,dt + h'(X)\,dX + \frac{1}{2}h''(X)\,dX\cdot dX \\ &= h'(X)\frac{(1-X)^2}{2i}\sqrt{\kappa}\,dB \\ &\quad + \frac{6-\kappa}{4}\left[\frac{(1-X)^4}{2}h''(X) - (1-X)^3h'(X)\right]\,dt \\ &\quad + \frac{h'(x)^2(1-x)^4}{4h(x)}\left[\frac{2}{i}\zeta(i\ln h(x)) - 3 - \frac{2\eta_1}{\pi}\ln h(x)\right]dt, \end{aligned} \tag{3.2}$$

and

$$\begin{aligned} \frac{dh'(X)}{h'(X)} &= \frac{h''(X)}{h'(X)}\frac{(1-X)^2}{2i}\sqrt{\kappa}\,dB \\ &\quad + \frac{3}{2}(1-X)^2\,dt - \frac{h''(X)^2(1-X)^4}{8h'(X)^2}\,dt \\ &\quad - \frac{6-\kappa}{4}(1-X)^3\frac{h''(X)}{h'(X)}\,dt + \frac{8/3-\kappa}{8}\frac{h'''(X)}{h'(X)}(1-X)^4\,dt \\ &\quad + \frac{h'(x)^2(1-x)^4}{4h(x)^2}\left[\frac{2}{i}\zeta(i\ln h(x)) - \frac{5}{6} - \frac{2\eta_1}{\pi}(1+\ln h(x))\right]dt, \end{aligned} \tag{3.3}$$

where the t subscript is suppressed for ease of notation.

Lemma 3.1 *Set* $Y_t = (u^{-1} \circ h_t \circ u)'(W_t)$ *and let* α *be a real number. Then*

$$\begin{aligned}
\frac{dY^\alpha}{\alpha Y^\alpha} &= i(1-X)\left[1 - \frac{h'(X)(1-X)}{1-h(X)} - \frac{h''(X)(1-X)}{2h'(X)}\right]\sqrt{\kappa}\, dB \\
&+ \left[\frac{1+h(X)}{1-h(X)}\left(\frac{2}{i}\zeta(i\ln h(X)) - \frac{2\eta_1}{\pi}\ln h(X)\right) - \frac{5}{6} - \frac{2\eta_1}{\pi}\right] d\ln q \\
&- (2\alpha+1)\kappa\frac{h(X)}{(1-h(X))^2}\, d\ln q \\
&+ \frac{(1-X)^4}{8}\left[(8/3-\kappa)\frac{h'''(X)}{h'(X)} - (1+(\alpha-1)\kappa)\frac{h''(X)^2}{h'(X)^2}\right] dt \\
&+ \frac{6-(2\alpha+1)\kappa}{4}\left[(1-X)^2 - \frac{h''(X)(1-X)^3}{h'(X)} + \frac{h''(X)(1-X)^4}{1-h(X)}\right] dt \\
&- \frac{6-(2\alpha+1)\kappa}{4}\left[\frac{2h'(X)(1-X)^3}{1-h(X)} + \frac{h'(X)(1-X)^4}{h(X)(1-h(X))}\right] dt. \qquad (3.4)
\end{aligned}$$

Proof First,

$$(u^{-1} \circ h_t \circ u)'(u^{-1}(z)) = h_t'(z)\frac{(1-z)^2}{(1-h_t(z))^2}.$$

Thus, suppressing the subscript t, $Y = h'(X)[(1-X)/(1-h(X))]^2$. Next,

$$dX_t = \frac{(1-X_t)^2}{2i}\sqrt{\kappa}\, dB_t + \frac{(1-X_t)^3}{4}\kappa\, dt,$$

and we find by application of an appropriate version of Itô's formula

$$\begin{aligned}
\frac{d\left(\frac{1-X}{1-h(X)}\right)^2}{\left(\frac{1-X}{1-h(X)}\right)^2} &= i(1-X)\sqrt{\kappa}\, dB + \frac{h'(X)(1-X)^3}{1-h(X)}\kappa\, dt \\
&\quad - \frac{3}{4}\kappa\left[(1-X)^2 + \frac{h'(X)^2}{(1-h(X))^2}\right] dt + \frac{2\, dh(X)}{1-h(X)}. \qquad (3.5)
\end{aligned}$$

Since

$$\frac{dY}{Y} = \frac{dh'(X)}{h'(X)} + \frac{d\left(\frac{1-X}{1-h(X)}\right)^2}{\left(\frac{1-X}{1-h(X)}\right)^2} + \frac{dh'(X)}{h'(X)} \cdot \frac{d\left(\frac{1-X}{1-h(X)}\right)^2}{\left(\frac{1-X}{1-h(X)}\right)^2}, \qquad (3.6)$$

we get

$$\begin{aligned}\frac{dY}{Y} &= i(1-X)\left[1-\frac{h'(X)(1-X)}{1-h(X)}-\frac{h''(X)(1-X)}{2h'(X)}\right]\sqrt{\kappa}\,dB\\&+\left[\frac{1+h(X)}{1-h(X)}\left(\frac{2}{i}\zeta(i\ln h(X))-\frac{2\eta_1}{\pi}\ln h(X)\right)-\frac{5}{6}-\frac{2\eta_1}{\pi}\right]d\ln q\\&-\frac{3}{2}\frac{h'(X)^2(1-X)^4}{h(X)(1-h(X))}\,dt\\&+\frac{3}{4}(2-\kappa)\left[(1-X)^2-\frac{h''(X)(1-X)^3}{h'(X)}+\frac{h''(X)(1-X)^4}{1-h(X)}\right]dt\\&-\frac{3}{2}(2-\kappa)\frac{h'(X)(1-X)^3}{1-h(X)}\,dt-\frac{h''(X)^2(1-X)^4}{8h'(X)^2}\,dt\\&-\frac{3\kappa}{4}\frac{h'(X)^2(1-X)^4}{(1-h(X))^2}\,dt+\frac{8/3-\kappa}{8}\frac{h'''(X)}{h'(X)}(1-X)^4\,dt,\end{aligned}\tag{3.7}$$

where we combined the two angular-bracket-terms involving the ζ-function using $1+2h/(1-h)=(1+h)/(1-h)$. Finally,

$$\frac{dY^\alpha}{\alpha Y^\alpha}=\frac{dY}{Y}+\frac{\alpha-1}{2}\frac{dY}{Y}\cdot\frac{dY}{Y}.$$

Using

$$\begin{aligned}\frac{\frac{dY}{Y}\cdot\frac{dY}{Y}}{\kappa\,dt} &= -(1-X)^2+\frac{h''(X)(1-X)^3}{h'(X)}+\frac{2h'(X)(1-X)^3}{1-h(X)}\\&-\frac{h''(X)^2(1-X)^4}{4h'(X)^2}-\frac{h''(X)(1-X)^4}{1-h(X)}-\frac{h'(X)^2(1-X)^4}{(1-h(X))^2},\end{aligned}\tag{3.8}$$

and

$$-\frac{3}{2}-\frac{(2\alpha+1)\kappa}{4}\frac{h}{1-h}=-\frac{(2\alpha+1)\kappa}{4}\frac{1}{1-h}-\frac{6-(2\alpha+1)\kappa}{4},$$

we get (3.4). □

4 A PDE for the non-intersection probability

Let $h(X)=e^{iV}$. Then

$$\begin{aligned}dV &= \frac{1}{i}\frac{dh(X)}{h(X)}-\frac{1}{i}\frac{dh(X)\cdot dh(X)}{2h(X)^2}\\&= -\frac{h'(X)(1-X)^2}{2h(X)}\sqrt{\kappa}\,dB+\left[2\zeta(\nu)-\frac{2\eta_1}{\pi}\nu\right]d\ln q\\&+\frac{6-\kappa}{8i}(1-X)^4\left[\frac{h''(X)}{h(X)}-\frac{h'(X)^2}{h(X)^2}-\frac{2h'(X)}{h(X)(1-X)}\right]dt.\end{aligned}\tag{4.1}$$

Theorem 4.1 *Let $(q,\nu)\mapsto F(\ln q,\nu)$ be smooth. Then*

$$
\begin{aligned}
&\frac{d[F(\ln q,V)Y^\alpha]}{Y^\alpha}\\
&=\partial_{\ln q}F\ d\ln q\\
&+\partial_\nu F\left\{-\sqrt{\kappa}\ dB_{\ln q}+\left[2\zeta(V)-\frac{2\eta_1}{\pi}V-\alpha\kappa\cot\frac{V}{2}\right]d\ln q\right.\\
&\qquad\left.+\frac{6-(2\alpha+1)\kappa}{8i}(1-X)^4\left[\frac{h''(X)}{h(X)}-\frac{2h'(X)}{h(X)(1-X)}-\frac{h'(X)^2}{h(X)^2}\right]\right\}\\
&+\frac{\kappa}{2}\partial_{\nu\nu}F\ d\ln q\\
&+\alpha F\left\{i(1-X)\left[1-\frac{h'(X)(1-X)}{1-h(X)}-\frac{h''(X)(1-X)}{2h'(X)}\right]\sqrt{\kappa}\ dB\right.\\
&\qquad-\left(\cot\frac{\nu}{2}\left[2\zeta(V)-\frac{2\eta_1}{\pi}V\right]+\frac{5}{6}+\frac{2\eta_1}{\pi}\right)d\ln q\\
&\qquad+\frac{(2\alpha+1)\kappa}{4}\csc^2\frac{V}{2}\ d\ln q\\
&\qquad+(1-X)^4\left[\frac{8/3-\kappa}{8}\frac{h'''(X)}{h'(X)}-\frac{1+(\alpha-1)\kappa}{8}\frac{h''(X)^2}{h'(X)^2}\right]dt\\
&\qquad+\frac{6-(2\alpha+1)\kappa}{4}\left[(1-X)^2-\frac{h''(X)(1-X)^3}{h'(X)}+\frac{h''(X)(1-X)^4}{1-h(X)}\right]dt\\
&\qquad\left.-\frac{6-(2\alpha+1)\kappa}{4}\left[\frac{2h'(X)(1-X)^3}{1-h(X)}+\frac{h'(X)(1-X)^4}{h(X)(1-h(X))}\right]dt\right\}. \qquad (4.2)
\end{aligned}
$$

Denote $\mathcal{S}h$ the Schwarzian derivative of h,

$$\mathcal{S}h(z)=\frac{h'''(z)}{h'(z)}-\frac{3}{2}\left(\frac{h''(z)}{h'(z)}\right)^2.$$

Corollary 4.1 *For $\kappa>0$ let α be such that $6-(2\alpha+1)\kappa=0$, i.e. $\alpha=3/\kappa-1/2$. Then*

$$
\begin{aligned}
&\frac{d[F(\ln q,V)Y^\alpha]}{Y^\alpha}\\
&=d(loc.\ mart.)+\partial_{\ln q}F\ d\ln q\\
&+\partial_\nu F\left[2\zeta(V)-\frac{2\eta_1}{\pi}V-(3-\kappa/2)\cot\frac{V}{2}\right]d\ln q\\
&+\frac{\kappa}{2}\partial_{\nu\nu}F\ d\ln q\\
&+\alpha F\left\{\left(\frac{3}{2}\csc^2\frac{V}{2}-\cot\frac{V}{2}\left[2\zeta(V)-\frac{2\eta_1}{\pi}\nu\right]-\frac{5}{6}-\frac{2\eta_1}{\pi}\right)d\ln q\right.\\
&\qquad\left.+(1-X)^4\frac{8/3-\kappa}{8}\mathcal{S}h(X)\ dt\right\}. \qquad (4.3)
\end{aligned}
$$

Corollary 4.2 *If $a = \ln q$, $\kappa = 8/3$, and $\alpha = 3/\kappa - 1/2 = 5/8$, then $F(a,V)Y^\alpha$ is a local martingale if and only if*

$$-\partial_a F = \partial_{\nu\nu} F \cdot \frac{4}{3} + \partial_\nu F \left[2\zeta(\nu) - \frac{2\eta_1}{\pi}\nu - \frac{5}{3}\cot\frac{\nu}{2}\right] + F\left\{\frac{15}{16}\csc^2\frac{\nu}{2} - \frac{5}{8}\left(\cot\frac{\nu}{2}\left[2\zeta(\nu) - \frac{2\eta_1}{\pi}\nu\right] + \frac{5}{6} + \frac{2\eta_1}{\pi}\right)\right\}. \tag{4.4}$$

If $F(a,\nu)$, is the probability that chordal SLE(8/3) in the unit disk from $e^{i\nu}$ to 1 avoids the disk of radius $q = e^a (< 1)$ centered at zero, then by a result of Beffara [3],

$$F(a,V)Y^{5/8}[(u^{-1}\circ h \circ u)'(\infty)]^{5/8}$$

is a martingale. Here the derivative of a conformal map f at infinity is defined as $[\lim_{z\to\infty} f'(z)]^{-1}$. By (2.10),

$$\begin{aligned}\dot{h}'(1) &= -2h'(1)\frac{h'(x)^2(1-x)^4}{4h'(x)^2}\left[\wp(i\ln h(x)) + \frac{\eta_1}{\pi}\right] \\ &= -[2\wp(i\ln h(x)) + \frac{2\eta_1}{\pi}]\partial_t \ln q. \end{aligned} \tag{4.5}$$

Thus, if the probability $F(a,\nu)$ is $C^{1,2}$, then

$$-\partial_a F = \partial_{\nu\nu} F \cdot \frac{4}{3} + \partial_\nu F \left[2\zeta(\nu) - \frac{2\eta_1}{\pi}\nu - \frac{5}{3}\cot\frac{\nu}{2}\right] + F\left\{\frac{15}{16}\csc^2\frac{\nu}{2} - \frac{5}{8}\left(\cot\frac{\nu}{2}\left[2\zeta(\nu) - \frac{2\eta_1}{\pi}\nu\right] + \frac{5}{6} + 2\wp(\nu) + \frac{4\eta_1}{\pi}\right)\right\}. \tag{4.6}$$

It can be shown from the stochastic representation of $F(a,\nu)$ derived in [1] that F is indeed $C^{1,2}$,

$$F(a,\nu) = \left[\prod_{n=1}^{\infty} \frac{1-2q^{2n}+q^{4n}}{1-2q^{2n}\cos\nu + q^{4n}}\right]^{3/4} \times \mathbb{E}\left[\exp\left(-\int_a^{a^*} \left(\wp(Y_b) - \frac{1}{4}\csc^2(Y_b/2)\right) db\right), a^* < 0\right], \tag{4.7}$$

where Y is a Legendre process ([11]) defined by $dY = -\sqrt{8/3}dB - \frac{2}{3}\cot(Y/2)\,db$ with $Y_a = \nu$ and $a^* = \inf\{b \geq a : Y_b \in \{0, 2\pi\}\}$. The derivation of the stochastic representation (4.7) rests on the estimate

$$F(a,\nu) \asymp \exp\left(-\frac{5\pi}{8}\cdot\frac{\nu}{1-q}\right), \quad q = e^a \nearrow 1, \tag{4.8}$$

where $\nu \in [0,\pi]$, and $f(a) \asymp g(a)$ as $a \nearrow 0$ means

$$\lim_{a\nearrow 0} \ln f(a)/\ln g(a) = 1.$$

Rather then sketching the proof of (4.8) given in [1] and, for an alternative derivation, in [15], we will obtain below a related estimate from an exact expression for Brownian excursions.

5 Brownian excursion in an annulus

If $\hat{B}$ denotes a Brownian excursion in the unit disk $\mathbb{D}$ from e^{ix} to 1, and B a standard 2-dimensional Brownian motion, then

$$P(\hat{B} \subset A_q) = \lim_{\epsilon \to 0} P^{(1-\epsilon)e^{ix}}(B(0,\tau_{\mathbb{D}}) \subset A_q | B_{\tau_{\mathbb{D}}} \in \{e^{is} : |s| < \epsilon\}),$$

where τ_D denotes the exit time of B from the domain D. Thus

$$P(\hat{B} \subset A_q) = \lim_{\epsilon \searrow 0} \frac{P^{(1-\epsilon)e^{ix}}(B_{\tau_{A_q}} \in \{e^{is} : |s| < \epsilon\})}{P^{(1-\epsilon)e^{ix}}(B_{\tau_{\mathbb{D}}} \in \{e^{is} : |s| < \epsilon\})}$$

$$= \frac{\frac{\partial^2}{\partial n_1 \partial n_2} G_{A_q}(e^{ix}, 1)}{\frac{\partial^2}{\partial n_1 \partial n_2} G_{\mathbb{D}}(e^{ix}, 1)} = \frac{H_{A_q}(e^{ix}, 1)}{H_{\mathbb{D}}(e^{ix}, 1)}, \tag{5.1}$$

where G_D and H_D denote the Green function and the Boundary Poisson kernel of the domain D, respectively, and $\partial/\partial n_1$, $\partial/\partial n_2$ denote the normal derivative with respect to the first and second argument. For more information on excursion measures and the boundary Poisson kernel see [7].

Using Villat's kernel, [6], we find the exit density with respect to normalized Lebesgue measure on the unit circle (i.e. so that the circle's length is 1) for Brownian motion in the annulus A_q at 1 and starting from re^{ix}, $(r < 1)$, to be

$$\frac{1 - r^2}{1 - 2r\cos x + r^2} - \frac{\ln r}{\ln q} - \sum_{n=1}^{\infty} \frac{2q^{2n}}{1 - q^{2n}}(1 - r^{2n})\frac{\cos nx}{r^n}. \tag{5.2}$$

Note that in the limit $q \searrow 0$, the punctured disk, we obtain the Poisson kernel of the unit disk,

$$\frac{1 - r^2}{1 - 2r\cos x + r^2}.$$

To get the normal derivative of (5.2), we divide by $1 - r$ and let $r \nearrow 1$, giving

$$H_{A_q}(e^{ix}, 1) = \frac{1}{1 - \cos x} + \frac{1}{\ln q} - \sum_{n=1}^{\infty} \frac{4nq^{2n}}{1 - q^{2n}} \cos nx, \tag{5.3}$$

and

$$H_{\mathbb{D}}(e^{ix}, 1) = \frac{1}{1 - \cos x}. \tag{5.4}$$

Thus, using the Weierstrass $\wp$-function,

$$\frac{H_{A_q}(e^{ix}, 1)}{H_{\mathbb{D}}(e^{ix}, 1)} = 2\sin^2(x/2)\left[\frac{1}{a} + 2\left(\wp(x) + \frac{\eta_1}{\pi}\right)\right].$$

If $\vartheta_1(v;\tau)$ is the Jacobi theta function, then

$$-\wp(x) - \frac{\eta_1}{\pi} = \frac{1}{4\pi^2}(\ln \vartheta_1)''\left(\frac{x}{2\pi}\right).$$

Here $i\pi\tau = a$. Our notation in this section follows [5]. By a classical transformation formula, see [5, (II,7.,§2,(7))],

$$\vartheta_1(v,\tau) = \vartheta_1\left(\frac{v}{\tau}; -\frac{1}{\tau}\right) i\sqrt{\frac{i}{\tau}} e^{-\frac{i\pi}{\tau}v^2}.$$

Thus

$$(\ln \vartheta_1)''(v;\tau) = (\ln \vartheta_1)''\left(\frac{v}{\tau}; -\frac{1}{\tau}\right) \cdot \frac{1}{\tau^2} - \frac{2\pi i}{\tau}. \tag{5.5}$$

Since

$$\vartheta_1(v;\tau) = f(\tau)\left(e^{i\pi v} - e^{-i\pi v}\right)\prod_{n=1}^{\infty}\left(1 - e^{2i\pi\tau n}e^{2i\pi v}\right)\left(1 - e^{2i\pi\tau n}e^{-2i\pi v}\right),$$

we have

$$\begin{aligned}\vartheta_1\left(\frac{v}{\tau};-\frac{1}{\tau}\right) &= f\left(-\frac{1}{\tau}\right)\left(e^{-\pi^2 v/a} - e^{\pi^2 v/a}\right)\\ &\quad\times\prod_{n=1}^{\infty}\left(1 - \left[e^{\frac{2\pi^2}{a}(n-v)} + e^{\frac{2\pi^2}{a}(n+v)}\right] + e^{\frac{2\pi^2}{a}n}\right).\end{aligned}$$

So, from (5.5) follows

$$\begin{aligned}&(\ln\vartheta_1)''(v;\tau)\\ &= \frac{2\pi^2}{a} - \frac{4\pi^4}{a^2}\cdot\frac{e^{\frac{2\pi^2}{a}v}}{\left(1 - e^{\frac{2\pi^2}{a}v}\right)^2}\\ &\quad - \frac{4\pi^4}{a^2}\sum_{n=1}^{\infty}\frac{e^{\frac{2\pi^2}{a}(n-v)} + e^{\frac{2\pi^2}{a}(3n-v)} + e^{\frac{2\pi^2}{a}(n+v)} + e^{\frac{2\pi^2}{a}(3n+v)} - 4e^{\frac{4\pi^2}{a}n}}{\left(1 - \left[e^{\frac{2\pi^2}{a}(n-v)} + e^{\frac{2\pi^2}{a}(n+v)}\right] + e^{\frac{4\pi^2}{a}n}\right)^2}\end{aligned} \tag{5.6}$$

Thus, with $v = \frac{x}{2\pi}$,

$$\begin{aligned}&\frac{H_{A_q}(e^{ix},1)}{H_{\mathbb{D}}(e^{ix},1)}\cdot\left(\frac{4\pi^2}{a^2}\sin^2\frac{x}{2}\right)^{-1}\\ &= \frac{e^{\frac{\pi}{a}x}}{\left(1 - e^{\frac{\pi}{a}x}\right)^2} + \sum_{n=1}^{\infty}\frac{e^{\frac{\pi}{a}(2\pi n - x)} + e^{\frac{\pi}{a}(6\pi n - x)} + e^{\frac{\pi}{a}(2\pi n + x)} + e^{\frac{\pi}{a}(6\pi n + x)} - 4e^{\frac{4\pi^2}{a}n}}{\left(1 - \left[e^{\frac{\pi}{a}(2\pi n - x)} + e^{\frac{\pi}{a}(2\pi n + x)}\right] + e^{\frac{4\pi^2}{a}n}\right)^2}\end{aligned} \tag{5.7}$$

From this explicit expression, we find by inspection,

$$\lim_{a\nearrow 0}\frac{P(\hat{B}\subset A_q)}{\frac{4\pi^2}{a^2}\sin^2\frac{x}{2}\ e^{\frac{\pi}{a}\min\{x,2\pi-x\}}} = \begin{cases}1 & \text{if } x\in(0,\pi)\cup(\pi,2\pi),\\ 2 & \text{if } x = \pi.\end{cases} \tag{5.8}$$

To derive the estimate (4.8) from (5.8), one needs to show that the probability that a Brownian excursion from e^{ix} to 1 in $\mathbb{D}$ disconnects the circle $\{|z| = q\}$ from the circle $\{|z| = 1\}$ decays even faster than (5.8).

References

[1] Bauer, Robert O., *Restricting SLE(8/3) to an annulus*, preprint, arXiv:math.PR/0602391.

[2] Bauer, Robert O., Friedrich, Roland M., *On chordal and bilateral stochastic Loewner evolution in multiply connected domains*, available at www.arXiv.org, math.PR/0503178.

[3] Beffara, Vincent, Thesis, University of Paris, Orsay, 2003.

[4] Dubédat, Julien, *Critical percolation in annuli and* SLE$_6$, Comm. Math. Phys., **245**, 2004, no. 3, 627–637.

[5] Hurwitz, Adolf, Vorlesungen über allgemeine Funktionentheorie und elliptische Funktionen, 4th edition, Springer-Verlag, Berlin, 1964.

[6] Komatu, Yûsaku, *Untersuchungen über konforme Abbildung von zweifach zusammenhängenden Gebieten*, Proc. Phys.-Math. Soc. Japan (3), **25**, 1943, 1–42.

[7] Lawler, Gregory F., Conformally invariant processes in the plane, Mathematical Surveys and Monographs, 114, American Mathematical Society, Providence, RI, 2005.

[8] Lawler, Gregory and Schramm, Oded and Werner, Wendelin, *Conformal restriction: the chordal case*, J. Amer. Math. Soc., **16**, 2003, no. 4, 917–955 (electronic).

[9] Lawler, Gregory F. and Schramm, Oded and Werner, Wendelin, *On the scaling limit of planar self-avoiding walk*, in "Fractal geometry and applications: a jubilee of Benoît Mandelbrot, Part 2," Proc. Sympos. Pure Math., vol. 72, 339–364, Amer. Math. Soc., Providence, RI, 2004.

[10] Löwner, Karl, *Untersuchungen über schlichte konforme Abbildungen des Einheitskreises. I*, Math. Ann., **89**, 1923, no. 1-2, 103–121.

[11] Revuz, Daniel and Yor, Marc, Continuous martingales and Brownian motion, Grundlehren der Mathematischen Wissenschaften [Fundamental Principles of Mathematical Sciences], 293, 3rd edition, Springer-Verlag, Berlin, 1999.

[12] Rohde, Steffen and Schramm, Oded, *Basic properties of SLE*, Ann. of Math. (2), **161**, no. 2, 2005, pp. 883–924,

[13] Schiffer, Menahem and Spencer, Donald C., Functionals of finite Riemann surfaces, Princeton University Press, Princeton, N. J., 1954.

[14] Schramm, Oded, *Scaling limits of loop-erased random walks and uniform spanning trees*, Israel J. Math., **118**, 2000, 221–288.

[15] Werner, Wendelin, *The conformally invariant measure on self-avoiding loops*, available at www.arxiv.org, math.PR/0511605.

[16] Zhan, Dapeng, *Stochastic Loewner evolution in doubly connected domains*, Probab. Theory Related Fields, **129**, 2004, no. 3, 340–380.

Fields Institute Communications
Volume **50**, 2007

Cardy's Formula on the Triangular Lattice, the Easy Way

Vincent Beffara
CNRS
UMPA – ENS-Lyon
46 allée d'Italie
69364 Lyon Cedex 07, France
vbeffara@ens-lyon.fr

Abstract. We give a simple derivation of Cardy's formula for site-percolation in the triangular lattice, as proved by Smirnov; our main goal, besides simplifying the proof, is to pinpoint why the triangular lattice is so special ...

Introduction

The problem of determining the scaling limit of critical percolation in two dimensions is long-standing, and although there has been substantial progress in the understanding of that limit *via* the study of SLE_6 processes, convergence is known only in the case of critical site-percolation on the triangular lattice, for which the essential estimate, known as Cardy's formula, was proved by Smirnov in a much celebrated paper (see [5, 6]).

In several steps of Smirnov's proof, the specific structure of the triangular lattice is used in a seemingly essential way, and it is a very natural question to try and determine where exactly it is necessary to be in this specific case, and where the construction can be unified — and thus simplified ...

The aim of this short note is to provide a mostly self-contained re-writing of Smirnov's proof, written in such a way as to emphasize the one key part of the proof where the precise geometry of the graph is used. As such, it does not contain any new result *per se* (except for the quite anecdotal Corollary 2.1), but we do believe that it will provide the reader with some insight into the model.

For general background on percolation, we refer the avid reader to the books of Grimmett [2] and Kesten [3]; and besides Smirnov's papers, related work on the scaling limit of percolation includes a paper of Camia and Newman [1], where they provide the full details about the convergence of percolation cluster boundaries to SLE processes, and a paper of Smirnov and Werner [7] where they derive the values of critical exponents for 2D percolation.

2000 *Mathematics Subject Classification.* Primary 60K35, 82B27, 82B43; Secondary 60D05.

We will keep the same notation as in Smirnov's paper as much as possible; at various points we will be a bit terse or even a bit sloppy, since writing the proof in full would take many more pages, be tedious to read and probably not be more informative.

Acknowledgments. Of course I would like to thank the organizers of the workshop, and especially Ilia Binder who convinced me to contribute to these proceedings. I would also like to mention that part of this construction was done independently by Tóth and his student Bálázs Rath, who first had the idea of replacing H in Smirnov's proof with the more symmetric version we use here.

1 Notations and setup

1.1 The graph. Let G_δ be the hexagonal lattice, embedded as usual to preserve its rotational symmetry of order 3, and scaled by a factor $\delta > 0$ (which we will refer to as the *mesh* of the lattice). The vertices of G_δ all have degree 3. In all the following discussion, e will always stand for an oriented edge of G_δ — which, by a slight abuse of notation, we will denote as $e \in G_\delta$.

Let G_δ^* be the dual graph of G_δ, defined as follows: The set of vertices (resp. faces) of G_δ^* is in bijection with the set of faces (resp. vertices) of G_δ, and there is an edge between two vertices of G_δ^* if and only if the two corresponding faces of G_δ share an edge. It can be embedded in the plane by choosing each of its vertices to be the center of the corresponding regular hexagon in G_δ, in which case it is exactly the usual triangular lattice, scaled by the same factor of δ. If e is an edge of G_δ, we will denote by e^* its dual edge, *i.e.* the unique edge of G_δ^* intersecting it, oriented in such a way that the frame (e, e^*) is direct.

For each vertex x of G_δ, there are three oriented edges having x as their origin. Letting $\tau = e^{2\pi i/3}$, if e is one of these edges, we will denote the other two by $\tau.e$ and $\tau^2.e$. Seeing each edge as a complex number (the difference between the positions of its endpoint and its origin), this corresponds to complex multiplication; however, we prefer seeing it as an action of the group $\{1, \tau, \tau^2\}$ on the set of oriented edges of G_δ, since this will make the constructions below easier. When needed we will also use the notation $-e$ for the edge sharing the same endpoints as e but with opposite orientation.

Notice the following, fundamental identity: For each oriented edge $e \in G_\delta$,

$$e^* + \tau(\tau.e)^* + \tau^2(\tau^2.e)^* = 0. \tag{1.1}$$

Here, $\tau(\tau.e)^*$ means the product of two complex numbers, namely τ and the difference between the endpoint and the origin of $(\tau.e)^*$ — we will stick to this convention, that a dot means the action of τ on edges while no dot means complex multiplication. This identity is actually trivial (using the fact that the action of τ on edges can itself be seen as complex multiplication, the sum above is equal to $(1+\tau^2+\tau^4)e^*$), but as it will appear it is essentially the only place in the construction where we use the fact that G_δ^* is the usual, equilateral triangular lattice.

It is to be noted that the above identity actually characterizes the triangular lattice; more precisely, the sum is 0 if and only if the triangle around the origin of e is equilateral. So, it seems that the triangular lattice is the only one (apart from trivial modifications of it) in which a fully combinatorial proof like follows is possible ...

1.2 The model. Let Ω be a smooth, bounded, simply connected domain in the complex plane. We will be interested in critical site-percolation on the intersection of Ω with G_δ^*. More specifically, the question we are interested in is the following. Let A, B, C and D be four points on the boundary of Ω, in that order. For every $\delta > 0$, let Ω_δ be the largest connected component (in terms of graph connectivity) of the intersection of Ω with G_δ, and let Ω_δ^* be its dual graph. Ω_δ should be seen as a discretization of Ω at scale δ. Let A_δ, B_δ, C_δ and D_δ be the vertices of Ω_δ that are closest to A, B, C and D respectively.

The model we are considering is now critical site-percolation on Ω_δ^*, *i.e.*, each vertex of Ω_δ^* is taken to be open with probability $p_c = 1/2$, independently of all the others. Let $C_\delta(\Omega, A, B, C, D)$ be the event that there is an open crossing in Ω_δ^*, between the intervals $A_\delta B_\delta$ and $C_\delta D_\delta$ of its boundary. Standard Russo-Seymour-Welsh estimates (cf. [2] for instance) tells us that at criticality, the probability of $C_\delta(\Omega, A, B, C, D)$ is bounded away from both 0 and 1 as δ goes to 0; The main result we prove here is the following:

Theorem 1.1 (Smirnov) *The probability of the event $C_\delta(\Omega, A, B, C, D)$ has a limit $f(\Omega, A, B, C, D)$ as δ goes to 0. Moreover, the limit is conformally invariant, in the following sense: If Φ is a conformal map from Ω to another simply connected domain $\Omega' = \Phi(\Omega)$, and extends continuously to $\partial\Omega$, then*

$$f(\Omega, A, B, C, D) = f(\Phi(\Omega), \Phi(A), \Phi(B), \Phi(C), \Phi(D)).$$

As will appear naturally in the proof and was first pointed to by Carleson, f has a most simple expression in the case where Ω itself is an equilateral triangle with vertices A, B and C and D on the interval (CA): Then $f(\Omega, A, B, C, D) = |CD|/|CA|$. By conformal invariance, this gives the value of f for every conformal rectangle.

2 The proof

2.1 General framework. The framework of the proof here is the same as in Smirnov's proof, but we describe it in some detail for sake of self-containedness. Recall that Ω_δ is a discrete approximation of Ω; for every point z be a point of the Ω_δ lying inside one of the triangles of Ω_δ^*, we will denote by $E_{A,\delta}(z)$ the event that there exists a simple path of open vertices in Ω_δ^*, separating A_δ and z from B_δ and C_δ — and $E_{B,\delta}(z)$, $E_{C,\delta}(z)$ similarly, with obvious circular permutations of the letters. Let $H_{A,\delta}(z)$ (resp. $H_{B,\delta}(z)$, $H_{C,\delta}(z)$) be the probability of $E_{A,\delta}(z)$ (resp. $E_{B,\delta}(z)$, $E_{C,\delta}(z)$). From now on, in many cases where this does not introduce confusion, we will omit the mention of δ in those notations.

From Russo-Seymour-Welsh theory, we obtain the following: There are two positive constants K_E and ε_E such that, for every $\delta > 0$ and any two points z and z' in Ω_δ (up to the boundary of the domain),

$$|H_{A,\delta}(z') - H_{A,\delta}(z)| \leqslant K_E \, |z' - z|^{\varepsilon_E} \tag{2.1}$$

and a similar bound for H_B and H_C. Hence, if we suitably extend these functions continuously to Ω, we obtain a family of uniformly Hölder maps from Ω to $\mathbb{R}$. The family is then relatively compact with respect to uniform convergence, and it is hence possible to extract a subsequence $(H_{A,\delta_n}, H_{B,\delta_n}, H_{C,\delta_n})_{n>0}$, with $\delta_n \to 0$, which converges uniformly to a triple of Hölder maps (h_A, h_B, h_C) from $\overline{\Omega}$ to $[0, 1]$, with the same exponent and norm as the H_A's.

If we can identify the triple (h_A, h_B, h_C) uniquely, we will obtain convergence of the family $(H_{A,\delta}, H_{B,\delta}, H_{C,\delta})$ as δ goes to 0. Then, making the additional remark that $E_{C,\delta}(D_\delta)$ is the same event as $C_\delta(\Omega, A, B, C, D)$, this will conclude the proof of the first statement in Theorem 1.1.

The way we will do that is slightly different from Smirnov's, the difference being essentially in the determination of the boundary conditions (see Subsection 2.4). Introduce the following two functions:

$$H_\delta(z) := H_{A,\delta}(z) + \tau H_{B,\delta}(z) + \tau^2 H_{C,\delta}(z) \qquad S_\delta(z) = H_{A,\delta}(z) + H_{B,\delta}(z) + H_{C,\delta}(z).$$

(This is slightly different from Smirnov's H, but the idea is exactly the same and the version we use here is more symmetric.) They also form a family of uniformly Hölder maps, and the subsequence (H_{δ_n}) (resp. S_{δ_n}), with the same (δ_n) as were introduced previously, converges to

$$h := h_A + \tau h_B + \tau^2 h_C, \qquad \text{resp.} \quad s := h_A + h_B + h_C.$$

The key step of the proof, to which the next two sections are devoted, is to prove that the functions h and s are holomorphic on Ω. Subsection 2.4 then concludes the proof. Notice though that since s is both holomorphic and real-valued, it has to be constant, and it is easy to see from boundary conditions that it is actually equal to 1:

Corollary 2.1 *As the mesh of the discretization vanishes, the sum of the probabilities of $E_{A,\delta}(z)$, $E_{B,\delta}(z)$ and $E_{C,\delta}(z)$ converges to 1, uniformly in $z \in \overline{\Omega}$.*

The exact same proof works for any triangulation in which Russo-Seymour-Welsh holds, so this fact seems to be a fundamental property of critical two-dimensional percolation (and *might* be the key to understanding universality in this particular, limited case, though this is hardly even speculative). As of this time, no direct, combinatorial proof of the corollary seems to be known.

2.2 Discrete integration. The theorem best adapted in the present case to prove that h is indeed holomorphic is Morera's theorem, which we recall without proof.

Theorem 2.1 (Morera) *Let Ω be a simply connected domain of the complex plane, and let $f : \Omega \to \mathbb{C}$ be a continuous function. Then f is holomorphic if, and only if, for every simple, closed, smooth curve γ contained in Ω, the integral of f along γ vanishes.*

So, let γ be a simple, closed, smooth curve contained in Ω, and for every $\delta > 0$ let γ_δ be a discretization of γ contained in Ω_δ, *i.e.* a finite chain $(\gamma_\delta(k))_{0 \leqslant k \leqslant N_\delta}$ of pairwise distinct vertices of G_δ, ordered in the positive direction, such that for every index k, $\gamma_\delta(k)$ and $\gamma_\delta(k+1)$ are nearest neighbors, and chosen in such a way that the Hausdorff distance between γ_δ and δ goes to 0 with δ. Notice that N_δ can be taken of order δ^{-1}, which we shall assume from now on.

Since the subsequence (δ_n) is chosen so that h_{δ_n} converges uniformly to h, we have the following fact: As n goes to infinity,

$$I_n(\gamma) := \sum_{k=0}^{N_{\delta_n}} \frac{H_{\delta_n}(\gamma_{\delta_n}(k)) + H_{\delta_n}(\gamma_{\delta_n}(k+1))}{2} (\gamma_{\delta_n}(k+1) - \gamma_{\delta_n}(k)) \to \oint_\gamma h(z) \mathrm{d}z. \tag{2.2}$$

So, to prove that h is holomorphic, it is sufficient to show that $I_n(\gamma)$ goes to 0 as n goes to infinity.

The discrete curve γ_δ surrounds a finite family of faces of G_δ, which we shall denote by $\mathring{\gamma}_\delta$. Besides, for every (oriented) edge $e = xy$ in G_δ, define the following notations:

$$H_\delta(e) := \frac{H_\delta(x) + H_\delta(y)}{2}, \qquad \partial_e H := H(y) - H(x)$$

(without dividing by the length of e). Last, if f is a face of G_δ, let ∂f be its oriented boundary, seen as a set of oriented edges. With these notations, we get a shorter notation for I_n (where $e \in \gamma_\delta$ means that e is of the form $\gamma_{\delta_n}(k)\gamma_{\delta_n}(k+1)$ for some k) and the following identity:

$$I_n(\gamma) = \sum_{e \in \gamma_\delta} eH_\delta(e) = \sum_{f \in \mathring{\gamma}_\delta} \sum_{e \in \partial f} eH_\delta(e). \tag{2.3}$$

Indeed, in the last equality, each boundary term is obtained exactly once with the correct sign, and each interior term appears twice with opposite signs — for every edge e, $H_\delta(-e) = H_\delta(e)$.

If f is a face of G_δ, let $c(f)$ be its center (it is a vertex of G_δ^*). The sum of $eH(e)$ around f can be rewritten in the following fashion: If $x_0, x_1, \dots, x_{|f|} = x_0$ are the vertices of f, we can "integrate by parts" (which is the same as re-indexing part of the sum — we do it explicitly this time, and shall do it quietly a few more times below) and obtain

$$\begin{aligned}
\sum_{e \in \partial f} eH_\delta(e) &= \sum_{k=0}^{|f|-1} (x_{k+1} - x_k) \frac{H(x_{k+1}) + H(x_k)}{2} \\
&= \sum_{k=0}^{|f|-1} (x_{k+1} - c(f)) \frac{H(x_{k+1}) + H(x_k)}{2} \\
&\qquad - \sum_{k=0}^{|f|-1} (x_k - c(f)) \frac{H(x_{k+1}) + H(x_k)}{2} \\
&= \sum_{k=0}^{|f|-1} (x_k - c(f)) \frac{[H(x_{k+1}) - H(x_k)] + [H(x_k) - H(x_{k-1})]}{2} \\
&= \sum_{k=0}^{|f|-1} \left(\frac{x_k + x_{k+1}}{2} - c(f) \right) [H(x_{k+1}) - H(x_k)].
\end{aligned}$$

Putting this back in the previous sum, notice that for x, y nearest neighbors in the interior of γ, the term $H(y) - H(x)$ appears twice, once for each face of G_δ adjacent to xy, and that the factors $(x+y)/2$ cancel between them, leaving only the difference between the centers of the faces, *i.e.* the dual edge of xy.

On the boundary, we obtain of the order of δ^{-1} terms, each being of order $\delta^{1+\varepsilon_H}$, so the contribution of the boundary terms goes to 0 with δ, and we finally obtain

$$I_n(\gamma) = \frac{1}{2} \sum_{e \subset \gamma_\delta} e^* \partial_e H + o(1), \tag{2.4}$$

with the slight abuse of notation that $e \subset \gamma_\delta$ means that e is an edge of $H\delta$ lying in the interior of γ. We will keep this notation from now on.

For every edge $e = xy$ of Ω_δ let $P_{A,\delta}(e)$ (and similarly P_B and P_C) be the probability that $E_{A,\delta}(y)$ is satisfied, but $E_{A,\delta}(x)$ is not. Again by Russo-Seymour-Welsh, we get the following estimate, uniformly in δ and e:

$$P_{A_\delta}(e) \leqslant K_P \delta^{\varepsilon_P}. \tag{2.5}$$

Besides, we have for every edge e:

$$\partial_e H_A = P_A(e) - P_A(-e).$$

So, replacing H by its definition in (2.4), and re-indexing the sum to obtain each oriented edge in exactly one term, we get the following expression for the discrete contour integral:

$$I_n(\gamma) = \sum_{e \subset \gamma_\delta} e^* \left[P_A(e) + \tau P_B(e) + \tau^2 P_C(e)\right] + o(1). \tag{2.6}$$

2.3 Color-swapping. The main combinatorial tool that we will use in what follows, and in fact the only place where the fact that the model is critical percolation at $p_c = 1/2$, is the following (stated here with our notations):

Proposition 2.1 (Smirnov) *For every edge e of Ω_δ, we have the following identities:*

$$P_A(e) = P_B(\tau.e) = P_C(\tau^2.e).$$

We refer the reader to [6] for the (elementary, but very clever) proof of this result; said proof does not actually use the fact that we are considering the triangular lattice, or even the fact that 1/2 is the critical parameter: It extends to site-percolation with parameter 1/2 on any planar triangulation. Using this identity in (2.6), we obtain

$$\begin{aligned} I_n(\gamma) &= \sum_{e \subset \gamma_\delta} e^* \left[P_A(e) + \tau P_A(\tau^2.e) + \tau^2 P_A(\tau.e)\right] + o(1) \\ &= \sum_{e \subset \gamma_\delta} (e^* + \tau(\tau.e)^* + \tau^2(\tau^2.e)^*) P_A(e) + o(1), \end{aligned}$$

where the last step is obtained by re-indexing the sum one last time. Each term in the sum is then equal to 0, by Equation (1.1), so we get $I_n(\gamma) = o(1)$ which is exactly what we needed. It follows that the limit h is holomorphic on Ω.

Replacing H with S in the above computation, it is easy to check that one gets the sum, over the same family of oriented edges, of $(e^* + (\tau.e)^* + (\tau^2.e)^*)P_A(e)$; but the identity

$$e^* + (\tau.e)^* + (\tau^2.e)^* = 0 \tag{2.7}$$

is true in any lattice, as it is the sum of the oriented edges of the triangle around the origin of e. Notice how this does not use the exact details of the triangulation anymore, only Russo-Seymour-Welsh estimates, so it holds on any triangulation for which those hold, as announced.

2.4 Boundary conditions. The last step we have to perform, now that we know that h is holomorphic, is to identify enough boundary conditions to specify it uniquely. This can be done in a very simple way, as follows.

Let z be a point on the boundary of Ω_δ lying between B and C. It is clear that $h_A(z) = 0$, and it is true that $h_B(z) + h_C(z) = 1$ (either from Corollary 2.1, or by looking at the discrete exploration process started at A and aiming at the boundary interval $[BC]$, for which $H_B(z)$ (resp. $H_C(z)$) is the probability that it touches $[zC]$ before (resp. after) $[Bz]$). Hence, $h(z)$ lies on the interval $[\tau, \tau^2]$ of the complex plane. Besides, $h(B) = \tau$ and $h(C) = \tau^2$, so h induces a continuous map from the boundary interval $[BC]$ of Ω onto $[\tau, \tau]$. By Russo-Seymour-Welsh yet again, h is one-to-one on this boundary interval.

Similarly, h induces a bijection between the boundary interval $[AB]$ (resp. $[CA]$) of Ω and the complex interval $[1, \tau]$ (resp. $[\tau^2, 1]$), so putting the pieces together we see that h is a holomorphic map from Ω to $\mathbb{C}$ which extends continuously to $\bar{\Omega}$ and induces a continuous bijection between $\partial\Omega$ and the boundary of the triangle with vertices at 1, τ and τ^2.

From standard results of complex analysis ("principle of corresponding boundaries", cf. for instance Theorem 4.3 in [4]), this implies that h is actually a conformal map from Ω to the interior of the same triangle. But we know that h maps A (resp. B, C) to 1 (resp. τ, τ^2), and this determines it uniquely. In other words, there is only one possible limit for the triple (H_A, H_B, H_C) as δ goes to 0, which gives conformal invariance for free and concludes the proof of Theorem 1.1.

As a corollary of Theorem 1.1, we get a nice expression for h_A: If $\Phi_{\Omega,A,B,C}$ is the conformal map from Ω to the triangle mapping A, B and C as previously (which means of course that $\Phi_{\Omega,A,B,C} = h$) then

$$H_{A,\delta}(z) \to \frac{2\Re(\Phi_{\Omega,A,B,C}(z)) + 1}{3}.$$

If in particular Ω is the equilateral triangle itself, then h is the identity map and we obtain Cardy's formula in Carleson's form: if $D \in [CA]$ then

$$P_{p_c}([AB] \leftrightarrow [CD]) = \frac{|CD|}{|AB|}.$$

References

[1] F. Camia and C. M. Newman, *The full scaling limit of two-dimensional critical percolation*, Preprint, (2005). `arXiv:math.PR/0504036`.

[2] G. Grimmett, *Percolation*, vol. 321 of Grundlehren der Mathematischen Wissenschaften, Springer-Verlag, Berlin, second ed., 1999.

[3] H. Kesten, *Percolation theory for mathematicians*, vol. 2 of Progress in Probability and Statistics, Birkhäuser, Boston, Mass., 1982.

[4] S. Lang, *Complex Analysis*, vol. 103 of Graduate Texts in Mathematics, Springer-Verlag, New York, fourth ed., 1999.

[5] S. Smirnov, *Critical percolation in the plane: Conformal invariance, Cardy's formula, scaling limits*, C. R. Acad. Sci. Paris Sér. I Math., 333 (2001), pp. 239–244.

[6] S. Smirnov, *Critical percolation in the plane. I. Conformal invariance and Cardy's formula. II. Continuum scaling limit*, Unpublished, (2001). `http://www.math.kth.se/~stas/papers/percol.ps`.

[7] S. Smirnov and W. Werner, *Critical exponents for two-dimensional percolation*, Mathematical Research Letters, 8 (2001), pp. 729–744.

Fields Institute Communications
Volume **50**, 2007

Rota–Baxter Algebras in Renormalization of Perturbative Quantum Field Theory

Kurusch Ebrahimi-Fard
I.H.É.S. Le Bois-Marie
35, Route de Chartres
F-91440 Bures-sur-Yvette, France
kurusch@ihes.fr

Li Guo
Department of Mathematics and Computer Science
Rutgers University
Newark, NJ 07102 USA
liguo@newark.rutgers.edu

Abstract. Recently, the theory of renormalization in perturbative quantum field theory underwent some exciting new developments. Kreimer discovered an organization of Feynman graphs into combinatorial Hopf algebras. The process of renormalization is captured by a factorization theorem for regularized Hopf algebra characters. In this context the notion of Rota–Baxter algebras enters the scene. We review several aspects of Rota–Baxter algebras as they appear in other sectors also relevant to perturbative renormalization, for instance multiple-zeta-values and matrix differential equations.

Contents

2000 *Mathematics Subject Classification.* Primary 81T15, 81T18, 16W30; Secondary 47A68, 37B55.

Key words and phrases. Rota–Baxter operators, Atkinson's theorem, Spitzer's identity, Baker–Campbell–Hausdorff formula, matrix factorization, renormalization, Hopf algebra of renormalization, Birkhoff decomposition.

1 Introduction

In this paper we extend the talk given by one of us (L. G.) at the workshop *Renormalization and Universality in Mathematical Physics* held at the Fields Institute for Research in Mathematical Sciences, October 18-22, 2005. This talk, as well as the following presentation, is based on our results in [EGK2004, EGK2005, EGGV2006, EG2005b] and explores in detail the algebraic structure underlying the Birkhoff decomposition in the theory of renormalization of perturbative quantum field theory. This decomposition result was discovered by Alain Connes and Dirk Kreimer using the minimal subtraction scheme in dimensional regularization in the context of their Hopf algebraic approach to renormalization [CK2000, CK2001, Krei1998, Krei1999a]. Recently, we extended our findings in joint work with D. Manchon in [EGM2006], presenting a unified picture of several apparently unrelated factorizations arisen from quantum field theory, vertex operator algebras, combinatorics and numerical methods in differential equations.

With a broad range of audience in mind, we summarize briefly some background in perturbative quantum field theory (pQFT) as well as in Hopf algebras and Rota–Baxter algebras, before presenting some of the recent developments. In doing so, we hope to reach a middle ground between mathematics and physics where the algebraic structures which emerged in the work of Connes and Kreimer can be appreciated and further studied by students and researchers in pure and applied areas. For other aspects and more details on the work of Kreimer and his collaborators on the Hopf algebra approach to renormalization in pQFT we refer the reader to [EK2005, FG2005, Krei2002, Man2001], as well as to the contributions of D. Kreimer and S. Weinzierl in this volume. For readable and brief sources on the basics of renormalization in pQFT using the standard terminology, that is, before the appearance of the more recent Hopf algebra approach, the reader might wish to take a look into J. Gracey's conference contribution, as well as [Cal1975, CaKe1982], and more recently [Col2006, Del2004]. For a more complete picture on renormalization theory in the context of applications, including calculational details, one should consult the standard textbooks in the field, e.g. [BoSh1959, Col1984, IzZu1980, Muta1987, Vas2004].

The new picture of renormalization in pQFT The very concept of renormalization has a long history [Br1993] going beyond its most famous appearance in perturbative quantum field theory, which started with a short paper on the Lamb shift in perturbative quantum electrodynamics (QED) by H. Bethe [Bet1947]. Eventually, in the context of pQFT, renormalization together with the gauge principle reached the status of a fundamental concept. Its development is marked by contributions by some of the most important figures in 20th century theoretical physics including the founding fathers of pQFT, to wit, Feynman, Dyson, Schwinger, and Tomonaga – see [Kai2005, Schw1994] for more details. Renormalization theory in pQFT reached a satisfying form, from the physics' point of view, with the formulation known as BPHZ renormalization prescription due to Bogoliubov and Parasiuk [BoPa1957], further improved by Hepp [Hepp1966] and Zimmermann [Zim1969].

As a matter of fact, in most of the interesting and relevant 4-dimensional quantum field theories, even simple perturbative calculations are plagued with ill-defined, i.e., ultraviolet divergent multidimensional integrals. The removal of these so-called short-distance singularities in a physically and mathematically sound way is the process of renormalization. Although the work of the aforementioned authors played a decisive role in the acceptance of the theory of renormalization in theoretical physics, the conceptual foundation as well as application of renormalization was nevertheless plagued by an ambivalent reputation. Let us cite the following statement from [Del2004], *Removing [these] divergencies has been during the decades, the nightmare and the delight of many physicists working in particle physics. [...] This matter of fact even participated to the nobility of the subject.* Moreover, renormalization theory suffered from its lack of a concise mathematical formulation. It was only recently that the foundational, i.e., mathematical structures of renormalization theory in pQFT experienced new and enlightening contributions. These developments started out with the Hopf algebraic formulation of its combinatorial-algebraic structures discovered by Kreimer in [Krei1998] – 50 years after Bethe's paper. This new approach found a satisfactory formulation in the work of Connes and Kreimer [CK1998, CK2000, CK2001], and Broadhurst and Kreimer [BK1998, BK1999, BK2000a, BK2000b, BK2001].

In fact, in the Hopf algebra picture the analytic and algebraic aspects of perturbative renormalization nicely decouple, hence making its structure more transparent. The process of renormalization in pQFT finds a surprisingly compact formulation in terms of a classical group theoretical factorization problem. This is one of the main results discovered by Connes and Kreimer in [CK2000, CK2001] using the minimal subtraction scheme in dimensional regularization. It was achieved by first organizing the set F of one-particle irreducible Feynman graphs coming from a perturbative renormalizable QFT (e.g. ϕ^3_6, ϕ^4_4, QED,...) into a commutative graded connected Hopf algebra $\mathcal{H}_{\mathcal{F}}$, where $\mathcal{F} := \mathbb{K}F$ denotes the vector space freely generated by F. Its non-cocommutative coproduct map essentially encodes the disentanglement of Feynman graphs into collections of ultraviolet divergent proper one-particle irreducible subgraphs and the corresponding cograph, just as it appears in the original BPHZ prescription – more precisely, Bogoliubov's $\bar{\mathrm{R}}$-map.

Furthermore, regularized Feynman rules seen as linear maps $\phi : \mathcal{F} \to A$ from the vector space of graphs into a commutative unital algebra A are extended to $\phi : \mathcal{H}_{\mathcal{F}} \to A$ in the group G_A of algebra homomorphisms from $\mathcal{H}_{\mathcal{F}}$ to A. This setting, as we will see, already gives rise to a factorization of such generalized, that is, A-valued algebra homomorphisms in the group G_A. In addition we will observe that when the target space algebra A carries a Rota–Baxter algebra structure, Connes–Kreimer's Birkhoff decomposition for G_A in terms of Bogoliubov's $\bar{\mathrm{R}}$-map follows immediately. The last ingredient, i.e. the Rota–Baxter structure is naturally provided by the choice of a renormalization scheme on the regularization space A, e.g. minimal subtraction in dimensional regularization.

It was the goal of this talk to present the algebraic underpinning for the result of Connes–Kreimer in terms of a general factorization theorem for complete filtered associative, not necessarily commutative algebras combined with other key results, such as Spitzer's identity and Atkinson's theorem for such algebras in the presence of a Rota–Baxter structure.

In fact, we would like to show how the above factorization problem for renormalization may be formulated more transparently as a classical decomposition problem for matrix Lie groups. This is achieved by establishing a representation of the group of regularized Hopf algebra characters by – infinite dimensional – unipotent lower triangular matrices with entries in a commutative unital Rota–Baxter algebra. The representation space is given by – a well-chosen vector subspace of – the vector space $\mathcal{F}$ of 1PI Feynman graphs corresponding to the pQFT. The demand for entries in a commutative Rota–Baxter algebra induces a non-commutative Rota–Baxter structure on the underlying matrix algebra such that Spitzer's theorem applies, and indeed, as we will see, this paves the way smoothly to Bogoliubov's $\bar{\mathrm{R}}$-map, though in matrix form.

Before giving a more detailed account on this purely algebraic structure we should underline that the initial proof of the Connes–Kreimer factorization of regularized Feynman rules was achieved using dimensional regularization, i.e., assuming A to be the field of Laurent series, hereby opening a hitherto hidden geometric viewpoint on perturbative renormalization in terms of a correspondence to the Riemann–Hilbert problem. For the minimal subtraction scheme it amounts to the multiplicative decomposition of the Laurent series associated to a Feynman graph using dimensionally regularized Feynman rules. Such a Laurent series has poles of finite order in the regulator parameter and the decomposition happens to be the one into a part holomorphic at the origin and a part holomorphic at the complex infinity. The latter moreover distinguishes itself by the particular property of locality. This has far reaching geometric implications, starting with the interpretation upon considering the Birkhoff decomposition of a loop around the origin, providing the clutching data for the two half-spheres defined by that very loop. Connes and Kreimer found a scattering type formula for the part holomorphic at complex infinity, that is, the counterterm, giving rise to its description in terms of iterated integrals.

Moreover, by some recent progress, which has been made in the mathematical context of number theory, and motivic aspects of Feynman integrals, this geometric point of view leads to motivic Galois theory upon studying the notion of equisingular connections in the related Riemann–Hilbert correspondence. This was used to explore Tannakian categories and Galois symmetries in the spirit of differential Galois theory in [CM2004a, CM2006, CM2004b]. Underlying the notion of an equisingular connection is the locality of counterterms, which itself results from Hochschild cohomology. The resulting Dyson–Schwinger equation allows for gradings similar to the weight- and Hodge filtrations for the polylogarithm [Krei2003a, Krei2004]. More concretely, the motivic nature of primitive graphs has been established very recently by Bloch, Esnault and Kreimer in [BEK2005].

We now outline the organization of the paper. In Section 2 we review some pieces from renormalization theory in pQFT up to the introduction of Bogoliubov's $\bar{\mathrm{R}}$-map. We use Section 3 for introducing most of the basic mathematical notions needed in the sequel, especially with respect to Hopf algebras. A general factorization theorem for complete filtered algebras is presented, based on a new type of recursion equation defined in terms of the Baker–Campbell–Hausdorff (BCH) formula. We establish Connes' and Kreimer's Hopf respectively Lie algebra of Feynman graphs as well as their Birkhoff decomposition of regularized Hopf algebra characters. Section 4 is devoted to a not-so-short review of the concept of Rota–Baxter algebra, including a list of instructive examples as well as Spitzer's identity and a link between the Magnus recursion known from matrix differential equations and the new BCH-recursion. Section 5 contains a presentation of the Rota–Baxter algebra structure underlying the Birkhoff decomposition of Connes–Kreimer. We establish a matrix representation of the group of regularized Feynman rules characters and recover Connes–Kreimer's Birkhoff decomposition in terms of a classical matrix factorization problem.

2 Renormalization in pQFT: Bogoliubov's $\bar{\mathrm{R}}$-operation

We recall some of the essential points from *classical* renormalization theory as applied in the context of pQFT. We follow Kreimer's book [Krei2000] and the recent brief and concise summary given by Collins in [Col2006].

2.1 Feynman rules and renormalizability. Let us begin with a statement which summarizes perturbative renormalization theory of QFT in its Lagrangian formulation. The quintessence of –multiplicative– renormalization says that under the severe constraint of maintaining the physical principles of locality, unitarity, and Lorentz invariance, it is possible to absorb to all orders in perturbation theory the (ultraviolet) divergencies in a redefinition of the parameters defining the QFT. Here, in fact, two very distinct concepts enter, to wit, that of *renormalizability*, and the *process of renormalization*. The former distinguishes those theories with only a finite number of parameters, lending them considerably more predictive power. However, the process of renormalization instead works order by order, indifferently of the number of parameters.

In the Lagrangian picture we start with a Lagrangian density defining the QFT, which we assume to be renormalizable. For instance, the so-called ϕ^4 and ϕ^3 theories are described by the following two Lagrangians

$$L=\frac{1}{2}(\partial_\mu\phi)^2-\frac{1}{2}m_1^2\phi^2+\frac{\lambda_4}{4!}\phi^4 \quad \text{resp.} \quad L=\frac{1}{2}(\partial_\mu\phi)^2-\frac{1}{2}m_2^2\phi^2+\frac{\lambda_3}{3!}\phi^3.$$

The classical calculus of variations applied to $L=\frac{1}{2}(\partial_\mu\phi)^2-\frac{1}{2}m_i^2\phi^2$, $i=1,2$ provides the equations of motion

$$\partial_\mu\frac{\delta L}{\delta\partial_\mu\phi}-\frac{\delta L}{\delta\phi}=0\Rightarrow(\Box+m_i^2)\phi(x)=0, \qquad i=1,2.$$

There is only one type of –scalar– field denoted by $\phi=\phi(x)$, parametrized by the D dimensional space-time point $x=(t,\mathbf{x})$. The parameters m_1,m_2 and λ_4,λ_3 are called the mass and coupling constants, respectively. The equations of motion as well as the Lagrangian are local in the sense that all fields are evaluated at the same space-time point. The reader interested in quantum field theories more relevant to physics, such as QED or quantum chromodynamics, should consult the standard literature, for instance [BoSh1959, Col1984, IzZu1980, Muta1987, Ryd1985, tHVel1973].

We may separate the above Lagrangians into a free and an interaction part, $L=L_f+L_i$, where in the both cases above $L_f=\frac{1}{2}(\partial_\mu\phi)^2-\frac{1}{2}m_i^2\phi^2$, $i=1,2$. The remaining quartic respectively cubic term represents the interaction part, L_i.

The general goal is to calculate the n-point Green's functions,

$$G(x_1,\dots,x_n):=\langle 0|\mathrm{T}[\phi(x_1)\cdots\phi(x_n)]|0\rangle$$

where the time-ordering T-operator is defined by

$$\mathrm{T}[\phi(x_1)\phi(x_2)]:=\begin{cases}\phi(x_1)\phi(x_2), & \text{if } t_1>t_2,\\ \phi(x_2)\phi(x_1), & \text{if } t_2>t_1,\end{cases}$$

since from these vacuum correlators of time-ordered products of fields one can obtain all S-matrix elements with the help of the LSZ-formalism, see for instance [IzZu1980] for more details.

So far, perturbation theory is the most successful approach to Lagrangian QFT with an overwhelming accordance between experimental and theoretical, that is, perturbative results.

In perturbation theory Green's functions are expanded in power series of a supposed to be small parameter like the coupling constant (or Planck's $\hbar$). Its coefficients are given by complicated iterated integrals of the external parameters defining the scattering process, respectively the participating particles. However, these coefficients turn out to show a highly organized structure. Indeed, Feynman discovered a graphical coding of such expressions in terms of graphs that later bore his name. A *Feynman graph* is a collection of internal and external lines, or edges, and vertices of several types. Proper subgraphs of a Feynman graph are determined by proper subsets of the set of internal edges and vertices.

Eventually, perturbative expansions in QFT are organized in terms of one-particle irreducible (1PI) Feynman graphs, i.e., connected graphs that stay connected upon removal of any of the internal edges, also called propagators. For example, we have such simple drawings as  or 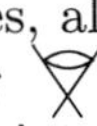 encoding Feynman amplitude integrals appearing in the expansion of the 4-point Green's function in ϕ^4-theory. The fact that there is only one type of –undirected– lines and only 4-valent vertices reflects the fact that the Lagrangian L only contains one type of –scalar– field ϕ with only a quartic interaction term in L_i. For the ϕ^3 case we would find drawings with only 3-valent vertices and one type of undirected lines.

Feynman rules give a mean to translate these drawings –back– into the corresponding amplitudes. The goal is to calculate them order by order in the coupling constants of the theory. Originally, a graph was somehow 'identified' with its amplitude. Let us remark that it is only in the Hopf algebraic picture that these objects get properly distinguished into the algebraic-combinatorial and analytic side of the same underlying pQFT picture. We will see that in this context graphs are studied as combinatorial objects organized into (pre-)Lie algebras and the corresponding Hopf algebras. Whereas Feynman rules provide particular maps assigning analytic expressions, the so-called Feynman amplitudes, to these graphs.

As an example for the latter we list here the set of Feynman rules related to the above Lagrangian of the ϕ^4-theory in four space-time dimensions [Krei2000].

- For each internal scalar field line, associate a propagator: $i\Delta(p) = \frac{i}{p^2-m^2+i\eta}$.
- At each vertex require momentum conservation and place a factor of $-i\lambda_4$.
- For each closed loop in the graph, integrate over $\int \frac{d^4q}{(2\pi)^4}$, where q is the momentum associated with the loop.
- For any Feynman graph Γ built from these propagators and vertices divide by the symmetry factor $sym(\Gamma)$ of the graph, which is assumed to contain r vertices. The symmetry factor is then given by the number of possibilities to connect r vertices to give Γ, divided by $(4!)^r$.

A few examples are presented now, see for instance [FG2005] and the standard literature. Let us write down the integral expression corresponding to the graph called *fish*, 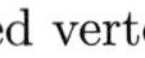, made out of two 4-valent vertices and two propagators. The unconnected vertex legs carry the external momenta of the in- and outgoing particles which are denoted by k_1, k_2, k_3, k_4. They are related by a delta function ensuring the overall momentum conservation.

$I($$) := (-i\lambda_4)^2 \int \frac{d^4q}{(2\pi)^4} \frac{i}{(q+k_1+k_2)^2 - m^2 + i\eta} \frac{i}{q^2 - m^2 + i\eta}. \qquad (2.1)$

For the *bikini* graph, , with three 4-valent vertices and four propagators we find after some algebra a product of two 'fishy' integrals

$I($$) \quad := \quad (-i\lambda_4)^3 \int \frac{d^4q}{(2\pi)^4} \frac{i}{(q+k_1+k_2)^2 - m^2 + i\eta} \frac{i}{q^2 - m^2 + i\eta} \times$

$$\int \frac{d^4p}{(2\pi)^4} \frac{i}{(p+k_1+k_2)^2 - m^2 + i\eta} \frac{i}{p^2 - m^2 + i\eta}. \tag{2.2}$$

Let us go one step further and write down the integral for the *winecup* graph, , of order three in the coupling constant $i\lambda_4$, also made out of four propagators

$I($$) \quad := \quad (-i\lambda_4)^3 \int \frac{d^4p}{(2\pi)^4} \int \frac{d^4q}{(2\pi)^4} \frac{i}{(q+k_1+k_2)^2 - m^2 + i\eta} \frac{i}{q^2 - m^2 + i\eta} \times$

$$\frac{i}{(q^2 - m^2 + i\eta)\,((p-q-k_3)^2 - m^2 + i\eta)}. \tag{2.3}$$

Unfortunately, all of the above integrations over the loop-momenta are not restricted by any constraint and some power-counting analysis of the integrands reveals a serious problem as we observe that the above integrals are not well-defined for high loop-momenta.

For the purpose of classifying such problematic integrals within a pQFT, we must introduce the notion of *superficial degree of divergence* associated to a Feynman graph Γ and denoted by $\omega(\Gamma)$. In general, $\omega(\Gamma)$ simply follows from counting powers of integral loop-momenta in the corresponding Feynman integral in the limit when they are large. The amplitude corresponding to the graph Γ is superficially *convergent* if $\omega(\Gamma) < 0$, and superficially *divergent* for $\omega(\Gamma) \geq 0$. For $\omega(\Gamma) = 0$ we call the diagram superficially *logarithmically* divergent and we say that it is superficially *linearly* divergent, if $\omega(\Gamma) = 1$. No doubt, we may continue this nomenclature. Superficially here reflects the very fact that the integral associated to a graph might appear to be convergent, but due to ill-defined subintegrals corresponding to ultraviolet divergent 1PI subgraphs it is divergent in fact.

Restricting ourselves to scalar theories such as the two examples at the beginning one can derive a simple expression for $\omega(\Gamma)$ depending on the structure of the graph, i.e., its vertices and edges, as well as the space-time dimension. We assume a Lagrangian of the general form $L = L_f + \sum_{k=3} L_{i_k}$ where L_f as usual consists of the free part and the sum contains the interaction field monomials of the form $L_{i_k} := \frac{\lambda_k}{k!}\phi^k$. Recall that the canonical dimension (inverse length) in D space-time dimensions of the scalar field ϕ is given by $[\phi] = (D-2)/2$ in natural units, following from the free part of the Lagrangian. Therefore we have $[\phi^k] = (D-2)k/2 =: \omega_k$ and the coupling constant must be of dimension $[\lambda_k] = D - \omega_k$.

For a given connected graph Γ consisting of $E = E(\Gamma)$ external lines, $I = I(\Gamma)$ internal lines and $V = V(\Gamma)$ vertices we find $L = I - V + 1$ loop integrations in the corresponding amplitude. By standard considerations it follows that $E - 2I = \sum_{j=1}^{V} n_j$, where n_j denotes the order, or valency, of the j-th vertex in Γ. Since each D dimensional loop integration contributes D loop momenta in the numerator and each propagator contributes two inverse loop momenta it follows that

$$\omega(\Gamma) = DL - 2I = D - \frac{D-2}{2}E + \sum_{j=1}^{V}\Big(\frac{D-2}{2}n_j - D\Big).$$

The last term is just $\frac{D-2}{2}n_j - D = \omega_{n_j} - D = [\lambda_{n_j}]$ and hence

$$\omega(\Gamma) = D - \frac{D-2}{2}E + \sum_{j=1}^{V}[\lambda_{n_j}].$$

For our ϕ^4-theory in $D = 4$ space-time dimensions we find a dimensionless coupling constant, $[\lambda_4] = 0$, which implies $\omega(\Gamma) = 4 - E$. In fact, from this formula it follows immediately that the above integrals are all superficially logarithmically divergent, rendering the perturbative series expansion of –the 4-point– Green's functions quite useless right from the start, to wit, for the 1-loop fish diagram  corresponding to the first non-trivial term in the perturbation expansion of the 4-point Green's functions in $D = 4$ space-time dimensions we find $\omega($$) = D - 4 = 4 - 4 = 0$, which indicates the existence of a so-called logarithmic *ultraviolet* (UV) *divergence*. The reader is referred to the literature for more details on the related dimensional analysis of pQFT revealing the role played by the number of space-time dimensions and the related classification into *non-renormalizable*, (just) *renormalizable* and *super-renormalizable* theories [Cole1988, IzZu1980, PS1995, Ryd1985]. Let us mention that in the ϕ^4 example the number of $D = 4$ space-time dimensions just happens to be the critical one, and that we would make a similar observation for the Lagrangian with cubic interaction in $D = 6$ space-time dimensions.

At this level and under the assumption that the perturbative QFT is *renormalizable*, the programm of –multiplicative– renormalization enters the stage. Its non-trivial operations will cure the ultraviolet deficiencies in a self-consistent and physically sound way order by order in the series expansion.

2.2 Multiplicative renormalization. The machinery of multiplicative renormalization consists of the following parts.

Regularization: first, one must regularize the theory, rendering integrals of the above type formally finite upon the introduction of new and nonphysical parameters. For instance, one might truncate, i.e., *cut-off* the integration limits at an upper bound Λ by introducing a step-function Θ

$$\int d^4q \frac{\Theta[\Lambda - q^2]}{(q + k_1 + k_2)^2 - m^2 + i\eta}\frac{1}{q^2 - m^2 + i\eta}.$$

Evaluating such an integral results in log-terms containing Λ so that naively removing the cut-off parameter by taking the limit $\Lambda \to \infty$ must be avoided since it gives back the original divergence. The cut-off approach usually interferes with gauge symmetry.

Another regularization scheme is *dimensional regularization* [Col1984, tH1973], which respects –almost– all symmetries of the theory under consideration. It introduces a complex parameter by changing the integral measure, that is, the space-time dimension $D \in \mathbb{N}$ to $\mathcal{D} \in \mathbb{C}$

$$\frac{d^Dq}{(2\pi)^D} \longrightarrow \mu^{2\varepsilon}\frac{d^{\mathcal{D}}q}{(2\pi)^{\mathcal{D}}},$$

where $\varepsilon = \frac{(D-\mathcal{D})}{2} \in \mathbb{C}$. The parameter μ (t'Hooft's mass) is introduced for dimensional reasons. Hence, forgetting about the other involved variables, that do not enter our considerations, dimensionally regularized Feynman rules associate to each graph an element from the field of Laurent series $\mathbb{C}[\varepsilon^{-1}, \varepsilon]]$. In general, one obtains a proper Laurent series of degree at most n in an n-loop calculation, which means that we have at most poles of order n in the regularized Feynman integral corresponding to a graph with n independent closed cycles. The pole terms parametrize the UV-deficiency of the integrals. Obviously, the limit of $\varepsilon \to 0$ results in the original divergences of the integral and must be postponed.

Let us emphasize that it is common knowledge that the choice of the regularization procedure is highly arbitrary and mainly driven by calculational conveniences and the wish to conserve symmetries of the original theory.

But, so far we have not achieved much more than a parametrization of the divergent structure of Feynman integrals. Though important and useful all this just indicates the need of some more sophisticated set of rules on how to handle the terms which diverge with $\Lambda \to \infty$ in the cut-off setting respectively $\varepsilon \to 0$ for dimensional regularization, since in the end we must remove these non-physical regularization parameters.

Renormalization: to render the picture complete we must fix a so-called *renormalization scheme* (or *condition*), which is supposed to isolate the divergent part of a regularized integral. For instance, in dimensional regularization a natural choice is the minimal subtraction scheme $R := R_{ms}$, mapping a Laurent series to its strict pole part, i.e., $R\big(\sum_{n\geq -N} a_n\varepsilon^n\big) := \sum_{n=-N}^{-1} a_n\varepsilon^n$.

The next and conceptually most demanding step is to provide a procedure which renders the integrals finite upon the removal of the regularization parameters order by order in a –mathematically and– physically sound way. This actually goes along with the proof of perturbative renormalizability of the underlying QFT. A convenient procedure which accomplishes this and which is commonly used nowadays was established by Bogoliubov and Parasiuk, further improved by Hepp and Zimmermann.

Let us go back to the statement we made at the beginning of this section, characterizing the concept of perturbative renormalization in a –renormalizable– QFT. The idea is to *redefine multiplicatively* the parameters appearing in the Lagrangian L so as to absorb all divergent contributions order by order in the perturbation expansion. We will outline this idea in the next step where we briefly indicate how to implement Bogoliubov's *subtraction* procedure for renormalization in a *multiplicative* way on the level of the Lagrangian by *adding* the so-called *counterterm* Lagrangian L_c to the original Lagrangian $L = L_f + L_i$. Indeed, the result is $L \xrightarrow{ren.} L_{ren} = L + L_c$ with L_{ren} describing the perturbatively renormalized quantum field theory. We achieve this by introducing so-called Z-factors together with the idea of renormalized respectively bare parameters. In the case of the ϕ^4-theory we have

$$L_{ren} = L_f + L_i + \underbrace{\frac{Z_\phi - 1}{2}(\partial_\mu\phi)^2 - \frac{(m_1^2 Z_{m_1} - m_1^2)}{2}\phi^2 + \frac{(\lambda_4 Z_{\lambda_4} - \lambda_4)}{4!}\phi^4}_{L_c}.$$

One can reorganize the right hand side into the *original form* upon introducing the notion of bare fields and bare parameters. First, we replace the field ϕ by the bare field

$$\phi_0 := Z_\phi^{\frac{1}{2}} \phi$$

defined in terms of the Z-factor $Z_\phi := 1 + \delta\phi$. Then, we introduce the Z-factors Z_{m_1}, Z_{λ_4} and replace the parameters m_1 and λ_4 by *bare* parameters

$$m_{1,0}^2 := m_1^2 Z_{m_1} Z_\phi^{-1} = m_1^2 \frac{(1+\delta m_1)}{(1+\delta\phi)}, \text{ and } \lambda_{4,0} := \lambda_4 Z_{\lambda_4} Z_\phi^{-2} = \lambda_4 \frac{(1+\delta\lambda_4)}{(1+\delta\phi)^2},$$

such that we can write the multiplicatively renormalized Lagrangian as

$$L_{ren} = \frac{1}{2}(\partial_\mu \phi_0)^2 - \frac{1}{2} m_{1,0}^2 \phi_0^2 + \frac{\lambda_{4,0}}{4!}\phi_0^4.$$

The property of the original ϕ^4 Lagrangian, $L = L_f + L_i$, to be renormalizable is encoded in the fact that the counterterm Lagrangian, L_c, only contains field monomials which also appear in L. This allows us to recover L_{ren} in the original form of L upon introducing bare parameters respectively the Z-factors, which implies the notion of *multiplicative renormalization.* We observe explicitly that Z_ϕ must be a numerical constant to assure *locality* ($[Z_\phi, \partial_\mu] = 0$). This is also true for the other Z-factors as well. However, they depend in a highly singular form on the regularization parameter. In fact, $Z_X = 1 + \delta X$ is an order by order determined infinite power series of *counterterms*, each of which is divergent when removing the regulator parameter. They are introduced to compensate –i.e. renormalize– the pole parts in the original perturbation expansion, and –one must prove that– such a process assures the cancellation of divergent contributions at each order in perturbation theory, see e.g. [Col1984, IzZu1980].

2.3 Preparation of Feynman graphs for simple subtraction. Recall that a proper connected UV divergent 1PI subgraph γ of a 1PI Feynman graph Γ is determined by proper subsets of the set of internal edges and vertices of Γ.

Bogoliubov's $\bar{\mathrm{R}}$-operation provides a recursive way to calculate the counterterms, which we mentioned above and denoted by $C(\Gamma)$ for each Feynman graph Γ as follows. The $\bar{\mathrm{R}}$-operation maps a graph Γ, that is, the corresponding regularized amplitude to a linear combination of suitably modified graphs including Γ itself

$$\bar{\mathrm{R}}(\Gamma) := \Gamma + \sum\nolimits'_{\{\gamma'\} \subsetneq \Gamma} \Gamma(\{\gamma'\} \to C(\gamma')). \tag{2.4}$$

We follow the notation of [Col2006]. The primed sum is over all unions $\{\gamma'\}$ of UV divergent 1PI subgraphs sitting inside Γ. More concretely, following the terminology used in [CaKe1982] we call $\mathfrak{s} := \{\gamma'\}$ a *proper spinney* of Γ if it consists of a nonempty union of disjoint *proper* UV divergent 1PI subgraphs, $\{\gamma'\} := \{\gamma'_1 \dots \gamma'_n\}$, $\gamma'_i \subsetneq \Gamma$, $\gamma'_i \cap \gamma'_j = \emptyset$ for $i \neq j$. We call the union of all such spinneys in the graph Γ a *proper wood* which we denote by

$$\mathfrak{W}(\Gamma) := \big\{\{\gamma'\} \subsetneq \Gamma \,\big|\{\gamma'\} = \{\text{union of disjoint proper 1PI subgraphs}\}\big\}.$$

Let us enlarge this set to $\bar{\mathfrak{W}}(\Gamma)$ which includes by definition the empty set, $\{\emptyset\}$, and the graph $\{\Gamma\}$ itself, i.e., $\{\emptyset\}, \{\Gamma\} \in \bar{\mathfrak{W}}(\Gamma)$.

For instance, the two loop self-energy graph $\Gamma =$ 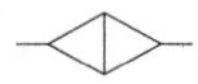 , borrowed from ϕ^3-theory in six dimensions, has two proper 1PI UV subgraphs $\gamma_l =$  and $\gamma_r =$  , but they are not disjoint. Instead, they happen to be overlapping, i.e.

$\gamma_r \cap \gamma_l \neq \emptyset$ in Γ. This is why we do not have $\{\gamma_l\ \gamma_r\} \notin \mathfrak{W}(\Gamma)$, that is, the set of proper spinneys corresponding to that graph is

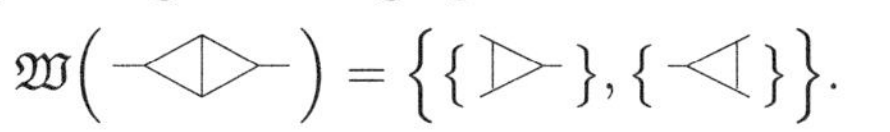
(2.5)

A commonly used way to denote such a wood is by putting each subgraph of a spinney into a box

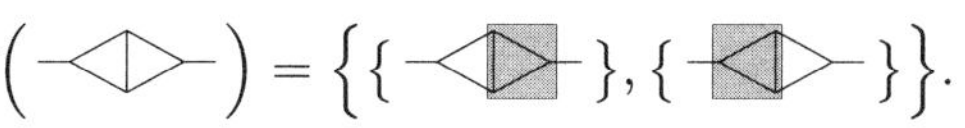
(2.6)

Let us remark that the notion of spinney respectively wood is particularly well-adapted to Bogoliubov's $\bar{\mathrm{R}}$-map [CaKe1982]. Zimmermann [Zim1969] gave a solution to Bogoliubov's recursion by using the combinatorial notion of *forests* of graphs, which differs from that of spinneys [Col1984].

For more examples, this time borrowed from ϕ^4-theory and QED in four space-time dimensions, we look at the woods of the divergent 4-point graphs 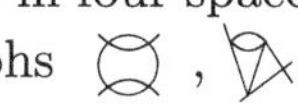

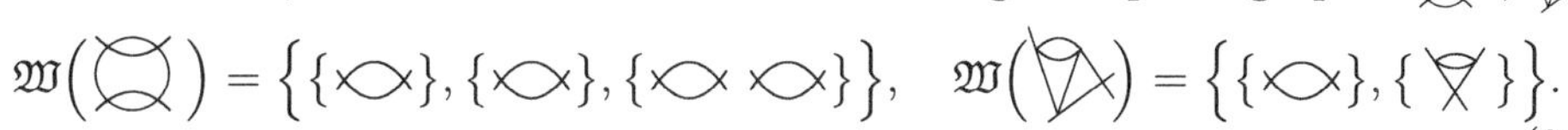
(2.7)

and the QED self-energy graphs 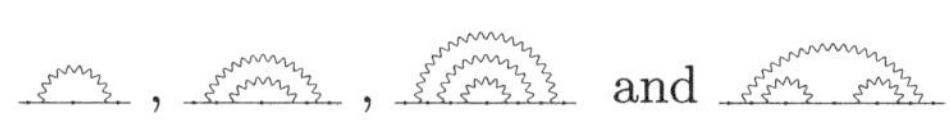

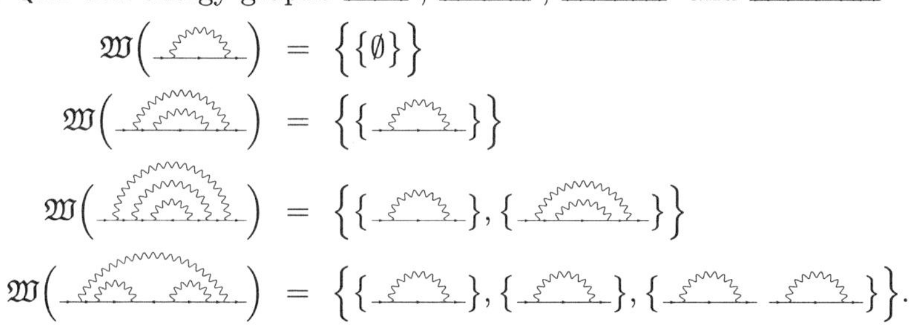
(2.8)
(2.9)
(2.10)
(2.11)

We will later see that the union of 1PI subgraphs, such as for instance in the spinney 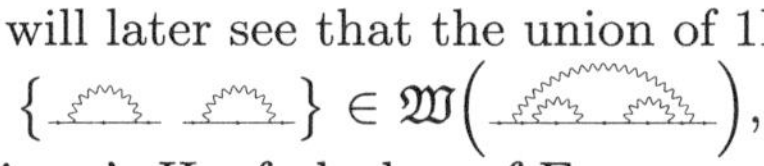, gives rise to the algebra part in Connes' and Kreimer's Hopf algebra of Feynman graphs in Section 3.5. Also, we should underline that the subgraphs in a spinney may posses nontrivial spinneys themselves, see for instance the graph 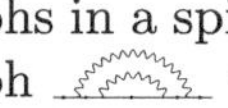 which appears as a spinney in the wood (2.10).

We call graphs such as the one-loop diagram 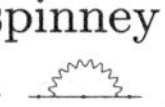 *primitive* if their woods contain only the empty set, i.e., they have no proper UV divergent 1PI subgraphs and hence the primed sum in (2.4) disappears.

To each element $\{\gamma'\} \in \mathfrak{W}(\Gamma)$ of a graph Γ corresponds a *cograph* denoted by $\Gamma/\{\gamma'\}$, which follows from the contraction of all graphs $\gamma \in \{\gamma'\}$ in Γ to a point at once. For instance, to the spinneys in the woods (2.9) - (2.11) correspond the cographs

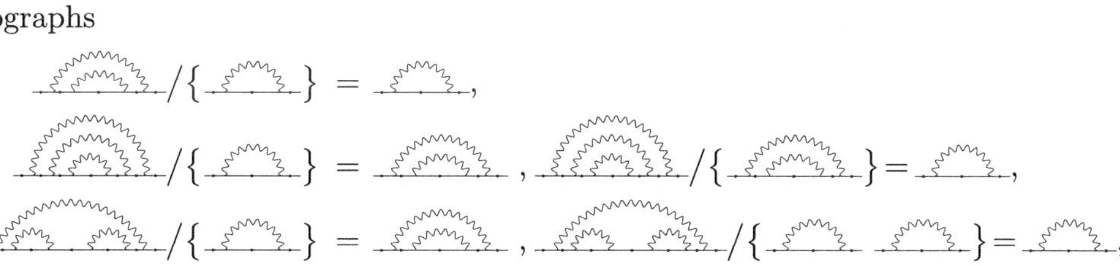

respectively. Remember, that we defined $\bar{\mathfrak{W}}(\Gamma)$ to contain the spinneys $\{\Gamma\}$ and $\{\emptyset\}$. Those spinneys have the following cographs, $\Gamma/\{\Gamma\} = \emptyset$ and $\Gamma/\{\emptyset\} = \Gamma$, respectively. Again, as a preliminary remark we mention here that in the Hopf algebra of Feynman graphs the coproduct of a graph consists simply of the sum of tensor products of each spinney with its cograph, see Eqs. (3.37) and (3.38).

We call a spinney *maximal* if its cograph is primitive. Each of the graphs in the QED example contains only one maximal spinney. The wood in (2.6) consists of two maximal spinneys, as the two UV divergent 1PI subgraphs overlap. In fact, Feynman graphs with overlapping UV divergent 1PI subgraphs always contain more than one maximal spinney. We will come back to this in subsection 3.5.3 where we remark about the correspondence between Feynman graphs and decorated non-planar rooted trees.

As an example for the sum in Eq. (2.4) we write it explicitly for the graph using its wood in (2.7)

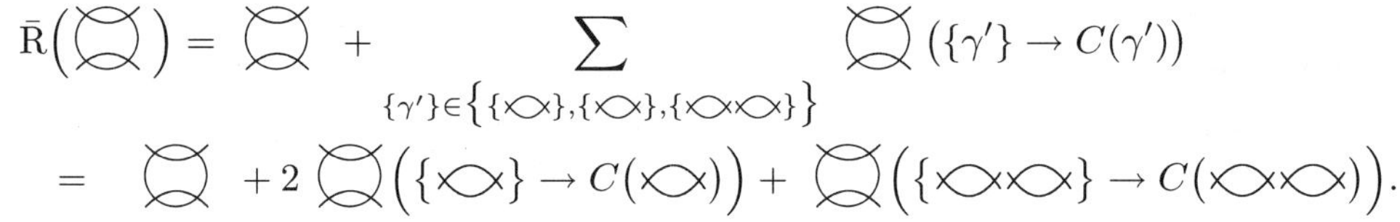

Recall the notational abuse by using a graph just as a mnemonic for an analytic expression, e.g. a Laurent series. We must now explain what happens inside the primed sum in Eq. (2.4)

$$\sum\nolimits'_{\{\gamma'\}\in\mathfrak{W}(\Gamma)}\Gamma\big(\{\gamma'\}\rightarrow C(\gamma')\big). \tag{2.12}$$

More precisely, the expression $\Gamma\big(\{\gamma'\}\rightarrow C(\gamma')\big)$ denotes the *replacement* of each subgraph $\gamma\in\{\gamma'\}$ in Γ by its *counterterm* $C(\gamma)$ defined in terms of the $\bar{\mathrm{R}}$-operation and the particular renormalization scheme map R

$$C(\gamma) := -R(\bar{\mathrm{R}}(\gamma)). \tag{2.13}$$

At this point the recursive nature of Bogoliubov's procedure becomes evident, making it a suitable tool for applying inductive arguments in the proof of perturbative renormalization to all orders. Zimmermann established a solution to Bogoliubov's recursive formula [Zim1969]. A subtle point arises from the evaluation of the counterterm map on the disjoint union of graphs, e.g. $C(\ )$ in the above example. In fact, it is natural to demand that C is an algebra morphism, that is, $C(\ \ ) = C(\ )C(\ )$. We should remark here that it is exactly at this point where Connes and Kreimer in the particular context of the minimal subtraction scheme in dimensional regularization needed the renormalization scheme map R_{ms} to satisfy the so-called multiplicativity constraint [CK2000, Kreil999a], i.e. R_{ms} must have the Rota–Baxter property (3.44), which in turn implies the property of multiplicativity for the counterterm map C. We will return to this problem later in more detail.

Graphically the replacement operation, $\big(\{\gamma'\}\rightarrow C(\gamma')\big)$, is represented by shrinking the UV divergent 1PI subgraph(s) from $\{\gamma'\}$ to a point –,i.e., becoming a vertex– in Γ and to mark that point with a cross or alike to indicate the counterterm. For a graph γ having no proper UV divergent 1PI subgraphs we find $C(\gamma) = -R(\gamma)$.

As we will see, the Hopf algebra approach provides a very convenient way to describe this recursive graph replacement operation on Feynman graphs in a combinatorial well-defined manner in terms of a convolution product.

From Bogoliubov's $\bar{\mathrm{R}}$-operation the renormalized expression for the graph Γ follows simply by a subtraction

$$\begin{aligned}\Gamma \xrightarrow{ren.} \Gamma_+ := (\mathrm{id} - R)\big(\bar{\mathrm{R}}(\Gamma)\big) &= \bar{\mathrm{R}}(\Gamma) + C(\Gamma) \\ &= \Gamma + \sum\nolimits'_{\{\gamma'\}\in\mathfrak{W}(\Gamma)} \Gamma\big(\{\gamma'\} \to C(\gamma')\big) + C(\Gamma).\end{aligned}$$

This simple subtraction is made possible since the $\bar{\mathrm{R}}$-operation applied to Γ, see Eq. (2.4), contains the graph Γ itself, such that adding to Γ the sum (2.12) where the proper subgraphs γ in Γ are replaced by there counterterms $C(\gamma)$ prepares the graph in such a particular way that a simple subtraction finally cancels its superficial divergence. Indeed, Bogoliubov's $\bar{\mathrm{R}}$-operation leads to the renormalization of all proper UV divergent 1PI subgraphs γ in Γ, i.e., they are replaced by their renormalized value γ_+ and we may write for Bogoliubov's $\bar{\mathrm{R}}$-operation

$$\bar{\mathrm{R}}(\Gamma) = \sum\nolimits'_{\{\gamma'\}\in\mathfrak{W}(\Gamma)} \Gamma\big(\{\gamma'\} \to (\gamma'_+)\big).$$

As before, we encounter a subtle point with respect to the multiplicativity of the renormalization map, that is, on the disjoint union of graphs $\{\gamma_1 \cdots \gamma_n\}$ we need, $(\gamma_1 \cdots \gamma_n)_+ = \gamma_{1+} \cdots \gamma_{n+}$.

Let us exemplify this intriguing operation by a little toy-model calculation. We choose the following integral $f_1(c) := \int_0^\infty (y+c)^{-1} dy \in \mathbb{R}$, which is ill-defined at the upper boundary. It is represented pictorially by the toy-model 'self-energy' graph . Here $c > 0$ is an external dimensional parameter. Its non-zero value avoids a so-called *infrared divergency* problem at the lower integral boundary. We may now iterate those integrals, representing them by rainbow graphs

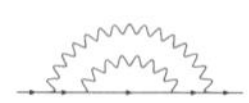 $$\sim f_2(c) := \int_0^\infty f_1(y)(y+c)^{-1} dy,$$

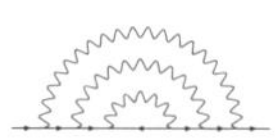 $$\sim f_3(c) := \int_0^\infty f_2(y)(y+c)^{-1} dy \cdots$$

Of course, these graphs just exemplify the simple iterated nature of the integral expression and we could have used instead rooted trees without side branchings [CK1998, Krei1999a]. Similar graphs appear in QED. However, there they represent different functions corresponding to the QED Feynman rules [IzZu1980, Krei2000]. Here and further below, in Section 5, we will use those graphs without any reference to QED. In fact, they will appear in the context of a simple matrix calculus capturing the combinatorics involved in the Bogoliubov formulae which was inspired by the Hopf algebra approach to renormalization.

The functions $f_n(c)$ actually appear as coefficients in the –formal– series expansion of the perturbative calculation of the physical quantity represented by the function $F(c)$

$$F(c) = 1 + \alpha f_1(c) + \alpha^2 f_2(c) + \alpha^3 f_3(c) + \cdots$$

where α is our perturbation parameter, which is supposed to be small and finite. We may write this as a recursive equation due to the simple iterated structure of the functions f_n

$$F(c) = 1 + \alpha \int_0^\infty F(y)\, \frac{dy}{(y+c)}.$$

This recursion is a highly simplified version of a Dyson–Schwinger equation for the quantity F (we refer the reader to Kreimer's contribution in this volume).

However we readily observe that the integrals f_n, $n > 0$ are not well-defined. Indeed, they are logarithmically divergent, $\omega(f_n) = 0$ as a simple power-counting immediately tells. Hence, we introduce a regularization parameter $\varepsilon \in \mathbb{C}$ to render them formally finite

$$f_n(c;\varepsilon) := \mu^\varepsilon \int_0^\infty f_{n-1}(y;\varepsilon)(y+c)^{-1-\varepsilon}dy \in \mathbb{C}[\varepsilon^{-1},\varepsilon]][[\ln(\mu/c)]], \qquad n > 0,$$

and $f_0 = 1$. Here, the parameter μ was introduced for dimensional reasons so as to keep the $f_n(c;\varepsilon)$ dimensionless. This implies a regularized, and hence formally finite Dyson–Schwinger equation

$$F(c;\varepsilon) = 1 + \alpha \int_0^\infty F(y;\varepsilon)\,\frac{\mu^\varepsilon\, dy}{(y+c)^{1+\varepsilon}}.$$

The –forbidden– limit $\varepsilon \to 0$ just brings back the original divergence. The first coefficient function is

$$f_1(c;\varepsilon) = \mu^\varepsilon c^{-\varepsilon}\varepsilon^{-1} = \varepsilon^{-1}\sum_{n\geq 0}\frac{(\varepsilon)^n \log^n(\mu/c)}{n!}.$$

For $n > 1$ those iterated ε-regularized integrals are solved in terms of the function

$$\int_0^\infty y^{-a\varepsilon}(y+c)^{-1-b\varepsilon}dy = B\big((a+b)\varepsilon, 1-a\varepsilon\big)c^{-(a+b)\varepsilon}\ \text{, where } B(a,b) := \frac{\Gamma(a)\Gamma(b)}{\Gamma(a+b)},$$

and $\Gamma(a)$ is the usual Gamma-function [Krei2000]. For $n = 2$ we find

$$\begin{aligned} f_2(c;\varepsilon) &= \mu^{2\varepsilon}\int_0^\infty\int_0^\infty (y+c)^{-1-\varepsilon}(z+y)^{-1-\varepsilon}dzdy \\ &= \frac{\mu^{2\varepsilon}}{\varepsilon}\int_0^\infty y^{-\varepsilon}(y+c)^{-1-\varepsilon}dy = \frac{\mu^{2\varepsilon}c^{-2\varepsilon}}{\varepsilon}B(2\varepsilon, 1-\varepsilon) \\ &= \sum_{n\geq 0}\frac{(2\varepsilon)^n \log^n(\mu/c)}{n!}\Big(\frac{1}{2\varepsilon^2} + h + O(\varepsilon)\Big), \end{aligned}$$

where h is a constant term. Now, first observe that,

$$f_1(c;\varepsilon) - f_1(\mu;\varepsilon) = f_1(c;\varepsilon) - \Big(\frac{1}{\varepsilon}\Big) \in \mathbb{C}[[\varepsilon]][[\ln(\mu/c)]] \xrightarrow{\varepsilon\to 0} f_{1,+} = \log \mu/c.$$

So, by subtracting $f_1(\mu;\varepsilon)$ from $f_1(c;\varepsilon)$ we have achieved a Laurent series without pole part, such that the limit $\varepsilon \to 0$ is allowed and gives a finite, that is, renormalized value of the first order coefficient, $f_1(c) \to f_{1,+}$.

Let us formalize the operation which provides the term to be subtracted by calling it the *renormalization scheme map* R, defined to be

$$R(g)(t) := g(t)_{|t=\mu},$$

for an arbitrary function g. In fact, we just truncated the Taylor expansion of $g(t)$, $\mathcal{T}_\mu[g](t) := \sum_{k\geq 0} g^{(k)}(\mu)\frac{(t-\mu)^k}{k!}$, at zeroth order. The order at which one truncates the Taylor expansion corresponds to the superficial degree of divergence, e.g. in our toy model example we find logarithmic divergencies, $\omega(f_n) = 0$, $n > 0$. The counterterm for the primitive divergent regularized one-loop graph  is

$C($ 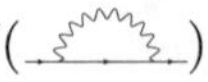 $) = -R(f_1)_{|t=\mu} = -f_1(\mu;\epsilon).$

Applying the same procedure to the second order term we find

$$f_2(c;\varepsilon) - f_2(\mu;\varepsilon) \to \frac{\log \mu/c}{\varepsilon} + h' + O(\varepsilon),$$

with a modified constant term h'. Hence, our subtraction ansatz fails already at second order. Indeed, what we find is an artefact of the –UV divergent 1PI– one-loop diagram  sitting inside the two loop diagram 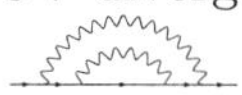 , reflected in the $1/\varepsilon$-term with logarithmic coefficient.

Instead, let us apply Bogoliubov's $\bar{\mathrm{R}}$-operation to the logarithmic divergent two-loop graph  with one-loop  subdivergence

$\bar{\mathrm{R}}($ 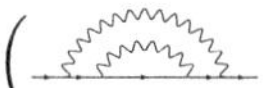 $) \; = $ 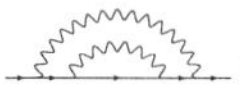 $+ \sum_{\{\gamma'\}\in\{\{$ 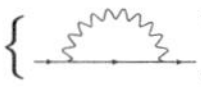 $\}\}}$ 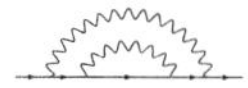 $(\{\gamma'\} \to C($ 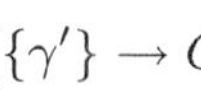 $))$

$=$  $+$ 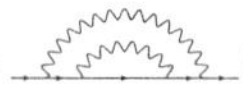 $(\{$  $\} \to C($ 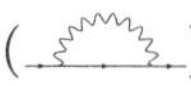 $))$

$=$ 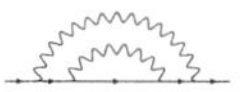 $+ C($  $)$  .

For the sake of notational transparency we omitted to mark the obvious place in the cograph  where the subgraph  of 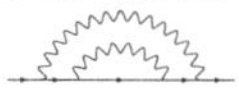 was contracted to a point. In fact, with the function f_1 and f_2 replacing the graphs, we find

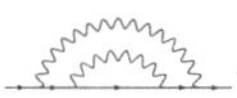 $+C($  $)$  $= \mu^\varepsilon \int_0^\infty \frac{f_1(y;\varepsilon)dy}{(y+c)^{1+\varepsilon}} + C($ 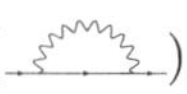 $)\mu^\varepsilon \int_0^\infty \frac{dy}{(y+c)^{1+\varepsilon}}$

$= \mu^\varepsilon \int_0^\infty \frac{\big(f_1(y) + C($  $)\big)dy}{(y+c)^{1+\varepsilon}}$

$$= \mu^\varepsilon \int_0^\infty \frac{\big(f_1(y;\varepsilon) - f_1(\mu;\varepsilon)\big)dy}{(y+c)^{1+\varepsilon}}$$

$$= \mu^\varepsilon \int_0^\infty \frac{f_{1,+}(y;\varepsilon)dy}{(y+c)^{1+\varepsilon}}.$$

And hence, we have replaced the subdivergent one-loop graph by its renormalized expression, see expression (2.12).

$\bar{\mathrm{R}}($ 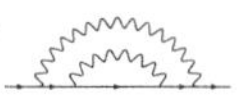 $) \; = \; f_2(c;\varepsilon) + C\big(f_1(c;\varepsilon)\big)f_1(c;\varepsilon) = f_2(c;\varepsilon) - R[f_1(c;\varepsilon)]f_1(c;\varepsilon)$

$$= \sum_{n\geq 0} \frac{(2\varepsilon)^n \log^n(\mu/c)}{n!}\Big(\frac{1}{2\varepsilon^2} + h + O(\varepsilon)\Big) - \frac{1}{\varepsilon^2}\sum_{n\geq 0}\frac{(\varepsilon)^n \log^n(\mu/c)}{n!}.$$

One then shows that $\bar{\mathrm{R}}($ 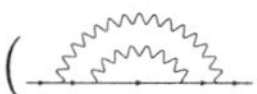 $) \in \mathbb{C}[\varepsilon^{-1},\varepsilon]]$ does not contain any $\log(\mu/c)$-terms as coefficients in the pole part and that the simple subtraction

$(\mathrm{id} - R)\bar{\mathrm{R}}($ 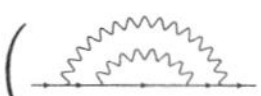 $) = \bar{\mathrm{R}}($ 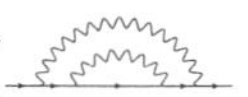 $) + C($ 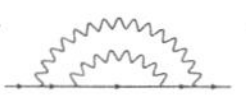 $) \xrightarrow{\varepsilon\to 0} f_{2,+}(c) < \infty.$

The renormalization scheme R, of course, is defined as a map only on the particular space of regularized amplitudes, e.g., the field of Laurent series. This provides a recursive way to calculate the counterterms as well as the renormalized amplitude corresponding to a graph order by order in the perturbation expansion. We may remark here that in the classical –pre-Hopfian– approach the notion of graph and amplitude are used in an interchangeable manner.

We have described a way to renormalize each coefficient in the perturbative expansion of F. We may now introduce a Z-factor, defined as an expansion in α with the counterterms $C(f_n) = C(f_n(c;\varepsilon))$ as coefficients

$$\mathrm{Z} = \mathrm{Z}(\varepsilon) := 1 + \sum_{n>0} \alpha^n C(f_n).$$

We then multiply the Dyson–Schwinger toy-equation for F with Z

$$F(c;\varepsilon) \xrightarrow{\mathrm{Z}} \bar{F}(c;\varepsilon) = \mathrm{Z} + \alpha \int_0^\infty \bar{F}(y;\varepsilon) \, \frac{\mu^\varepsilon \, dy}{(y+c)^{1+\varepsilon}}.$$

Upon expanding Z we find up to second order

$$\bar{F}(c;\varepsilon) = 1 + \alpha^1\big(C(f_1) + f_1(c;\varepsilon)\big) + \alpha^2\big(C(f_2) + C(f_1)f_1(c;\varepsilon) + f_2(c;\varepsilon)\big) + O(\alpha^3).$$

Comparing with our calculations of the renormalized amplitudes $f_{1,+}$ and $f_{2,+}$ above, we readily observe that the multiplication of the Dyson–Schwinger toy-equation with Z has renormalized the perturbative expansion of F order by order in the coupling α

$$F(c;\varepsilon) \xrightarrow{\mathrm{Z}} \bar{F}(c;\varepsilon) \xrightarrow{\varepsilon \to 0} F_{ren}(c).$$

This is a simplified picture of the process of multiplicative renormalization via constant Z-factors, $\mathrm{Z}F(c;\varepsilon) \xrightarrow{\varepsilon \to 0} F_{ren}(c)$, which is a power series in α with finite coefficients, see [Krei1999a, Krei2000]. Later we will establish a more involved but similar result known as Connes–Kreimer's Birkhoff decomposition of renormalization.

Let us summarize and see what we have learned so far. Feynman amplitudes appear as the coefficients in the perturbative expansion of physical quantities in powers of a coupling constant. In general, they are ill-defined objects. The process of regularization renders them formally finite upon introducing nonphysical parameters into the theory which allow for a decomposition of such amplitudes into a finite part, although ambiguously defined, and one that contains the divergent contributions. Renormalization of such regularized amplitudes is mainly guided by constraints coming from fundamental physical principles, such as locality, unitarity, and Lorentz invariance and consists of careful manipulations eliminating, i.e., subtracting, though in a highly non-trivial manner, the illness causing parts, together with a renormalization hypotheses which fixes the value of the remaining finite parts. Upon introducing multiplicative Z-factors respectively bare parameters m_0, λ_0 and bare fields in the Lagrangian, such that $L \xrightarrow{\phi_0, m_0, \lambda_0} L_{ren}$ we arrive at a local perturbative renormalization of a –renormalizable– quantum field theory in the Lagrangian approach. However, it is crucial for the Z-factors to be independent of external momenta, as otherwise such a dependence would obstruct the equations of motion derived from the Lagrangian. After renormalization one can remove consistently the regularization parameter in the perturbative expansion.

Bogoliubov's $\bar{\mathrm{R}}$-map was exactly invented to extract the finite parts of –regularized– Feynman integrals corresponding to Feynman graphs. The simple strategy used in this process consists of preparing a superficially divergent Feynman graph respectively its proper ultraviolet divergent one-particle irreducible subgraphs in such a way that a final 'naive' subtraction renders the whole amplitude associated with that particular graph finite. Today this is summarized in the BPHZ renormalization prescription refereing to Bogoliubov, Parasiuk, Hepp and Zimmermann

[BoPa1957, Hepp1966, Zim1969], see [BoSh1959, CaKe1982, Col1984] for more details.

3 Connes–Kreimer's Hopf algebra of Feynman graphs

Before recasting adequately the above classical setting of perturbative renormalization in Hopf algebraic terms we would like to recall some basic notions of connected graded Hopf algebras.

In fact, we will provide most of the algebraic notions we need in the sequel most of which are well-known from the classical literature, with only two exceptions, namely, we will explore Rota–Baxter algebras in an extra section. We do this simply because as we will see much of the algebraic reasoning for Connes–Kreimer's Birkhoff decomposition follows from key properties characterizing algebras especially associative ones with a Rota–Baxter structure. Furthermore, in subsection 3.1.3 we present a general factorization theorem for complete filtered algebras, which we discovered together with D. Kreimer in [EGK2004, EGK2005] and which was further explored jointly with D. Manchon in [EGM2006].

The concept of a Hopf algebra originated from the work of Hopf [Hop1941] by algebraic topologists [MM1965]. Comprehensive treatments of Hopf algebras can be found in [Abe1980, Swe1969]. Some readers may find Bergman's paper [Berg1985] a useful reference. Since the 1980s Hopf algebras became familiar objects in the realm of quantum groups [ChPr1995, FGV2001, Kas1995, Maj1995, StSh1993], and more recently in noncommutative geometry [CoMo1998]. The class of Hopf algebras we have in mind appeared originally in the late 1970s in the context of combinatorics, where they were introduced essentially by Gian-Carlo Rota. The seminal references in this field are the work by Rota [Rota1978], and in somewhat disguised form Joni and Rota [JoRo1979]. Actually, the latter article essentially presents a long list of bialgebras, whereas the Hopf algebraic extra structure, i.e., the antipode map was largely ignored. Later, one of Rota's students, W. Schmitt, further extended the subject considerably into the field of incidence Hopf algebras [Schm1987, Schm1993, Schm1994]. Other useful references on Hopf algebras related to combinatorial structures are [FG2005, Maj1995, NiSw1982, SpDo1997].

In the sequel, $\mathbb{K}$ denotes the ground field of $char(\mathbb{K}) = 0$ over which all algebraic structures are defined. The following results have all appeared in the literature and we refer the reader to the above cited references, especially [FG2005, Man2001] for more details.

3.1 Complete filtered algebra. In this section we establish general results to be applied in later sections. We obtain from the Baker–Campbell–Hausdorff (BCH) series a non-linear map χ on a complete filtered algebra A which we called *BCH*-recursion. This recursion gives a decomposition on the exponential level (see Theorem 3.1), and a one-sided inverse of the Baker–Campbell–Hausdorff series with the later regarded as a map from $A \times A \to A$.

3.1.1 *Algebra.* We denote associative $\mathbb{K}$-algebras by the triple (A, m_A, η_A) where A is a $\mathbb{K}$-vector space with a product $m_A : A \otimes A \to A$, supposed to be associative, $m_A \circ (m_A \otimes \mathrm{id}_A) = m_A \circ (\mathrm{id}_A \otimes m_A)$, and $\eta_A : \mathbb{K} \to A$ is the unit map. Often we denote the unit element in the algebra by 1_A. A $\mathbb{K}$-subalgebra in the algebra A is a $\mathbb{K}$-vector subspace $B \subset A$, such that for all $b, b' \in B$ we have $m_A(b \otimes b') \in B$.

A $\mathbb{K}$-subalgebra $I \subset A$ is called (right-) left-ideal if for all $i \in I$ and $a \in A$, $(m_A(i \otimes a) \in I)$ $m_A(a \otimes i) \in I$. $I \subset A$ is called a bilateral ideal, or just an ideal if it is a left- and right-ideal.

In order to motivate the concept of a $\mathbb{K}$-coalgebra to be introduced below, we rephrase the definition of a $\mathbb{K}$-algebra A as a $\mathbb{K}$-vector space A together with a $\mathbb{K}$-vector space morphism $m_A : A \otimes A \to A$ such that the diagram

$$\begin{array}{ccc} A \otimes A \otimes A & \xrightarrow{m_A \otimes \mathrm{id}_A} & A \otimes A \\ {\scriptstyle \mathrm{id}_A \otimes m_A}\Big\downarrow & & \Big\downarrow{\scriptstyle m_A} \\ A \otimes A & \xrightarrow{\quad m_A \quad} & A \end{array} \tag{3.1}$$

is commutative. A is a unital $\mathbb{K}$-algebra if there is furthermore a $\mathbb{K}$-vector space map $\eta_A : \mathbb{K} \to A$ such that the diagram

$$\begin{array}{ccccc} \mathbb{K} \otimes A & \xrightarrow{\eta_A \otimes \mathrm{id}_A} & A \otimes A & \xleftarrow{\mathrm{id}_A \otimes \eta_A} & A \otimes \mathbb{K} \\ & {\scriptstyle \alpha_l}\searrow & \Big\downarrow{\scriptstyle m_A} & \swarrow{\scriptstyle \alpha_r} & \\ & & A & & \end{array} \tag{3.2}$$

is commutative. Here α_l (resp. α_r) is the isomorphism sending $k \otimes a$ (resp. $a \otimes k$) to ka, for $k \in \mathbb{K}$, $a \in A$.

Let $\tau := \tau_{A,A} : A \otimes A \to A \otimes A$ be the flip map defined by $\tau_{A,A}(x \otimes y) := y \otimes x$. A is a commutative $\mathbb{K}$-algebra if the next diagram commutes,

$$\begin{array}{ccc} A \otimes A & \xrightarrow{\ \tau\ } & A \otimes A \\ {\scriptstyle m_A}\Big\downarrow & \swarrow{\scriptstyle m_A} & \\ A & & \end{array} \tag{3.3}$$

We denote by $\mathfrak{L}(A)$ the Lie algebra associated with a $\mathbb{K}$-algebra A by anti-symmetrization of the product m_A. At this point the reader may wish to recall the definition and construction of the universal enveloping algebra $\mathfrak{U}(\mathfrak{L})$ of a Lie algebra $\mathfrak{L}$ [FG2001, Reu1993], and the fact that for a Lie algebra $\mathfrak{L}$ with an ordered basis $l_1, l_2, \cdots, l_n, \cdots$ we have an explicit basis $\{l_{i_1}^{n_1} \ldots l_{i_s}^{n_s} | i_1 < \cdots < i_s,\ n_i > 0,\ s \geq i \geq 1\}$ for $\mathfrak{U}(\mathfrak{L})$.

3.1.2 *Filtered algebra.* A *filtered* $\mathbb{K}$*-algebra* is a $\mathbb{K}$-algebra A together with a *decreasing* filtration, i.e., there are nonunitary $\mathbb{K}$-subalgebras $A_{n+1} \subseteq A_n$, $n \geq 0$ of A such that

$$A = \bigcup_{n \geq 0} A_n\ , \qquad m_A(A_n \otimes A_m) \subseteq A_{n+m}.$$

It follows that A_n is an ideal of A. In a filtered $\mathbb{K}$-algebra A, we can use the subsets $\{A_n\}$ to define a *metric* on A in the standard way. Define, for $a \in A$,

$$l(a) = \begin{cases} \max\{k \mid a \in A_k\}, & a \notin \cap_n A_n, \\ \infty, & \text{otherwise} \end{cases}$$

and, for $a, b \in A$, $d(a,b) = 2^{-l(b-a)}$.

A filtered algebra is called *complete* if A is a complete metric space with metric $d(a,b)$, that is, every Cauchy sequence in A converges. Equivalently, we also record that a filtered $\mathbb{K}$-algebra A with $\{A_n\}$ is complete if $\cap_n A_n = 0$ and if the resulting embedding

$$A \to \bar{A} := \varprojlim A/A_n$$

is an isomorphism. When A is a complete filtered algebra, the functions

$$\exp : A_1 \to 1_A + A_1, \qquad \exp(a) := \sum_{n=0}^{\infty} \frac{a^n}{n!}, \tag{3.4}$$

$$\log : 1_A + A_1 \to A_1, \qquad \log(1_A + a) := -\sum_{n=1}^{\infty} \frac{(-a)^n}{n} \tag{3.5}$$

are well-defined and are the inverse of each other.

Let us mention two examples of complete filtered associative algebra that will cross our ways again further below. First, consider for A being an arbitrary associative $\mathbb{K}$-algebra, the power series ring $\mathcal{A} := A[[t]]$ in one (commuting) variable t. This is a complete filtered algebra with the filtration given by powers of t, $\mathcal{A}_n := t^n A[[t]]$, $n > 0$.

Example 3.1 *Triangular matrices:* Another example is given by the subalgebra $\mathcal{M}_n^{\ell}(A) \subset \mathcal{M}_n(A)$ of (upper) lower triangular matrices in the algebra of $n \times n$ matrices with entries in the associative algebra A, and with n finite or infinite. $\mathcal{M}_n^{\ell}(A)_k$ is the ideal of strictly lower triangular matrices with zero on the main diagonal and on the first $k-1$ subdiagonals, $k > 1$. We then have the decreasing filtration

$$\mathcal{M}_n^{\ell}(A) \supset \mathcal{M}_n^{\ell}(A)_1 \supset \cdots \supset \mathcal{M}_n^{\ell}(A)_{k-1} \supset \mathcal{M}_n^{\ell}(A)_k \supset \cdots, k < n,$$

with

$$\mathcal{M}_n^{\ell}(A)_k \, \mathcal{M}_n^{\ell}(A)_m \subset \mathcal{M}_n^{\ell}(A)_{k+m}.$$

For A being commutative we denote by $\mathfrak{M}_n(A)$ the group of lower triangular matrices with unit diagonal which is $\mathfrak{M}_n(A) = \mathbf{1} + \mathcal{M}_n^{\ell}(A)_1$. Here the $n \times n$, $n \leq \infty$, unit matrix is given by

$$\mathbf{1} := (\delta_{ij} 1_A)_{1 \leq i,j \leq n}. \tag{3.6}$$

3.1.3 *Baker–Campbell–Hausdorff recursion.* Formulae (3.4) and (3.5) naturally lead to the question whether the underlying complete filtered algebra A is commutative or not. In general, we must work with the Baker–Campbell–Hausdorff (BCH) formula for products of two exponentials

$$\exp(x)\exp(y) = \exp\big(x + y + \mathrm{BCH}(x,y)\big).$$

$\mathrm{BCH}(x,y)$ is a power series in the non-commutative power series algebra $Q := \mathbb{Q}\langle\langle x, y\rangle\rangle$. Let us recall the first few terms of [Reu1993, Var1984]

$$\mathrm{BCH}(x,y) = \frac{1}{2}[x,y] + \frac{1}{12}[x,[x,y]] - \frac{1}{12}[y,[x,y]] - \frac{1}{24}[x,[y,[x,y]]] + \cdots$$

where $[x,y] := xy - yx$ is the commutator of x and y in Q. Also, we denote $C(x,y) := x + y + \mathrm{BCH}(x,y)$. So we have

$$C(x,y) = \log\big(\exp(x)\exp(y)\big). \tag{3.7}$$

Then for any complete filtered algebra A and $u, v \in A_1$, $C(u,v) \in A_1$ is well-defined and we get a map

$$C : A_1 \times A_1 \to A_1.$$

Now let $P : A \to A$ be any linear map preserving the filtration of A, $P(A_n) \subseteq A_n$. We define $\tilde{P}$ to be $\mathrm{id}_A - P$. For $a \in A_1$, define $\chi(a) = \lim_{n\to\infty} \chi_{(n)}(a)$ where $\chi_{(n)}(a)$ is given by the *BCH*-recursion

$$\begin{aligned}
\chi_{(0)}(a) &:= a, \\
\chi_{(n+1)}(a) &= a - \mathrm{BCH}\big(P(\chi_{(n)}(a)), (\mathrm{id}_A - P)(\chi_{(n)}(a))\big),
\end{aligned} \tag{3.8}$$

and where the limit is taken with respect to the topology given by the filtration. Then the map $\chi : A_1 \to A_1$ satisfies

$$\chi(a) = a - \mathrm{BCH}\big(P(\chi(a)), \tilde{P}(\chi(a))\big). \tag{3.9}$$

This map appeared in [EGK2004, EGK2005, EG2005b], see also [EGM2006] for more details.

We observe that for any linear map $P : A \to A$ preserving the filtration of A the (usually non-linear) map $\chi : A_1 \to A_1$ is unique and such that $(\chi - \mathrm{id}_A)(A_i) \subset A_{2i}$ for any $i \geq 1$. Further, with $\tilde{P} := \mathrm{id}_A - P$ we have

$$\forall a \in A_1, \quad a = C\Big(P\big(\chi(a)\big), \tilde{P}\big(\chi(a)\big)\Big). \tag{3.10}$$

This map is bijective, and its inverse is given by

$$\chi^{-1}(a) = C\big(P(a), \tilde{P}(a)\big) = a + \mathrm{BCH}\big(P(a), \tilde{P}(a)\big). \tag{3.11}$$

Now follows the first theorem, which contains the key result and which appeared already in [EGK2004, EGK2005]. It states a general decomposition on A implied by the map χ. In fact, it applies to associative as well as Lie algebras.

Theorem 3.1 *Let A be a complete filtered associative (or Lie) algebra with a linear, filtration preserving map $P : A \to A$. For any $a \in A_1$, we have*

$$\exp(a) = \exp\big(P(\chi(a))\big) \exp\big(\tilde{P}(\chi(a))\big). \tag{3.12}$$

In our recent work [EGM2006] with D. Manchon we established a unified approach of several apparently unrelated factorizations arisen from quantum field theory [CK2000], vertex operator algebras [BHL2000], combinatorics [ABS2003] and numerical methods in differential equations [MQZ2001].

In Theorem 3.1 we would like to emphasize the particular case when the map P is idempotent, $P^2 = P$. Hence, let $P : A \to A$ be such an *idempotent* linear filtration preserving map. Let $A = A_- \oplus A_+$ be the corresponding vector space decomposition, with $A_- := P(A)$ and $A_+ := \tilde{P}(A)$. We define $A_{1,-} := P(A_1)$ and $A_{1,+} := \tilde{P}(A_1)$, and $\chi : A_1 \to A_1$ is the *BCH*-recursion map associated to the map P. Then under these hypotheses we find that for any $\eta \in 1_A + A_1$ there are *unique* $\eta_- \in \exp\big(A_{1,-}\big)$ and $\eta_+ \in \exp\big(A_{1,+}\big)$ such that $\eta = \eta_- \eta_+$.

The factorization in Theorem 3.1 gives rise to a simpler recursion for the map χ, without the appearance of $\tilde{P}$. Indeed, for A being a complete filtered algebra with a filtration preserving map $P : A \to A$ the map χ in (3.9) solves the following recursion for $u \in A_1$

$$\chi(u) := u + \mathrm{BCH}\big(-P(\chi(u)), u\big). \tag{3.13}$$

As a particularly simple but useful remark we mention the case of an idempotent algebra morphism on a complete filtered associative algebra A, which deserves some attention especially in the context of renormalization. In fact, such a map R on A

satisfies the weight one Rota–Baxter relation, see Section 4. Moreover, in this case the map χ in Eq. (3.13) simplifies considerably, to wit,

$$\chi(u) = u + \mathrm{BCH}\big(- R(u), u\big), \tag{3.14}$$

for any element $u \in A_1$.

The proof follows from $R(\chi(u)) = R(u)$ which results from the multiplicativity of R, i.e., applying R to Eq. (3.9) we obtain

$$R\big(\chi(u)\big) = R(u) + \mathrm{BCH}\big(R^2(\chi(u)), (R \circ \tilde{R})(u))\big).$$

and, since R is idempotent, we have $R \circ \tilde{R} = R - R^2 = 0$. Thus $R(\chi(u)) = R(u)$.

With the foregoing assumptions on R the factorization in Theorem 3.1 reduces to

$$\exp(a) = \exp\big(R(a)\big) \exp\big(\tilde{R}(a) + \mathrm{BCH}\big(- R(a), a\big)\big), \qquad a \in A_1. \tag{3.15}$$

An example for such a map is given by the evaluation map R_a on $\mathcal{C} := \mathbf{Cont}(\mathbb{R})$ which evaluates a function $f \in \mathcal{C}$ at the point $a \in \mathbb{R}$, $R_a(f) := f(a)$.

3.2 Coalgebra and bialgebra. A $\mathbb{K}$-coalgebra is obtained by reversing the arrows in the relations (3.1) and (3.2). Thus a $\mathbb{K}$-*coalgebra* is a triple $(C, \Delta_C, \epsilon_C)$, where the coproduct map $\Delta_C : C \to C \otimes C$ is coassociative, i.e. $(\Delta_C \otimes \mathrm{id}_C) \circ \Delta_C = (\mathrm{id}_C \otimes \Delta_C) \circ \Delta_C$, and $\epsilon_C : C \to \mathbb{K}$ is the counit map which satisfies $(\epsilon_C \otimes \mathrm{id}_C) \circ \Delta_C(x) = x = (\mathrm{id}_C \otimes \epsilon_C) \circ \Delta_C(x)$. We call its kernel $\ker(\epsilon_C)$ the *augmentation ideal*.

For $x \in C$, we use the notation $\Delta_C(x) = \sum_{(x)} x_{(1)} \otimes x_{(2)}$. Then the coassociativity of the coproduct map Δ_C just means

$$\sum_{(x)} \Big(\sum_{(x_{(1)})} x_{(1)(1)} \otimes x_{(1)(2)} \Big) \otimes x_{(2)} = \sum_{(x)} x_{(1)} \otimes \Big(\sum_{(x_{(2)})} x_{(2)(1)} \otimes x_{(2)(2)} \Big)$$

which allows us to use another short hand notation

$$\Delta_C^{(2)}(x) := (\mathrm{id}_C \otimes \Delta_C) \circ \Delta_C(x) = (\Delta_C \otimes \mathrm{id}_C) \circ \Delta_C(x) = \sum_{(x)} x_{(1)} \otimes x_{(2)} \otimes x_{(3)}.$$

We speak of a cocommutative coalgebra if $\tau \circ \Delta_C = \Delta_C$, that is, the arrows in (3.3) are reversed.

A simple example of a coalgebra is provided by the field $\mathbb{K}$ itself, with $\Delta_{\mathbb{K}} : \mathbb{K} \to \mathbb{K} \otimes \mathbb{K}$, $c \to c \otimes 1$, $c \in \mathbb{K}$ and $\epsilon_{\mathbb{K}} := \mathrm{id}_{\mathbb{K}} : \mathbb{K} \to \mathbb{K}$. We also denote the multiplication in $\mathbb{K}$ by $m_{\mathbb{K}}$.

Let $(C, \Delta_C, \epsilon_C)$ be a $\mathbb{K}$-coalgebra. A subspace $I \subseteq C$ is a *subcoalgebra* if $\Delta_C(I) \subseteq I \otimes I$. A subspace $I \subseteq C$ is called a (left-, right-) *coideal* if $(\Delta_C(I) \subseteq I \otimes C$, $\Delta_C(I) \subseteq C \otimes I)$ $\Delta_C(I) \subseteq I \otimes C + C \otimes I$.

An element x in a coalgebra $(C, \Delta_C, \epsilon_C)$ is called *primitive* if $\Delta_C(x) = x \otimes 1_C + 1_C \otimes x$. We will denote the set of primitive elements in C by $P(C)$.

A $\mathbb{K}$-*bialgebra* consists of a compatible pair of a $\mathbb{K}$-algebra structure and a $\mathbb{K}$-coalgebra structure. More precisely, a $\mathbb{K}$-bialgebra is a quintuple $(B, m_B, \eta_B, \Delta_B, \epsilon_B)$, where (B, m_B, η_B) is a $\mathbb{K}$-algebra, and $(B, \Delta_B, \epsilon_B)$ is a $\mathbb{K}$-coalgebra, such that m_B and η_B are morphisms of $\mathbb{K}$-coalgebras, with the natural coalgebra structure on

$B \otimes B$. In other words, we have the commutativity of the following diagrams.

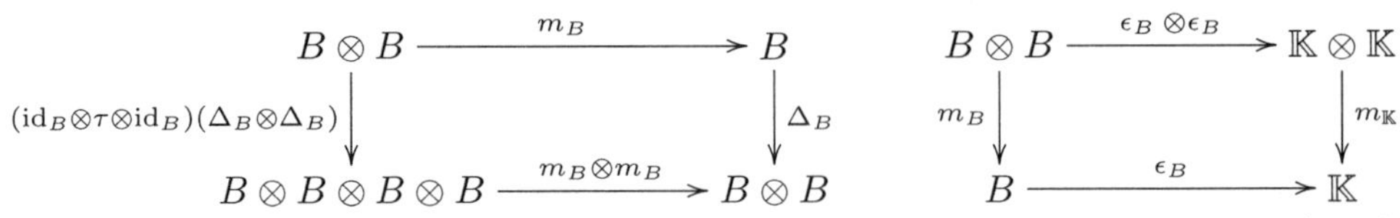

(3.16)

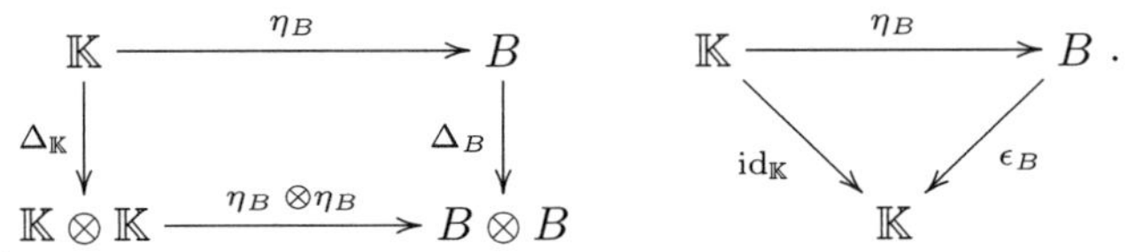

(3.17)

Here the flip map τ is defined as before. We can equivalently require that Δ_B and ϵ_B are morphisms of $\mathbb{K}$-algebras, with the natural algebra structure on $B \otimes B$. One often uses a slight abuse of notation and writes the compatibility condition as

$$\Delta_B(bb') = \Delta_B(b)\Delta_B(b'), \qquad b, b' \in B,$$

saying that the *coproduct of the product is the product of the coproducts.* The identity element in B will be denoted by 1_B. All algebra morphisms are supposed to be unital. The set $P(B)$ of primitive elements of a bialgebra B is a Lie subalgebra of the Lie algebra $\mathcal{L}(B)$.

A bialgebra B is called a *graded bialgebra* if there are $\mathbb{K}$-vector subspaces $B^{(n)}$, $n \geq 0$ of B such that

1. $B = \bigoplus_{n \geq 0} B^{(n)}$,
2. $m_B(B^{(n)} \otimes B^{(m)}) \subseteq B^{(n+m)}$,
3. $\Delta_B(B^{(n)}) \subseteq \bigoplus_{p+q=n} B^{(p)} \otimes B^{(q)}$.

Elements $x \in B^{(n)}$ are given a degree $\deg(x) = n$. Moreover, B is called *connected* if $B^{(0)} = \mathbb{K}$. A graded bialgebra $B = \bigoplus_{n \geq 0} B^{(n)}$ is said to be of *finite type* if each of its homogeneous components $B^{(n)}$ is a $\mathbb{K}$-vector space of finite dimension.

Let B be a connected graded $\mathbb{K}$-bialgebra. The key observation for such objects is the following result describing the coproduct for any element $x \in B$

$$\Delta_B(x) = x \otimes 1_B + 1_B \otimes x + \sum x' \otimes x'',$$

where $\sum x' \otimes x'' \in \ker(\epsilon_B) \otimes \ker(\epsilon_B)$. Hence, for any $x \in B^{(n)}$, the element

$$\bar{\Delta}_B(x) := \Delta_B(x) - x \otimes 1_B - 1_B \otimes x$$

is in $\oplus_{p+q=n,\ p>0,\ q>0} B^{(p)} \otimes B^{(q)}$. The map $\bar{\Delta}_B$ is coassociative on the augmentation ideal, which is $\ker(\epsilon_B) = \bigoplus_{n>0} B^{(n)}$ in the case of a connected graded bialgebra. Elements in the kernel of $\bar{\Delta}_B$ are just the primitive elements in B.

As an example we mention the *divided power bialgebra.* It is defined by the quintuple $(\mathcal{D}, m_{\mathcal{D}}, \eta_{\mathcal{D}}, \Delta_{\mathcal{D}}, \epsilon_{\mathcal{D}})$ where $\mathcal{D}$ is the free $\mathbb{K}$-module $\bigoplus_{n=0}^{\infty} \mathbb{K} d_n$ with basis d_n, $n \geq 0$. The multiplication is given by $m_{\mathcal{D}} : \mathcal{D} \otimes \mathcal{D} \to \mathcal{D}$, $d_m \otimes d_n \mapsto \binom{m+n}{m} d_{m+n}$, and the unital map $\eta_{\mathcal{D}} : \mathbb{K} \to \mathcal{D}$, $1_{\mathbb{K}} \mapsto d_0 := 1_{\mathcal{D}}$. For the coproduct we have $\Delta_{\mathcal{D}} : \mathcal{D} \to \mathcal{D} \otimes \mathcal{D}$, $d_n \mapsto \sum_{k=0}^{n} d_k \otimes d_{n-k}$, and $\epsilon_{\mathcal{D}} : \mathcal{D} \to \mathbb{K}$, $d_n \mapsto \delta_{0,n} 1_{\mathbb{K}}$ where $\delta_{0,n}$ is the Kronecker delta. Other examples of similar type are the *binomial bialgebra*, and the *shuffle* [Swe1969, Reu1993] respectively *quasi-shuffle bialgebra* [Hof2005].

3.3 Convolution product and Hopf algebra. For a $\mathbb{K}$-algebra A and a $\mathbb{K}$-coalgebra C, we define the *convolution product* of two linear maps f, g in $\mathrm{Hom}(C, A)$ to be the linear map $f \star g \in \mathrm{Hom}(C, A)$ given by the composition

$$C \xrightarrow{\Delta_C} C \otimes C \xrightarrow{f \otimes g} A \otimes A \xrightarrow{m_A} A.$$

In other words, for $a \in C$, we define

$$(f \star g)(a) = \sum_{(a)} f(a_{(1)})\, g(a_{(2)}).$$

For the maps $f_i \in \mathcal{A} := \mathrm{Hom}(C, A)$, $i = 1, \cdots, n$, $n > 1$, we define multiple convolution products by

$$f_1 \star \cdots \star f_n := m_A \circ (f_1 \otimes f_2 \otimes \cdots \otimes f_n) \circ \Delta_C^{(n-1)}, \tag{3.18}$$

where we define inductively $\Delta_C^{(0)} := \mathrm{id}_C$ and, for $n > 0$, $\Delta_C^{(n)} := (\Delta_C^{(n-1)} \otimes \mathrm{id}_C) \circ \Delta_C$.

Let $(\mathcal{H}, m_{\mathcal{H}}, \eta_{\mathcal{H}}, \Delta_{\mathcal{H}}, \epsilon_{\mathcal{H}})$ be a $\mathbb{K}$-bialgebra. A $\mathbb{K}$-linear endomorphism S of $\mathcal{H}$ is called an *antipode* for $\mathcal{H}$ if it is the inverse of $\mathrm{id}_{\mathcal{H}}$ under the convolution product

$$S \star \mathrm{id}_{\mathcal{H}} = m \circ (S \otimes \mathrm{id}_{\mathcal{H}}) \circ \Delta_{\mathcal{H}} = \eta_{\mathcal{H}} \circ \epsilon_{\mathcal{H}} = m_{\mathcal{H}} \circ (\mathrm{id}_{\mathcal{H}} \otimes S) \circ \Delta_{\mathcal{H}} = \mathrm{id}_{\mathcal{H}} \star S. \tag{3.19}$$

A *Hopf algebra* is a $\mathbb{K}$-bialgebra $(\mathcal{H}, m_{\mathcal{H}}, \eta_{\mathcal{H}}, \Delta_{\mathcal{H}}, \epsilon_{\mathcal{H}}, S)$ with an antipode S, which is unique. The algebra unit in $\mathcal{H}$ is denoted by $1_{\mathcal{H}}$.

As an example we mention the universal enveloping algebra $\mathfrak{U}(\mathcal{L})$ of a Lie algebra $\mathcal{L}$ which has the structure of a Hopf algebra.

Let $(\mathcal{H}, m_{\mathcal{H}}, \eta_{\mathcal{H}}, \Delta_{\mathcal{H}}, \epsilon_{\mathcal{H}}, S)$ be a Hopf algebra. The antipode is an algebra anti-morphism and coalgebra anti-morphism

$$m_{\mathcal{H}}(S \otimes S) \circ \tau = S \circ m_{\mathcal{H}} \qquad \Delta_{\mathcal{H}} \circ S = \tau \circ S \otimes S \circ \Delta_{\mathcal{H}}.$$

If the Hopf algebra $\mathcal{H}$ is commutative or cocommutative, then $S \circ S = \mathrm{id}_{\mathcal{H}}$.

Let A be an $\mathbb{K}$-algebra, $\mathcal{H}$ a Hopf algebra. By abuse of language we call an element $\phi \in \mathrm{Hom}(\mathcal{H}, A)$ a *character* if ϕ is an algebra morphism, that is, if it respects multiplication, $\phi\big(m_{\mathcal{H}}(x \otimes y)\big) = m_A\big(\phi(x) \otimes \phi(y)\big)$. An element $Z \in \mathrm{Hom}(\mathcal{H}, A)$ is called a *derivation* (or *infinitesimal character*) if

$$Z\big(m_{\mathcal{H}}(x \otimes y)\big) = m_A\big(e_A(x) \otimes Z(y)\big) + m_A\big(Z(x) \otimes e_A(y)\big), \tag{3.20}$$

for all $x, y \in \mathcal{H}$ and with $e_A := \eta_A \circ \epsilon_{\mathcal{H}}$. The set of characters (respectively derivations) is denoted by $G_A := \mathrm{char}(\mathcal{H}, A) \subset \mathrm{Hom}(\mathcal{H}, A)$ (respectively $g_A := \partial\mathrm{char}(\mathcal{H}, A) \subset \mathrm{Hom}(\mathcal{H}, A)$). We remark that 'proper' (infinitesimal) characters live in $\mathrm{Hom}(\mathcal{H}, \mathbb{K})$. We note that $\phi(1_{\mathcal{H}}) = 1_A$ if ϕ is a character and $Z(1_{\mathcal{H}}) = 0$ for Z being a derivation.

In [Man2001] one may find the proof of the following statements which will be important later.

Proposition 3.2 *Let (A, m_A, η_A) be a unital $\mathbb{K}$-algebra.*

1. *Let $(C, \Delta_C, \epsilon_C)$ be a $\mathbb{K}$-coalgebra. Then the triple $(\mathrm{Hom}(C, A), \star, e_A)$ is a unital $\mathbb{K}$-algebra, with $e_A := \eta_A \circ \epsilon_C$ as the unit.*
2. *Let $B = \oplus_{n \geq 0} B^{(n)}$ be a connected graded bialgebra. Let $\mathcal{A} := \mathrm{Hom}(B, A)$, and define*

$$\mathcal{A}_n = \big\{ f \in \mathrm{Hom}(B, A) \big|\ f(B^{(k)}) = 0, k \leq n - 1 \big\}$$

for $n \geq 0$ with the convention that $B^{(-1)} = \emptyset$. Then $\mathcal{A}$ is a complete filtered unital $\mathbb{K}$-algebra.

Of course, we can replace the target space algebra A in item (1) or (2) by the base field $\mathbb{K}$. And in case that B is a bialgebra we have the convolution algebra structure on $\mathrm{Hom}(B, B)$ with unit $e_B := \eta_B \circ \epsilon_B$.

For a connected graded bialgebra, we have the well-known result that any such $\mathbb{K}$-bialgebra $\mathcal{H}$ is a Hopf algebra. The antipode is defined by the geometric series $\mathrm{id}_{\mathcal{H}}^{\star(-1)} = \big(\eta_{\mathcal{H}} \circ \epsilon_{\mathcal{H}} - (\eta_{\mathcal{H}} \circ \epsilon_{\mathcal{H}} - \mathrm{id}_{\mathcal{H}})\big)^{\star(-1)}$

$$S(x) = \sum_{k \geq 0} (\eta_{\mathcal{H}} \circ \epsilon_{\mathcal{H}} - \mathrm{id}_{\mathcal{H}})^{\star k}(x), \tag{3.21}$$

well-defined because of Proposition 3.2. The proof of this result is straightforward, see [FG2001] for more details. The antipode preserves the grading, $S(\mathcal{H}^{(n)}) \subseteq \mathcal{H}^{(n)}$. The projector $P := \mathrm{id}_{\mathcal{H}} - \eta_{\mathcal{H}} \circ \epsilon_{\mathcal{H}}$ maps $\mathcal{H}$ to its augmentation ideal, $\ker(\epsilon_{\mathcal{H}})$.

The antipode S for connected graded Hopf algebras may also be defined recursively in terms of either of the following two formulae

$$\begin{aligned} S(x) = -S \star \mathrm{id}_{\mathcal{H}} \circ P(x) &= -x - \sum_{(x)} S(x_{(1)}) x_{(2)}, \qquad (3.22) \\ S(x) = -\mathrm{id}_{\mathcal{H}} \circ P \star S(x) &= -x - \sum_{(x)} x_{(1)} S(x_{(2)}), \end{aligned}$$

for $x \in \ker(\epsilon_{\mathcal{H}})$, following readily from (3.19) by recalling that $\ker(\epsilon_{\mathcal{H}}) = \bigoplus_{n>0} \mathcal{H}^{(n)}$, and $S(1_{\mathcal{H}}) := 1_{\mathcal{H}}$. The first formula will cross our way in disguised form in later sections.

An important fact due to Milnor and Moore [MM1965] concerning the structure of cocommutative connected graded Hopf algebras of finite type states that any such Hopf algebra $\mathcal{H}$ is isomorphic to the universal enveloping algebra of its primitive elements, $\mathcal{H} \cong \mathfrak{U}(P(\mathcal{H}))$, see also [FG2001].

Let A be a commutative $\mathbb{K}$-algebra and $\mathcal{H}$ a connected graded Hopf algebra. One can show that the subalgebra $\bar{g} := \big\{ f \in \mathrm{Hom}(\mathcal{H}, A) \;\big|\; f(1_{\mathcal{H}}) = 0 \big\}$ in the filtered algebra $\mathcal{A} = (\mathrm{Hom}(\mathcal{H}, A), \star, e_A)$, endowed with Lie brackets defined by anti-symmetrization of the convolution product is a Lie algebra.

Moreover, $\mathcal{G} := \big\{ f \in \mathrm{Hom}(\mathcal{H}, A) \;\big|\; f(1_{\mathcal{H}}) = 1_A \big\} \subset \mathcal{A}$ endowed with the convolution product forms a group in $\mathcal{A}$. The inverse of $f \in \mathcal{G}$ is given by composition with the antipode of $\mathcal{H}$, $f^{-1} = f \circ S$, and we have that $\mathcal{G} = e_A + \bar{g}$. As a matter of fact we find that

1. A-valued characters, $G_A = \mathrm{char}(\mathcal{H}, A)$, form a subgroup of $\mathcal{G} = e_A + \bar{g}$ under convolution;
2. A-valued derivations, $g_A = \partial\mathrm{char}(\mathcal{H}, A)$ form a Lie subalgebra of $\bar{g}$;
3. The bijection $\exp : \bar{g} \to e_A + \bar{g}$ defined by its power series with respect to convolution restricts to a bijection $\exp : g_A \to G_A$.

As $\mathcal{A} = (\mathrm{Hom}(\mathcal{H}, A), \star, e_A)$ is a complete filtered associative algebra by Proposition 3.2, we may immediately apply Theorem 3.1 giving rise to a factorization in the group $G_A = \mathrm{char}(\mathcal{H}, A)$ of A-valued characters.

Theorem 3.3 *Let A be a commutative $\mathbb{K}$-algebra and $\mathcal{H}$ be a connected graded commutative Hopf algebra. Let $\mathcal{A} = (\mathrm{Hom}(\mathcal{H}, A), \star, e_A)$. Let $P : \mathcal{A} \to \mathcal{A}$ be any* *filtration preserving linear map. Then we have for all* $\phi = \exp(Z) \in G_A$, $Z \in g_A$ *the characters* $\phi_-^{-1} := \exp\big(P(\chi(Z))\big)$ *and* $\phi_+ := \exp\big(\tilde{P}(\chi(Z))\big)$ *such that*

$$\phi = \phi_-^{-1} \star \phi_+. \tag{3.23}$$

If P *is idempotent this decomposition is unique.*

As a simple example take the even-odd decomposition described in [ABS2003], see also [EGM2006]. Take an arbitrary connected graded Hopf algebra $\mathcal{H}$ and let Y denote the grading operator, $Y(h) = deg(h)h = nh$ for a homogeneous element $h \in \mathcal{H}^{(n)}$. We may define an involutive automorphism on $\mathcal{H}$, denoted by $^{-} : \mathcal{H} \to \mathcal{H}$, $\overline{h} := (-1)^Y h = (-1)^{deg(h)}h$, for $h \in \mathcal{H}^{(n)}$. It induces by duality an involution on $\mathrm{Hom}(\mathcal{H},\mathbb{K})$, $\overline{\phi}(h) := \phi(\bar{h})$ for $\phi \in \mathrm{Hom}(\mathcal{H},\mathbb{K})$, $h \in \mathcal{H}$. $\mathcal{H}$ naturally decomposes on the level of vector spaces into elements of odd respectively even degree

$$\mathcal{H} = \mathcal{H}_- \oplus \mathcal{H}_+,$$

with projectors $\pi_\pm : \mathcal{H} \to \mathcal{H}_\pm$. Such that for $\overline{\pi_+(h)} = \pi_+(h) =: h_+$ and $\overline{\pi_-(h)} = \overline{h_-} = -h_-$, and $h = h_- + h_+ \in \mathcal{H}$. Let ϕ be a character in the group G. It is called even if it is a fixed point of the involution, $\overline{\phi} = \phi$, and is called odd if it is an anti-fixed point, $\overline{\phi} = \phi^{-1} = \phi \circ S$. The set of odd and even characters is denoted by G_-, G_+, respectively. Even characters form a subgroup in G. Whereas the set of odd characters forms a symmetric space. Theorem 3.3 says that any $\phi \in G$ has a unique decomposition

$$\phi = \exp(Z) = \exp\big(\pi_-(Z) + \pi_+(Z)\big) = \exp\big(\pi_-(\chi(Z))\big) \star \exp\big(\pi_+(\chi(Z))\big), \tag{3.24}$$

with

$$\phi_-^{-1} := \exp\big(\pi_-(\chi(Z))\big) \in G_-$$

being an odd character, and

$$\phi_+ := \exp\big(\pi_+(\chi(Z))\big) \in G_+$$

being an even character. From the factorization we derive a closed form for the BCH-recursion [EGM2006]

$$\chi(Z) = Z + \mathrm{BCH}\Big(-\pi_-(Z) - \frac{1}{2}\mathrm{BCH}\big(Z, Z - 2\pi_-(Z)\big), Z\Big).$$

Finally, by abuse of language we shall call a connected graded commutative bialgebra $(\mathcal{H} = \bigoplus_{n\geq 0}\mathcal{H}^{(n)}, m_{\mathcal{H}}, \eta_{\mathcal{H}}, \Delta_{\mathcal{H}}, \epsilon_{\mathcal{H}})$ of finite type a *renormalization Hopf algebra* if it is polynomially generated by a graded vector space $\mathcal{H}_{\ell in}$ which is also a right-coideal. This implies that the right hand side of $\bar{\Delta}_{\mathcal{H}}(x) \in \ker(\epsilon_{\mathcal{H}}) \otimes \ker(\epsilon_{\mathcal{H}})$ is linear for $x \in \mathcal{H}_{\ell in}$

$$\mathcal{H}_{\ell in} \xrightarrow{\Delta_{\mathcal{H}}} \mathcal{H} \otimes \mathcal{H}_{\ell in}. \tag{3.25}$$

In the sequel we will also use the name *combinatorial Hopf algebra* if we want to underline its combinatorial, i.e. graphical structure.

3.4 Hopf algebra of non-decorated non-planar rooted trees. As a key-example for such a type of Hopf algebra we mention briefly the Hopf algebra of non-decorated non-planar rooted trees. Connes and Kreimer introduced it in [CK1998]. For more details on its relation to renormalization we refer the reader so Section 3.5.3. The reader may also consult the references [BeKr2005a, FGV2001, Fois2002, Hof2003, Holt2003].

A rooted tree t is a connected and simply-connected set of vertices $V(t)$ and oriented edges $E(t)$ such that there is precisely one distinguished vertex, called the root, with no incoming edge. We draw the root on top of the tree with its outgoing edges oriented towards the root. The empty tree is denoted by $1_{\mathcal{T}}$

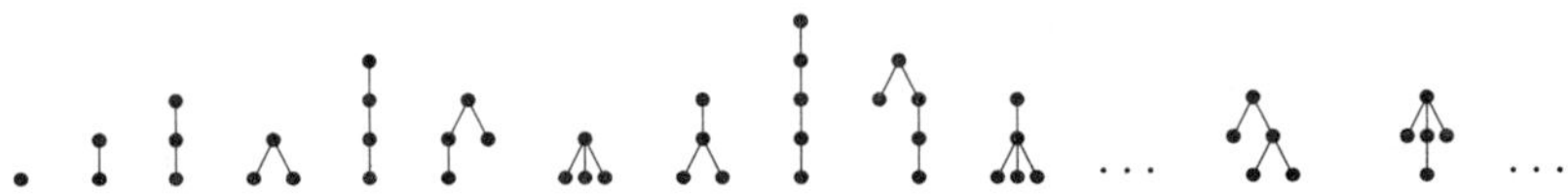

Let T be the set of isomorphic classes of rooted trees. Let $\mathcal{T}$ be the $\mathbb{K}$-vector space generated by T, which is graded by the number of vertices, denoted by $\deg(t) := |V(t)|$ with the convention that $\deg(1_{\mathcal{T}}) = 0$. Let $\mathcal{H}_{\mathcal{T}}$ be the graded commutative polynomial algebra of finite type over $\mathbb{K}$ generated by $\mathcal{T}$, $\mathcal{H}_{\mathcal{T}} := \mathbb{K}[\mathcal{T}] = \bigoplus_{n\geq 0} \mathbb{K}\mathcal{H}^{(n)}$. Monomials of trees are called forests. We will define a coalgebra structure on $\mathcal{T}$. The counit is defined by

$$\epsilon(t_1\cdots t_n) := \begin{cases} 0, & t_1\cdots t_n \neq 1_{\mathcal{T}} \\ 1, & t_1\cdots t_n = 1_{\mathcal{T}}. \end{cases} \tag{3.26}$$

The coproduct is defined in terms of *cuts* $c(t) \subset E(t)$ on a tree $t \in \mathcal{T}$. A *primitive cut* is the removal of a single edge, $|c(t)| = 1$, from the tree t. The tree t decomposes into two parts, denoted by the pruned part $P_c(t)$ and the rooted part $R_c(t)$, where the latter contains the original root vertex. An *admissible cut* of a rooted tree t is a set of primitive cuts, $|c(t)| > 1$, such that any path from any vertex of t to its root has at most one cut.

The *coproduct* is then defined as follows. Let C_t be the set of all admissible cuts of the rooted tree $t \in \mathcal{T}$. We exclude the empty cut $c^{(0)}(t)$, $P_{c^{(0)}}(t) = \emptyset$, $R_{c^{(0)}}(t) = t$ and the full cut $c^{(1)}(t)$, $P_{c^{(1)}}(t) = t$, $R_{c^{(1)}}(T) = \emptyset$. Also let C'_t be $C_t \cup \{c^{(0)}(t), c^{(1)}(t)\}$. Define

$$\Delta(t) := t \otimes 1_{\mathcal{T}} + 1_{\mathcal{T}} \otimes t + \sum_{c_t \in C_t} P_c(t) \otimes R_c(t) = \sum_{c_t \in C'_t} P_c(t) \otimes R_c(t). \tag{3.27}$$

We see easily, that $\deg(t) = \deg(P_c(t)) + \deg(R_c(t))$, for all admissible cuts $c_t \in C_t$, and therefore

$$\sum_{c_t \in C_t} P_c(t) \otimes R_c(t) \in \sum_{p+q=\deg(t),\ p,\,q>0} \mathcal{H}^{(p)} \otimes \mathcal{H}^{(q)}.$$

Furthermore, this map is extended by definition to an algebra morphism on $\mathcal{H}_{\mathcal{T}}$

$$\Delta\Big(\prod_{i=1}^{n} t_i\Big) := \prod_{i=1}^{n} \Delta(t_i).$$

We see here a very concrete instance of the coproduct map of a Hopf algebra. The best way to get use to this particular coproduct is to present some examples,

$$\begin{aligned}
\Delta(\bullet) &= \bullet \otimes 1_{\mathcal{T}} + 1_{\mathcal{T}} \otimes \bullet \\
\Delta(\substack{\bullet\\ \bullet}) &= \substack{\bullet\\ \bullet} \otimes 1_{\mathcal{T}} + 1_{\mathcal{T}} \otimes \substack{\bullet\\ \bullet} + \bullet \otimes \bullet \qquad (3.28)\\
\Delta(\bullet\bullet) &= \Delta(\bullet)\Delta(\bullet) = (\bullet \otimes 1_{\mathcal{T}} + 1_{\mathcal{T}} \otimes \bullet)(\bullet \otimes 1_{\mathcal{T}} + 1_{\mathcal{T}} \otimes \bullet) \\
&= \bullet\bullet \otimes 1_{\mathcal{T}} + 1_{\mathcal{T}} \otimes \bullet\bullet + 2 \bullet \otimes \bullet \qquad (3.29)\\
\Delta\Big(\substack{\bullet\\ \bullet\\ \bullet}\Big) &= \substack{\bullet\\ \bullet\\ \bullet} \otimes 1_{\mathcal{T}} + 1_{\mathcal{T}} \otimes \substack{\bullet\\ \bullet\\ \bullet} + \bullet \otimes \substack{\bullet\\ \bullet} + \substack{\bullet\\ \bullet} \otimes \bullet \qquad (3.30)\\
\Delta(\substack{\bullet\\ \bullet\,\bullet}) &= \substack{\bullet\\ \bullet\,\bullet} \otimes 1_{\mathcal{T}} + 1_{\mathcal{T}} \otimes \substack{\bullet\\ \bullet\,\bullet} + 2 \bullet \otimes \substack{\bullet\\ \bullet} + \bullet\bullet \otimes \bullet \qquad (3.31)
\end{aligned}$$

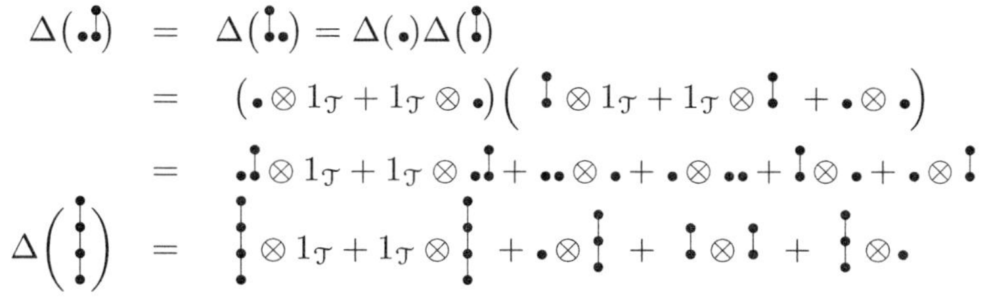

(3.32) (3.33) (3.34)

Connes and Kreimer showed that $(\mathcal{H}_{\mathcal{T}}, m, \eta, \Delta, \epsilon)$ with coproduct $\Delta : \mathcal{H}_{\mathcal{T}} \to \mathcal{H}_{\mathcal{T}} \otimes \mathcal{H}_{\mathcal{T}}$ defined by (3.27), and counit $\epsilon : \mathcal{H}_{\mathcal{T}} \to \mathbb{K}$ (3.26) is a $\mathbb{Z}_{\geq 0}$ connected graded commutative, but non-cocommutative bialgebra of finite type. $\mathcal{H}_{\mathcal{T}}$ is connected since $\mathcal{H}^{(0)} \simeq \mathbb{K}$, and hence a Hopf algebra with antipode S. See Eqs. (3.21), (3.22), defined recursively by $S(1_{\mathcal{T}}) = 1_{\mathcal{T}}$ and

$$S(t) := -t - \sum_{c_t \in C_t} S(P_c(t)) R_c(t).$$

Again, a couple of examples might be helpful here.

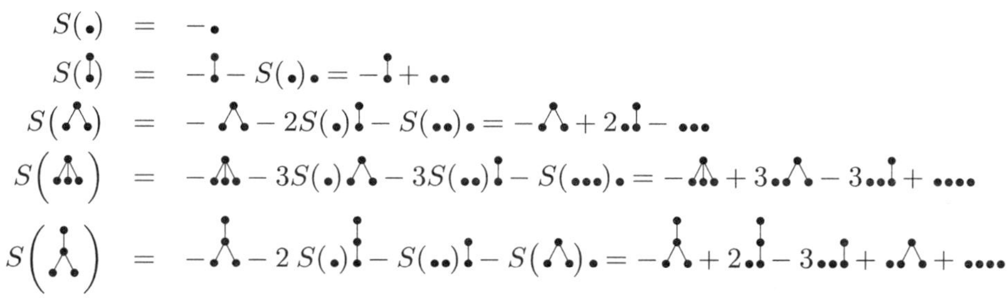

Remark 3.4 1. The coproduct (3.27) can be written in a recursive way, using the B_+ operator, which is a closed but not exact Hochschild 1-cocycle [Krei1999a, FGV2001, BeKr2005a], hence

$$\Delta(t) = \Delta(B_+(t_{i_1} \cdots t_{i_n})) = t \otimes 1_{\mathcal{T}} + \{\mathrm{id}_{\mathcal{T}} \otimes B_+\}\Delta(t_{i_1} \cdots t_{i_n}). \tag{3.35}$$

$B_+ : \mathcal{H}_{\mathcal{T}} \to \mathcal{H}_{\mathcal{T}}$, in fact $B_+ : \mathcal{H}_{\mathcal{T}} \to \mathcal{T}$, is a linear operator, mapping a (forest of) rooted tree(s) to a rooted tree, by putting a new root on top of the (forrest) tree and connecting the old root(s) to this new adjoined root. A couple of examples tell everything

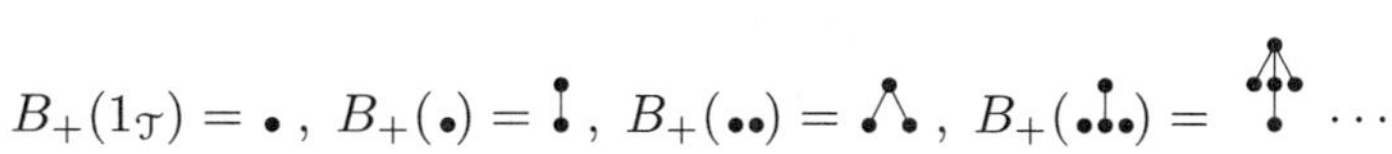

It therefore raises the degree by 1, $\deg\left(B_+(t_{i_1} \cdots t_{i_n})\right) = \deg\left(t_{i_1} \cdots t_{i_n}\right) + 1$. The inductive proof of coassociativity of the coproduct (3.27) formulates easily in terms this map. Every rooted tree lies in the image of the B_+ operator. The notion of subtrees becomes evident from this fact. The conceptual importance of the B_+ map with respect to fundamental notions of physics was further elaborated in recent work [BeKr05b, Krei2003a, Krei2005].

2. It is important to notice that the right hand side of $\Delta(t) \in \mathcal{H}_{\mathcal{T}} \otimes \mathcal{H}_{\mathcal{T}}$ is linear for $t \in \mathcal{T}$. Therefore we may write

$$\mathcal{T} \xrightarrow{\Delta} \mathcal{H}_{\mathcal{T}} \otimes \mathcal{T}. \tag{3.36}$$

This is of course not true for the coproduct of proper forests of rooted trees, $t = t_1 \cdots t_n$, $n > 1$.

3.5 Hopf and Lie algebra of Feynman graphs. In the work of Kreimer [Krei1998], and Connes and Kreimer [CK2000, CK2001] Feynman graphs as the main building blocks of perturbative QFT are organized into a Hopf algebra. In fact it is this Hopf algebra of Feynman graphs that one calls the renormalization Hopf algebra corresponding to the pQFT. Theorem 3.3 establishes the Birkhoff decomposition upon replacing the linear filtration preserving map P by the regularization prescription and the corresponding renormalization scheme map R.

3.5.1 *Hopf algebra of Feynman graphs.* We first give a more detailed description of a *connected Feynman graph* Γ. It is a collection of several types of internal and external lines, and vertices. We denote by $\Gamma^{[0]}$ its set of vertices and by $\Gamma^{[1]} := \Gamma^{[1]}_{\rm int} \cup \Gamma^{[1]}_{\rm ext}$ its set of possibly oriented internal and external edges. Internal lines are also called propagators, whereas external ones are called legs. For instance, for QED we have just two types of edges,  and , together with one type of vertex, 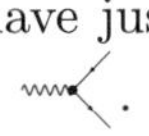. A proper subgraph of a Feynman graph is determined by proper subsets of the set of internal edges and vertices.

Of vital importance is the class of so-called *one-particle irreducible* (1PI) Feynman graphs, which consists of connected graphs that cannot be made disconnected by removing any of its internal edges. In general, a 1PI Feynman graph Γ is parameterized by attaching to the set of external edges $\Gamma^{[1]}_{\rm ext}$ the quantum numbers, such as masses, momenta, and spin, corresponding to the particles that enter respectively exit the scattering process described by the graph. In fact, the set of such quantum numbers specifies in a precise manner the external leg structure, denoted by $\underline{r}$, of a 1PI graph $\Gamma = \Gamma^{\underline{r}}$, that is, the physical process to which that graph respectively its amplitude contributes. We denote the set of all external leg structures $\underline{r}$ by $\mathfrak{R}$ and observe that for a renormalizable QFT it consists of those edges and vertices corresponding to the monomials in the defining Lagrangian.

Beyond one-loop order, the process of the renormalization of a 1PI graph is characterized essentially by the appearance of its ultraviolet (UV) divergent 1PI Feynman subgraphs $\gamma_i \subset \Gamma$. For instance, two proper Feynman subgraphs $\gamma_1, \gamma_2 \subset \Gamma$ might be strictly *nested*, $\gamma_1 \subset \gamma_2 \subset \Gamma$, or *disjoint*, $\gamma_1 \cap \gamma_2 = \emptyset$. This hierarchy in which subgraphs are either located inside another subgraph or appear to be disjoint is best represented by a decorated rooted tree, see Section 3.5.3. However, there is a third possibility, consisting of subgraphs which might be *overlapping*, see Figure 1. In fact, such Feynman graphs are represented by linear combinations of decorated rooted trees. For a detailed treatment of this important case of overlapping graph structures in pQFT we refer to [Krei1999b].

We denote by F the set of all equivalence classes of one-particle irreducible Feynman graphs. Let $\mathcal{F} := \mathbb{K}F$ be the $\mathbb{K}$-vector space with basis F. Let $\mathcal{H}_{\mathcal{F}}$ be the commutative polynomial algebra $\mathbb{K}[\mathcal{F}]$ with the product denoted by the disjoint union and with the empty graph $1_{\mathcal{F}}$ as the unit. Those 1PI Feynman graphs come with a grading by the number of loops. This gives rise to a grading on $\mathcal{H}_{\mathcal{F}}$ by $\deg(\prod_{i=1}^k \Gamma_i) = \sum_{i=1}^k \deg(\Gamma_i)$, making it into a connected graded commutative

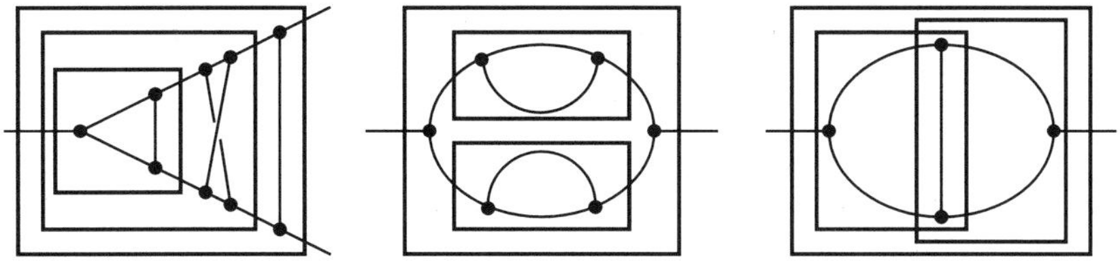

Figure 1 [Krei2000] There are various ways subdivergences can appear. They can be nested as in (a), disjoint as in (b) or overlapping as in (c). We put every subdivergence in a box. These boxes are called forests.

algebra $\mathcal{H}_{\mathcal{F}} = \oplus_{n=0}^{\infty}\mathcal{H}^{(n)}$ with $\mathcal{H}^{(n)}$ the subspace spanned by degree n graphs. We have $\mathcal{H}^{(0)} \simeq \mathbb{K}$.

Define a counit ϵ on $\mathcal{H}_{\mathcal{F}}$ by $\epsilon(1_{\mathcal{F}}) = 1_{\mathbb{K}}$ and zero else. The key map is the coproduct $\Delta : \mathcal{H}_{\mathcal{F}} \to \mathcal{H}_{\mathcal{F}} \otimes \mathcal{H}_{\mathcal{F}}$ defined on 1PI Feynman graphs $\Gamma \in F$ by

$$\Delta(\Gamma) = \Gamma \otimes 1_{\mathcal{F}} + 1_{\mathcal{F}} \otimes \Gamma + \sum_{\gamma \subset \Gamma} \gamma \otimes \Gamma/\gamma, \tag{3.37}$$

where by abuse of notation the sum is over all unions of disjoint divergent proper 1PI subgraphs $\gamma := \{\gamma_1 \cdots \gamma_n\}$, $\gamma_i \cap \gamma_j = \emptyset$, $i \neq j$, of Γ, and Γ/γ denotes the cograph which follows from the contraction of γ in Γ.

Remark 3.5 In fact, one may define the coproduct of a graph Γ directly by the use of its spinneys in the wood $\bar{\mathfrak{W}}(\Gamma)$

$$\Delta(\Gamma) = \sum_{\{\gamma'\} \in \bar{\mathfrak{W}}(\Gamma)} \gamma' \otimes \Gamma/\{\gamma'\}. \tag{3.38}$$

Where the first two terms $\Gamma \otimes 1_{\mathcal{F}}$ and $1_{\mathcal{F}} \otimes \Gamma$ follow from the spinneys $\{\Gamma\}$ and $\{\emptyset\} =: 1_{\mathcal{F}}$, respectively. Hence, the coproduct can be seen as a tensor list, systematically storing term by term the spinneys and associated cographs for each Feynman graph.

We extend this definition to products of graphs (forests) in $\mathcal{H}_{\mathcal{F}}$, $\Delta\big(\prod_{i=1}^{k} \Gamma_i\big) = \prod_{i=1}^{k} \Delta(\Gamma_i)$, so that we get a connected graded commutative non-cocommutative bialgebra, hence a Hopf algebra of that type with antipode map S, as defined in Eq. (3.21) . This result was first established by Connes and Kreimer in [CK2000]. By the definition of the coproduct in (3.37) we may call $\mathcal{H}_{\mathcal{F}}$ a *renormalization Hopf algebra of Feynman graphs*, since for $\Gamma \in F$ the cograph Γ/γ in (3.37) always lies in F, such that

$$\mathcal{F} \xrightarrow{\Delta} \mathcal{H}_{\mathcal{F}} \otimes \mathcal{F}. \tag{3.39}$$

As an example we calculate the coproduct of the graph 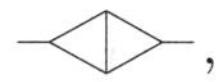, borrowed from ϕ_6^3 theory, which is the same as the graph in case (c) in Figure 1. We hope that the way we draw this two-loop Feynman self-energy graph now helps to identify its two one-loop vertex subgraphs, 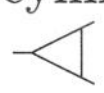 and 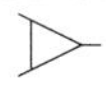.

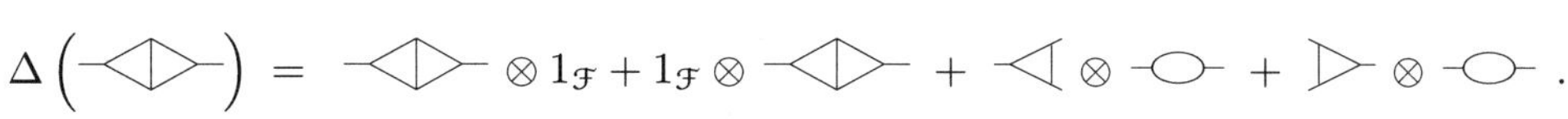

Compare this with the wood containing the spinneys, $\bar{\mathfrak{W}}($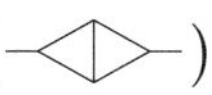$)$, which is just given by Eq.(2.6) enlarged by the empty graph $\{\emptyset\}$ and the graph itself $\{$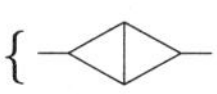$\}$.

Using Eq. (3.22) the antipode of this graph is given by

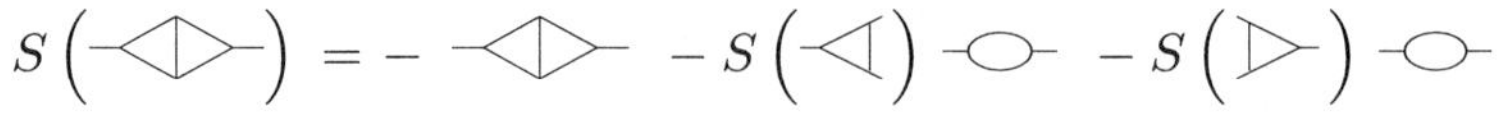

We must compare this with Bogoliubov's classical formula for the counterterm $C(\Gamma)$ of a Feynman graph Γ, see Eq. (2.13) of Section 2.

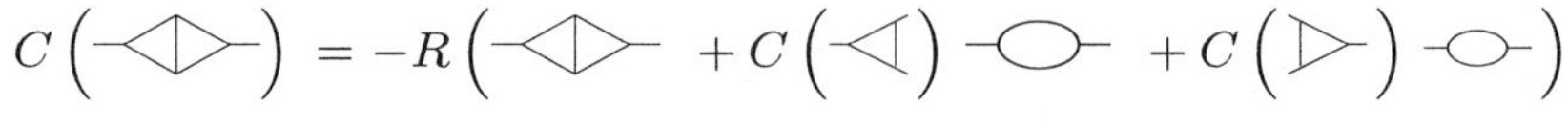

Here, of course, each graph represents its corresponding –regularized– Feynman amplitude, and R denotes the renormalization scheme map.

3.5.2 *Lie and pre-Lie algebra of Feynman graphs.* The combinatorial Hopf algebra of Feynman graphs has a dual counter part. To see this let us go back to the coproduct and antipode respectively the counterterm expression in the above example. In fact, we readily observe that the Connes-Kreimer coproduct just *disentangles* a graph analogously to Bogoliubov's $\bar{\mathrm{R}}$-map which in turn gives the counterterm map C.

One might therefore think about an opposite operation in terms of gluing graphs into another graph. Let us define such a bilinear *gluing operation* on the vector space $\mathcal{F}$ of 1PI Feynman graphs

$$\Gamma_1 * \Gamma_2 = \sum_{\Gamma \in F} n(\Gamma_1, \Gamma_2; \Gamma)\Gamma. \tag{3.40}$$

The sum on the right hand side runs over all 1PI graphs Γ in F. The number $n(\Gamma_1, \Gamma_2; \Gamma) \in \mathbb{N}$ for $\Gamma, \Gamma_1, \Gamma_2 \in F$ is a section coefficient which counts the number of ways a subgraph Γ_2 in Γ can be reduced to a point such that Γ_1 is obtained. The above sum is finite as long as Γ_1 and Γ_2 are finite graphs. The graphs which contribute to this sum necessarily fulfill $\deg(\Gamma) = \deg(\Gamma_1) + \deg(\Gamma_2)$ and $\mathbf{res}(\Gamma) = \mathbf{res}(\Gamma_1)$, where we let $\mathbf{res}(\Gamma)$, called the *residue of* Γ, be the graph obtained when all internal edges are shrunk to a point. The graph we obtain in this manner consists of a single vertex with $\mathbf{res}(\Gamma)^{[0]}_{\text{ext}} = \Gamma^{[0]}_{\text{ext}}$. In case the initial graph was a self-energy graph, we regard its residue as a single edge.

As a matter of fact it was shown in several references [ChLi2001, CK1998, CK2002] that the operation $*$ on $\mathcal{F}$ defines a (right) *pre-Lie* product, satisfying the *pre-Lie relation*

$$(\Gamma_1 * \Gamma_2) * \Gamma_3 - \Gamma_1 * (\Gamma_2 * \Gamma_3) = (\Gamma_1 * \Gamma_3) * \Gamma_2 - \Gamma_1 * (\Gamma_3 * \Gamma_2) \tag{3.41}$$

and hence a pre-Lie algebra $\mathcal{P} := \mathcal{P}(\mathcal{F})$. This pre-Lie algebra structure on Feynman graphs was further explored by Mencattini and Kreimer [MeKr2004, MeKr2005]. See also [EMK2003] for more details. The pre-Lie relation is sufficient for the anti-symmetrization of this product to fulfil the Jacobi identity. Hence, we get a graded Lie algebra $\mathcal{L}(\mathcal{P}) := \bigoplus_{n \geq 0} \mathcal{L}^{(n)}$ of finite type, with commutator bracket

$$[\Gamma_1,\ \Gamma_2] := \Gamma_1 * \Gamma_2 - \Gamma_2 * \Gamma_1, \tag{3.42}$$

for $\Gamma_1, \Gamma_2 \in \mathcal{F}$. The above pre-Lie respectively Lie algebra structures are available once one has decided on the set of 1PI graphs F of interest. The Milnor–Moore theorem [MM1965] then implies the structure of a corresponding Hopf algebra of Feynman graphs and vice versa.

So far we have two complementary operations on Feynman graphs forming either a (pre-)Lie or a Hopf algebra. In the former case we have a composition of Feynman graphs in terms of a pre-Lie gluing product, where we replace vertices by Feynman graphs with compatible external leg structure. Dually, we have the decomposition of graphs in terms of a Hopf algebra coproduct, i.e., replacing non-trivial UV divergent 1PI subgraphs by their residues.

3.5.3 *Correspondence between non-planar rooted trees and Feynman graphs.* Connes and Kreimer developed in [CK1998] a detailed picture of a connected graded commutative Hopf algebra structure of finite type on non-planar decorated rooted trees giving rise to an interesting interplay between perturbative renormalization and non-commutative geometry. This Hopf algebra of rooted trees serves as the role model for combinatorial Hopf algebras of the above type due to its universal property. In Section 3.4 we gave a brief account of its main properties. Here we would like to indicate its link to Feynman graphs and renormalization.

In the renormalization problem we associated to each graph Γ its wood, i.e. the set of its spinneys, denoted by $\mathfrak{W}(\Gamma)$. We have seen that the coproduct is a storing list of those spinneys and the corresponding cographs. Indeed, the only information needed in this disentanglement problem of a graph is the hierarchical structure in which the spinneys appear to sit inside the graph. Each spinney of a graph consists of a union of disjoint UV divergent 1PI subgraphs $\gamma \subsetneq \Gamma$ which themselves may carry a nontrivial spinney structure, e.g. . Its set of proper spinneys was given in (2.7)

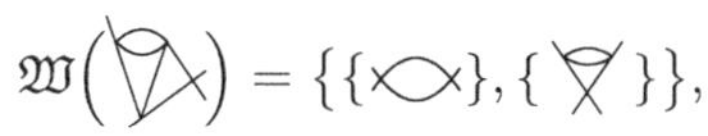

where  is a maximal spinney. We observe that this UV divergent 1PI subgraph  contains itself one –maximal– (sub)spinney

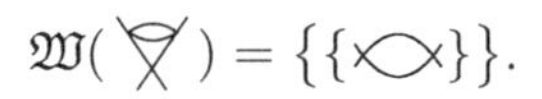

For the graph  we found

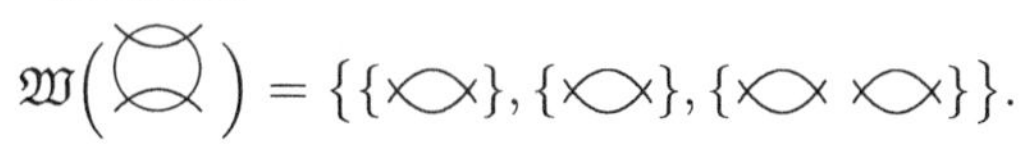

Each of its 1PI subgraphs is primitive, and we have one maximal spinney which is the union 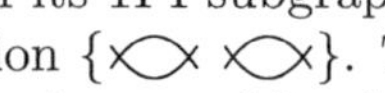. These hierarchies are best represented by rooted trees with vertices decorated by the primitive UV divergent 1PI 1-loop graph 

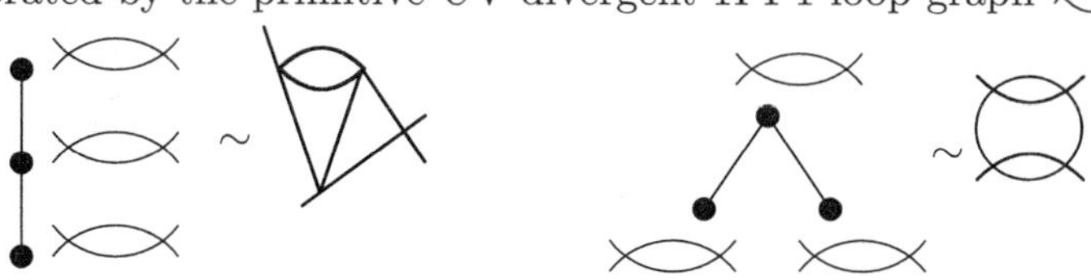

The orientation of the graphs is of no importance at this level. The branching of the tree on the right hand side reflects the disjoint location of the subgraphs in , i.e. the spinney 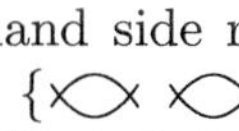. Of course, different decorations may appear naturally in more complicated situations, like for instance in QED.

The property that both graphs have exactly one maximal spinney, that is, they contain no overlapping subgraphs, is reflected by the fact that their subdivergences hierarchy can be properly represented by a single decorated rooted tree. For graphs with overlapping subgraphs we have in general several maximal spinneys. For instance, recall the wood, i.e. the set of proper spinneys corresponding to the 3-loop QED graph 

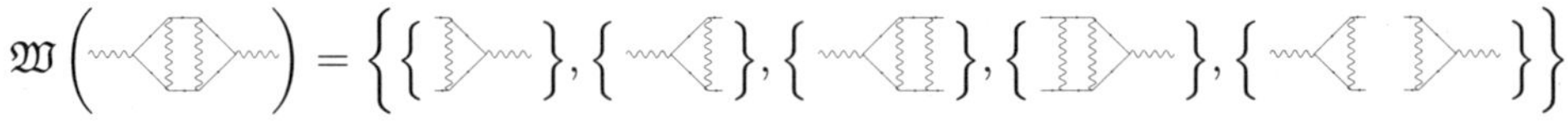

with the following cographs

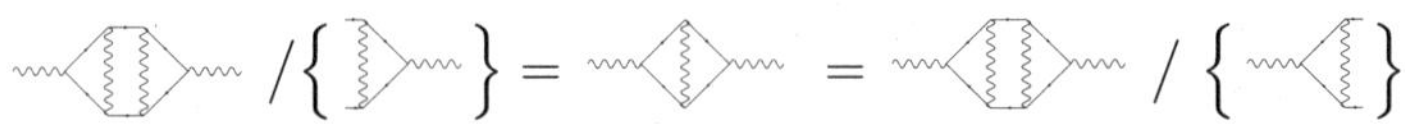

and

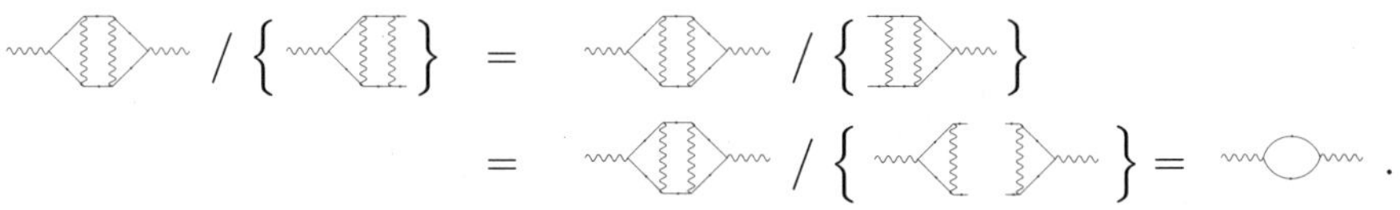

From the last line we see immediately that we have three maximal spinneys in 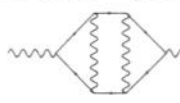, $\mathfrak{W}_{max}($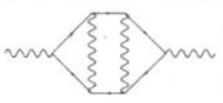$) :=$ 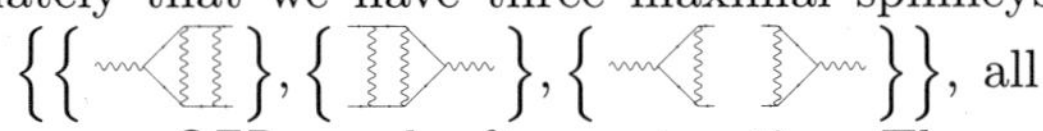, all resulting in the primitive 1-loop self energy QED graph after contraction. The corresponding tree representation consists of a linear combination of three rooted trees each of degree three

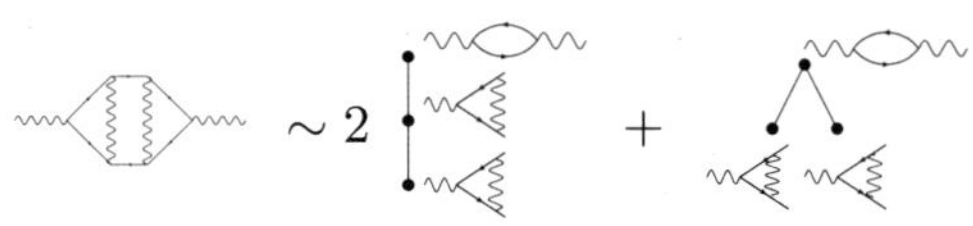

For the sake of completeness we should mention that the QED 2-loop self-energy graph  has two overlapping divergent 1-loop vertex subgraphs. Its union of maximal spinneys is given by $\mathfrak{W}_{max}($$) :=$ 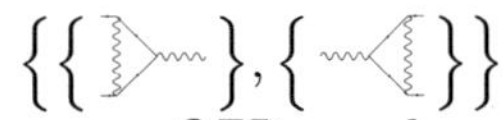, such that all contractions result in the primitive 1-loop self energy QED graph and

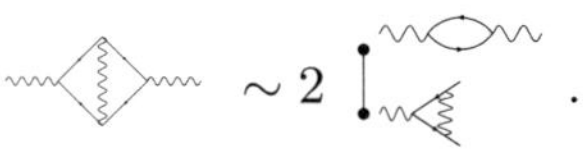

The rule may be summarized as follows. To each maximal spinney $\{\gamma'\} \in \mathfrak{W}_{max}(\Gamma)$ of a graph Γ corresponds a decorated rooted tree $t(\{\gamma'\})$, with as many vertices as the graph has loops and the root decorated by the primitive graph $\Gamma/\{\gamma'\}$ resulting from the contraction of the maximal spinney $\{\gamma'\}$ in Γ. The other vertices of this tree, representing the hierarchy of subgraphs in that spinney, are decorated by those spinney elements. Eventually, the graph Γ is represented by the linear combination of those trees,

$$\Gamma \sim \sum_{\{\gamma'\} \in \mathfrak{W}_{max}(\Gamma)} t(\{\gamma'\}).$$

One verifies readily that the coproduct on both objects, defined in $\mathcal{H}_{\mathcal{F}}$ respectively $\mathcal{H}_{\mathcal{T}}$, agree modulo the identification between graphs and decorated rooted trees. As an example compare the coproduct of the Feynman graph 

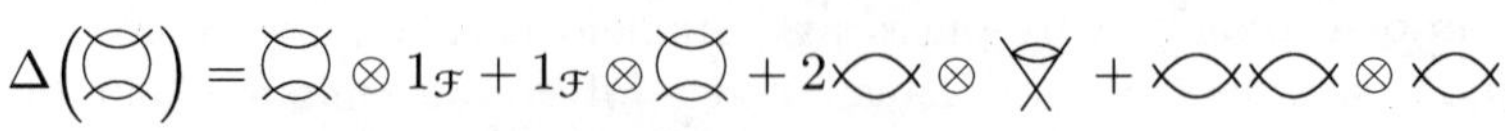

with that of the cherry tree  in (3.31).

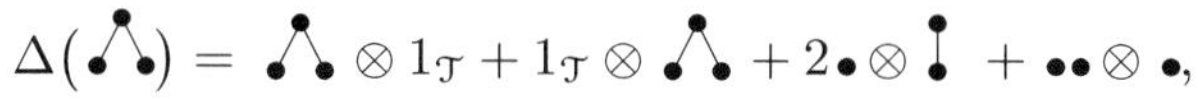

where one needs to put the decoration 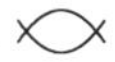 on each tree vertex. Observe that the winecup graph  contains  as the only UV divergent 1PI graph and hence is represented by the ladder tree  decorated by  . To phrase it differently, the linear combinations of decorated rooted trees representing graphs form a Hopf subalgebra in $\mathcal{H}_{\mathcal{T}}$ [Krei1999b].

3.6 Feynman characters and Birkhoff decomposition. Recall from the mathematical preliminaries in an earlier section that the space $\mathrm{Hom}(\mathcal{H}_{\mathcal{F}},\mathbb{K})$ together with the convolution product $\star$ and the counit map $\epsilon : \mathcal{H}_{\mathcal{F}} \to \mathbb{K}$ as unit forms a unital, associative and non-commutative $\mathbb{K}$-algebra, which contains the group of characters, $G := \mathrm{char}(\mathcal{H}_{\mathcal{F}},\mathbb{K})$, i.e., linear functionals ϕ from $\mathcal{H}_{\mathcal{F}}$ to $\mathbb{K}$ respecting multiplication, $\phi(\Gamma_1\Gamma_2) = \phi(\Gamma_1)\phi(\Gamma_2)$, $\Gamma_1,\ \Gamma_2 \in \mathcal{H}_{\mathcal{F}}$. This group of multiplicative maps possesses a corresponding Lie algebra, $\mathcal{L} = \partial\mathrm{char}(\mathcal{H}_{\mathcal{F}},\mathbb{K})$, of derivations, or infinitesimal characters, i.e., linear maps Z satisfying the Leibniz rule

$$Z(\Gamma_1\Gamma_2) = Z(\Gamma_1)\epsilon(\Gamma_2) + \epsilon(\Gamma_1)Z(\Gamma_2)$$

for all $\Gamma_1,\ \Gamma_2 \in \mathcal{H}_{\mathcal{F}}$. The grading of $\mathcal{H}_{\mathcal{F}}$ implies a decreasing filtration on the algebra $\mathrm{Hom}(\mathcal{H}_{\mathcal{F}},\mathbb{K})$, making it a unital complete filtered algebra. The exponential map $\exp^\star$ gives a bijection between the Lie algebra $\mathcal{L}$ and its corresponding group G.

3.6.1 *Feynman rules as characters.* As a matter of fact, general Feynman rules for any interesting perturbative QFT give a linear map $\phi \in \mathrm{Hom}(\mathcal{H}_{\mathcal{F}},\mathbb{K})$, which can be extended multiplicativity to form a subclass of characters in G.

The reader may wish to recall Theorem 3.3 which already implies at this level a –unique– factorization of the group $G := \mathrm{char}(\mathcal{H}_{\mathcal{F}},\mathbb{K})$ upon the choice of an arbitrary –idempotent– linear map P on $\mathrm{Hom}(\mathcal{H}_{\mathcal{F}},\mathbb{K})$. Eventually, the map P turns out to be naturally given in renormalization.

Amplitudes, viz those Feynman integrals associated with a Feynman graph via Feynman rules, are given by in general ill-defined integrals plagued with ultraviolet divergences coming from high momenta integrations. These divergences demand for a regularization. In general, a *regularization prescription* introduces extra non-physical parameters into the theory rendering such Feynman integrals finite but changing the nature of the target space of Feynman rules.

The particular choice of such a prescription is largely arbitrary, and mainly guided by two desires. One is the more practical interest, especially to practitioners in pQFT, for calculational convenience driven by the fact that multiloop calculations are in general very complicated problems. The other desire, though of more fundamental nature, is to maintain as many –if not all– of the physical properties of the original –non-regularized– theory as possible. However, what one must assure is that the final physical result is completely independent of such nonphysical intermediate steps.

Motivated by the need for regularization of pQFT, due to ultraviolet divergencies we take here the more general point of view, replacing the base field $\mathbb{K}$ as target space by a suitable commutative and unital algebra A. As an example we mention $A = \mathbb{C}[\varepsilon^{-1},\varepsilon]]$, the field of Laurent series that enters in dimensional regularization.

Let us denote by $G_A := \mathrm{char}(\mathcal{H}_{\mathcal{F}}, A) \subset \mathrm{Hom}(\mathcal{H}_{\mathcal{F}}, A)$ the group of A-valued, or regularized, algebra morphisms. The group law is given by the convolution product

$$\phi_1 \star \phi_2 := m_A \circ (\phi_1 \otimes \phi_2) \circ \Delta : \ \mathcal{H}_{\mathcal{F}} \xrightarrow{\Delta} \mathcal{H}_{\mathcal{F}} \otimes \mathcal{H}_{\mathcal{F}} \xrightarrow{\phi_1 \otimes \phi_2} A \otimes A \xrightarrow{m_A} A \quad (3.43)$$

With the now regularized Feynman rules understood as canonical A-valued characters ϕ we will have to make one further choice: a *renormalization scheme*. We do this by demanding the existence of a linear idempotent map R satisfying the *Rota–Baxter relation* of weight one

$$R(x)R(y) + R(xy) = R\big(R(x)y\big) + R\big(xR(y)\big) \quad (3.44)$$

for all $x, y \in A$. For R being such a Rota–Baxter map, $\tilde{R} := \mathrm{id}_A - R$ also satisfies relation (3.44). This equation turns out to lie at the heart of Connes–Kreimer's Birkhoff decomposition to be explored below. It appeared in [CK1998, Krei1999a] under the name *multiplicativity constraint* and tells us that the algebra A splits into two parallel subalgebras given by the image and kernel of R.

As a paradigm we mention again dimensional regularization together with the minimal subtraction scheme, that is, the pole part projection $R := R_{ms}$, which is a Rota–Baxter map. At this point we must postpone further comments on a complete census of renormalization schemes used in physics in the light of Rota–Baxter algebras.

3.6.2 *Connes–Kreimer's Birkhoff decomposition of Feynman rules.* In the sequel it is our goal to shed more light on the meaning of this operator relation in the context of Connes–Kreimer's work.

But before this, let us take a shortcut for the moment and see briefly how all the above structure comes together. Starting with regularized Feynman rules characters ϕ dictated by the pQFT and taking values in the field of Laurent series $A := \mathbb{C}[\varepsilon^{-1}, \varepsilon]]$ with

$$R := R_{ms} : A \to A, \qquad \sum_{i=-n}^{\infty} a_i \varepsilon^i \mapsto \sum_{i=-n}^{-1} a_i \varepsilon^i$$

defining the renormalization scheme map on A, Connes and Kreimer observed that Bogoliubov's recursive formula for the counterterm in renormalization has a Hopf algebraic expression given inductively by the map

$$\phi_-(\Gamma) = -R\big(\phi(\Gamma) + \sum_{\gamma \subset \Gamma} \phi_-(\gamma)\phi(\Gamma/\gamma)\big), \qquad \Gamma \in \ker(\epsilon), \quad (3.45)$$

with $\phi_-(1_{\mathcal{F}}) := 1_A$ and $\phi_-(\Gamma) = -R(\Gamma)$ if Γ is a primitive element in $\mathcal{H}_{\mathcal{F}}$, i.e., contains no subdivergences. We denoted this map earlier by C, the counterterm. The argument of R

$$\bar{\mathrm{R}}[\phi](\Gamma) := \phi(\Gamma) + \sum_{\gamma \subset \Gamma} \phi_-(\gamma)\phi(\Gamma/\gamma) \quad (3.46)$$

for $\Gamma \in \ker(\epsilon)$, is Bogoliubov's preparation map or the $\bar{\mathrm{R}}$-map. This leads to the Birkhoff decomposition of Feynman rules found by Connes and Kreimer [CK1998, CK2000, CK2001, Krei1999a], described in the following theorem.

Theorem 3.6 *The renormalization of the dimensionally regularized Feynman rules character* $\phi \xrightarrow{ren.} \phi_+$ *follows from the convolution product of the counterterm* $\phi_- := S_R^\phi$ *with* ϕ, $\phi_+ := S_R^\phi \star \phi$, *implying the inductive formula for* ϕ_+

$$\phi_+(\Gamma) = \phi(\Gamma) + \phi_-(\Gamma) + \sum_{\gamma \subset \Gamma} \phi_-(\gamma)\phi(\Gamma/\gamma), \qquad \Gamma \in \ker(\epsilon).$$

Furthermore, the maps ϕ_- *and* ϕ_+ *are the unique characters such that* $\phi = \phi_-^{-1} \star \phi_+$ *gives the algebraic Birkhoff decomposition of the regularized Feynman rules character* $\phi \in G_A$.

In the proof of this theorem relation (3.44) enters in the last statement, that is, in showing that S_R^ϕ is a character. We will see that its appearance has deeper reasons. Let us postpone to give more details. In Section 5.2, we will capture this theorem in the context of a Rota–Baxter matrix calculus.

4 Rota–Baxter algebras

Relation (3.44) plays a particular role in the decomposition result of Connes and Kreimer. We know already from Theorem 3.3 that we have a natural factorization of A-valued (or regularized) characters in the complete filtered algebra $\mathcal{A} := \mathrm{Hom}(\mathcal{H}_\mathcal{F}, A)$ upon the arbitrary choice of a filtration preserving idempotent map P on $\mathcal{A}$. Now, we explore the implications due to the Rota–Baxter relation.

4.1 Definition, history and examples. As a matter of fact, relation (3.44) is a well-known object in mathematics. In the 1950s and early 1960s, several interesting results were obtained in the fluctuation theory of probability. One of the most well-known is Spitzer's classical identity [Spit1956] in the theory of sums of independent random variables, see also [Spit1976].

In an important 1960 work [Bax1960] the American mathematician Glen Baxter (1930-1983) deduced Spitzer's identity from the above operator identity

$$R(x)R(y) = R\big(R(x)y + xR(y) - \theta xy\big), \tag{4.1}$$

for all $x, y \in A$, where R is a $\mathbb{K}$-linear endomorphism on a $\mathbb{K}$-algebra A, which he assumed to be associative, unital and commutative. Here θ is a fixed element in the base field $\mathbb{K}$ called the weight of the algebra A.

Gian-Carlo Rota (1932-1999) started a careful in depth elaboration of Baxter's article in his 1969 papers [Rota1969], where he solved the crucial 'word problem'. Together with one of his students, David Smith, he remarked –in the context of the Hilbert transform of a function– in a 1972 paper [RoSm1972] that Baxter's identity is equivalent, in characteristic zero, to the modified Rota–Baxter relation

$$B(x)B(y) = B\big(B(x)y + xB(y)\big) - \theta^2 xy, \qquad \forall x, y \in A, \tag{4.2}$$

where $B := \theta \mathrm{id}_A - 2R$. During the early 1960s and 1970s, further algebraic, combinatorial and analytic aspects of Baxter's identity were studied by several people [Car1972, Kin1962, Mil1966, Mil1969, Tho1977, Vog1963].

Acknowledging Rota's seminal contributions to the subject as well as his efforts to promote its further development and application in several fields of mathematics [Rota1995, Rota1998] we call a pair (A, R) where A is a not necessarily associative $\mathbb{K}$-algebra and R is a $\mathbb{K}$-linear operator $R : A \to A$ satisfying the *Rota–Baxter relation* (4.1) a *Rota–Baxter* $\mathbb{K}$*-algebra* of weight θ.

4.1.1 *Riemann and Jackson integration.* Let $A = \mathbf{Cont}(\mathbb{R})$ be the ring of continuous functions on $\mathbb{R}$. The indefinite Riemann integral is seen as a linear map

$$I : A \to A, \quad I[f](x) := \int_0^x f(t)dt. \tag{4.3}$$

Then I is a weight zero Rota–Baxter operator. Indeed, let

$$F(x) := I[f](x) = \int_0^x f(t)\,dt, \quad G(x) := I[g](x) = \int_0^x g(t)\,dt.$$

Then the integration-by-parts formula for the Riemann integral states that

$$\int_0^x F(t)G'(t)dt = F(x)G(x) - \int_0^x F'(t)G(t)dt,$$

that is,

$$I\big[I[f]g\big](x) = I[f](x)I[g](x) - I\big[fI[g]\big](x).$$

Thus the weight *0* Rota–Baxter relation (4.1) may be interpreted as a generalization of the integration-by-parts rule for integral-like operators on suitable function spaces.

Such a generalization of Riemann's integral map away from weight zero is given by the *q-analog* of the Riemann integral, also known as *Jackson's integral.* By this we mean the following operator on a well-chosen function algebra $\mathcal{F}$, defined for $0 < q < 1$

$$J[f](x) := \int_0^x f(y)d_qy = (1-q)\sum_{n\geq 0} f(xq^n)xq^n.$$

This may be written in a more algebraic way, using the operator

$$P_q[f] := \sum_{n>0} E_q^n[f],$$

where the algebra endomorphism (*q-dilatation*) $E_q[f](x) := f(qx)$, $f \in \mathcal{F}$. The map P_q is a Rota–Baxter operator of weight -1 and hence $\mathrm{id}_{\mathcal{F}} + P_q =: \hat{P}_q$ is of weight $+1$.

Jackson's integral is given in terms of the above operators P_q and the multiplication operator $M_{\mathrm{id}}(f)(x) := xf(x)$, $f \in \mathcal{F}$ as follows

$$J[f](x) = (1-q)\hat{P}_q\, M_{\mathrm{id}}\,[f](x)$$

satisfying the following 'mixed' Rota–Baxter relation of weight $(1-q)$,

$$J[f]\,J[g] + (1-q)J\,M_{\mathrm{id}}[f\,g] = J\Big[J[f]\,g + f\,J[g]\Big], \qquad f, g \in \mathcal{F},$$

replacing the integration-by-parts rule for the Riemann integral.

4.1.2 *Summation.* On a suitable class of functions, we define

$$S(f)(x) := \sum_{n\geq 1} f(x+n).$$

Then S is a Rota–Baxter operator of weight -1, since

$$\Big(\sum_{n\geq 1} f(x+n)\Big)\Big(\sum_{m\geq 1} g(x+m)\Big) = \sum_{n\geq 1, m\geq 1} f(x+n)g(x+m)$$

$$= \Big(\sum_{n>m\geq 1} + \sum_{m>n\geq 1} + \sum_{m=n\geq 1}\Big) f(x+n)g(x+m)$$

$$= \sum_{m\geq 1}\Big(\sum_{k\geq 1} f(x+\underbrace{k+m}_{=n})\Big)g(x+m) + \sum_{n\geq 1}\Big(\sum_{k\geq 1} g(x+\underbrace{k+n}_{=m})\Big)f(x+n) + \sum_{n\geq 1} f(x+n)g(x+n)$$

$$= S\big(S(f)g\big)(x) + S\big(fS(g)\big)(x) + S(fg)(x).$$

4.1.3 *Partial sums.* Let A be the set of sequences $\{a_n\}$ with values in $\mathbb{K}$. Then A is a $\mathbb{K}$-algebra with termwise sum, product and scalar product. Define

$$R: A \to A, \quad R(a_1, a_2, \cdots) := (a_1, a_1 + a_2, \cdots).$$

Then R is a Rota–Baxter operator of weight one.

Let A be a unitary algebra and let $\mathbb{N} := \mathbb{N}_{>0}$. Define $\mathcal{A} := \mathrm{map}(\mathbb{N}, A) = A^{\mathbb{N}}$ to be the algebra of maps $f: \mathbb{N} \to A$ with point-wise operations. Define the linear operator

$$Z := Z_A : \mathcal{A} \to \mathcal{A}, \quad Z[f](k) := \begin{cases} \sum_{i=1}^{k-1} f(i), & k > 1, \\ 0, & k = 1. \end{cases} \tag{4.4}$$

The map Z is a Rota–Baxter operator on $\mathcal{A}$ of weight -1. Furthermore, we have, for $f_1, \cdots, f_n \in \mathcal{A}$, the iteration

$$Z\Big[f_1 Z\big[f_2 \cdots Z[f_n]\cdots\big]\Big](k) = \sum_{k>i_1>\cdots>i_n>0} f_1(i_1)\cdots f_n(i_n)$$

and thus in the limit

$$\lim_{k\to\infty} Z\Big[f_1 Z\big[f_2 \cdots Z[f_n]\cdots\big]\Big](k) = \sum_{i_1>\cdots>i_n>0} f_1(i_1)\cdots f_n(i_n) \tag{4.5}$$

if the nested infinite sum on the righthand side exists. For the particular choice of functions $f_i(x) \in \mathcal{A}' := \{f_s(x) := x^{-s} \mid s \in \mathbb{N}\} \subset \mathcal{A}$, we obtain from the summation map Z in (4.4) respectively its limit (4.5) the multiple-zeta-value

$$\zeta(s_1, \cdots, s_k) := \sum_{s_1>s_2>\cdots>s_k\geq 1} \frac{1}{n_1^{s_1}\cdots n_k^{s_k}} = \lim_{l\to\infty} Z\big[x^{-s_1} Z\big[x^{-s_2}\cdots Z[x^{-s_k}]\cdots\big]\big](l). \tag{4.6}$$

of length $\ell(\zeta(s_1, \cdots, s_k)) := k$ and weight $w(\zeta(s_1, \cdots, s_k)) := \sum_{i=1}^k s_i$. The reader may take a glimpse into the proceedings contribution by Weinzierl for the link to multiple-zeta-values and their generalizations, see also [EG2005c, MUW2002]. The *shuffle* and *quasi-shuffle* relation mentioned for these numbers [Hof2005] are simply a consequence of the Rota–Baxter relation for the Riemann integral operator (4.3) and the summation operator Z in (4.4). More details about these generalized shuffle products and its relation to free commutative Rota–Baxter algebra [GK2000] can be found in [EG2005a].

4.1.4 *Matrix Rota–Baxter maps.* The next example was introduced in [Mil1969] and is related to Rota's original (free) Rota–Baxter algebra in [Rota1969]. Let A be a finite dimensional $\mathbb{K}$-vector space with basis $e_1, \ldots, e_n$. We make it into a commutative $\mathbb{K}$-algebra by defining the product componentwise on the column vectors of $n = s + t$ components. Then the following matrix $R \in \mathcal{M}_n(\mathbb{K})$ defines a Rota–Baxter operator of weight $\theta = 1$ on A

$$R := \begin{pmatrix} S_s & 0 \\ 0 & T_t \end{pmatrix}, \tag{4.7}$$

where the matrices S, T of size s respectively t are given by

$$S_s := \begin{pmatrix} 1 & 1 & \dots & 1 \\ 0 & 1 & \dots & 1 \\ \vdots & \ddots & \ddots & \vdots \\ 0 & \dots & 0 & 1 \end{pmatrix}_{s\times s}, \quad T_t := \begin{pmatrix} 0 & 0 & \dots & 0 \\ -1 & 0 & \dots & 0 \\ \vdots & \ddots & \ddots & \vdots \\ -1 & \dots & -1 & 0 \end{pmatrix}_{t\times t}. \tag{4.8}$$

4.1.5 *Triangular matrices.* On the algebra of upper triangular matrices $\mathcal{M}_n^u(\mathbb{K})$, define the linear operator D

$$\big(D(\alpha_{kl})\big)_{ij} := \theta\delta_{ij}\sum_{k\geq i}\alpha_{ik}, \qquad \alpha_{ik}\in\mathcal{M}_n^u(\mathbb{K}).$$

Then D is a Rota–Baxter operator on $\mathcal{M}_n^u(\mathbb{K})$ of weight $\theta\in\mathbb{K}$ [Le2004].

4.1.6 *Idempotent Rota–Baxter maps.* If the $\mathbb{K}$-algebra A, for instance Lie or associative, decomposes directly into subalgebras A_1 and A_2 , $A = A_1 \oplus A_2$, then the projection to A_1, $R : A \to A$, $R(a_1, a_2) = a_1$, is an idempotent Rota–Baxter operator. Let us verify this briefly for $a, b \in A = A_1 \oplus A_2$

$$\begin{aligned} R(a)b + aR(b) - ab &= R(a)\big(R(b) + (\mathrm{id}_A - R)(b)\big) \\ &\quad - \big(R(a) + (\mathrm{id}_A - R)(a)\big)(\mathrm{id}_A - R)(b) \\ &= R(a)R(b) - (\mathrm{id}_A - R)(a)(\mathrm{id}_A - R)(b) \end{aligned}$$

such that applying R on both sides gives

$$R\big(R(a)b + aR(b) - ab\big) = R(a)R(b),$$

since it kills the term $(\mathrm{id}_A - R)(a)(\mathrm{id}_A - R)(b)$ without changing the term $R(a)R(b)$, as $R(\mathrm{id}_A - R)(a) = 0$ and A_1, A_2 are subalgebras.

As an example we saw already the *minimal subtraction scheme* in *dimensional regularization*, where $A = \mathbb{C}[\varepsilon^{-1}, \varepsilon]] = \varepsilon^{-1}\mathbb{C}[\varepsilon^{-1}] \oplus \mathbb{C}[[\varepsilon]]$. So $R\big(\sum_{n\geq -N} a_n\varepsilon^n\big) := \sum_{n=-N}^{-1} a_n\varepsilon^n$ is a Rota–Baxter operator. In general if R is multiplicative and idempotent, then R is a Rota–Baxter operator.

4.2 Properties of Rota–Baxter algebras. Let (A, R), (B, P) be two Rota–Baxter algebras of the same weight. We call an algebra morphism f between A and B a *Rota–Baxter homomorphism*, if $f \circ R = P \circ f$. A *Rota–Baxter ideal* of a Rota–Baxter algebra (A, R) is an ideal I of A such that $R(I) \subseteq I$. The following basic properties of a Rota–Baxter algebra (A, R) are readily verified. If R has weight θ, then $\theta^{-1}R$ has weight one. We already remarked that $\tilde{R} := \theta\mathrm{id}_A - R$ is a Rota-Baxter operator. The images $A_- := R(A)$ as well as $A_+ := \tilde{R}(A)$ are subalgebras in A. The so-called *double Rota–Baxter product*

$$x *_R y := xR(y) + R(x)y - \theta xy, \qquad x, y \in A, \tag{4.9}$$

equips the vector space underlying A with another Rota–Baxter structure denoted by $(A, *_R, R)$ where R satisfies the Rota–Baxter relation for this product. One readily verifies that

$$R(x *_R y) = R(x)R(y) \quad \text{and} \quad \tilde{R}(x *_R y) = -\tilde{R}(x)\tilde{R}(y), \qquad x, y \in A. \tag{4.10}$$

This construction might be continued giving a hierarchy of Rota–Baxter algebras. Later, we will need the following useful identity

$$a_1 *_R \cdots *_R a_n = \frac{1}{\theta}\Big(\prod_{i=1}^{n} R(a_i) - (-1)^n \prod_{i=1}^{n} \tilde{R}(a_i)\Big), \qquad a_i \in A. \tag{4.11}$$

If (A, R) is a Rota-Baxter algebra with idempotent Rota–Baxter map ($R^2 = R$), then $A = A_- \oplus A_+$. More generally, we have that a linear operator $R : A \to A$ is a Rota–Baxter operator if and only if the following is true, first, A_+ and A_- are both closed under multiplication, secondly, if for $x, y \in A$ we have that for a $z \in A$, $R(x)R(y) = R(z)$ implies $\tilde{R}(x)\tilde{R}(y) = -\tilde{R}(z)$. We find moreover that $K_+ := \ker(R)$ and $K_- := \ker(\tilde{R})$ are ideals in A_R and therefore $A_\pm = A_R/K_\pm$. They are also ideals in $A_\mp$ and the map $\xi : A_+/K_+ \to A_-/K_-$, defined by $\xi\big(R(x)\big) := \tilde{R}(x)$ is an algebra isomorphism.

From a mathematical as well as physics point of view it is interesting to mention that the Rota–Baxter relation was independently (re)discovered in the 1980s by several Russian physicists, including Semenov-Tian-Shansky, and Belavin and Drinfeld [BelDri1982, Sem1983] (see [BBT2003] for more details). In the particular context of Lie algebras they studied r-matrix solutions to the *classical Yang–Baxter equation*

$$[r_{13}, r_{12}] + [r_{23}, r_{12}] + [r_{23}, r_{13}] = 0 \tag{4.12}$$

named after the physicists C.N. Yang from China and the Australian Rodney Baxter. Under suitable circumstances a related operator R satisfies

$$[R(x), R(y)] + R([x, y]) = R\big([R(x), y] + [x, R(y)]\big).$$

Especially Semenov-Tian-Shansky found a link between the modified classical Yang–Baxter equation and the Riemann–Hilbert problem [Sem1983, Sem2000]. His results are a generalization of the Kostant–Adler scheme.

4.3 Spitzer's identity for commutative Rota–Baxter algebras. We mentioned Spitzer's paper [Spit1956] playing an important role in Baxter's work. A transparent way to understand this identity may be seen by following Baxter's original approach. Let $A = \mathbf{Cont}(\mathbb{R})$ be the ring of continuous functions on $\mathbb{R}$ with the Riemann integral map $I(f)(x) = \int_0^x f(u)du$ (4.3). It is well-known that the initial value problem

$$\frac{d}{dt}y(t) = a(t)y(t), \quad y(0) = 1, \qquad a \in A \tag{4.13}$$

has a unique solution, $y(t) = \exp\left(\int_0^t a(u)du\right)$. As we may transform the differential equation into an integral equation by applying I to (4.13)

$$y(t) = 1 + I(ay)(t) \tag{4.14}$$

we arrive naturally at the non-trivial identity

$$\exp\left(\int_0^t a(u)du\right) = 1 + \sum_{n=1}^{\infty} \underbrace{I\big(I(I(\cdots I}_{n-times}(a)a)\ldots a)a\big)(t) \tag{4.15}$$

which follows from the identity

$$\big(I(a)(t)\big)^n = n!\,\underbrace{I\big(I(I(\cdots I}_{n-times}(a)a)\ldots a)a\big)(t). \tag{4.16}$$

This relation is the weight zero case of the Bohnenblust–Spitzer identity, see Eq. (4.18) below. Baxter's simple question was to replace within the *commutative* setting the weight zero Rota–Baxter map I in (4.14) by an arbitrary Rota–Baxter map R of weight $\theta \neq 0$ and to find the analog of the exponential solution respectively the corresponding relation (4.15). The result is the *classical Spitzer identity* for a commutative Rota–Baxter algebra (A, R) of weight $\theta \neq 0$,

$$\exp\Big(-R\Big(\frac{\log(1_A-\theta ax)}{\theta}\Big)\Big) \;=\; \sum_{n=0}^{\infty} x^n \underbrace{R(R(R(\cdots R(a)a)\ldots a)a)}_{n-times}, \quad (4.17)$$

in the ring of power series $A[[x]]$. We may interpret this identity as an infinite set of identities in A followed by comparison of coefficients of powers of x [RoSm1972]. Observe that $\frac{\log(1_A-\theta ax)}{\theta}$ reduces to $-ax$ in the limit $\theta \to 0$, which gives back the classical Riemann integral identity (4.15).

To get ourselves acquainted with this identity let us consider a simple, but interesting example of its use. We go back to the Z-summation Rota–Baxter map in (4.4) and its link to multiple-zeta-values (4.6). Observe that a simple calculation gives for the function $f_k := 1/x^k$

$$Z\big[\log(1+f_k t)\big](m) \;=\; Z\left[\sum_{i=1}^{\infty}\frac{(-1)^{i-1}}{i}\Big(\frac{t}{x^k}\Big)^i\right](m) = \sum_{i=1}^{\infty}\frac{(-1)^{i-1}t^i}{i}Z\left[\frac{1}{x^{ki}}\right](m).$$

Hence, in the limit

$$\lim_{m\to\infty} Z\big[\log(1+f_k t)\big](m) = \sum_{i=1}^{\infty}\frac{(-1)^{i-1}t^i}{i}\zeta(ki).$$

Therefore, Spitzer's classical identity for the Z-summation map gives

$$\begin{aligned}\exp\left(\sum_{i=1}^{\infty}(-1)^{i-1}\zeta(ik)\frac{t^i}{i}\right) &= 1+\sum_{n>0}^{\infty} t^i\, \underbrace{Z[Z[\cdots Z[Z[f_k]f_k]f_k\cdots]f_k]}_{n\text{-times}} \\ &= 1+\sum_{n>0}^{\infty} t^i\zeta(\underbrace{k,\cdots,k}_{n-times}).\end{aligned}$$

Another example is given for the case where R is the –trivial weight one Rota–Baxter map– id_A. Then the left hand side of Eq. (4.17) becomes the power series

$$\exp\big(-\log(1_A-ax)\big) = \exp(\log(1_A-ax)^{-1}) = \frac{1}{1_A-ax}$$

and the right hand side is $\sum_{n=0}^{\infty} x^n a^n = \sum_{n=0}^{\infty}(ax)^n$. So we have the familiar geometric expansion.

Let us remark here, that there exist several proofs of Spitzer's classical identity in the commutative setting. Other than the original proof of Spitzer in the case of probability distributions, there are the combinatorial proofs of Baxter and Kingman [Kin1962], the analytic proofs of Atkinson [Atk1963] and Wendel [Wen1962], as well as the algebraic proofs of Cartier and Rota–Smith working with free commutative Rota–Baxter algebras and making use of Waring's formula.

The Rota–Baxter point of view opens simple ways to recognize particular identities, e.g. for multiple-zeta-values (4.6) [EG2005c], where Hoffman found the so-called *Partition Identity* [Hof2005], which is just a particular example of the Bohnenblust–Spitzer formula for commutative Rota–Baxter algebras.

Let us briefly recall the exact form of the *Bohnenblust–Spitzer formula* [Gio2002, RoSm1972] of weight θ. Let (A, R) be a commutative Rota–Baxter algebra of weight θ and fix $s_1, \ldots, s_n \in A$, $n > 0$. Let S_n be the set of permutations of $\{1, \cdots, n\}$. Then

$$\sum_{\sigma \in S_n} R\Big(s_{\sigma(1)} R\big(s_{\sigma(2)} \cdots R(s_{\sigma(n)}) \cdots\big)\Big) = \sum_{\mathcal{T}} \theta^{n-|\mathcal{T}|} \prod_{T \in \mathcal{T}} (|T|-1)!\, R\big(\prod_{j \in T} s_j\big). \quad (4.18)$$

Here $\mathcal{T}$ runs through all unordered set partitions of $\{1, \cdots, n\}$. The weight $\theta = 0$ case reduces the sum over $\mathcal{T}$ to $|\mathcal{T}| = n$. The Rota–Baxter relation itself appears as a particular case for $n = 2$.

In the next part, we will see that the *noncommutative* version of Spitzer's identity naturally follows from Theorem 3.1 in the context of a filtered associative Rota–Baxter algebra of weight θ. The limit $\theta \to 0$ retrieves Magnus' solution to the initial value problem (4.13) in a noncommutative setting, say, for matrix valued functions. The results presented here can be found in [EGM2006] and were achieved together with D. Manchon.

4.4 Spitzer's identity for non-commutative Rota–Baxter algebras. We will derive the noncommutative version of Spitzer's identity from Theorem 3.1 by comparison with the following fundamental result for associative Rota–Baxter algebras due to F. V. Atkinson. His comprehensive 1963 paper [Atk1963] is important in the understanding of the role played by Rota–Baxter algebras in the description of renormalization as a factorization problem. For convenience we put ourselves into the realm of complete filtered associative unital Rota–Baxter algebras $(\mathcal{A}, R)$ with filtration preserving Rota–Baxter map R of weight $\theta \neq 0$ over the field $\mathbb{K}$ of characteristic zero.

For instance, recall Example 3.1 of triangular matrices. More generally, let (A, R) be a Rota–Baxter algebra of weight θ. Define a Rota–Baxter map R on $\mathcal{M}_n^\ell(A)$ by extending the Rota–Baxter map R entrywise,

$$\mathrm{R}(\alpha) = \big(R(\alpha_{ij})\big).$$

We have the following important, though simple theorem

Theorem 4.1 [EG2005b, EGGV2006] *The triple* $\big(\mathcal{M}_n^\ell(A), \mathrm{R}, \{\mathcal{M}_n^\ell(A)_k\}_{k\geq 1}\big)$ *forms a complete filtered Rota–Baxter algebra of weight* θ.

Atkinson observed in [Atk1963] a multiplicative decomposition theorem for associative unital Rota–Baxter algebras of weight θ. Let $a \in \mathcal{A}_1$. For elements X and Y in $\mathcal{A}$ such that

$$X = 1_{\mathcal{A}} - R(X\, a) \quad \text{respectively} \quad Y = 1_{\mathcal{A}} - \tilde{R}(a\, Y) \qquad (4.19)$$

we find by simple Rota–Baxter gymnastic, that

$$X(1_{\mathcal{A}} + \theta a)Y = 1_{\mathcal{A}}. \qquad (4.20)$$

We therefore obtain the factorization

$$(1_{\mathcal{A}} + \theta a) = X^{-1} Y^{-1} \qquad (4.21)$$

for elements in $1_{\mathcal{A}} + \theta\mathcal{A}_1$. If R is idempotent, then this is the unique decomposition of $1_{\mathcal{A}} + \theta a$ into a product of an element in $\mathfrak{A}_- := 1_{\mathcal{A}} + R(\mathcal{A}_1)$ with an element in

$\mathfrak{A}_+ := 1_\mathcal{A} + \tilde{R}(\mathcal{A}_1)$. Both, $\mathfrak{A}_\pm$ are subgroups, due to the Rota–Baxter relation. The inverses $X' := X^{-1}$ and $Y' := Y^{-1}$ satisfy the equations

$$X' = 1_\mathcal{A} - R(\breve{a}\, X') \quad \text{respectively} \quad Y' = 1_\mathcal{A} - \tilde{R}(Y'\, \breve{a}), \tag{4.22}$$

where $\breve{a} := (1_\mathcal{A} + \theta a)^{-1} - 1_\mathcal{A}$ in $\mathcal{A}_1$.

Now, recall Theorem 3.1, which we generalize to linear filtration preserving maps P on $\mathcal{A}$, such that $P + \tilde{P} = \theta \mathrm{id}_\mathcal{A}$. This is easily achieved upon the generalization of the BCH-recursion from χ to χ_θ.

Theorem 4.2 *Let $\mathcal{A}$ be a complete filtered $\mathbb{K}$-algebra and P a linear filtration preserving map, such that $P + \tilde{P} = \theta \mathrm{id}_\mathcal{A}$. The map χ in factorization (3.9) generalizes to*

$$\chi_\theta(u) = u - \frac{1}{\theta}\mathrm{BCH}\Big(P\big(\chi_\theta(u)\big), \tilde{P}\big(\chi_\theta(u)\big)\Big). \tag{4.23}$$

Similarly the recursion in Eq. (3.13) transposes into

$$\chi_\theta(u) = u + \frac{1}{\theta}\mathrm{BCH}\Big(- P\big(\chi_\theta(u)\big), \theta u\Big), \qquad u \in \mathcal{A}_1. \tag{4.24}$$

Such that for all $\exp(\theta u) \in 1_\mathcal{A} + \theta\mathcal{A}_1$, $u \in \mathcal{A}_1$, *we have the decomposition*

$$\exp(\theta u) = \exp\big(P(\chi_\theta(u))\big) \exp\big(\tilde{P}(\chi_\theta(u))\big). \tag{4.25}$$

We call χ_θ the BCH-recursion of weight $\theta \in \mathbb{K}$, or simply θ-BCH-recursion. In the light of Atkinson's decomposition (4.21) and the factorization in Eq. (4.25) of Theorem 4.2 we observe by taking into account Eqs. (4.10) that

$$x := \exp\big(- R(\chi_\theta(u))\big) = 1_\mathcal{A} + R\big(\exp^{*_R}\big(- \chi_\theta(u)\big) - 1_\mathcal{A}\big), \tag{4.26}$$

for $1_\mathcal{A} + \theta a = \exp(\theta u)$, $u \in \mathcal{A}_1$. One readily verifies using the identity in Eq. (4.11) that

$$\begin{aligned} \exp^{*_R}\big(- \chi_\theta(u)\big) - 1_\mathcal{A} &= \sum_{n>0} \frac{(-\chi_\theta(u))^{*_R n}}{n!} \\ &= \sum_{n>0} \frac{(-1)^n}{n!\theta}\big(R(\chi_\theta(u))^n - (-\tilde{R}(\chi_\theta(u))^n)\big) \\ &= \frac{1}{\theta}\exp\big(- R(\chi_\theta(u)\big) - \frac{1}{\theta}\exp\big(\tilde{R}(\chi_\theta(u)\big) \\ &= -\exp\big(- R(\chi_\theta(u)\big)a. \end{aligned} \tag{4.27}$$

In the last equality we used (4.25). Hence we arrive at Spitzer's identity for a complete filtered noncommutative Rota–Baxter algebras $(\mathcal{A}, R)$ of weight $\theta \neq 0$

$$\exp\Big(- R\Big(\chi_\theta\Big(\frac{\log(1_\mathcal{A} + \theta a)}{\theta}\Big)\Big)\Big) = \sum_{n=0}^{\infty} (-1)^n \underbrace{R(R(R(\cdots R(a)a)\ldots a)a)}_{n-times}, \tag{4.28}$$

for $a \in \mathcal{A}_1$. It is obvious that χ_θ reduces to the identity for commutative algebras, giving back Spitzer's classical identity (4.17).

We clearly see that in Theorem 4.2 no Rota–Baxter property is needed. Instead, in identity (4.28) respectively (4.26) it becomes evident that in the presence of a

Rota–Baxter map we may express the exponential factors in Eq.(4.25) in terms of Atkinson's recursions (4.19)

$$X = \exp\Big(-R\Big(\chi_\theta\Big(\frac{\log(1_{\mathcal{A}}+\theta a)}{\theta}\Big)\Big)\Big) \qquad Y = \exp\Big(-\tilde{R}\Big(\chi_\theta\Big(\frac{\log(1_{\mathcal{A}}+\theta a)}{\theta}\Big)\Big)\Big).$$

As we will see in the following part, the particular appearance of the weight θ in Eqs. (4.23,4.24) reflects the fact that in the case of weight $\theta = 0$ we find $\tilde{R} = -R$ such that Atkinson's factorization formula (4.20) collapses to

$$XY = \big(1_{\mathcal{A}} - R(X\,a)\big)\big(1_{\mathcal{A}} + R(a\,Y)\big) = 1_{\mathcal{A}} \tag{4.29}$$

for any $a \in \mathcal{A}_1$ which is compatible with the $\theta = 0$ limit in Eq. (4.25).

4.5 Magnus' expansion and the weight zero *BCH*-recursion. In the light of Baxter's approach to Spitzer's classical identity and its generalization to noncommutative associative Rota–Baxter algebras of weight θ it seems to be natural to look at the initial value problem in Section 4.3 in a noncommutative setting, say, for matrix valued functions, $\frac{d}{dt}\alpha(t) = \beta(t)\alpha(t)$, $\alpha(0) = 1$, where $\alpha(t)$ and $\beta(t)$ take values in the matrices $\mathcal{M}_n(\mathbb{K})$ of size $n \times n$, $1 < n < \infty$. Again, we can traverse the differential equation of the initial value problem into an integral equation, which we write more generally like

$$\alpha(t) = \mathbf{1} + I(\beta\,\alpha)(t). \tag{4.30}$$

But, under this assumption we can not write its solution as a simple exponential.

Instead, recall the seminal work of Magnus [Mag1954] on initial value problems of the above type. He proposed an exponential solution

$$\alpha(t) = \exp\big(\Omega[\beta](t)\big)$$

with $\Omega[\beta](0) = 0$. For $\Omega[\beta](t)$ we assume an expansion, $\Omega[\beta](t) := \sum_{n>0}\Omega^{(n)}[\beta](t)$, in terms of multiple Riemann integrals of nested commutators of the matrix $\beta(t)$. Magnus established a recursive equation for the terms $\Omega^{(n)}[\beta](t)$ defined in terms of the differential equation

$$\frac{d}{dt}\Omega[\beta](t) = \frac{\mathrm{ad}\,\Omega[\beta]}{\mathrm{e}^{\mathrm{ad}\,\Omega[\beta]} - 1}(\beta)(t). \tag{4.31}$$

Let us go back to Eq. (4.28) for a moment. It happens to be useful to write Eq.(3.7) as a sum [Reu1993]

$$C(a,b) = a + b + \mathrm{BCH}(a,b) = \sum_{n\geq 0} H_n(a,b),$$

where each $H_n(a,b)$ is homogenous of degree n with respect to b. Especially, $H_0(a,b) = a$. For $n = 1$ we have

$$H_1(a,b) = \frac{\mathrm{ad}\,a}{1 - \mathrm{e}^{-\mathrm{ad}\,a}}(b).$$

Hence, for Eq.(4.24) we get in the limit $\theta \to 0$ a non-linear map χ_0 inductively defined on the pro-nilpotent Lie algebra $\mathcal{A}_1$ by the formula

$$\chi_0(a) \;=\; -\frac{\mathrm{ad}R\big(\chi_0(a)\big)}{\mathrm{id}_{\mathcal{A}} - \mathrm{e}^{\mathrm{ad}R(\chi_0(a))}}(a) = \Big(\mathrm{id}_{\mathcal{A}} + \sum_{n>0} b_n\Big[\mathrm{ad}R\big(\chi_0(a)\big)\Big]^n\Big)(a) \tag{4.32}$$

where R is now a weight zero Rota–Baxter operator. We call this the weight zero *BCH*-recursion. The coefficients $b_n := \frac{B_n}{n!}$ where B_n are the Bernoulli numbers. For

$n = 1, 2, 3, 4$ we find the numbers $b_1 = -1/2$, $b_2 = 1/12$, $b_3 = 0$ and $b_4 = -1/720$. The first three terms in (4.32) are

$$\chi_0(a) = a - \frac{1}{2}[R(a),\, a] + \Big(\frac{1}{4}\big[R([R(a),\, a]),\, a\big] + \frac{1}{12}\big[R(a),\, [R(a),\, a]\big]\Big) + \cdots \quad (4.33)$$

Lemma 4.3 *Let $(\mathcal{A}, R)$ be a complete filtered Rota–Baxter algebra of weight zero. For $a \in \mathcal{A}_1$ the weight zero BCH-recursion $\chi_0 : \mathcal{A}_1 \to \mathcal{A}_1$ is given by the recursion in Eq. (4.32)*

$$\chi_0(a) = -\frac{adR\big(\chi_0(a)\big)}{\mathrm{id}_{\mathcal{A}} - \mathrm{e}^{\mathrm{ad}R(\chi_0(a))}}(a).$$

1. *The equation $x = 1_{\mathcal{A}} - R(x\ a)$ has a unique solution $x = \exp\big(-R(\chi_0(a))\big)$.*
2. *The equation $y = 1_{\mathcal{A}} + R(a\ y)$ has a unique solution $y = \exp\big(R(\chi_0(a))\big)$.*

Atkinson's factorization for the weight zero case, Eq. (4.29), follows immediately from the preceding lemma.

Comparison of (4.32) with (4.31) reveals the link between Magnus' recursion and the *BCH*-recursion in the context of a vanishing Rota–Baxter weight, namely

Corollary 4.4 *Let A be a function algebra over $\mathbb{R}$ with values in an operator algebra. Let I denote the indefinite Riemann integral operator. Magnus' Ω expansion is given by the formula*

$$\Omega[a](x) = I\big(\chi_0(a)\big)(x). \quad (4.34)$$

Hence, the θ-*BCH*-recursion (4.23) generalizes Magnus' expansion to general weight $\theta \neq 0$ Rota–Baxter operators R by replacing the weight zero Riemann integral in $F = 1 + I\{aF\}$.

The following commutative diagram captures the picture established. We observe on the right wing of (4.35) that starting from a complete filtered associative Rota–Baxter algebra $(\mathcal{A}, R)$ of weight $\theta \neq 0$, we arrive at Spitzer's classical identity assuming a commutative algebra. Taking the limit $\theta \to 0$ gives the classical expression for integral-like equations. Instead, following the left wing (4.35) and first taking the limit $\theta \to 0$ while maintaining noncommutativity we retrieve Magnus' expansion. It reduces to the classical exponential solution for the initial value problem in (4.13) when the underlying algebra is commutative.

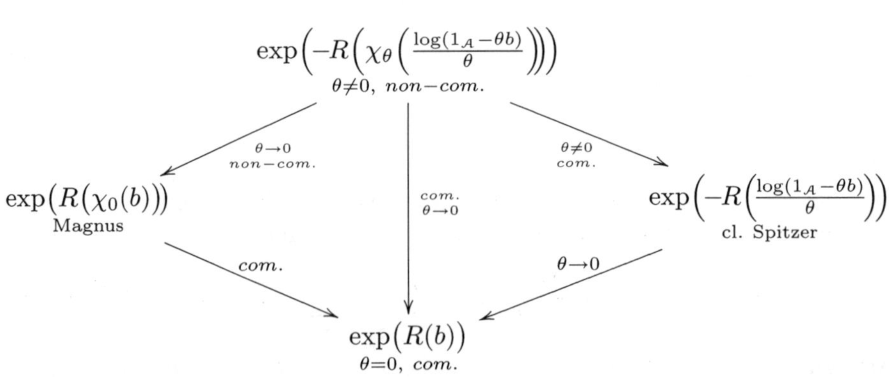

(4.35)

5 The Rota–Baxter structure of renormalization in pQFT

Let us briefly recapitulate Section 3. We assume a renormalizable QFT treated perturbatively. Recall that for a proper –Hopf algebraic– renormalization picture in pQFT we need the combinatorial Hopf algebra of 1PI Feynman graphs, $\mathcal{H}_{\mathcal{F}}$, as well as Feynman rules ϕ which are identified as particular multiplicative maps on $\mathcal{H}_{\mathcal{F}}$ into the base field $\mathbb{K}$. Both, the set of 1PI Feynman graphs F and Feynman rules are dictated by the QFT itself upon physical constraints. Renormalization enters with the choice of a regularization prescription rendering ill-defined Feynman amplitudes formally finite, together with a renormalization scheme, the latter may reflect some input from physics (physical renormalization scheme). The last two objects enter the setting via the regularization target space $A \neq \mathbb{K}$ in $\mathrm{Hom}(\mathcal{H}_{\mathcal{F}}, A)$ additionally equipped with a particular linear map, denoted by R, which extracts divergent parts of regularized Feynman amplitudes. The choice of A, that is, the regularization prescription for the Feynman rules, is mainly guided by practical reasons and severe demands for conserving as many as possible of the physical properties of the unregulated QFT.

Motivated by the example of the minimal subtraction scheme in dimensional regularization, leading to regularized Hopf algebra characters with values in $A = \mathbb{C}[\varepsilon^{-1}, \varepsilon]] = \varepsilon^{-1}\mathbb{C}[\varepsilon^{-1}] \oplus \mathbb{C}[[\varepsilon]]$ forming a commutative unital Rota–Baxter algebra (A, R), where R is the projection of a Laurent series onto its pole part, we assume in general that regularization prescriptions and the related renormalization schemes form a commutative unital Rota–Baxter algebra denoted by the tuple (A, R) with the renormalization scheme given by the Rota–Baxter map R.

5.1 Birkhoff decomposition in Rota–Baxter algebras. We will now see how in the light of the last sections this purely algebraic *Rota–Baxter* picture naturally leads to an algebraic Birkhoff type decomposition of Feynman rules capturing the process of renormalization. Minimal subtraction in dimensional regularization as renormalization prescription simply corresponds to the choice of a particular Rota–Baxter algebra with an idempotent Rota–Baxter map. In principle any other 'sensible' choice of such an algebra would be as good, though dimensional regularization has proven itself to be a veritable setting reflecting the state of the art in several areas of high energy particle physics.

Eventually we will cast the process of renormalization into a simple calculus on upper or lower triangular matrices with entries in the regularization algebra (A, R). This gives rise to a linear representation of Connes–Kreimer's group G_A of A-valued characters and its Birkhoff decomposition, which is recovered in terms of a factorization of such matrices implied by the underlying Rota–Baxter structure.

The following theorem describes the Birkhoff decomposition of Connes and Kreimer in Theorem 3.6 using the algebraic setting developed in the earlier sections, to wit, the Rota–Baxter structure on the target space implied by the choice of the renormalization scheme.

Remember that the Hopf algebra of Feynman graphs $\mathcal{H}_{\mathcal{F}}$, which is the polynomial algebra generated by the set F of 1PI Feynman graphs, has the coproduct defined in (3.37) which is of combinatorial type, i.e., for $\Gamma \in F$ the cograph Γ/γ in (3.37) always lies in F. Recall $\mathcal{F} := \mathbb{K}F$. We have

$$\mathcal{F} \xrightarrow{\Delta} \mathcal{H}_{\mathcal{F}} \otimes \mathcal{F}. \tag{5.1}$$

Remember that we called $\ker(\epsilon)$ the augmentation ideal of $\mathcal{H}_{\mathcal{F}}$ and that we denoted by P the projection $\mathcal{H}_{\mathcal{F}} \to \ker(\epsilon)$ onto the augmentation ideal, $P := \mathrm{id} - \eta \circ \epsilon$.

Recall from Proposition 3.2 that the increasing filtration on $\mathcal{H}_{\mathcal{F}}$ by its loop number grading implies in duality a complete decreasing filtration on the associative $\mathbb{K}$-algebra $\mathcal{A} := (\mathrm{Hom}(\mathcal{H}_{\mathcal{F}}, A), \star, e_A)$ with unit $e_A := \eta_A \circ \epsilon$. Here (A, R, η_A) is a unital commutative Rota–Baxter algebra with idempotent weight one Rota–Baxter map R dictated by the regularization prescription. In $\mathcal{A}$ we have the group $G_A := \mathrm{char}(\mathcal{H}_{\mathcal{F}}, \mathrm{A})$ of multiplicative maps and its corresponding Lie algebra of infinitesimal characters $\mathcal{L}_A$, related bijectively via the exponential map $\exp^\star$.

The crucial point is the observation that the linearity of R on A allows for a lifting of the Rota–Baxter structure to $\mathcal{A}$, making it into a unital associative non-commutative Rota–Baxter algebra $(\mathcal{A}, \mathcal{R})$ of weight one. Having such a Rota–Baxter algebra on $\mathcal{A}$ at hand we arrive at the following theorem.

Theorem 5.1 [EGK2004, EGK2005] $(\mathrm{Hom}(\mathcal{H}_{\mathcal{F}}, A), \mathcal{R})$ *is a complete filtered Rota–Baxter algebra with Rota–Baxter operator* $\mathcal{R}(\phi) := R \circ \phi$ *and filtration from* $\mathcal{H}_{\mathcal{F}}$. *We denote its unit by* $e_A := \eta_A \circ \epsilon$. *For an A-valued character* $\phi = \exp^\star(b) \in \mathrm{char}(\mathcal{H}_{\mathcal{F}}, \mathrm{A})$ *take* $a := \phi - e_A$. *Then* a *is in* $\mathcal{A}_1$ *and Theorem 4.2 implies that*

1. *the equations in (4.19) are the recursive formulae for* $X =: \phi_- \in G_A$ *and* $Y =: \phi_+^{-1} \in G_A$ *in Theorem 3.6 of Connes–Kreimer. They take the following form for* $\Gamma \in \ker(\epsilon)$
$$\begin{aligned} \phi_-(\Gamma) &= e_A(\Gamma) - \mathcal{R}\big(\phi_- \star a\big)(\Gamma) \qquad (5.2)\\ &= -R\Big(\phi(\Gamma) + \sum_{(\gamma\subset\Gamma)} \phi_-(\gamma)\phi(\Gamma/\gamma)\Big), \\ \phi_+(\Gamma) &= e_A(\Gamma) - \tilde{\mathcal{R}}(\phi_+ \star (\phi^{-1} - e_A))(\Gamma) \qquad (5.3)\\ &= \tilde{R}\Big(\phi(\Gamma) + \sum_{(\gamma\subset\Gamma)} \phi_-(\gamma)\phi(\Gamma/\gamma)\Big). \end{aligned}$$

2. *Spitzer's identity for non-commutative complete filtered Rota–Baxter algebras (4.28) implies that the above recursions for the linear maps* ϕ_- *and* ϕ_+ *can be written as the exponential factors in Eq. (3.12) giving the unique explicit formulae for* $X = \phi_-$ *and* $Y^{-1} = \phi_+$. *They are algebra homomorphisms, i.e., A-valued characters in* $G_A^- := e_A + \mathcal{R}(\mathcal{A}_1)$ *respectively in* $G_A^+ := e_A + \tilde{\mathcal{R}}(\mathcal{A}_1)$
$$\phi_-^{-1} = \exp^\star\Big(\mathcal{R}\big(\chi(b)\big)\Big) \text{ resp. } \phi_+ = \exp^\star\Big(\tilde{\mathcal{R}}\big(\chi(b)\big)\Big). \qquad (5.4)$$

3. *equation (3.12) gives the unique Birkhoff decomposition of* $\phi = \phi_-^{-1} \star \phi_+$ *found in Theorem 3.6 of Connes–Kreimer;*
4. *Bogoliubov's* $\bar{\mathrm{R}}$*-map,* $\bar{\mathrm{R}}[\phi] : \mathcal{H}_{\mathcal{F}} \to A$, *is given by* $\bar{\mathrm{R}}[\phi] = \exp^{\star_{\mathcal{R}}}\big(-\chi(\log^\star(\phi))\big)$, *for* $\phi \in G_A$, *such that* $\tilde{\mathcal{R}}(\bar{\mathrm{R}}[\phi]) = 2e_A - \phi_+$ *and* $\mathcal{R}(\bar{\mathrm{R}}[\phi]) = \phi_- - e_A$. *Here, the double Rota–Baxter product (4.9) is defined in terms of* $\mathcal{R}$
$$\phi_1 \star_{\mathcal{R}} \phi_2 := \mathcal{R}(\phi_1) \star \phi_2 + \phi_1 \star \mathcal{R}(\phi_2) - \phi_1 \star \phi_2, \quad \phi_1, \phi_2 \in \mathcal{A}.$$

Observe that Eq. (5.3) for $Y^{-1} = \phi_+$ in item (1) of Theorem 5.1 does not contain the counterterm ϕ_- explicitly
$$\phi_+ = e_A - \tilde{\mathcal{R}}\big(\phi_+ \star (\phi^{-1} - e_A)\big).$$

Recall that the inverse of $\phi \in G_A$ is given by the composition with the antipode of $\mathcal{H}_\mathcal{F}$, $\phi^{-1} = \phi \circ S$. The reader will easily check that $\phi_+ \star (\phi^{-1} - e_A) = -\phi_- \star (\phi - e_A)$ in accordance with the classical expression in terms of Bogoliubov's $\bar{\mathrm{R}}$-operation (3.46).

In the light of the last remark we may emphasize the observation in item (4) of the above theorem, where we see that Bogoliubov's $\bar{\mathrm{R}}$-operation (3.46) (see also Eq. (2.4)) can be written as an exponential, defined with respect to the double Rota–Baxter convolution product $\star_\mathcal{R}$ defined on $(\mathcal{A}, \mathcal{R})$, and the BCH-recursion χ in (3.9)

$$\bar{\mathrm{R}}[\phi] = \phi_- \star (\phi - e_A) = -\exp^{\star_\mathcal{R}}\big(-\chi(\log^\star(\phi))\big)$$

for elements in the augmentation ideal of $\mathcal{H}_\mathcal{F}$.

The above theorem presents a purely algebraic setting for the formulation of renormalization as a factorization problem in the group of regularized Hopf algebra characters, situated in the theory of non-commutative Rota–Baxter algebras with idempotent Rota–Baxter map. The formulae for the counterterm (5.2) and renormalized character (5.3) are completely dictated by a general decomposition structure, which characterizes Rota–Baxter algebras. The additional property of R being a projector implies a direct decomposition of the algebra A, hence the uniqueness of the factorization in item (3). We would like to emphasis the necessary freedom in the choice of the regularization prescription, encoded in the particular structure of the commutative Rota–Baxter algebra A as target space of linear Hopf algebra functionals in $\mathrm{Hom}(\mathcal{H}_\mathcal{F}, A)$.

Geometric picture for dimensional regularization: Specializing the target space Rota–Baxter algebra A to the field of Laurent series, i.e., using dimensional regularization together with minimal subtraction scheme, the above theorem amounts to the decomposition of the Laurent series $\phi(\Gamma)(\varepsilon)$, which has poles of finite order in the regulator parameter ε, into a part holomorphic at the origin and a part holomorphic at complex infinity. This has a geometric interpretation upon considering the Birkhoff decomposition of a loop around the origin, which is central in the work of Connes and Kreimer [CK2000, CK2001].

In this special case the theorem opens a hitherto hidden geometric viewpoint on perturbative renormalization and puts the mathematics in the theory of renormalization in a broader context of the Riemann–Hilbert correspondence which has its origin in Hilbert's 21st problem and has been extended into several areas of analysis, geometry, mathematical physics [Itz2003].

Finally, let us mention that the above notion of complete Rota–Baxter algebra and the statements in Theorem 5.1 become very transparent for (pro) uni- respectively nilpotent upper (or lower) triangular matrices with entries in a unital commutative Rota–Baxter algebra. This will be our last point in the following final section. It is extracted from our recent work [EGGV2006, EG2005b], which was conducted together with J. M. Gracia-Bondía and J. C. Várilly. It shows how the combinatorics of perturbative renormalization can be represented by matrix factorization of unipotent triangular matrices with entries in a commutative Rota–Baxter algebra. As we have seen above such triangular matrices provide a simple example of a complete filtered Rota–Baxter algebra.

5.2 Matrix calculus for the renormalization process in pQFT. Recall that $\mathcal{H}_{\mathcal{F}}$ is defined to be the graded polynomial algebra generated by F. Connes and Kreimer's coproduct $\Delta : \mathcal{H}_{\mathcal{F}} \to \mathcal{H}_{\mathcal{F}} \otimes \mathcal{H}_{\mathcal{F}}$ on $\mathcal{H}_{\mathcal{F}}$, making it into a combinatorial Hopf algebra, is simply given by (3.37) which we repeat here for $\Gamma \in \mathcal{H}_{\mathcal{F}}$

$$\Delta(\Gamma) = \Gamma \otimes 1_{\mathcal{F}} + 1_{\mathcal{F}} \otimes \Gamma + \sum_{\gamma \subset \Gamma} \gamma \otimes \Gamma/\gamma, \tag{5.5}$$

where the sum is over superficially divergent subgraphs γ of Γ and $\Gamma/\gamma \in \mathcal{F}$ was the corresponding 1PI cograph. Recall the Remark 3.5 after (3.37) on the spinney structure of graphs and the terms in the coproduct.

Let A be a commutative Rota–Baxter algebra with Rota–Baxter operator R and define $\mathcal{A} := \mathrm{Hom}(\mathcal{H}_{\mathcal{F}}, A)$ which is associative with respect to convolution and has a unit $e_A := \eta_A \circ \epsilon$. The grading of $\mathcal{H}_{\mathcal{F}}$ and the Rota–Baxter structure on A imply that $\mathcal{A}$ is a complete filtered non-commutative unital Rota–Baxter algebra.

In this final section we will establish a representation of $\mathcal{A}$ in terms of lower triangular matrices with entries in A. The group $G_A := e_A + \mathcal{A}_1 \in \mathcal{A}$ and its corresponding Lie algebra $\mathcal{L}_A$ are associated to (pro) unipotent and nilpotent lower triangular matrices, respectively.

Let us choose a subset $F' \subseteq F$ of graphs and order them, $\Gamma_1 := 1_{\mathcal{F}}, \Gamma_2, \Gamma_3, \ldots$ in accordance to the grading by degree. For graphs having the same degree we choose an arbitrary order. By the definition of the coproduct Δ in (5.5) we have

$$\Delta(\Gamma_i) = \Gamma_i \otimes 1_{\mathcal{F}} + 1_{\mathcal{F}} \otimes \Gamma_i + \sum_{j=2}^{i-1} \Gamma_{ij} \otimes \Gamma_j = \sum_{j=1}^{i} \Gamma_{ij} \otimes \Gamma_j, \tag{5.6}$$

where $\Gamma_{ij} \in \mathcal{H}_{\mathcal{F}}$ and $\deg(\Gamma_j) < \deg(\Gamma_i)$ for $j < i$. Here we see that to assure that the restriction of the coproduct becomes a good comodule map, $\Delta|_{\mathcal{F}'} : \mathcal{F}' \to \mathcal{H}_{\mathcal{F}} \otimes \mathcal{F}'$ where $\mathcal{F}' = \mathbb{K}\, F'$, we must have that for any $\Gamma_i \in F'$ all corresponding cographs in (5.6) appear in F', too.

Define the following possibly infinite lower triangular matrix

$$M_{\mathcal{H}_{\mathcal{F}}}(F') = (\Gamma_{ij})_{1 \le j \le i \le |F'|}$$

to be the *coproduct matrix* of $\mathcal{H}_{\mathcal{F}}$ corresponding to F'. $M_{\mathcal{H}_{\mathcal{F}}}(F')$ is an $|F'| \times |F'|$ lower triangular matrix ($\infty \times \infty$ if $F' \subseteq F$ contains infinitely many graphs) with entries in $\mathcal{H}_{\mathcal{F}}$ and with unit on the diagonal.

For example, take the following simple set of electron self-energy graphs borrowed from QED in four space-time dimensions

$F' := \Big\{ \Gamma_1 := 1_{\mathcal{F}},\ \Gamma_2 :=$ 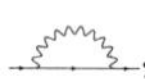, $\Gamma_3 :=$ 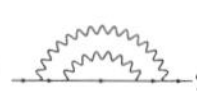, $\Gamma_4 :=$ 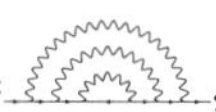, $\Gamma_5 :=$ 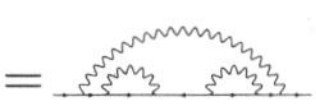 $\Big\}$ (5.7)

and with the prescribed order. Then $\Delta(\Gamma_i) = \sum_{j \le i} \Gamma_{ij} \otimes \Gamma_j$. So, that by a slight abuse of the tensor product notation as matrix product we have

$$\Delta \begin{pmatrix} 1_{\mathcal{F}} \\ \Gamma_2 \\ \Gamma_3 \\ \Gamma_4 \\ \Gamma_5 \end{pmatrix} = (\Gamma_{ij}) \otimes \begin{pmatrix} 1_{\mathcal{F}} \\ \Gamma_2 \\ \Gamma_3 \\ \Gamma_4 \\ \Gamma_5 \end{pmatrix} = \begin{pmatrix} 1_{\mathcal{F}} & 0 & 0 & 0 & 0 \\ \Gamma_2 & 1_{\mathcal{F}} & 0 & 0 & 0 \\ \Gamma_3 & \Gamma_2 & 1_{\mathcal{F}} & 0 & 0 \\ \Gamma_4 & \Gamma_3 & \Gamma_2 & 1_{\mathcal{F}} & 0 \\ \Gamma_5 & \Gamma_2\Gamma_2 & 2\Gamma_2 & 0 & 1_{\mathcal{F}} \end{pmatrix} \otimes \begin{pmatrix} 1_{\mathcal{F}} \\ \Gamma_2 \\ \Gamma_3 \\ \Gamma_4 \\ \Gamma_5 \end{pmatrix},$$

where the coproduct matrix $(\Gamma_{ij}) \in \mathcal{M}_{\mathcal{H}_\mathcal{F}}(F')$ is lower 5×5 triangular with unit diagonal and entries in $\mathcal{H}_\mathcal{F}$. In the last row we used that

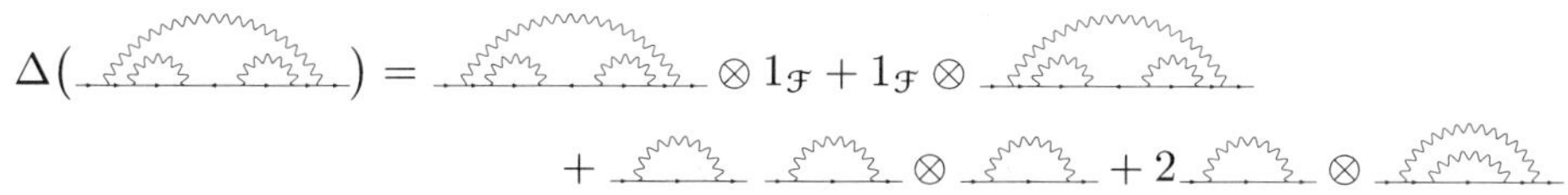

As a remark we should underline that the entries in the above coproduct matrix can also be seen directly from the spinneys in the woods $\bar{\mathfrak{W}}(\Gamma_i)$, $i = 1, 2, 3, 4, 5$, see Eqs. (2.8-2.11) from Section 2.3. Indeed, this is in accordance with the remark after the definition of the coproduct on Feynman graphs in Eq. (3.37) of Section 3, since we observe that the k-th row of (Γ_{ij}) corresponding to the graph Γ_k simply contains the spinneys in $\bar{\mathfrak{W}}(\Gamma_k)$. More precisely, the entry in (Γ_{kj}) for $j = 1, 2, 3, 4, 5$ is given by the particular spinney in $\bar{\mathfrak{W}}(\Gamma_k)$ which produces the cograph Γ_j when it is contracted in Γ_k. Recall that $\bar{\mathfrak{W}}(\Gamma_k)$ contains the spinneys $\{\Gamma_k\}$ and $\{\emptyset\}$ with cographs $\Gamma_k/\{\emptyset\} = \Gamma_k$ and $\Gamma_k/\{\Gamma_k\} = \emptyset = 1_\mathcal{F}$, leading to the entries (Γ_{k1}) and (Γ_{kk}), respectively.

We must put some emphasis on the fact that F' can be an infinite subset of F, or even F itself, in the latter case we denote the coproduct matrix by $M_{\mathcal{H}_\mathcal{F}}$. Both cases would imply that the matrix $M_{\mathcal{H}_\mathcal{F}}(F')$ is triangular of infinite size. Such matrices form a very well-behaved subalgebra.

Let us define $n := |F'|$, $n \leq \infty$. Now let $f : \mathcal{H}_\mathcal{F} \to A$ denote an arbitrary regularized linear functional on $\mathcal{H}_\mathcal{F}$. Simply applying f to either $M_{\mathcal{H}_\mathcal{F}}(F')$ or, in the full case, to $M_{\mathcal{H}_\mathcal{F}}$ entry by entry gives a lower triangular matrix

$$\widehat{f} := \widehat{f}_{F'} := f\big(M_{\mathcal{H}_\mathcal{F}}(F')\big) = \big(f(\Gamma_{ij})\big)_{1 \leq j \leq i \leq |F'|} \in \mathcal{M}^{\ell}_{|F'|}(A)$$

with $f(\Gamma_1), f(\Gamma_2), f(\Gamma_3), \cdots, f(\Gamma_{|F'|})$ for all graphs Γ_i, $i = 1, \ldots, n$ (possibly infinite), as the first column. The matrix $\widehat{f}$ is now in the algebra of lower triangular matrices of size $|F'|$ with entries in the commutative Rota–Baxter algebra A.

For a Feynman rules character $\phi \in G_A$ we obtain the unipotent *Feynman rules matrix* $\widehat{\phi}$ with unit diagonal. For example, with our F' in (5.7), we have

$$\widehat{\phi} = \begin{pmatrix} 1_A & 0 & 0 & 0 & 0 \\ \phi(\Gamma_2) & 1_A & 0 & 0 & 0 \\ \phi(\Gamma_3) & \phi(\Gamma_2) & 1_A & 0 & 0 \\ \phi(\Gamma_4) & \phi(\Gamma_3) & \phi(\Gamma_2) & 1_A & 0 \\ \phi(\Gamma_5) & \phi(\Gamma_2)\phi(\Gamma_2) & 2\phi(\Gamma_2) & 0 & 1_A \end{pmatrix} \tag{5.8}$$

The unit $e_A \in A$ obviously maps to the unit matrix $\widehat{e_A} = \mathbf{1}$, where $\mathbf{1} := (\delta_{ij} 1_A)_{1 \leq i,j \leq n}$ is the identity matrix. Compare this with the matrix representation of the identity map $\mathrm{id}_{\mathcal{H}_\mathcal{F}}$, which is just the coproduct matrix

$$\widehat{\mathrm{id}_{\mathcal{H}_\mathcal{F}}} = M_{\mathcal{H}_\mathcal{F}}(F').$$

Since A is assumed to be commutative the space of those triangular matrices with unit diagonal forms the Lie group $\widehat{G_A} := \mathbf{1} + \mathcal{M}^{\ell}_n(A)_1 \subset \mathcal{M}^{\ell}_{|F'|}(A)$.

On the other hand, for elements in the corresponding Lie algebra $\mathcal{L}_A$ of infinitesimal characters we find that $Z \in \mathcal{L}_A$ applied to $M_{\mathcal{H}_\mathcal{F}}(F')$ maps the unit diagonal and non-linear entries to zero due to relation (3.20). For F' in (5.7), this gives

$$\widehat{Z} = \begin{pmatrix} 0 & 0 & 0 & 0 & 0 \\ Z(\Gamma_2) & 0 & 0 & 0 & 0 \\ Z(\Gamma_3) & Z(\Gamma_2) & 0 & 0 & 0 \\ Z(\Gamma_4) & Z(\Gamma_3) & Z(\Gamma_2) & 0 & 0 \\ Z(\Gamma_5) & 0 & 2Z(\Gamma_2) & 0 & 0 \end{pmatrix}$$

Hence, derivations correspond to nilpotent matrices forming the matrix Lie algebra $\widehat{\mathcal{L}_A} \subset \mathcal{M}^{\ell}_{|F'|}(A)$ bijectively corresponding to $\widehat{G_A}$ via the matrix exponential map respectively logarithmic map. Let us underline that we are just in the setting described in Example 3.1, i.e., the complete filtered (non-commutative) algebra of lower triangular matrices with entries in a commutative unital algebra A. It contains the group $\widehat{G_A}$ of unit diagonal triangular matrices as well as its Lie algebra $\widehat{\mathcal{L}_A}$. In fact, this picture represents most transparently the pro-nilpotent structure of the Connes–Kreimer group of combinatorial Hopf algebra characters. For any finite subset $F' \subseteq F$, $n = |F'| < \infty$ we see that finite powers of derivations $\widehat{Z}^m = 0$, $Z \in \widehat{\mathcal{L}_A}$ and $m > n$.

Observe that for the Feynman rules character ϕ the first column of the associated matrix $\widehat{\phi}$ contains the Feynman amplitudes corresponding to the elements in F' (resp. $\mathcal{F}$). Therefore it is natural to introduce a *bra-ket notation*, i.e., we identify the elements in $\Gamma_i \in \mathcal{F}'$, $1 \leq i \leq n$ as column vectors $\{|\Gamma_i\rangle\}_{1\leq i\leq |F'|}$ with unit entry at position i and zero else. Such that for $1 \leq i \leq |F'|$

$$\langle \Gamma_i|\ \widehat{\phi}\ |\Gamma_i\rangle = \phi(\Gamma_i) \in A.$$

An intriguing simple observation is the following. Let $\phi \in G_A$ respectively $\log^{\star}(\phi) \in \mathcal{L}_A$. Then, in terms of matrices we have immediately

$$\widehat{\phi} = \exp\big(\log(\widehat{\phi})\big) = \exp\Big(\phi\big(\log\big(\widehat{\mathrm{id}_{\mathcal{H}_{\mathcal{F}}}}\big)\big)\Big),$$

since ϕ is multiplicative. The matrix $\log\big(\widehat{\mathrm{id}_{\mathcal{H}_{\mathcal{F}}}}\big) \in \mathcal{M}_{|F'|}(\mathcal{H}_{\mathcal{F}})$ is easily calculated

$$\log\big(\widehat{\mathrm{id}_{\mathcal{H}_{\mathcal{F}}}}\big) = \log\big(\mathbf{1} + (\widehat{\mathrm{id}_{\mathcal{H}_{\mathcal{F}}}} - \mathbf{1})\big) = -\sum_{k>0} \frac{\big(-(\widehat{\mathrm{id}_{\mathcal{H}_{\mathcal{F}}}} - \mathbf{1})\big)^k}{k}.$$

In the above QED example F' in (5.7) we find the following concrete matrix

$$\log\big(\widehat{\mathrm{id}_{\mathcal{H}_{\mathcal{F}}}}\big) = \begin{pmatrix} 0 & 0 & 0 & 0 & 0 \\ \Gamma_2 & 0 & 0 & 0 & 0 \\ \Gamma_3 - \frac{1}{2}{\Gamma_2}^2 & \Gamma_2 & 0 & 0 & 0 \\ \Gamma_4 - \Gamma_2\Gamma_3 + \frac{1}{3}{\Gamma_2}^3 & \Gamma_3 - \frac{1}{2}{\Gamma_2}^2 & \Gamma_2 & 0 & 0 \\ \Gamma_5 - \Gamma_2\Gamma_3 + \frac{1}{6}{\Gamma_2}^3 & 0 & 2\Gamma_2 & 0 & 0 \end{pmatrix}$$

This matrix just contains the *normal coordinates* used in the exponential representation of ϕ, see [CQRV2002, FG2005]. In the same spirit we may calculate the inverse of an element in $\widehat{G_A}$. Recall that for $\phi \in G_A$ the inverse was given by composition with the Hopf algebra antipode, $\phi^{-1} := \phi \circ S$. Since S is supposed to be the convolution inverse of the identity on $\mathcal{H}_{\mathcal{F}}$, we readily see that

$$\widehat{\phi^{-1}} = \phi \circ \widehat{\mathrm{id}^{-1}_{\mathcal{H}_{\mathcal{F}}}} = \phi \circ \Big(\sum_{k\geq 0} \big(-(\widehat{\mathrm{id}_{\mathcal{H}_{\mathcal{F}}}} - \mathbf{1})\big)^k\Big)$$

where again in our QED example we find

$$\mathbf{1}+\sum_{k>0}\big(-\,(\widehat{\mathrm{id}_{\mathcal{H}_{\mathcal{F}}}-\mathbf{1}})\big)^{k}=\begin{pmatrix} 1_{\mathcal{F}} & 0 & 0 & 0 & 0\\ -\Gamma_2 & 1_{\mathcal{F}} & 0 & 0 & 0\\ -\Gamma_3-{\Gamma_2}^2 & -\Gamma_2 & 1_{\mathcal{F}} & 0 & 0\\ -\Gamma_4+2\Gamma_2\Gamma_3-{\Gamma_2}^3 & \Gamma_3-{\Gamma_2}^2 & -\Gamma_2 & 1_{\mathcal{F}} & 0\\ -\Gamma_5+2\Gamma_2\Gamma_3-{\Gamma_2}^3 & {\Gamma_2}^2 & -\Gamma_2 & 0 & 1_{\mathcal{F}} \end{pmatrix}$$

One confirms without difficulty that $\widehat{\phi^{-1}}=\widehat{\phi}^{-1}$. In the light of Theorem 4.1 and Theorem 5.1 the reader may guess the next step. Indeed, we lift the Rota–Baxter structure from (A,R) to $\mathcal{M}^{\ell}_{|\mathcal{F}'|}(A)$ in the usual way by defining for $\alpha\in\mathcal{M}^{\ell}_{|\mathcal{F}'|}(A)$

$$\mathrm{R}(\alpha):=\big(R(\alpha_{ij})\big)_{1\le j\le i\le |F'|}.$$

All this is readily summarized in [EG2005b, EGGV2006]. In fact, denote by $A\mathcal{F}':=A\otimes\mathcal{F}'$ the free A-module with basis F'. Recall that $\mathcal{A}:=\mathrm{Hom}(\mathcal{H}_{\mathcal{F}},A)$ is a complete filtered unital non-commutative Rota–Baxter algebra of weight one with idempotent Rota–Baxter map $\mathcal{R}(f):=R\circ f$. Define a map

$$\Psi:=\Psi_{A,\mathcal{F}'}:\mathrm{Hom}(\mathcal{H}_{\mathcal{F}},A)\to\mathrm{End}(A\otimes\mathcal{F}')$$

by the composition

$$\Psi[f]:A\otimes\mathcal{F}'\xrightarrow{\mathrm{id}_A\otimes\Delta}A\otimes\mathcal{H}_{\mathcal{F}}\otimes\mathcal{F}'\xrightarrow{\mathrm{id}_A\otimes f\otimes\mathrm{id}_{\mathcal{F}'}}A\otimes A\otimes\mathcal{F}'\xrightarrow{m_A\otimes\mathrm{id}_{\mathcal{F}'}}A\otimes\mathcal{F}'.$$

So that for $\Psi[f]$, $f\in\mathcal{A}$ applied to $\Gamma_i\in F'$ we find

$$\Psi[f](\Gamma_i)=f\star\mathrm{id}_{\mathcal{F}}(\Gamma_i)=\sum_{j=1}^{i}f(\Gamma_{ij})\Gamma_j\in A\mathcal{F}'.$$

It was shown in [EG2005b] that Ψ is an anti-homomorphism of algebras. We then obtain an algebra homomorphism $\mathrm{Hom}(\mathcal{H}_{\mathcal{F}},A)\to\mathcal{M}^{\ell}_{|F'|}(A)$, still denoted by Ψ, when composing this with the anti-homomorphism

$$\mathrm{End}(A\otimes\mathcal{F}')\to\mathcal{M}^{\ell}_{|F'|}(A)$$

sending $P:A\otimes\mathcal{F}'\to A\otimes\mathcal{F}'$ to its standard matrix $M=(m_{ij})$ with respect to the ordered basis F': $P(\Gamma_i)=\sum_j m_{ij}\Gamma_j$. Altogether, we have the algebra morphism

$$\Psi_{A,\mathcal{F}'}:\big(\mathrm{Hom}(\mathcal{H}_{\mathcal{F}},A),\mathcal{R}\big)\to\big(\mathcal{M}^{\ell}_{|F'|}(A),\mathrm{R}\big),\qquad \Psi_{A,\mathcal{F}'}[f]=\widehat{f}$$

of complete filtered Rota–Baxter algebras. We omit the indices now for notational transparency. So in particular for $f,g\in\mathcal{A}$

$$\Psi[f\star g]=\Psi[f]\ \Psi[g]=\widehat{f}\,\widehat{g},\qquad \Psi[\mathcal{R}(f)]=\mathrm{R}(\Psi[f])=\mathrm{R}(\widehat{f}).$$

The map Ψ gives a matrix representation of $\mathrm{Hom}(\mathcal{H}_{\mathcal{F}},A)$ in terms of lower triangular matrices. In a more abstract context, Ψ gives a morphism between the affine group scheme of $\mathcal{H}_{\mathcal{F}}$ and the matrix group scheme.

Let us go back to Theorem 5.1. Applying Ψ to item (3) we have for the matrix representation of the character ϕ respectively its Birkhoff decomposition

$$\widehat{\phi}=\widehat{\phi_-}^{-1}\,\widehat{\phi_+},\tag{5.9}$$

where $\widehat{\phi_-}=\Psi[\phi_-]$ and $\widehat{\phi_+}=\Psi[\phi_+]$. Hence, we recover $\widehat{\phi_+}$ and $\widehat{\phi_-}$ for each element in F' from the first column of the corresponding matrices.

Further the two matrix factors, $\widehat{\phi_+}^{-1}$, $\widehat{\phi_-}$, are unique solutions to the matrix Spitzer identities, i.e., Bogoliubov's formulae for the counterterm and renormalized Feynman rules, respectively

$$\widehat{\phi_-} = \mathbf{1} - \mathrm{R}\Big(\widehat{\phi_-}\ (\widehat{\phi} - \mathbf{1})\Big), \tag{5.10}$$

$$\widehat{\phi_+}^{-1} = \mathbf{1} - \tilde{\mathrm{R}}\Big((\widehat{\phi} - \mathbf{1})\ \widehat{\phi_+}^{-1}\Big) \quad \text{and} \quad \widehat{\phi_+} = \mathbf{1} - \tilde{\mathrm{R}}\Big(\widehat{\phi_+}\ (\widehat{\phi}^{-1} - \mathbf{1})\Big) \tag{5.11}$$

In [EG2005b, EGGV2006] we worked with an upper triangular representation which makes Ψ is an anti-morphism, exchanging the order of products when comparing with the 'scalar' formulae in the Hopf algebra formalism. The matrix entries can be calculated without recursions using $\alpha = \widehat{\phi}$ from the equations

$$(\widehat{\phi_-})_{ij} = -R(\alpha_{ij}) + \sum_{k=2}^{j-i} \sum_{i>l_1>\cdots>l_{k-1}>j} (-1)^k R\big(R(\cdots R(\alpha_{il_1})\alpha_{l_1l_2})\cdots\alpha_{l_{k-1}j}\big) \tag{5.12}$$

$$(\widehat{\phi_+}^{-1})_{ij} = -\tilde{R}(\alpha_{ij}) + \sum_{k=2}^{j-i} \sum_{i>l_1>\cdots>l_{k-1}>j} (-1)^k \tilde{R}\big(\alpha_{il_1}\tilde{R}(\alpha_{l_1l_2}\cdots\tilde{R}(\alpha_{l_{k-1}j})\cdots)\big) \tag{5.13}$$

Let us finish this section and the paper with the simple example from QED. We should emphasize that the calculations of the matrix Birkhoff factorization are just meant to point up the combinatorial structure involved in the renormalization of matrix Feynman rules characters.

The Feynman rules matrix (5.8) of the graphs up to 3 loops from the QED example decomposes into the counterterm

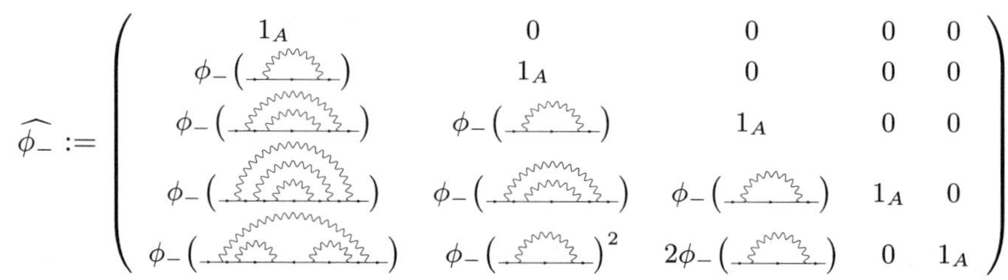

(5.14)

Of course, this follows immediately from general reasoning by applying the abstract Hopf algebra character ϕ_- to the coproduct matrix $M_{\mathcal{H}_\mathcal{F}}(F')$. Instead, the goal is to calculate the matrix entries directly either by use of the matrix Equation (5.10), which terminates after 3 iterations, or entrywise with help of Eq. (5.12).

In this example the only nontrivial counterterm matrix entry we need to calculate is position $(5,1)$ in $\widehat{\phi}_-$. Applying formula (5.12) we find

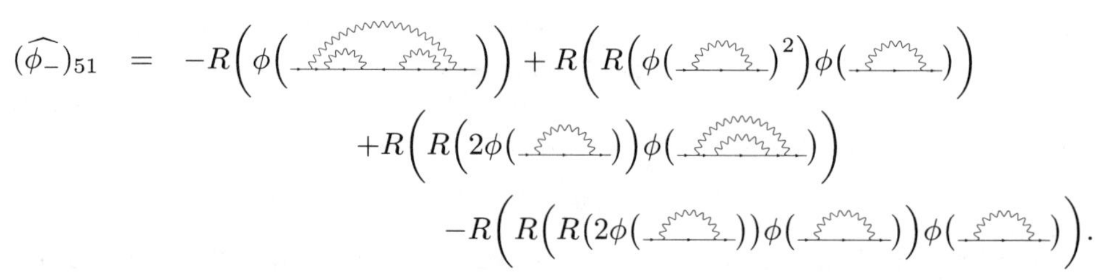

The identity $R(a)^2 = 2R(aR(a)) - R(a^2)$ which follows from the Rota–Baxter identity, and which is true only for commutative algebras, immediately implies

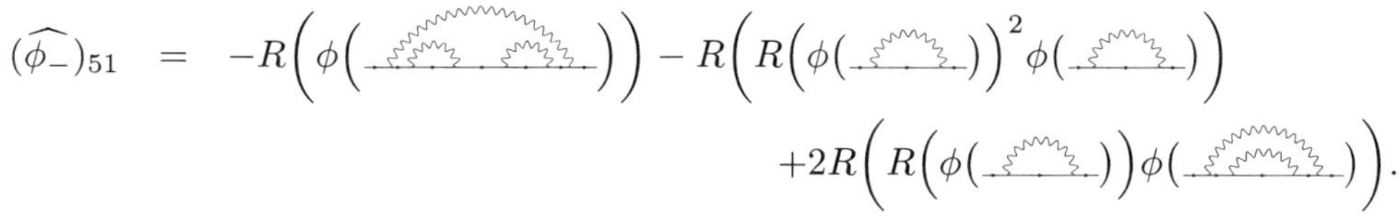

Likewise for the corresponding entry in the renormalized matrix $\widehat{\phi_+} = \widehat{\phi_-}\ \widehat{\phi}$ we find

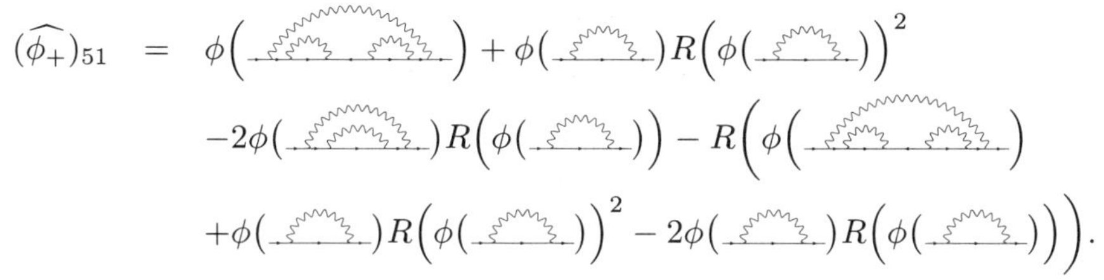

More examples can be found in [EGGV2006, EG2005b].

6 Conclusion and outlook

The algebraic-combinatorial structure of renormalization in perturbative quantum field theory has found a concise formulation in terms of Hopf algebras of Feynman graphs. The process of renormalization is captured by an algebraic Birkhoff decomposition of regularized Feynman characters discovered by Connes and Kreimer. Associative Rota–Baxter algebras naturally provide a suitable general underpinning for such factorizations in terms of Atkinson's theorem and Spitzer's identity. This approach gives a different perspective on the original result due to Connes and Kreimer in the context of other renormalization schemes. It enabled us to establish a simple matrix calculus for renormalization. However, a general characterization of the algebraic structure underlying the notion of renormalization schemes needs to be explored in future work. We should underline that our results presented here go beyond the particular application in the context of perturbative renormalization as we presented a general factorization theorem for filtered algebras in terms of a recursion defined using the Baker–Campbell–Hausdorff formula for which we found a closed formula under suitable circumstances. Moreover, we showed its natural link to a classical result of Magnus known from matrix differential equations. The examples of Rota–Baxter algebras, especially its appearance in classical integrable systems makes it a rich field to be further explored.

Acknowledgements. We would like to thank the organizers for giving us the opportunity to present parts of our research results which we achieved in joint work with D. Kreimer, and partly with J. M. Gracia-Bondía, J. Várilly and more recently D. Manchon. The stimulating atmosphere we found at this memorable workshop affected us considerably and helped a lot in shaping our understanding of the concept and applications of renormalization. The generous support, relaxing atmosphere and warm hospitality we experienced at the Fields Institute is greatly acknowledged. Thanks goes especially to its staff for the efficient organization. The first author acknowledges greatly the support by the European Post-Doctoral

Institute (E.P.D.I.) and Institute des Hautes Études Scientifiques (I.H.É.S.), and last but not least D. Kreimer's ongoing collaboration. He would also like to thank the department of theoretical physics at the University of Zaragoza (Spain) where this work was finished for warm hospitality and support. The second author thanks NSF and Rutgers University Research Council for support. Dirk Kreimer is thanked for his corrections, comments and remarks.

URL of KEF: http://www.th.physik.uni-bonn.de/th/People/fard/

References

[Abe1980] E. Abe, *Hopf algebras*, Cambridge University Press, Cambridge, (1980).

[Agu2000] M. Aguiar, *Prepoisson algebras*, Lett. Math. Phys. **54** no. 4 (2000), 263–277.

[ABS2003] M. Aguiar, N. Bergeron and F. Sottile, *Combinatorial Hopf algebras and generalized Dehn–Sommerville relations*, Composito Mathematica **142** (2006), 1–30. [arXiv:math.CO/0310016]

[Atk1963] F. V. Atkinson, *Some aspects of Baxter's functional equation*, J. Math. Anal. Appl. **7** (1963), 1–30.

[BBT2003] O. Babelon, D. Bernard and M. Talon, *Introduction to classical integrable systems*, Cambridge Monographs on Mathematical Physics. Cambridge University Press, Cambridge, (2003).

[BHL2000] K. Barron, Y. Huang and J. Lepowsky, *Factorization of formal exponentials and uniformazation*, J. Algebra **228** (2000), 551–579.

[Bax1960] G. Baxter, *An analytic problem whose solution follows from a simple algebraic identity*, Pacific J. Math. **10** (1960), 731–742.

[BelDri1982] A. A. Belavin and V. G. Drinfeld, *Solutions of the classical Yang–Baxter equation for simple Lie algebras*, Funct. Anal. Appl. **16** (1982), 159–180.

[BeKr2005a] C. Bergbauer and D. Kreimer, *The Hopf algebra of rooted trees in Epstein–Glaser renormalization*, Ann. Inst. Henri Poincaré **6** (2005), 343–367. [arXiv:hep-th/0403207]

[BeKr05b] C. Bergbauer and D. Kreimer, *Hopf Algebras in Renormalization Theory: Locality and Dyson–Schwinger Equations from Hochschild Cohomology*, IRMA lectures in Mathematics and Theoretical Physics Vol.**10**, Physics and Number Theory, European Mathematical Society, Eds. V. Turaev, L. Nyssen. [arXiv:hep-th/0506190]

[Berg1985] G. M. Bergman *Everybody knows what a Hopf algebra is*, Group actions on rings (Brunswick, Maine, 1984), Contemp. Math. **43**, (1985), 25–48.

[Bet1947] H. Bethe, *The electromagnetic shift of energy levels*, Phys. Rev. **72** (1947), 339–341.

[BoPa1957] N. N. Bogoliubov and O. S. Parasiuk, *On the multiplication of causal functions in the quantum theory of fields*. Acta Math. **97** (1957), 227–266.

[BoSh1959] N. N. Bogoliubov and D. V. Shrikov, *Introduction to the theory of Quantized Fields*, Intersci. Monogr. Phys. Astron. **3** (1959), 1–720.

[BK1998] D. J. Broadhurst and D. Kreimer, *Association of multiple zeta values with positive knots via Feynman diagrams up to* 9 *loops*, Phys. Lett. B **426** no. 3-4 (1998), 339–346. [arXiv:hep-th/9609128]

[BK1999] D. J. Broadhurst and D. Kreimer, *Renormalization automated by Hopf algebra*, J. Symb. Comput. **27** (1999), 581–600. [arXiv:hep-th/9810087]

[BK2000a] D. J. Broadhurst and D. Kreimer, *Combinatoric explosion of renormalization tamed by Hopf algebra: 30-loop Pade–Borel resummation*, Phys. Lett. B **475** (2000), 63–70. [arXiv:hep-th/9912093]

[BK2000b] D. J. Broadhurst and D. Kreimer, *Towards cohomology of renormalization: bigrading the combinatorial Hopf algebra of rooted trees*, Comm. in Math. Phys. **215** (2000), 217–236. [arXiv:hep-th/0001202]

[BK2001] D. J. Broadhurst and D. Kreimer, *Exact solutions of Dyson–Schwinger equations for iterated one-loop integrals and propagator-coupling duality*, Nucl. Phys. B **600** (2001), 403–422. [arXiv:hep-th/0012146]

[BEK2005] S. Bloch, H. Esnault and D. Kreimer, *Motives associated to graph polynomials*, preprint: May 2005, arXiv:math.ag/0510011.

[Br1993] L. Brown (ed.), *Renormalization. From Lorentz to Landau (and beyond)*, Springer-Verlag, New York, (1993).

[Cal1975] C. Callan, *Introduction to Renormalization Theory*, in "Methods in Field Theory", 1975, Les Houches Lect., R. Ballan, J. Zinn-Justin (eds.), North-Holland, Amsterdam, (1976).

[Car1972] P. Cartier, *On the structure of free Baxter algebras*, Advances in Math. **9** (1972), 253–265.

[CaKe1982] W. E. Caswell and A. D. Kennedy, *A Simple approach to renormalization theory*, Phys. Rev. D **25** (1982), 392–408.

[ChLi2001] F. Chapoton and M. Livernet, *Pre-Lie algebras and the rooted trees operad*, Internat. Math. Res. Notices no. **8** (2001), 395–408. [arXiv:math.QA/0002069]

[ChPr1995] V. Chari and A. Pressley, *A guide to Quantum Groups*, Cambridge Univeristy Press, Cambridge, (1995).

[CK1998] A. Connes and D. Kreimer, *Hopf algebras, Renormalization and Noncommutative Geometry*, Comm. in Math. Phys. **199** (1998), 203–242. [arXiv:hep-th/9808042]

[CK2000] A. Connes and D. Kreimer, *Renormalization in quantum field theory and the Riemann-Hilbert problem. I. The Hopf algebra structure of graphs and the main theorem*, Comm. in Math. Phys. **210** no. 1 (2000), 249–273. [arXiv:hep-th/0003188]

[CK2001] A. Connes and D. Kreimer, *Renormalization in quantum field theory and the Riemann–Hilbert problem. II. The β-function, diffeomorphisms and the renormalization group*, Comm. in Math. Phys. **216** (2001), 215–241. [arXiv:hep-th/0003188]

[CK2002] A. Connes and D. Kreimer, *Insertion and elimination: the doubly infinite Lie algebra of Feynman graphs*, Ann. Inst. Henri Poincaré **3** no. 3 (2002), 411–433. [arXiv:hep-th/0201157]

[Cole1988] S. Coleman, *Aspects of Symmetry Selected Erice Lectures* Cambridge University Press, Cambridge, (1988).

[Col1984] J. C. Collins, *Renormalization*, Cambridge Monographs on Mathematical Physics, Cambridge University Press, Cambridge, (1984).

[Col2006] J. C. Collins, *Renormalization: general theory* preprint: Februray 2006 arXiv:hep-th/0602121.

[CM2004a] A. Connes and M. Marcolli, *Renormalization and motivic Galois theory*, Internat. Math. Res. Not. no. 76 (2004), 4073–4091. [arXiv:math.NT/0409306]

[CM2006] A. Connes and M. Marcolli, *Quantum Fields and Motives*, J. Geom. Phys. **56** (2006), 55–85. [arXiv:hep-th/0504085]

[CM2004b] A. Connes and M. Marcolli, *From Physics to Number Theory via Noncommutative Geometry, Part II: Renormalization, the Riemann-Hilbert correspondence, and motivic Galois theory*, to appear in the volume "Frontiers in Number Theory, Physics, and Geometry", preprint: Nov. 2004, arXiv:hep-th/0411114.

[CoMo1998] A. Connes and H. Moscovici, *Hopf algebras, cyclic cohomology and the transverse index theorem*, Comm. in Math. Phys. **198** no. 1 (1998), 199–246.

[CQRV2002] C. Chryssomalakos, H. Quevedo and M. Rosenbaum and J. D. Vergara, *Normal Coordinates and Primitive Elements in the Hopf Algebra of Renormalization*, Comm. in Math. Phys. **225** no. 3 (2002), 465–485. [arXiv:hep-th/0105252]

[Del2004] B. Delamotte, *A hint of renormalization*, Am. J. Phys. **72** (2004), 170–184. [arXiv:hep-th/0212049]

[EGK2004] K. Ebrahimi-Fard, L. Guo and D. Kreimer, *Spitzer's Identity and the Algebraic Birkhoff Decomposition in pQFT*, J. Phys. A: Math. Gen. **37** (2004), 11037–11052. [arXiv:hep-th/0407082]

[EGK2005] K. Ebrahimi-Fard, L. Guo and D. Kreimer, *Integrable Renormalization II: the General case*, Ann. H. Poincaré **6** (2005), 369–395. [arXiv:hep-th/0403118]

[EGGV2006] K. Ebrahimi-Fard, J. M. Gracia-Bondía, L. Guo and J.C. Várilly, *Combinatorics of renormalization as matrix calculus*, Phys. Lett. B. **632** no 4 (2006), 552–558. [arXiv:hep-th/0508154]

[EG2005a] K. Ebrahimi-Fard and L. Guo, *Mixable Shuffles, Quasi-shuffles and Hopf Algebras*, Journal of Algebraic Combinatorics **24** no 1 (2006), 83–101. [arXiv:hep-th/0506418]

[EG2005b] K. Ebrahimi-Fard and L. Guo, *Matrix Representation of Renormalization in Perturbative Quantum Field Theory*, submitted, preprint: August 2005. [arXiv:hep-th/0508155]

[EK2005] K. Ebrahimi-Fard and D. Kreimer, *Hopf algebra approach to Feynman diagram calculations*, J. Phys. A: Math. Gen., **38**, R385-R406, 2005. [arXiv:hep-th/0510202]

[EG2005c] K. Ebrahimi-Fard and L. Guo, *Multiple zeta values and Rota–Baxter algebras*, submitted, preprint: August 2005. [arXiv:math.NT/0601558]

[EGM2006] K. Ebrahimi-Fard, L. Guo and D. Manchon *Birkhoff type decompositions and the Baker–Campbell–Hausdorff recursion*, Comm. in Math. Phys., **267** (2006), 821. [arXiv:math-ph/0602004]

[EMK2003] K. Ebrahimi-Fard, I. Mencattini and D. Kreimer, *On the Insertion-Elimination Lie algebra of Feynman graphs*, prepared for 5th International Workshop on Lie Theory and Its Applications in Physics, Varna, Bulgaria, 16-22 Jun 2003. Published in *Varna 2003, Lie theory and its applications in physics* (2003), 124–134.

[FG2001] H. Figueroa and J. M. Gracia-Bondía, *On the antipode of Kreimer's Hopf algebra*, Mod. Phys. Lett. **A16** (2001), 1427–1434. [arXiv:hep-th/9912170]

[FG2005] H. Figueroa and J. M. Gracia-Bondía, *Combinatorial Hopf algebras in quantum field theory I*, Reviews of Mathematical Physics **17** (2005), 881–976. [arXiv:hep-th/0408145]

[FGV2001] H. Figueroa, J. M. Gracia-Bondía and J. C. Várilly, *Elements of Noncommutative Geometry*, Birkhäuser, (2001).

[Fois2002] L. Foissy, *Les algèbres de Hopf des arbres enracinés décorés I, II*, Bull. Sci. Math. **126** no. 3 (2002), 193–239; ibid. **126** no. 4 (2002), 249–288.

[Gio2002] D. Gioev, *Generalized Hunt–Dyson Formula and Bohnenblust–Spitzer Theorem*, Internat. Math. Res. Not. **32** (2002), 1703–1722.

[GK2000] L. Guo and W. Keigher, *Baxter algebras and shuffle products*, Advances in Math. **150** (2000), 117–149.

[Hepp1966] K. Hepp, *Proof of the Bogoliubov–Parasiuk theorem on renormalization*, Comm. in Math. Phys. **2** (1966), 301–326.

[Hof2003] M. E. Hoffman, *Combinatorics of rooted trees and Hopf algebras*, Trans. Amer. Math. Soc. **355** (2003), 3795–3811. [arXiv:math.CO/0201253]

[Hof2005] M. E. Hoffman, *Algebraic aspects of multiple zeta values*, in "Zeta functions, topology and quantum physics", 51–73, Dev. Math. **14** Springer, New York, (2005). [arXiv:math.QA/0309425]

[Holt2003] R. Holtkamp, *Comparison of Hopf algebras on trees*, Arch. Math. (Basel) **80** (2003), 368–383.

[Hop1941] H. Hopf, *Über die Topologie der Gruppen-Mannigfaltigkeiten und ihren Verallgemeinerungen*, Ann. of Math. **42** (1941), 22–52.

[Itz2003] A. R. Its, *The Riemann–Hilbert problem and integrable systems*, Notices AMS **50** (2003), 1389–1400.

[IzZu1980] C. Itzykson and J.-B. Zuber, *Quantum Field Theory*, McGraw-Hill, New-York, (1980).

[JoRo1979] S. A. Joni and G.-C. Rota, *Coalgebras and bialgebras in combinatorics*, Studies in Applied Mathematics **61** (1979), 93–139. Reprinted in: "Gian-Carlo Rota on

Combinatorics: Introductory papers and commentaries", J. P. S. Kung, Ed., Contemp. Mathematicians, Birkhäuser Boston, Boston, MA, 1995.

[Kai2005] D. Kaiser, *Drawing Theories Apart: The Dispersion of Feynman Diagrams in Postwar Physics*, The University of Chicago Press, Chicago, (2005).

[Kas1995] C. Kassel, *Quantum Groups*, Springer-Verlag, New York, (1995).

[Kin1962] J. F. C. Kingman, *Spitzer's identity and its use in probability theory*, J. London Math. Soc. **37** (1962), 309–316.

[Krei1998] D. Kreimer, *On the Hopf algebra structure of perturbative quantum field theories*, Adv. Theor. Math. Phys. **2** (1998), 303–334. [arXiv:q-alg/9707029]

[Krei1999a] D. Kreimer, *Chen's iterated integral represents the operator product expansion*, Adv. Theor. Math. Phys. **3** no. 3 (1999), 627–670. [arXiv:hep-th/9901099]

[Krei1999b] D. Kreimer, *On Overlapping Divergencies*, Comm. in Math. Phys. **204** (1999), 669–689. [arXiv:hep-th/9810022]

[Krei2001] D. Kreimer, *Structures in Feynman graphs: Hopf algebras and symmetries*, *Dennisfest* Proceedings, Stony Brook, June 2001. [arXiv:hep-th/0202110]

[Krei2002] D. Kreimer, *Combinatorics of (perturbative) Quantum Field Theory*, Phys. Rep. **363** (2002), 387–424. [arXiv:hep-th/00110059]

[Krei2003a] D. Kreimer, *Factorization in quantum field theory: An exercise in Hopf algebras and local singularities*, contributed to Les Houches School of Physics: *Frontiers in Number Theory, Physics and Geometry*, Les Houches, France, March 2003, 9–21. [arXiv:hep-th/0306020]

[Krei2003b] D. Kreimer, *New Mathematical Structures in Renormalizable Quantum Field Theory*, Annals of Phys. **303** (2003), 179–202.

[Krei2004] D. Kreimer, *The Residues of Quantum Field Theory – Numbers we should know*, contributed to the proceedings of the *Workshop on Noncommutative Geometry and Number Theory*, August 18-22 2003, Max Planck Institut für Mathematik, Bonn. [arXiv:hep-th/0407016]

[Krei2000] D. Kreimer, *Knots and Feynman Diagrams*, Cambridge Monographs on Mathematical Physics, Cambridge University Press, Cambridge, (2000).

[Krei2005] D. Kreimer, *Anatomy of a gauge theory*, preprint: Sept. 2005, arXiv:hep-th/0509135.

[Le2004] P. Leroux, *Ennea-algebras*, J. Algebra **281** no. 1 (2004), 287–302. [arXiv:math.QA/0309213]

[Lod1994] J.-L. Loday, *Série de Hausdorff, idempotents Eulériens et algèbres de Hopf*, Expo. Math. **12** (1994), 165–178.

[Mag1954] W. Magnus, *On the exponential solution of differential equations for a linear operator*, Comm. Pure Appl. Math. **7** (1954), 649–673.

[Maj1995] S. Majid, *Foundations of quantum group theory*, Cambridge University Press, Cambridge, (1995).

[Man2001] D. Manchon, *Hopf algebras, from basics to applications to renormalization*, Comptes-rendus des Rencontres mathématiques de Glanon 2001. [arxiv:math.QA/0408405]

[MeKr2004] I. Mencattini, D. Kreimer, *Insertion and elimination Lie algebra: the ladder case*, Lett. Math. Phys., **67**, 61–74, (2004). [arXiv:math.QA/0308042]

[MeKr2005] I. Mencattini, D. Kreimer, *The structure of the ladder Insertion-Elimination Lie algebra*, Comm. in Math. Phys., **259**, 413–432, (2005). [arXiv:mat-th/0408053]

[Mil1966] B. Miller, *Some properties of Baxter operators*, Acta Math. Acad. Sci. Hungar. **17** (1966), 387–400.

[Mil1969] B. Miller, *Baxter Operators and Endomorphisms on Banach Algebras*, J. Math. Anal. Appl. **25** (1969), 503–520.

[MM1965] J. W. Milnor and J. C. Moore, *On the structure of Hopf algebras*, Annals of Math. **81** (1965), 211–264.

[MUW2002] S. Moch, P. Uwer and S. Weinzierl, *Nested sums, expansion of transcendental functions and multiple-loop integrals*, J. Math. Phys. **43** no. 6 (2002), 3363–3386. [arXiv:hep-ph/0110083]

[MQZ2001] H. Z. Munthe-Kaas, G. R. W. Quispel and A. Zanna, *Generalized polar decompositions on Lie groups with involutive automorphisms*, Found. Comput. Math. **1** no. 3 (2001), 297–324.

[Muta1987] T. Muta *Foundations of quantum chromodynamics: an introduction to perturbative methods in gauge theories*, World Sci. Lect. Notes Phys. **5** (1987), 1–409.

[NiSw1982] W. Nichols and M. Sweedler, *Hopf algebras and combinatorics*, Umbral calculus and Hopf algebras (Norman, Okla., 1978), Contemp. Math. **6** (1982), 49–84.

[PS1995] M. E. Peskin and D. V. Schroeder, *An introduction to Quantum Field Theory*, Addison–Wesley, Reading, (1995).

[Reu1993] C. Reutenauer, *Free Lie Algebras*, Oxford University Press, Oxford, (1993).

[Rota1969] G.-C. Rota, *Baxter algebras and combinatorial identities. I, II.*, Bull. Amer. Math. Soc. **75** (1969), 325–329; ibid. **75** (1969), 330–334.

[RoSm1972] G.-C. Rota and D. Smith, *Fluctuation theory and Baxter algebras*, Istituto Nazionale di Alta Matematica **IX** (1972), 179–. Reprinted in: "Gian-Carlo Rota on Combinatorics: Introductory papers and commentaries", J. P. S. Kung, Ed., Contemp. Mathematicians, Birkhäuser Boston, Boston, MA, 1995.

[Rota1978] G.-C. Rota, *Hopf algebras in combinatorics*, in "Hopf algebra methods in combinatorics. Problmes combinatoires et thorie des graphes", (Colloq. Internat. CNRS, Univ. Orsay, Orsay, (1976), 363–365, Colloq. Internat. CNRS, 260, CNRS, Paris, 1978. Reprinted in: "Gian-Carlo Rota on Combinatorics: Introductory papers and commentaries", J.P.S. Kung Ed., Contemp. Mathematicians, Birkhäuser Boston, Boston, MA, 1995.

[Rota1995] G.-C. Rota, *Baxter operators, an introduction*, In: "Gian-Carlo Rota on Combinatorics, Introductory papers and commentaries", J. P. S. Kung, Ed., Contemp. Mathematicians, Birkhäuser Boston, Boston, MA, 1995.

[Rota1998] G.-C. Rota, *Ten mathematics problems I will never solve*, Invited address at the joint meeting of the American Mathematical Society and the Mexican Mathematical Society, Oaxaca, Mexico, December 6, 1997. DMV Mittellungen Heft **2** (1998), 45–52.

[Ryd1985] L. H. Ryder, *Quantum Field Theory*, Cambridge University Press, Cambridge, (1985).

[Schm1987] W. Schmitt, *Antipodes and incidence coalgebras*, J. of Comb. Theory A **46** (1987), 264–290.

[Schm1993] W. Schmitt, *Hopf algebras in combinatorial structures*, Canadian J. of Mathematics **45** (1993), 412–428.

[Schm1994] W. Schmitt, *Incidence Hopf Algebras*, J. of Pure and Appl. Alg. **96** (1994), 299–330.

[Schw1994] S. Schweber, *QED and the Men who Made It: Dyson, Feynman, Schwinger and Tomonoga*, Princeton: Princeton University Press, (1994).

[Sem1983] M. A. Semenov-Tian-Shansky, *What is a classical r-matrix?*, Funct. Ana. Appl. **17** no.4 (1983), 254–272.

[Sem2000] M. A. Semenov-Tian-Shansky, *Integrable Systems and Factorization Problems*, Lectures given at the "Faro International Summer School on Factorization and Integrable Systems" (Sept. 2000), Birkhäuser 2003. [arXiv:nlin.SI/0209057]

[SpDo1997] E. Spiegel and C. J. O'Donnell, *Incidence Algebras*, Dekker Incorporated, Marcel, Vol. 206, (1997).

[Spit1956] F. Spitzer, *A combinatorial lemma and its application to probability theory*, Trans. Amer. Math. Soc. **82** (1956), 323–339.

[Spit1976] F. Spitzer, *Principles of random walks*, Second edition. Graduate Texts in Mathematics, Vol. **34**. Springer-Verlag, New York-Heidelberg, (1976).

[StSh1993] S. Shnider and S. Sternberg, *Quantum groups. From coalgebras to Drinfel'd algebras. A guided tour*, Graduate Texts in Mathematical Physics, II. International Press, Cambridge, MA, (1993).

[Swe1969] M. Sweedler, *Hopf algebras*, W. A. Benjamin, Inc., New York, (1969).

[Tho1977] G. P. Thomas, *Frames, Young tableaux and Baxter sequences*, Advances in Math. **26** (1977), 275–289.

[tH1973] G. 't Hooft, *Dimensional regularization and the renormalization group*, Nucl. Phys. B **41** (1973), 455–468.

[tHVel1973] G. 't Hooft and M. J. G. Veltman, *Diagrammar*, Louvain 1973, Particle Interactions At Very High Energies, Part B*, New York 1973, 177–322 and CERN Geneva - CERN 73-9 (73,REC.OCT).

[Var1984] V. S. Varadarajan, *Lie Groups, Lie Algebras, and Their Representations*, Springer-Verlag, (1984).

[Vas2004] A. N. Vasilev, *The field theoretic renormalization group in critical behavior theory and stochastic dynamics*, Chapman & Hall/CRC, Boca Raton, FL, (2004).

[Vog1963] W. Vogel, *Die kombinatorische Lösung einer Operator-Gleichung*, Z. f. Wahrscheinlichkeitstheorie und verw. Gebiete **2** (1963), 122–134.

[Wen1962] J. G. Wendel, *A brief proof of a theorem of Baxter*, Math. Scand. **11** (1962), 107–108.

[Zim1969] W. Zimmermann, *Convergence of Bogoliubov's method of renormalization in momentum space*, Comm. in Math. Phys. **15** (1969), 208–234.

Fields Institute Communications
Volume **50**, 2007

Practicalities of Renormalizing Quantum Field Theories

J.A. Gracey
Theoretical Physics Division
Department of Mathematical Sciences
University of Liverpool
P.O. Box 147
Liverpool, L69 3BX UK
jag@amtp.liv.ac.uk

Abstract. We review the techniques used to renormalize quantum field theories at several loop orders. This includes the techniques to systematically extract the infinities in a Feynman integral and the implementation of the algorithm within computer algebra. To illustrate the method we discuss the renormalization of ϕ^4 theory and QCD including the application of the critical point large N technique as a check on the anomalous dimensions. The renormalization of non-local operators in QCD is also discussed including the derivation of the two loop correction to the Gribov mass gap equation in the Landau gauge.

1 Introduction

Our theoretical understanding of quantum phenomena in particle physics and condensed matter is guided by the underlying quantum field theory. This is formulated in terms of a Lagrangian of fields which have symmetries motivated by experimental observation. For instance, the electromagnetic field governing light leads us to the gauge principle and gauge field theories, and in particular quantum electrodynamics. The properties of this field theory have been generalized to the current theory describing all elementary particles which is the standard model. Making predictions from these quantum field theories requires one to develop the loop expansion of perturbation theory where the interactions of the fields are represented by Feynman diagrams. One of the main properties of such diagrams is that when one evaluates them they are divergent. However, as has been established for a long time there is a systematic and mathematical procedure for handling the resulting infinities which goes under the general title of renormalization. For example, a comprehensive survey of the area is provided in [1]. The topic is by nature a technical one and if one wishes to extract meaningful and accurate predictions from a quantum field theory, one needs to be able to perform the renormalization

2000 *Mathematics Subject Classification.* Primary 81T15; Secondary 70S15.

at a large number of loops. In this article we review some of the practicalities of achieving this in several field theories which are of main interest. These are scalar ϕ^4 theory which is relevant to phase transitions in various spacetime dimensions and quantum chromodynamics (QCD) which is the quantum field theory describing the quarks and gluons associated with the strong nuclear force. Our aim is to address the basics in the first part of the article where we discuss various techniques for evaluating the complicated and divergent Feynman integrals which arise in renormalizable field theories, and the way they are handled in practice by using computer algebra and symbolic manipulation packages. Since renormalization is a complicated procedure we also indicate the rudiments of the important task of how one actually verifies that a calculation is in fact correct. The second part of the article discusses recent problems of interest which apply the techniques discussed in the first part and illustrate their application from a practical point of view. These will primarily be in the context of QCD where, for instance, its three loop renormalization is discussed in a particular non-linear gauge. We will also consider the problem of renormalizing operators which have a particular degree of non-locality.

The article is organized as follows. We outline the main issues concerning renormalization in section 2 before discussing various calculational techniques in section 3. Section 4 surveys the main checks one has on the derivation of the renormalization group functions which lead on to more specific checks for QCD in the large N_f limit in section 5. We discuss more recent aspects of renormalization of QCD in section 6, before examining the issue of how one can treat a particular class of non-local operators in section 7. Finally, various concluding remarks are provided in section 8.

2 Statement of problem

We commence by summarizing from a general point of view the main issues underlying the renormalization of a quantum field theory and the main terminology of the subject used in this article. At the outset it is worth doing this with several basic field theories in mind but initially we will focus on the scalar quartic interaction, ϕ^4, in four spacetime dimensions. Later we will consider theories with gauge symmetries and in particular QCD. However, the general remarks and comments will apply equally to all field theories which are renormalizable in four or other dimensions. For ϕ^4 theory the basic Lagrangian is

$$L = \frac{1}{2} (\partial \phi_o)^2 + \frac{g_o}{4!} \phi_o^4 \tag{2.1}$$

where the subscript $_o$ denotes that the field ϕ and coupling constant g are *bare* quantities. This observation is essentially the foundation of the problem of the need for renormalization. If one were to naively compute quantum corrections to any Green's function with (2.1) as the starting point, then the resulting Feynman diagrams would be infinite in four dimensions. This is not surprising as one is using a (local) Lagrangian involving quantum fields defined at the same spacetime point. In other words the variables, parameters or fields of the initial Lagrangian can be regarded as being insufficient or not the correct variables in which to define the problem. To circumvent this, one defines a new set of (renormalized) variables by scaling the original variables with a (multiplicative) factor, known as the *renormalization constant.* For (2.1) we define these formally as $\phi_o = \sqrt{Z_\phi}\phi$ and $g_o = Z_g g$

which leads to the Lagrangian in terms of renormalized variables

$$L = \frac{1}{2} Z_\phi (\partial \phi)^2 + \frac{g}{4!} Z_\phi^2 Z_g \phi^4 . \tag{2.2}$$

This establishes the framework for renormalization. If one subsequently computes the previously infinite Green's functions with these, as yet undetermined renormalization constants, then by choosing their value appropriately the infinities can be absorbed into the renormalization constants, Z_ϕ and Z_g. As the quantum field theory is renormalizable in four spacetime dimensions then the choice one makes for say the divergent 2-point and 4-point Green's functions of ϕ^4 theory, means that *all* other n-point functions are finite. Non-renormalizable theories require additional operators in the Lagrangian over and above the original ones to retain finiteness. By contrast a superrenormalizable theory either does not require all the available renormalization constants or these actually have a finite number of terms when expanded in a coupling constant expansion. Whilst this is the general procedure, one ordinarily fixes the renormalization constants order by order in a coupling constant or perturbative expansion. Though one can still follow this prescription non-perturbatively such as in a lattice regularization.

Whilst this is the overall essence of renormalization, there are clearly several technical issues to be addressed which we now briefly discuss. The choice in redefining the variables may seem ad hoc and has little connection with the real physical world. Indeed one may not only absorb the infinities of the Green's functions but also an arbitrary finite part. How one practically removes the divergences is known as the renormalization scheme. Although there are a large number of such schemes the most widely used is the modified minimal subtraction scheme denoted by $\overline{\text{MS}}$. To ensure that the same physics emerges independently of how the infinities are absorbed, the information contained in the renormalization constants are encoded in renormalization group functions such as $\gamma_\phi(g)$ and $\beta(g)$ which appear in the renormalization group equation. These functions determine properties of the quantum theory. For instance, in field theories underlying condensed matter problems the critical points relating to the phase transitions are given by the non-trivial values of the coupling constant, g_c, where $\beta(g_c) = 0$. In particle physics applications, for example, the solution of the differential equation defined by the β-function determines how the running coupling constant depends on the renormalization scale, which is defined later. For instance, in QCD since the β-function is negative for small values of the coupling constant then the theory is asymptotically free, [2]. In other words the (confined) quarks and gluons behave at very large energies as if they were free particles.

In alluding to infinities in general terms so far, for practical calculations one must have a mathematical way of handling them and determining their nature. Moreover, for theories with a classical symmetry which is assumed to be preserved in the quantum theory, the procedure for quantifying or *regularizing* the infinities must respect the symmetries of the theory. In this context we briefly mention the algebraic renormalization technology developed primarily to determine the general consequences for the renormalization of gauge theories but also for theories with supersymmetry. For example, see [3]. In essence algebraic renormalization determines the form of the renormalization constants consistent with the Slavnov-Taylor identities. Two of the most widely used regularization procedures are lattice regularization and dimensional regularization. In the former one replaces or approximates the continuum of spacetime by a discrete set of points regularly spaced a distance a apart[1]. The presence of a is a feature of any regularization which is that every regularization introduces an arbitrary scale. For lattice regularization one wishes to take a as small as possible in order to be as close as possible to the continuum. Though in practical (and financial) terms this is very costly from the point of view of computer running time. However, the point is that whatever physical quantity one is computing the result has to be independent of the arbitrary regularization scale.

The same feature is true in the other regularization we consider. By contrast, one retains a continuum spacetime but changes the dimension away from four by an infinitesimal amount by defining the spacetime dimension as $d = 4 - 2\epsilon$ where $\epsilon \ll 1$ and $\epsilon > 0$. In this regularization the infinities in Green's functions manifest themselves as poles in ϵ and after the poles are removed by some criterion in some scheme, then the $\epsilon \to 0$ limit can be taken in a non-singular way. However, in changing the spacetime dimension an arbitrary scale is introduced by requiring that the coupling constant remains dimensionless in d-dimensions. We now take

$$g_o = \mu^{2\epsilon} Z_g g \tag{2.3}$$

where μ is the arbitrary renormalization scale. Since the Green's function after renormalization, $\Gamma^{(n)}(\mu, p, g, \ldots)$, will necessarily depend on μ in the combination p^2/μ^2 where p is a momentum, we can quantify the essence of the renormalization group as

$$\mu \frac{d}{d\mu} \Gamma^{(n)}(\mu, p, g, \ldots) = 0 \tag{2.4}$$

so that results are independent of the arbitrary scale μ.

One point of clarification needs to be made in the context of defining (dimensional) regularization which is that infinities fall into two classes. Those which are ultraviolet arising from divergences in Feynman integrals at large momenta and those which are infrared coming from divergences in integrals at low loop momenta. In massless theories dimensional regularization regularizes both types of infinities so that it is never clear where the poles in ϵ originate from. However, infrared infinities can be regularized by a non-zero mass which clearly acts as a low momentum cutoff. This mass can either be present in the original Lagrangian or introduced by hand. In the latter case when the theory is renormalized it can be smoothly set to zero. However, in gauge and supersymmetric theories such extra masses can break the symmetries and if such regularizing masses are introduced, care is required to preserve the symmetry.

We close this section by briefly summarizing several renormalization schemes which are in common use. We have alluded to $\overline{\text{MS}}$ already and we note that the original minimal subtraction scheme, MS, on which it is based requires the removal of the poles in ϵ only. (Though it is worth noting that MS and $\overline{\text{MS}}$ are not solely tied to dimensional regularization. One can minimally subtract divergences in lattice regularization.) The $\overline{\text{MS}}$ scheme is a modification of MS in that a common finite term, $4\pi e^{-\bar{\gamma}}$ where $\bar{\gamma}$ is the Euler-Mascheroni constant, is removed in addition to the poles as it was observed that in practical calculations that the convergence of perturbative expansions was improved, [4]. These two schemes fall into the class of mass independent renormalization schemes. By contrast mass dependent

[1] This is known as a regular or square lattice which is the most widely used. However, lattice regularization is not restricted to square lattices. One can define triangular or random lattices.

schemes are those where, aside from the poles in ϵ being removed certain finite parts are also removed which involve masses of the fields introduced by putting the external particle masses on-shell or setting a Green's function to a particular mass independent number at some value of the renormalization scale. Examples of such mass dependent schemes are MOM, [5], on-shell and the modified regularization (RI$'$), [6], schemes. In lattice regularization RI$'$ is one of the main choices as it is constructed in such a way as to minimize computer calculation time. In a similar way to requiring that physical quantities are independent of the renormalization scale introduced through the regularization, they also have to be independent of the renormalization scheme. However, it is possible to convert between different renormalization schemes. Indeed the $\overline{\text{MS}}$ scheme is used as the reference in this context primarily because one can compute much further in the coupling constant expansion than in a mass dependent renormalization scheme. Therefore, on the assumption that with more terms the series is more accurate, it should be closer to the real physical value of the quantity being computed. So, for example, in lattice computations which are non-perturbative but computed in the RI$'$ scheme, one has to convert the results to the $\overline{\text{MS}}$ scheme and match the large momentum behaviour to the perturbative $\overline{\text{MS}}$ result at whichever perturbative order is available. Thus the larger the number of terms in the coupling constant expansion that are available then this will reduce the error in the final numerical estimate. For certain problems this has been achieved at three and four loops, [7, 8].

3 Calculational methods

Having discussed the method of renormalization in general terms, we now detail some of the techniques used to extract regularized infinities from Feynman integrals. We will concentrate on the application of dimensional regularization to massless and massive Feynman diagrams. Throughout we will regard the integrals as expanded near four dimensions with $d = 4 - 2\epsilon$. First, we recall a simple but powerful technique to treat one loop diagrams which is based on a simple integral and attributed to Feynman. Within a Feynman graph one has products of propagators and the basic idea is to write these as an integral using

$$\frac{1}{ab} = \int_0^1 \frac{dx}{[ax + (1-x)b]^2} \tag{3.1}$$

where notionally $a = k^2 - m_1^2$ and $b = (k-p)^2 - m_2^2$ with k regarded as an internal momentum and p as external. With the two momenta within one factor, one can use Lorentz symmetry before performing the k-integration. This leaves a function of the Feynman parameter, x, which at one loop can be related to known functions. For example, when the masses m_1 and m_2 are equal the integral can be written in terms of a hypergeometric function. Whilst this is a powerful approach at one loop, it ceases to be practical at higher loops especially when there are more than one mass and external momenta. In practice one chooses a method of calculating the Feynman integral which is tailored to the overall (renormalization) problem of interest.

We will focus on one of these and consider the situation where the field theory involves particles of one non-zero mass and is renormalizable. Then to renormalize the divergent Green's functions and to extract the infinities, one expands the set of (scalar) Feynman integrals in powers of p^2 where p is the external momentum. This method is known as the vacuum bubble expansion. In regarding all integrals as scalars, we have assumed that one has first decomposed Lorentz tensor integrals into their scalar pieces. To focus on the particulars of the vacuum bubble expansion we consider the simple example of a self energy type integral in a 2-point function which gives

$$\int_k \frac{1}{[k^2 - m^2][(k-p)^2 - m^2]} = \int_k \frac{1}{[k^2 - m^2]^2} - \int_k \frac{p^2}{[k^2 - m^2]^3} + \frac{4}{d} \int_k \frac{k^2 p^2}{[k^2 - m^2]^4} + O((p^2)^2) \, . \quad (3.2)$$

In the final term the $(kp)^2$ numerator factor has been simplified by using Lorentz symmetry. The expansion truncates due to the renormalizability condition which states that since the theory is renormalizable the $O((p^2)^2)$ terms are finite since otherwise one would require a fourth derivative 2-point term which is forbidden by renormalizability. The resulting integrals on the right hand side of (3.2) are simple vacuum loops which can be evaluated. More appropriately the method can be easily extended to higher loops. For instance, in four dimensions three loop single mass scale vacuum bubbles are known to their finite part and at two loops, three mass scale vacuum graphs are also known to their finite part with respect to ϵ. See, for example, [9]. One of the benefits of this algorithm is that it can be implemented in computer algebra in an automatic way. Also the method is applicable to higher leg Green's functions if one wants to renormalize them too. However, in certain problems where the finite part is required exactly in, say, scattering problems, the more recent technique of [10] is appropriate. There each propagator is represented by a Mellin-Barnes integral and contour integration used to evaluate the two loop 4-point functions, for example, to their finite parts. Though currently results have yet to be derived for the physically relevant case of all possible massive propagator combinations.

Next, we note that one distinct advantage of the massive vacuum bubble expansion is that there are no infrared problems as the inherent mass provides a *natural* infrared regulator. For theories where there are massless fields it may seem that the vacuum bubble approach is inapplicable. However, [11], one can manufacture a fictitious mass $\bar{\mu}$ which acts as an infrared regulator via the identity

$$\frac{1}{(k-p)^2} = \frac{1}{[k^2 - \bar{\mu}^2]} + \frac{[2kp - p^2 - \bar{\mu}^2]}{(k-p)^2 [k^2 - \bar{\mu}^2]} \, . \quad (3.3)$$

Within a Feynman diagram this identity can be used repeatedly with the truncation criterion based on Weinberg's theorem, [12], for the overall finiteness of a Feynman integral. Like the completely massive case this algorithm can be implemented automatically in computer algebra. Though one ought to be aware of the potential problem of breaking an inherent symmetry of the theory when a non-zero $\bar{\mu}$ is introduced. For non-abelian gauge theories this is discussed in [11].

Another equally powerful approach for massless field theories, such as QCD, is the use of integration by parts based on the identity, in d-dimensions, [13],

$$0 = \int \frac{d^d k}{(2\pi)^d} \frac{\partial}{\partial k^\mu} \left[k^\mu I(p, k, ...) \right] \quad (3.4)$$

where $I(p, k, ...)$ is the integrand derived from the propagators and vertices. Although such identities can be used for massive integrals, the power lies in the observation that the differentiation can introduce numerator propagator factors to simplify the momentum topology. By momentum topology we mean that diagram which represents all the scalar propagators. This is not necessarily the same as the actual Feynman diagram topology itself as the cancellation of a denominator factor means that that line would be omitted in the momentum topology. This cancellation is one of the principles which underlies the MINCER algorithm, [14]. This is a package for evaluating massless three loop 2-point functions in dimensional regularization to their finite parts. Moreover, it has been encoded in the symbolic manipulation language FORM, [15, 16]. Essentially at three loops there are fourteen basic integration topologies each with its own integration by parts routine. Though these momentum topologies are not all independent. Whilst primarily used for 2-point functions, MINCER can be applied to 3-point functions when one of the external momenta is set to zero or nullified. However, in this case one must ensure that the nullification does not introduce spurious infrared singularities such as $\int_k \frac{1}{(k^2)^2}$ which result in poles in ϵ which being infrared in nature, cannot be distinguished from ultraviolet ones. Despite this technicality there are methods for handliing nullification known as infrared rearrangement, [17]. Though this has not been implemented automatically in computer algebra. The main application of MINCER is to the renormalization of four dimensional gauge theories at three loops and in particular QCD.

Given this emphasis on automatic computer algebra we make some specific remarks. The need for such machinery can be gauged from the fact that when one increases the loop order of a calculation, the number of Feynman diagrams increases almost exponentially. As examples we note that the recent full three loop renormalization of QCD in the maximal abelian gauge required the evaluation of 37322 diagrams, [18]. Also the four loop QCD β-function of [19] in a linear covariant gauge required of the order of 50000 diagrams. Clearly computers are necessary to not only implement the computational algorithm, such as MINCER or the vacuum bubble method, but also to handle the sum of the individual results. Packages such as MINCER, which have been optimized, are essential to having as short a computation time as possible. Indeed a publicly available four loop MINCER would be equally as useful. For such automatic computations the Feynman diagrams themselves need to be generated electonically and QGRAF, [20], has been developed specifically for this purpose. It has various output formats which can be readily converted to the notation used by MINCER before applying the algorithm itself. The implementation of the renormalization procedure can also be performed automatically without the need for the traditional method of subtractions. This method determines the absolute divergence of a diagram by subtracting all subgraph divergences. Instead in the automatic approach, [21], the Green's function is computed as a function of the *bare* parameters. Then the counterterms are introduced naturally (and equivalently to the subtraction method) by rescaling by the perturbatively expanded renormalization constants $g_o = Z_g g$ and so on. Once the counterterms have been implemented at a particular loop order, the divergence remaining is then that associated with the renormalization constant of that particular Green's function.

Finally, we briefly comment on other recent techniques of evaluating Feynman diagrams. One which is also appropriate to n-point functions at two loops for nonzero external momenta is the differential equation method of [22]. The basic idea is to derive a complete set of differential equations at a particular loop order for relevant momentum topologies. These are then solved with a basic set of master integrals as the boundary conditions which have to be evaluated by direct methods. For completely massless integrals in problems where conformal symmetry is present, such as at a fixed point, the method of uniqueness, [23, 24], is also powerful. Essentially when the sum of powers of three momenta in a loop integral satisfy a particular condition depending on the spacetime dimension, then the integral can simply be replaced by products of related propagators and Euler Γ-functions of the original propagator exponents. For large N_f methods, which will be discussed later, this has proved to be extremely powerful in computing information on the renormalization group functions beyond the leading large N_f order and to all orders in perturbation theory.

4 Checks

Performing a renormalization even of a simple quantum field theory to several loop orders can lead to the computation of a large number of Feynman diagrams which are determined by one or other of the methods previously discussed. However, one natural question arises in the ultimate derivation of the renormalization group functions and that is whether the results are correct. For major loop calculations in four dimensional gauge theories to check by repeating the work using independent computer algebra programmes will not only be time consuming but make a large demand in both human and computer resources, which could be used for other problems. The exception to this is the situation where one is extending existing anomalous dimensions and β-functions to the next loop order. This requires the counterterms which depend on the finite part of the Green's functions which were renormalized in the previous loop calculation but which are not ordinarily computed to construct the established renormalization constants. Therefore, one invariably has to reconstruct the previous calculation prior to tackling the new loop order. Aside from this check we now address one standard way of assessing whether the calculation one has performed is correct. By this we mean that it satisfies a set of internal consistency checks at the very least and others tailored to the problem in hand. First, the renormalization group provides a clear insight into the structure of the renormalization constants. To illustrate this, if we work from the form of the $\overline{\text{MS}}$ β-function and anomalous dimension $\gamma_\phi(g)$ given by

$$\begin{aligned} \beta(g) &= (d-4)g + Ag^2 + Bg^3 + Cg^4 + O(g^5) \\ \gamma_\phi(g) &= ag + bg^2 + cg^3 + O(g^4) \end{aligned} \tag{4.1}$$

in $d = 4 - 2\epsilon$ dimensions where g is a generic coupling constant then the corresponding renormalization constants must be of the form

$$\begin{aligned} Z_g &= 1 - \frac{Ag}{(d-4)} + \left(\frac{A^2}{(d-4)^2} - \frac{B}{2(d-4)} \right) g^2 \\ &\quad + \left[-\frac{A^3}{(d-4)^3} + \frac{7AB}{6(d-4)^2} - \frac{C}{3(d-4)} \right] g^3 + O(g^4) \\ Z_\phi &= 1 + \frac{ag}{(d-4)} + \left(\frac{(a^2 - aA)}{2(d-4)^2} + \frac{b}{2(d-4)} \right) g^2 \\ &\quad + \left[\frac{(2aA^2 - 3a^2A + a^3)}{6(d-4)^3} + \frac{(3ab - 2aB - 2bA)}{6(d-4)^2} - \frac{c}{3(d-4)} \right] g^3 \\ &\quad + O(g^4) \end{aligned} \tag{4.2}$$

to three loops. Clearly the residues of the simple poles are in a one to one correspondence with the coefficients of the renormalization group functions since the $\overline{\text{MS}}$ scheme is used. However, the residues of the other poles depend only on the structure of the *previous* loop calculations. Therefore, in a new loop calculation these poles are already predetermined and *must* emerge from the new one loop calculation. This is important in automatic calculations since one determines the full renormalization constant without using the subtraction method. Another internal check is that provided by symmetries of the original theory. Ignoring the technicalities produced by anomalies, in a gauge theory certain renormalization group functions are independent of the choice of gauge in mass independent renormalization schemes. Therefore, working in arbitrary covariant gauges means that the gauge parameter must be absent in the final result. By contrast, when renormalization group functions which depend on the gauge parameter, α, are calculated, the residues of the triple and double poles of the renormalization constants will depend on α and also be constrained by the conditions above. Briefly, in supersymmetric theories similar conditions emerge. For example, when supersymmetry is unbroken, the anomalous dimensions of the fields in the same supermultiplet have to be equal when the component field version of the Lagrangian is considered. This is a stringent check not only on the renormalization procedure but also on the regularization which must preserve the supersymmetry. Whilst this ensures the renormalization constants are checked to an extent, the residues of the simple poles are not. However, there are partial checks available in some theories from the large N_f methods developed for scalar field theories, [24, 25, 26].

This can be illustrated in the case of $O(N)$ ϕ^4 theory where the scalar field is a vector in $O(N)$. In $\overline{\text{MS}}$ in d-dimensions the β-function has the form

$$\begin{aligned}\beta(g) &= (d-4)\frac{g}{2} + [N+8]\frac{g^2}{6} - [3N+14]\frac{g^3}{6} \\ &+ \left[33N^2 + 922N + 2960 + 96(5N+22)\zeta(3)\right]\frac{g^4}{432} + O(g^5) \quad (4.3)\end{aligned}$$

where $\zeta(n)$ is the Riemann zeta function. In d-dimensions there is a non-trivial fixed point of the β-function, g_c, defined by $\beta(g_c) = 0$ known as the Wilson-Fisher fixed point. Clearly g_c will be a function of d and N. Though near four dimensions with $d = 4 - 2\epsilon$ we have $g_c = g_c(\epsilon, N)$, which can be expanded in powers of $1/N$ as $N \to \infty$ giving

$$g_c = \frac{6\epsilon}{N} + \left[-48\epsilon + 108\epsilon^2 - 99\epsilon^3 + O(\epsilon^4)\right]\frac{1}{N^2} + O\left(\frac{1}{N^3}\right). \quad (4.4)$$

Evaluating the renormalization group invariant universal critical exponent $\beta'(g_c)$ in the same limit gives

$$\beta'(g_c) = -\epsilon + \left[18\epsilon^2 - 33\epsilon^3 - \frac{5}{2}\epsilon^4 + O(\epsilon^5)\right]\frac{1}{N} + O\left(\frac{1}{N^2}\right) \quad (4.5)$$

which encodes the information of the original β-function, albeit in a different way. However, if one can compute $\beta'(g_c)$ in d-dimensions order by order in the $1/N$ expansion then the coefficients in the polynomials of N in (4.3) can be read off directly. In [24, 25, 26, 27] the d-dimensional large N technique was developed for the $O(N)$ non-linear σ model in $d = 2 + \varepsilon$ dimensions and the first three terms in the $1/N$ series for the exponents η, ν and ω have been determined in d-dimensions.

As this model is in the same universality class as $O(N)$ ϕ^4 theory at the Wilson-Fisher fixed point then one can partially check off the known coefficients in the corresponding renormalization group functions. For instance, writing the leading large N wave function anomalous dimension as

$$\gamma(g) = \sum_{r=1}^{\infty} (c_r N^2 + d_r N + e_r) N^{r-2} g^{r+1} \tag{4.6}$$

then with $U_{6,2} = \sum_{n>m>0}^{\infty} \frac{(-1)^{n-m}}{n^6 m^2}$ as the $(3,4)$ torus knot number, [28], one finds at *ten* loops and $O(1/N^3)$ that

$$\begin{aligned} e_9 = \; & [1560674304\zeta(10) - 12534896640\zeta(9) + 11070010560\zeta(8) \\ & + 1732018176\zeta(7)\zeta(3) + 581961984\zeta(7) - 3411394560\zeta(6)\zeta(3) \\ & - 2684240640\zeta(6) + 209534976\zeta^2(5) - 1567752192\zeta(5)\zeta(4) \\ & + 1754664960\zeta(5)\zeta(3) - 975533568\zeta(5) - 9289728\zeta(4)\zeta^2(3) \\ & + 1310201856\zeta(4)\zeta(3) + 1636615872\zeta(4) - 137158656\zeta^3(3) \\ & - 1708996608\zeta^2(3) + 294403968\zeta(3) \\ & - 89800704 U_{62} - 341350433]/1950396973056 \, . \end{aligned} \tag{4.7}$$

This requires not only knowledge of η at $O(1/N^3)$ but also ω at $O(1/N^2)$ as this encodes the value of g_c required for $\eta = \gamma(g_c)$. By way of illustration as to the form of such large N exponents we note that with

$$\omega = \mu - 2 + \sum_{i=1}^{\infty} \frac{\omega_i}{N^i} \tag{4.8}$$

then

$$\omega_1 = (2\mu - 1)^2 \eta_1 \tag{4.9}$$

and, [27],

$$\begin{aligned} \omega_2 = & \Bigg[- \frac{4(\mu^2 - 5\mu + 5)(2\mu - 3)^2(\mu - 1)\mu^2(\bar{\Phi}(\mu) + \bar{\Psi}^2(\mu))}{(\mu - 2)^3(\mu - 3)} \\ & - \frac{16\mu(2\mu - 3)^2}{(\mu - 2)^3(\mu - 3)^2\eta_1} \\ & - \frac{3(4\mu^5 - 48\mu^4 + 241\mu^3 - 549\mu^2 + 566\mu - 216)(\mu - 1)\mu^2\hat{\Theta}(\mu)}{2(\mu - 2)^3(\mu - 3)} \\ & - [16\mu^{10} - 240\mu^9 + 1608\mu^8 - 6316\mu^7 + 15861\mu^6 \\ & - 25804\mu^5 + 26111\mu^4 - 14508\mu^3 + 2756\mu^2 \\ & + 672\mu - 144)]/[(\mu - 2)^4(\mu - 3)^2]\bar{\Psi}(\mu) \\ & + [144\mu^{14} - 2816\mu^{13} + 24792\mu^{12} - 130032\mu^{11} + 452961\mu^{10} \\ & - 1105060\mu^9 + 1936168\mu^8 - 2447910\mu^7 + 2194071\mu^6 \\ & - 1320318\mu^5 + 460364\mu^4 - 43444\mu^3 - 26280\mu^2 \\ & + 8208\mu - 864]/[2(2\mu - 3)(\mu - 1)(\mu - 2)^5(\mu - 3)^2\mu] \Bigg] \eta_1^2 \end{aligned} \tag{4.10}$$

where $\eta_1 = - 4\Gamma(2\mu - 2)/[\Gamma(2 - \mu)\Gamma(\mu - 1)\Gamma(\mu - 2)\Gamma(\mu + 1)]$ and we have set $d = \mu/2$. The expression for ω_2 contains derivatives of the Γ-function which are denoted by

$$\begin{aligned}\bar{\Psi}(\mu) &= \psi(2\mu-3) + \psi(3-\mu) - \psi(\mu-1) - \psi(1) \\ \hat{\Theta}(\mu) &= \psi'(\mu-1) - \psi'(1) \\ \bar{\Phi}(\mu) &= \psi'(2\mu-3) - \psi'(3-\mu) - \psi'(\mu-1) + \psi'(1) \, . \end{aligned} \tag{4.11}$$

Whilst $O(N)$ ϕ^4 theory has been examined to several orders in $1/N$, the same technique has been developed for QCD in [29, 30]. There the expansion is in terms of the number of quark flavours, N_f, and *not* N_c which is the number of colours. The latter $1/N_c$ expansion deals with the structure of QCD from a completely different point of view.

5 Large N_f QCD

Having outlined the critical point approach for $O(N)$ ϕ^4 theory, we now concentrate on how one develops the same formalism for QCD in the large N_f limit. Examining the d-dimensional $\overline{\text{MS}}$ β-function for QCD, one finds that there is an equivalent Wilson-Fisher fixed point which can be accessed in powers of $1/N_f$ as a function of ϵ. However, to compute critical exponents in QCD in d-dimensions and relate them to the associated anomalous dimensions in perturbation theory at whatever order they are available, one does not compute with the QCD Lagrangian itself. Instead one exploits the properties of the d-dimensional fixed point by realising that at this fixed point QCD is in the same universality class as the non-abelian Thirring model (NATM), [31]. This is a four-fermi theory which is renormalizable in two dimensions and plays the role of the non-linear σ model in the previous ϕ^4 theory example. The Lagrangian of the NATM is

$$L^{\text{NATM}} = i\bar{\psi}^{iI}\partial\!\!\!/\psi^{iI} + \frac{\lambda^2}{2}\left(\bar{\psi}^{iI}T^a_{IJ}\gamma^\mu\psi^{iJ}\right)^2 \tag{5.1}$$

where λ is the coupling constant of the NATM and is dimensionless in two dimensions. Introducing an *auxiliary* spin-1 field the Lagrangian can be written as

$$L^{\text{NATM}} = i\bar{\psi}^{iI}\partial\!\!\!/\psi^{iI} + A^a_\mu\bar{\psi}^{iI}T^a_{IJ}\gamma^\mu\psi^{iJ} - \frac{A^{a\,2}_\mu}{2\lambda} \, . \tag{5.2}$$

Clearly there is no field strength term in the non-abelian Thirring model and from a critical point of view this is due to the fact that at this infrared stable fixed point that operator is irrelevant. The critical behaviour is driven by the common quark gluon vertex when one compares with the QCD Lagrangian

$$L^{\text{QCD}} = -\frac{1}{4}G^a_{\mu\nu}G^{a\,\mu\nu} - \frac{1}{2\tilde{\alpha}}(\partial^\mu A^a_\mu)^2 - \bar{c}^a\partial^\mu D_\mu c^a + i\bar{\psi}^{iI}D\!\!\!\!/\,\psi^{iI} \tag{5.3}$$

where in this section we use $\tilde{\alpha}$ as the covariant gauge parameter with $\tilde{\alpha} = 0$ corresponding to the Landau gauge. Within the large N_f critical point formalism the triple and quartic gluon vertices are embedded inside quark loops with respectively three or four external auxiliary spin-1 fields, [31].

As an example of the large N_f critical point formalism, the $O(1/N_f)$ correction to the β-function can be computed by considering the critical point structure of the gluon 2-point function with the insertion of the operator $G^a_{\mu\nu}G^{a\,\mu\nu}$, [29]. The anomalous dimension of this operator will be related to the critical exponent of the associated coupling constant which is clearly the QCD β-function when expanded in powers of ϵ near four dimensions since $\omega = -\,\beta'(g_c)$. At criticality the propagators of the fields of the non-abelian Thirring model or QCD Lagrangian obey asymptotic

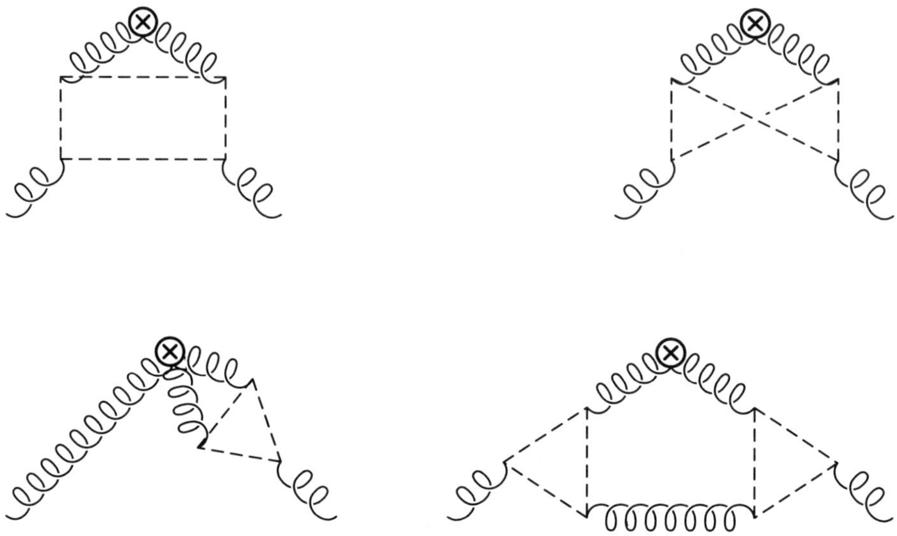

Figure 1 Feynman diagrams for $O(1/N_f)$ correction to ω.

scaling forms where the canonical exponents are derived from the dimensionality of each term in the d-dimensional Lagrangian. Thus we have, in the Landau gauge, [29, 30],

$$\psi(k) \sim \frac{A\not{k}}{(k^2)^{\mu-\alpha}} \; , \; A_{\mu\nu}(k) \sim \frac{B}{(k^2)^{\mu-\beta}} \left[\eta_{\mu\nu} - \frac{k_\mu k_\nu}{k^2} \right] \tag{5.4}$$

where A and B are momentum independent amplitudes and

$$\alpha = \mu - 1 + \tfrac{1}{2}\eta \; , \; \beta = 1 - \eta - \chi \; . \tag{5.5}$$

The exponents η and χ are respectively the quark anomalous dimension and the quark gluon vertex anomalous dimension at criticality. In large N_f there are four diagrams contributing to the gluon 2-point function with the operator inserted which are illustrated in Figure 1. The three loop diagram is included since it is of the same order in large N_f as the two loop ones because in the large N_f counting one regards the amplitudes as $A = O(1)$ and $B = O(1/N_f)$. These diagrams are infinite but can be regularized by shifting β to $\beta - \Delta$ where Δ is the regularizing parameter, [24, 25, 26]. It plays a role akin to ϵ in dimensional regularization and it is important to appreciate that within the critical point large N_f method ϵ is not the regularization. The calculations are performed in *fixed* d-dimensions. After subtraction of the poles in Δ, using the renormalization procedure of [32], the scaling behaviour of the remaining finite Green's function determines ω and we have, [29],

$$\begin{aligned} \omega &= (\mu - 2) - \Big[(2\mu-3)(\mu-3)C_F \\ &\quad - \frac{(4\mu^4 - 18\mu^3 + 44\mu^2 - 45\mu + 14)C_A}{4(2\mu-1)(\mu-1)} \Big] \frac{\eta_0}{T(R)N_f} \end{aligned} \tag{5.6}$$

where $\eta_0 = (2\mu-1)(\mu-2)\Gamma(2\mu)/[4\Gamma^2(\mu)\Gamma(\mu+1)\Gamma(2-\mu)]$. The factors deriving from the group theory are defined by

$$T^aT^a = C_F I \; , \; f^{acd}f^{bcd} = C_A\delta^{ab} \; , \; \mathrm{Tr}\left(T^aT^b\right) = T_F\delta^{ab} \; . \tag{5.7}$$

Consequently one can verify that the known leading $1/N_f$ coefficients agree with the $\overline{\text{MS}}$ four loop β-function of [19]. Moreover, the unknown higher loop coefficients at $O(1/N_f)$ are encoded in (5.6). For instance, if we write the leading large N_f part of the QCD β-function as

$$\beta(g) = \beta_0 g^2 + \sum_{n=1}^{\infty} a_{n+1} [T_F N_f]^n g^{n+2} \tag{5.8}$$

then at *five* loops we have

$$a_5 = [(288\zeta(3) + 214)C_F + (480\zeta(3) - 229)C_A]/31104 . \tag{5.9}$$

For $O(N)$ ϕ^4 theory, it was possible with the critical point method to go to several orders beyond the leading $1/N$ order. For QCD the same is possible, in principle, but one has the added complication of the Lorentz structure associated with fermions and the fact that the structure of the quark gluon vertex is such that the method of uniqueness cannot be applied immediately. Instead one has to use integration by parts and other methods to produce vertices which satisfy the uniqueness condition before that integration technique can be used. Nevertheless the quark anomalous dimension and quark mass anomalous dimensions have been determined at $O(1/N_f^2)$. For the quark anomalous dimension in arbitrary covariant gauge we have, [33],

$$\eta_1 = \frac{[(2\mu - 1)(\mu - 2) + \tilde{\alpha}\mu]C_F \eta_0}{[(2\mu - 1)(\mu - 2)T_F]} \tag{5.10}$$

and

$$\begin{aligned}
\eta_2 = \Big[& - \mu(2\mu^2 + \mu\tilde{\alpha} - 5\mu + 2)(\mu - 1) \left[\bar{\Psi}^2(\mu) + \bar{\Phi}(\mu)\right] C_A \\
& + (8\mu^5 - 92\mu^4 + 270\mu^3 - 301\mu^2 + 124\mu - 12)((2\mu - 1)(\mu - 2) + \mu\tilde{\alpha}) \\
& \times \frac{\bar{\Psi}(\mu) C_A}{2(2\mu - 1)(2\mu - 3)(\mu - 2)} \\
& + 3\mu(\mu - 1)[\mu\tilde{\alpha}C_A + 2(2\mu^2 + \mu\tilde{\alpha} - 5\mu + 2)C_F]\hat{\Theta}(\mu) \\
& - [(32\mu^7\tilde{\alpha} - 96\mu^7 + 8\mu^6\tilde{\alpha}^2 - 224\mu^6\tilde{\alpha} + 912\mu^6 - 4\mu^5\tilde{\alpha}^3 - 84\mu^5\tilde{\alpha}^2 \\
& + 704\mu^5\tilde{\alpha} - 3360\mu^5 + 16\mu^4\tilde{\alpha}^3 + 278\mu^4\tilde{\alpha}^2 - 1124\mu^4\tilde{\alpha} \\
& + 6240\mu^4 - 19\mu^3\tilde{\alpha}^3 - 387\mu^3\tilde{\alpha}^2 + 846\mu^3\tilde{\alpha} - 6292\mu^3 \\
& + 6\mu^2\tilde{\alpha}^3 + 230\mu^2\tilde{\alpha}^2 - 222\mu^2\tilde{\alpha} + 3416\mu^2 - 48\mu\tilde{\alpha}^2 \\
& - 4\mu\tilde{\alpha} - 908\mu + 88)C_A\mu \\
& - 8(4\mu^5 + 4\mu^4\tilde{\alpha} - 32\mu^4 - 13\mu^3\tilde{\alpha} + 75\mu^3 + 8\mu^2\tilde{\alpha} - 70\mu^2 - 2\mu\tilde{\alpha} \\
& + 32\mu - 6)(2\mu - 1)(2\mu - 3)(\mu - 2)C_F] \\
& /[4(2\mu - 1)(2\mu - 3)(\mu - 1)(\mu - 2)\mu] \Big] \frac{C_F \eta_0^2}{2(2\mu - 1)^2(\mu - 2)^2 T_F^2}
\end{aligned} \tag{5.11}$$

where

$$\eta = \sum_{i=1}^{\infty} \frac{\eta_i}{N_f^i} . \tag{5.12}$$

The expression for the quark mass anomalous dimension is independent of $\tilde{\alpha}$ and in the same notation,

$$\eta_{\bar{\psi}\psi,1} = - \frac{2C_F \eta_0}{(\mu - 2)T_F} \tag{5.13}$$

and, [33],

$$\begin{aligned}\eta_{\bar{\psi}\psi,2} &= -\Bigg[2\left(3\hat{\Theta}(\mu) + \frac{(4\mu^3 - 13\mu^2 + 9\mu - 3)}{\mu^2(\mu-1)^2}\right) C_F \\ &+ \Bigg(\frac{(8\mu^5 - 92\mu^4 + 270\mu^3 - 301\mu^2 + 124\mu - 12)\bar{\Psi}(\mu)}{2\mu(\mu-1)(2\mu-1)(2\mu-3)(\mu-2)} \\ &- [16\mu^6 - 128\mu^5 + 480\mu^4 - 900\mu^3 + 831\mu^2 \\ &- 344\mu + 44)]/[4\mu(2\mu-1)(2\mu-3)(\mu-1)^2(\mu-2)] \\ &- \bar{\Psi}^2(\mu) - \bar{\Phi}(\mu)\Bigg) C_A\Bigg] \frac{\mu(\mu-1)(2\mu-1)C_F\eta_0^2}{(2\mu-1)(\mu-2)^2T_F^2} . \qquad (5.14)\end{aligned}$$

Having focused on the operator associated with the coupling constant of QCD, one can also repeat the same analysis for the analogous operator of the non-abelian Thirring model which is the dimension two gluon mass operator. Whilst it clearly is not a gauge invariant operator its anomalous dimension has an interesting structure. Inserting $\frac{1}{2}A^a_\mu A^{a\,\mu}$ into the same Green's function as $G^a_{\mu\nu}G^{a\,\mu\nu}$, the $O(1/N_f)$ exponent in the Landau gauge is, [34],

$$\eta_{A^2} = -\frac{C_A\eta_1^{\rm O}}{4(\mu-2)T_FN_f} + O\left(\frac{1}{N_f^2}\right) . \qquad (5.15)$$

Interestingly this can be rewritten as

$$\gamma_{A^2}(g_c) = \gamma_A(g_c) + \gamma_c(g_c) \qquad (5.16)$$

in *all* dimensions d. It turns out that this is a general property of this operator in the Landau gauge. In [35] it was shown that in QCD

$$\gamma_{A^2}(g) = \gamma_A(g) + \gamma_c(g) \qquad (5.17)$$

to all orders in perturbation theory having first been shown at three loops in $\overline{\rm MS}$ by explicit computation, [36]. The explicit four loop value is also now available in $SU(N_c)$, [37]. Although this operator is gauge variant, it has been the subject of intense study in recent years as an effective gluon mass term. See, for example, [38]. Moreover, it has been studied in other gauges such as the maximal abelian gauge (MAG), [39, 40, 18]. In this latter gauge the gauge field is written in terms of its diagonal (centre) and off-diagonal fields

$$A^A_\mu T^A = A^a_\mu T^a + A^i_\mu T^i \qquad (5.18)$$

where $[T^i, T^j] = 0$, $T^i \in \{\text{centre}\}$, $1 \leq i \leq N^d_A$, $1 \leq a \leq N^o_A$ and $1 \leq A \leq N_A$ with N^d_A and N^o_A the respective dimensions of the centre of the group and its complement. (Here and in the next section we use the index A to denote the whole colour group, using the notation usually used in discussing the maximal abelian gauge.) Then in this gauge it turns out that the off-diagonal BRST invariant mass operator

$$O = \tfrac{1}{2}A^a_\mu A^{a\,\mu} + \alpha\bar{c}^a c^a \qquad (5.19)$$

satisfies a Slavnov-Taylor identity, similar to that for the analogous operator in the Landau gauge, which is, [39],

$$\gamma_O(g) = \gamma_{A^i}(g) - \gamma_{c^i}(g) \qquad (5.20)$$

where c^i is the ghost in the centre of the colour group. Whilst this identity has been established on general grounds, for practical calculations, such as studying the condensation of the operator O of (5.19), one needs the *explicit* values of the anomalous dimensions. This requires the application of much of the earlier general discussion on renormalization and the maximal abelian gauge provides a comprehensive example to illustrate these general remarks.

6 QCD in non-linear gauges

Given this interest in studying QCD in covariant gauges other than the canonical linear ones, we now focus on the explicit renormalization in the MAG. First, we define the gauge fixing for the maximal abelian gauge by recalling that for the Landau gauge. With (5.18) the gauge fixing is given by

$$L_{\mathrm{gf}}^{\mathrm{Landau}} = \delta\bar{\delta}\left[\frac{1}{2}A^A_\mu A^{A\,\mu} + \frac{1}{2}\alpha\bar{c}^A c^A\right] . \tag{6.1}$$

By contrast, the maximal abelian gauge is such that one minimizes the quadratic form in the off-diagonal sector only by fixing, [39],

$$L_{\mathrm{gf}}^{\mathrm{MAG}} = \delta\bar{\delta}\left[\frac{1}{2}A^a_\mu A^{a\,\mu} + \frac{1}{2}\alpha\bar{c}^a c^a\right] + \delta\left[\bar{c}^i\partial^\mu A^i_\mu\right] \tag{6.2}$$

where again δ and $\bar{\delta}$ are the respective BRST and anti-BRST variations with

$$\begin{aligned}
\delta A^a_\mu &= -\left(\partial_\mu c^a + g f^{ajc} A^j_\mu c^c + g f^{abc} A^b_\mu c^c + g f^{abk} A^b_\mu c^k\right) \\
\delta c^a &= g f^{abk} c^b c^k + \frac{1}{2} f^{abc} c^b c^c \\
\delta \bar{c}^a &= b^a \ , \ \delta b^a = 0 \\
\delta A^i_\mu &= -\left(\partial_\mu c^i + g f^{ibc} A^b_\mu c^c\right) \\
\delta c^i &= \frac{1}{2} g f^{ibc} c^b c^c \\
\delta \bar{c}^i &= b^i \ , \ \delta b^i = 0
\end{aligned} \tag{6.3}$$

and

$$\begin{aligned}
\bar{\delta} A^a_\mu &= -\left(\partial_\mu c^a + g f^{ajc} A^j_\mu c^c + g f^{abc} A^b_\mu c^c + g f^{abk} A^b_\mu c^k\right) \\
\bar{\delta} c^a &= -b^a + g f^{abc} c^b \bar{c}^c + g f^{abk} c^b \bar{c}^k + g f^{abk} \bar{c}^b c^k \\
\bar{\delta} \bar{c}^a &= g f^{abk} \bar{c}^b \bar{c}^k + \frac{1}{2} g f^{abc} \bar{c}^b \bar{c}^c \\
\bar{\delta} b^a &= -g f^{abc} b^b \bar{c}^c - g f^{abk} b^b \bar{c}^k + g f^{abk} \bar{c}^b b^k \\
\bar{\delta} A^i_\mu &= -\left(\partial_\mu \bar{c}^i + g f^{ibc} A^b_\mu \bar{c}^c\right) \ , \ \bar{\delta} c^i = -b^i + g f^{ibc} c^b \bar{c}^c \\
\bar{\delta} \bar{c}^i &= \frac{1}{2} g f^{ibc} \bar{c}^b \bar{c}^c \ , \ \bar{\delta} b^i = -g f^{ibc} b^b \bar{c}^c .
\end{aligned} \tag{6.4}$$

The final term in (6.2) is required to avoid a singular ghost propagator. The residual gauge freedom in the diagonal sector is fixed by applying the usual Landau gauge. Though one can introduce an extra gauge parameter, $\bar{\alpha}$, for the inversion to obtain a centre gluon propagator. Varying δ and $\bar{\delta}$ gives contributions to the interaction

Lagrangian with the remaining terms derived from $(G^A_{\mu\nu})^2$. Thus,

$$\begin{aligned}
L_{\mathrm{gf}} &= -\frac{1}{2\alpha}\left(\partial^\mu A^a_\mu\right)^2 - \frac{1}{2\bar{\alpha}}\left(\partial^\mu A^i_\mu\right)^2 + \bar{c}^a\partial^\mu\partial_\mu c^a + \bar{c}^i\partial^\mu\partial_\mu c^i \\
&+ \Big[f^{abk} A^a_\mu \bar{c}^k \partial^\mu c^b - f^{abc} A^a_\mu \bar{c}^b \partial^\mu c^c - \frac{1}{\alpha} f^{abk} \partial^\mu A^a_\mu A^b_\nu A^{k\,\nu} \\
&- f^{abk}\partial^\mu A^a_\mu c^b \bar{c}^k - \frac{1}{2} f^{abc}\partial^\mu A^a_\mu \bar{c}^b c^c - 2 f^{abk} A^k_\mu \bar{c}^a \partial^\mu \bar{c}^b - f^{abk}\partial^\mu A^k_\mu \bar{c}^b c^c \Big] g \\
&+ \Big[f_d^{acbd} A^a_\mu A^{b\,\mu} \bar{c}^c c^d - \frac{1}{2\alpha} f_o^{akbl} A^a_\mu A^{b\,\mu} A^k_\nu A^{l\,\nu} + f_o^{adcj} A^a_\mu A^{j\,\mu} \bar{c}^c c^d \\
&- \frac{1}{2} f_o^{ajcd} A^a_\mu A^{j\,\mu} \bar{c}^c c^d + f_o^{ajcl} A^a_\mu A^{j\,\mu} \bar{c}^c c^l + f_o^{alcj} A^a_\mu A^{j\,\mu} \bar{c}^c c^l \\
&- f_o^{cjdi} A^i_\mu A^{j\,\mu} \bar{c}^c c^d - \frac{\alpha}{4} f_d^{abcd} \bar{c}^a \bar{c}^b c^c c^d - \frac{\alpha}{8} f_o^{abcd} \bar{c}^a \bar{c}^b c^c c^d \\
&+ \frac{\alpha}{8} f_o^{acbd} \bar{c}^a \bar{c}^b c^c c^d - \frac{\alpha}{4} f_o^{abcl} \bar{c}^a \bar{c}^b c^c c^l + \frac{\alpha}{4} f_o^{acbl} \bar{c}^a \bar{c}^b c^c c^l \\
&- \frac{\alpha}{4} f_o^{albc} \bar{c}^a \bar{c}^b c^c c^l + \frac{\alpha}{2} f_o^{akbl} \bar{c}^a \bar{c}^b c^k c^l \Big] g^2
\end{aligned} \tag{6.5}$$

where $f_d^{ABCD} = f^{ABi} f^{CDi}$ and $f_o^{ABCD} = f^{ABe} f^{CDe}$. Although this is a much more involved Lagrangian than that which results in the Landau gauge where there are a handful of interactions, it can be shown that it contains the usual covariant gauge fixed Lagrangian, [39]. Moreover, in the limit where the fields deriving from the centre of the colour group are set to zero one recovers the non-linear Curci-Ferrari gauge, [41]. In other words the off-diagonal sector of the maximal abelian gauge corresponds to QCD fixed in the Curci-Ferrari gauge. Clearly the Lagrangian includes quartic ghost interactions which are always a feature of a non-linear gauge fixing. Though they do not destroy the renormalizability of the maximal abelian gauge Lagrangian which has been established by the algebraic renormalization technology, [39, 40]. This is crucial to formulating the renormalization of QCD in the maximal abelian gauge as it produces the Slavnov-Taylor identities originating from the BRST transformations. More importantly it determines the *form* of the renormalization of the fields and parameters as

$$\begin{aligned}
A^{a\,\mu}_{\mathrm{o}} &= \sqrt{Z_A}\, A^{a\,\mu} \;,\; A^{i\,\mu}_{\mathrm{o}} = \sqrt{Z_{A^i}}\, A^{i\,\mu} \\
c^a_{\mathrm{o}} &= \sqrt{Z_c}\, c^a \;,\; \bar{c}^a_{\mathrm{o}} = \sqrt{Z_c}\, \bar{c}^a \\
c^i_{\mathrm{o}} &= \sqrt{Z_{c^i}}\, c^i \;,\; \bar{c}^i_{\mathrm{o}} = \frac{\bar{c}^i}{\sqrt{Z_{c^i}}} \\
\psi_{\mathrm{o}} &= \sqrt{Z_\psi}\psi \;,\; g_{\mathrm{o}} = \mu^\epsilon Z_g\, g \\
\alpha_{\mathrm{o}} &= Z_\alpha^{-1} Z_A\, \alpha \;,\; \bar{\alpha}_{\mathrm{o}} = Z_{\alpha^i}^{-1} Z_{A^i}\, \bar{\alpha} \,.
\end{aligned} \tag{6.6}$$

Consequently, one observes that the diagonal ghost 2-point function is finite. If this had not been deduced via algebraic renormalization, then in an explicit renormalization of the diagonal ghost 2-point function one would have erroneously deduced that Z_{c^i} was unity. Instead to determine the correct Z_{c^i} one has to renormalize a Green's function containing one Z_{c^i} where all the other associated renormalization constants are known. For this case one choice would be the $A^a_\mu c^b \bar{c}^i$ vertex renormalization. Another useful feature is that the diagonal gluon wave function renormalization constant is not independent but is in fact related to that of the coupling constant. Hence, the β-function can be determined without resort to the renormalization of a vertex function. Moreover, this simplification is similar to what happens in QCD fixed in the background field gauge, [42]. Though in practical calculations it reduces substantially the number of diagrams to be determined as well as computation time which can be significant at high loop orders. The key point is that the preliminary analysis from algebraic renormalization not only determines the structure of the renormalization constants consistent with the underlying symmetry, but also provides an efficient route for computing the anomalous dimensions themselves.

Given that the maximal abelian gauge gluon mass operator does not renormalize independently, to have its explicit anomalous dimension requires only knowledge of the diagonal ghost anomalous dimension. Clearly given the large number of interactions in (6.5), a three loop renormalization can only proceed with an automatic computer algebra approach for which the MINCER algorithm is the ideal tool, using the FORM version, [16], and QGRAF, [20]. In completing the full three loop maximal abelian gauge renormalization of QCD, 37322 Feynman diagrams had to be considered, [18]. Though not all of these are non-zero. Some vanish trivially by the group theory structure of a diagram. For instance, where one has a self-energy insertion in a propagator line with a diagonal and off-diagonal field as external to this subgraph, then it will vanish due to

$$f^{ijk} = 0 \,, \quad f^{ijc} = 0 \,, \quad f^{ibc} \neq 0 \,, \quad f^{abc} \neq 0 \,. \tag{6.7}$$

The result of the renormalization procedure yields the anomalous dimensions, [18], which have the structure illustrated by the example,

$$\begin{aligned}
\gamma_O(a) &= \frac{1}{12N^o_A} \left[N^o_A \left((-\,3\alpha + 35) C_A - 16 T_f N_f \right) + N^d_A \left((-\,6\alpha - 18) C_A \right) \right] a \\
&\quad + \frac{1}{96 N^{o\,2}_A} \Big[N^{o\,2}_A \left((-\,6\alpha^2 - 66\alpha + 898) C_A^2 - 560 C_A T_f N_f \right. \\
&\quad \left. -\,384 C_F T_f N_f \right) + N^o_A N^d_A \left((-\,54\alpha^2 - 354\alpha - 323) C_A^2 + 160 C_A T_f N_f \right) \\
&\quad + N^{d\,2}_A \left((-\,60\alpha^2 - 372\alpha + 510) C_A^2 \right) \Big] a^2 \\
&\quad + \frac{1}{6912 N^{o\,3}_A} \Big[N^{o\,3}_A ((-\,162\alpha^3 - 2727\alpha^2 - 2592\zeta(3)\alpha - 18036\alpha \\
&\quad -\,1944\zeta(3) + 302428) C_A^3 + (6912\alpha + 62208\zeta(3) - 356032) C_A^2 T_F N_f \\
&\quad + (-\,82944\zeta(3) - 79680) C_A C_F T_F N_f + 49408 C_A T_F^2 N_f^2 \\
&\quad + 13824 C_F^2 T_F N_f + 33792 C_F T_F^2 N_f^2) \\
&\quad + N^{o\,2}_A N^d_A ((-\,2754\alpha^3 + 648\alpha^2\zeta(3) - 28917\alpha^2 - 4212\alpha\zeta(3) - 69309\alpha \\
&\quad + 37260\zeta(3) - 64544) C_A^3 + (25488\alpha + 103680\zeta(3) \\
&\quad -\,13072) C_A^2 T_F N_f + (-\,165888\zeta(3) + 155520) C_A C_F T_F N_f \\
&\quad + 17920 C_A T_F^2 N_f^2) + N^o_A N^{d\,2}_A ((-\,7884\alpha^3 + 22680\alpha^2\zeta(3) - 84564\alpha^2 \\
&\quad + 97524\alpha\zeta(3) - 47142\alpha + 433836\zeta(3) - 56430) C_A^3 \\
&\quad + (25056\alpha - 124416\zeta(3) - 18144) C_A^2 T_F N_f) \\
&\quad + N^{d\,3}_A ((-\,6480\alpha^3 + 34992\alpha^2\zeta(3) - 70092\alpha^2 + 8424\alpha\zeta(3) \\
&\quad + 114912\alpha + 77112\zeta(3) - 161028) C_A^3) \Big] a^3 \; + \; O(a^4)
\end{aligned} \tag{6.8}$$

where $a = g^2/(16\pi^2)$ is conventionally used as the coupling constant. To be confident that the anomalous dimensions are in fact correct, the standard internal renormalization group checks we discussed above have been shown to hold. Moreover, since the β-function is a gauge independent object then the cancellation of α and the parameters N_A^d and N_A^o, provided not only a strong check on the calculation but on the group theory FORM module which handles the consequences of splitting the gauge group into two sectors. Additional checks for this specific computation are provided by noting that the anomalous dimensions of the off-diagonal fields are correctly equivalent for all α to those of QCD fixed in the non-linear Curci-Ferrari gauge, which was introduced in [41], in the limit $N_A^d/N_A^o \to 0$ where the explicit results were given in [43, 36].

7 Renormalization and non-locality

We now turn to more recent aspects of renormalization and that is the renormalization of quantum field theories where a non-locality is present. The general properties of renormalization theory as discussed, for example, in [44], is based on the assumption that the Lagrangian is local. In other words there are no non-local interactions or operators. However, in certain field theories of interest, such as QCD, one encounters important operators which are non-local but whose properties require investigation. We now summarize the status of two such studies where although there is a degree of non-locality present, it falls into a class which does not prevent calculations from being performed. By this we mean that the non-local operators can be rewritten in terms of a *finite* number of local fields and operators. We refer to this as a localizable non-locality. By contrast, there are operators which, whilst being non-local, do not allow for a *finite* number of auxiliary fields to lead to a local Lagrangian. We term this a non-localizable non-locality.

The first example is that of the Gribov problem which relates to the impossibility of *globally* fixing the gauge in a non-abelian gauge theory. This was first pointed out by Gribov, [45], who proceeded to construct a path integral to study the problem. There the non-abelian gauge field is restricted to the region defined by the first zero of the Faddeev-Popov operator. This region, known as the Gribov volume, is of finite size and characterised by the Gribov mass, γ. The resulting path integral, [45],

$$Z = \int DA \, \delta(\partial^\mu A^a_\mu) \det\left(-\partial^\nu D^a_\nu\right) \, e^{-S} \tag{7.1}$$

essentially leads to a non-local Lagrangian

$$L = -\frac{1}{4} G^a_{\mu\nu} G^{a\,\mu\nu} + \frac{C_A \gamma^4}{2} A^a_\mu \frac{1}{\partial^\nu D_\nu} A^{a\,\mu} - \frac{d N_A \gamma^4}{2g^2} \,. \tag{7.2}$$

The parameter γ is not independent and satisfies the Gribov gap equation which is, at one loop, [45],

$$1 = C_A \left[\frac{5}{8} - \frac{3}{8} \ln \left(\frac{C_A \gamma^4}{\mu^4} \right) \right] a + O(a^2) \,. \tag{7.3}$$

Subsequently, in [46, 47] Zwanziger managed to localize the original Gribov path integral by introducing a set of extra ghost fields $\{\phi^{ab}_\mu, \bar{\phi}^{ab}_\mu, \omega^{ab}_\mu, \bar{\omega}^{ab}_\mu\}$ where the last two fields are anti-commuting. This led to the Gribov-Zwanziger Lagrangian in the

Landau gauge, [46, 47],

$$\begin{aligned} L^{\mathrm{GZ}} &= L^{\mathrm{QCD}} + \bar{\phi}^{ab\,\mu}\partial^\nu \left(D_\nu\phi_\mu\right)^{ab} - \bar{\omega}^{ab\,\mu}\partial^\nu \left(D_\nu\omega_\mu\right)^{ab} \\ &- g f^{abc}\partial^\nu\bar{\omega}^{ae}_\mu \left(D_\nu c\right)^b \phi^{ec\,\mu} \\ &- \frac{\gamma^2}{\sqrt{2}}\left(f^{abc}A^{a\,\mu}\phi^{bc}_\mu + f^{abc}A^{a\,\mu}\bar{\phi}^{bc}_\mu\right) + \frac{dN_A\gamma^4}{2g^2} \end{aligned} \tag{7.4}$$

which clearly involves only a finite number of local interactions. Not only is the Lagrangian local, it is also renormalizable, [48, 49], which implies that one can perform loop computations and study the implications the Gribov parameter has on the infrared structure of QCD. The renormalization structure of (7.4) is interesting in that in the Landau gauge the renormalization constants of the extra ghost fields are not independent with, [48, 49],

$$Z_\phi = Z_\omega = Z_c \, . \tag{7.5}$$

Moreover, the renormalization of the Gribov parameter is not independent in the Landau gauge, satisfying, [49],

$$\gamma_\gamma(a) = \frac{1}{4}\left[\gamma_A(a) + \gamma_c(a)\right] \tag{7.6}$$

which is similar to the renormalization of $\frac{1}{2}A^a_\mu A^{a\,\mu}$. Moreover, the quark, gluon, Faddeev-Popov ghost anomalous dimensions and the β-function are unaltered by the presence of γ and the extra Zwanziger ghost fields in $\overline{\mathrm{MS}}$. In a general linear covariant gauge one has the additional two loop $\overline{\mathrm{MS}}$ results for the anomalous dimensions that

$$\begin{aligned} \gamma_\phi(a) &= \gamma_\omega(a) = \gamma_c(a) \\ \gamma_\gamma(a) &= (16T_FN_f - (35+3\alpha)C_A)\frac{a}{48} \\ &+ (192C_FT_FN_f + 280C_AT_FN_f - (449-3\alpha))\frac{a^2}{192} + O(a^3) \end{aligned} \tag{7.7}$$

which was deduced using MINCER.

Equipped with these properties one can compute the corrections to the one loop mass gap equation which Gribov originally derived. This corresponds to the horizon condition definition of [47], which is equivalent to

$$f^{abc}\langle A^{a\,\mu}(x)\phi^{bc}_\mu(x)\rangle = \frac{dN_A\gamma^2}{\sqrt{2}g^2} \tag{7.8}$$

which can be evaluated order by order in perturbation theory using the propagators, [46, 47, 50],

$$\begin{aligned} \langle A^a_\mu(p)A^b_\nu(-p)\rangle &= -\frac{\delta^{ab}p^2}{[(p^2)^2 + C_A\gamma^4]}P_{\mu\nu}(p) \\ \langle A^a_\mu(p)\bar{\phi}^{bc}_\nu(-p)\rangle &= -\frac{f^{abc}\gamma^2}{\sqrt{2}[(p^2)^2 + C_A\gamma^4]}P_{\mu\nu}(p) \\ \langle \phi^{ab}_\mu(p)\bar{\phi}^{cd}_\nu(-p)\rangle &= -\frac{\delta^{ac}\delta^{bd}}{p^2}\eta_{\mu\nu} + \frac{f^{abe}f^{cde}\gamma^4}{p^2[(p^2)^2 + C_A\gamma^4]}P_{\mu\nu}(p) \end{aligned} \tag{7.9}$$

where the gluon propagator is suppressed in the infrared. Hence, at two loops in $\overline{\text{MS}}$, [50],

$$
\begin{aligned}
1 &= C_A \left[\frac{5}{8} - \frac{3}{8} \ln \left(\frac{C_A \gamma^4}{\mu^4} \right) \right] a \\
&+ \left[C_A^2 \left(\frac{2017}{768} - \frac{11097}{2048} s_2 + \frac{95}{256} \zeta(2) - \frac{65}{48} \ln \left(\frac{C_A \gamma^4}{\mu^4} \right) \right.\right. \\
&+ \left. \frac{35}{128} \left(\ln \left(\frac{C_A \gamma^4}{\mu^4} \right) \right)^2 + \frac{1137}{2560} \sqrt{5} \zeta(2) - \frac{205 \pi^2}{512} \right) \\
&+ C_A T_F N_f \left(- \frac{25}{24} - \zeta(2) + \frac{7}{12} \ln \left(\frac{C_A \gamma^4}{\mu^4} \right) \right. \\
&- \left.\left. \frac{1}{8} \left(\ln \left(\frac{C_A \gamma^4}{\mu^4} \right) \right)^2 + \frac{\pi^2}{8} \right) \right] a^2 + O(a^3)
\end{aligned}
\tag{7.10}
$$

where $s_2 = (2\sqrt{3}/9)\mathrm{Cl}_2(2\pi/3)$. The one and two loop corrections are evaluated using the vacuum bubble approach discussed earlier. Though we note that the basic two loop multiscale vacuum bubble integral $I(m_1^2, m_2^2, m_3^2)$, where

$$
I(m_1^2, m_2^2, m_3^2) = \int_{kl} \frac{1}{[k^2 - m_1^2][l^2 - m_2^2][(k-l)^2 - m_3^2]} \tag{7.11}
$$

has to be determined for $m_i^2 \in \{0, i\sqrt{C_A}\gamma^2, -i\sqrt{C_A}\gamma^2\}$. For instance, the quantities s_2 and $\sqrt{5}\zeta(2)$ arise from the finite parts of $I(i\sqrt{C_A}\gamma^2, i\sqrt{C_A}\gamma^2, i\sqrt{C_A}\gamma^2)$ and $I(i\sqrt{C_A}\gamma^2, i\sqrt{C_A}\gamma^2, -i\sqrt{C_A}\gamma^2)$ respectively.

One interesting consequence of this gap equation is that it ensures the Kugo-Ojima confinement condition, [51], of Faddeev-Popov ghost propagator enhancement is preserved at two loops. If one considers the full ghost propagator to have the form

$$
G_c(p^2) = \frac{\delta^{ab}}{p^2[1 + u(p^2)]} \tag{7.12}
$$

where $u(p^2)$ represents the radiative corrections, then the Kugo-Ojima confinement criterion is that $u(0) = -1$, [51]. Hence, the ghost propagator behaves as $1/(p^2)^2$ as $p^2 \to 0$. As $u(p^2)$ corresponds to the loop corrections of the ghost 2-point function, then computing $u(p^2)$ in the vacuum bubble expansion to two loops one can check if the Kugo-Ojima criterion holds at this order. Using the Gribov-Zwanziger Lagrangian, (7.4), it transpires that the gap equation emerges as the $u(0) + 1$ term at $O(p^2)$ so that $u(0) = -1$ precisely at two loops, [50].

Finally, we discuss a more recent study of non-locality in QCD, [52]. Earlier we considered the gauge variant dimension two operator $\frac{1}{2} A^a_\mu A^{a\,\mu}$ given the current interest in it as a potential mass operator for the gluon. However, if one wishes to have a gauge invariant Lagrangian with gluon mass then this operator is excluded. Instead to preserve gauge invariance and insist on a mass operator, one has to allow for a non-local mass operator. In [52], it was pointed out that aside from the one usually associated with the (non-renormalizable) Stueckelberg term,

$$
\tilde{A}^2_\mu = \min_{\{U\}} \int \left(A^U_\mu \right)^2 \tag{7.13}
$$

there is another independent operator which is

$$G^a_{\mu\nu} \frac{1}{D^2} G^{a\,\mu\nu} \tag{7.14}$$

which can be added to the gauge fixed Lagrangian as

$$L ~=~ L^{\mathrm{gf}} ~-~ \frac{m^2}{4} G^a_{\mu\nu} \frac{1}{D^2} G^{a\,\mu\nu} ~. \tag{7.15}$$

This is a localizable operator in the sense we defined earlier. Consequently, [52], it can be localized by introducing the ghost fields $\{ B^a_{\mu\nu}, \bar{B}^a_{\mu\nu}, H^a_{\mu\nu}, \bar{H}^a_{\mu\nu} \}$ where the last two are anti-commuting, to give

$$L ~=~ L^{\mathrm{gf}} + \frac{im}{4} (B^a_{\mu\nu} ~-~ \bar{B}^a_{\mu\nu}) G^{a\,\mu\nu} ~+~ \frac{1}{4} \bar{B}^a_{\mu\nu} \left(D^\sigma D_\sigma B^{\mu\nu} \right)^a ~-~ \frac{1}{4} \bar{H}^a_{\mu\nu} \left(D^\sigma D_\sigma H^{\mu\nu} \right)^a ~. \tag{7.16}$$

This Lagrangian has been analysed by algebraic renormalization and whilst it is renormalizable, it is not multiplicatively renormalizable since new quartic ghost terms are generated through quantum corrections, [52]. Moreover, the localized dimension three operator $(B^a_{\mu\nu} - \bar{B}^a_{\mu\nu}) G^{a\,\mu\nu}$ is itself mixed into the lower dimensional mass operators $\left(\bar{B}^a_{\mu\nu} B^{a\,\mu\nu} ~-~ \bar{H}^a_{\mu\nu} H^{a\,\mu\nu} \right)$ and $(\bar{B}^a_{\mu\nu} ~-~ B^a_{\mu\nu})^2$. Given that (7.16) is renormalizable, the one loop anomalous dimensions have been determined. Clearly the addition of such an operator ought not to affect the established anomalous dimensions of the gluon, Faddeev-Popov ghost and quark fields and explicit calculations verify this. Essentially within the respective 2-point functions the contributions from the extra ghost fields cancel. Moreover,

$$\gamma_B(a) ~=~ \gamma_H(a) ~=~ (\alpha - 3) C_A a ~+~ O(a^2) \tag{7.17}$$

and the anomalous dimension of the mass operator itself can be deduced by inserting the gauge invariant equivalent operator $(B^a_{\mu\nu} ~-~ \bar{B}^a_{\mu\nu}) G^{a\,\mu\nu}$ into the $B^a_{\mu\nu}$-gluon 2-point function. The explicit MINCER calculation yields the result, [52],

$$\gamma_{BG}(a) ~=~ -~ \left(\frac{11}{6} C_A - \frac{2}{3} T_F N_f \right) a^2 ~+~ O(a^3) \tag{7.18}$$

which is not only independent of the gauge parameter as it ought to be, but is equivalent to the one loop QCD β-function. This is the same as the one loop anomalous dimension of higher dimensional operators with similar Lorentz structure, [53]. In other words the operators $G^a_{\mu\nu} G^{a\,\mu\nu}$, $D_\mu G^a_{\nu\sigma} D^\mu G^{a\,\nu\sigma}$, $D_\mu D_\nu G^a_{\sigma\rho} D^\mu D^\nu G^{a\,\sigma\rho}$ and $D_\mu D_\nu D_\sigma G^a_{\rho\theta} D^\mu D^\nu D^\sigma G^{a\,\rho\theta}$ all have *same* one loop anomalous dimension as the non-local operator. We complete this section by remarking that it would be interesting to determine the higher loop corrections to (7.18).

8 Conclusions

We close with general remarks. First, in the initial part of the article we have reviewed the techniques and general structure associated with the renormalization of quantum field theories in the context of $O(N)$ ϕ^4 theory and QCD. One of the interesting aspects of renormalization is the rich interplay between the abstract and mundane exercise of evaluating renormalization group functions and the implications such results have on the underlying physics. For example, the anomalous dimensions when evaluated at a critical point of the β-function lead to predictions for the scaling behaviour of Green's functions which can be measured in experiments. However, from a technical point of view the critical exponents, when studied in the large N expansion, also actually complement the checking of the explicit perturbative results. For large loop order computations this plays a role in establishing the correctness of the result as well as providing *new* information on the as yet undetermined terms in the series at several orders in large N. This can be important due to the fact that such high order computations can presently only be undertaken with intense use of computer algebra and symbolic manipulation techniques. The latter part of the article dealt with the application of renormalization to more current computations including the renormalization of QCD in non-linear gauges. Although this follows the application of established techniques, such calculations do provide additional checks on the already determined three loop gauge independent renormalization group functions such as the β-function. Whilst much of the theory of renormalization is based on the assumption of locality of the initial Lagrangian, we have also touched on very recent calculations of operators which are *non-local* in structure. Although these fall in the class of localizable non-local operators, it is possible that such studies might open the possibility of studying problems in QCD where such operators are important in probing the infrared structure of the theory. This is more evident, for example, in the Gribov problem where the structure of the ghost propagators satisfies the Kugo-Ojima confinement condition at *two* loops.

Acknowledgement

The author thanks the Fields Institute for financial support in attending the meeting.

References

[1] Zinn-Justin, J. *Quantum field theory and critical phenomena*, 2nd edition International Series of Monographs on Physics 85, Clarendon Press, Oxford, 1993.

[2] Gross, D. J. & Wilczek, F. *Ultraviolet behavior of nonabelian gauge theories*, Phys. Rev. Lett. **30** (1973), 1343–1346; Politzer, H. D. *Reliable results for strong interactions?*, Phys. Rev. Lett. **30** (1973), 1346-1349.

[3] Piguet. O. & Sorella, S. P. *Algebraic renormalization: perturbative renormalization, symmetries and anomalies*, Lect. Notes Phys. **M128** (1995), 1–134.

[4] Bardeen, W. A., Buras, A. J., Duke, D. W. & Muta, T. *Deep inelastic scattering beyond the leading order in asymptotically free gauge theories*, Phys. Rev. **D18** (1978), 3998–4017.

[5] Celmaster, W. & Gonsalves, R. J. *The renormalization prescription dependence of the QCD coupling constant*, Phys. Rev. **D20** (1979), 1420–1434.

[6] Martinelli, G., Pittori, C., Sachrajda, C. T., Testa, M. & Vladikas, A. *A general method for nonperturbative renormalizaton of lattice operators*, Nucl. Phys. **B445** (1995), 81–108.

[7] Chetyrkin, K. G. & Rétey, A. *Renormalization and running of quark mass and field in the regularization invariant and $\overline{MS}$ schemes at three and four loops*, Nucl. Phys. **B583** (2000), 3–34.

[8] Gracey, J. A. *Three loop anomalous dimension of the second moment of the transversity operator in the $\overline{MS}$ and RI′ schemes*, Nucl. Phys. **B667** (2003), 242–260.

[9] Ford, C., Jack, I. & Jones, D. R. T. *The standard model effective potential at two loops*, Nucl. Phys. **B387** (1992), 373–390.

[10] Smirnov, V. A. *Analytical result for dimensionally regularized massless on shell double box*, Phys. Lett. **B460** (1999), 397–404.

[11] Chetyrkin, K. G., Misiak, M. & Münz, M. *Beta functions and anomalous dimensions up to three loops*, Nucl. Phys. **B518** (1998), 473–494.

[12] Weinberg, S. *High energy behavior in quantum field theory* Phys. Rev. **118** (1960), 838–849.

[13] Chetyrkin, K. G. & Tkachov, F. V. *Integration by parts: the algorithm to calculate β-functions in 4 loops*, Nucl. Phys. **B192** (1981), 159–204.

[14] Gorishny, S. G., Larin, S. A., Surguladze, L. R. & Tkachov, F. K. *Mincer: program for multiloop calculations in quantum field theory for the Schoonschip system*, Comput. Phys. Commun. **55** (1989), 381–408.

[15] Vermaseren, J. A. M. *New features of form*, math-ph/0010025.

[16] Larin, S. A., Tkackov, F. V. & Vermaseren, J. A. M. *The form version of mincer*, NIKHEF-H-91-18.

[17] Vladimirov, A. A. *Method for computing renormalization group functions in dimensional renormalization scheme*, Theor. Math. Phys. **43** (1980), 417–422; Chetyrkin, K. G., Kataev, A. L. & Tkachov, F. V. *New approach to evaluation of multiloop Feynman integrals: the Gegenbauer polynomial x space technique*, Nucl. Phys. **B174** (1980), 345–377.

[18] Gracey, J. A. *Three loop* $\overline{MS}$ *renormalization of QCD in the maximal abelian gauge*, JHEP **04** (2005), 012 0–29.

[19] van Ritbergen, T., Vermaseren, J. A. M. & Larin, S. A. *The four loop* β*-function in quantum chromodynamics*, Phys. Lett. **B400** (1997), 379-384.

[20] Nogueira, P. *Automatic Feynman graph generation*, J. Comput. Phys. **105** (1993), 279-289.

[21] Larin, S. A. & Vermaseren, J. A. M. *The three loop QCD* β*-function and anomalous dimensions*, Phys. Lett. **B303** (1993), 334–336.

[22] Remiddi, E. *Differential equations for Feynman graph amplitudes*, Nuovo Cim. **A110** (1997), 1435–1452; Gehrmann, T. & Remiddi, E. *Differential equations for two loop four point functions*, Nucl. Phys. **B580** (2000), 485–518.

[23] d'Eramo, M., Peliti, L. & Parisi, G. *Theoretical predictions for critical exponents at the* λ*-point of Bose liquids*, Lett. Nuovo Cim. **2** (1971), 878–880.

[24] Vasil'ev, A. N., Pismak, Y. M. & Honkonen, J. R. *Simple method of calculating the critical indices in the* $1/N$ *expansion*, Theor. Math. Phys. **46** (1981), 104–113.

[25] Vasil'ev, A. N., Pismak, Y. M. & Honkonen, J. R. $1/N$ *expansion: calculation of the exponents* η *and* ν *in the order* $1/N^2$ *for arbitrary number of dimensions*, Theor. Math. Phys. **47** (1981), 465–475.

[26] Vasil'ev, A. N., Pismak, Y. M. & Honkonen, J. R. $1/N$ *expansion: calculation of the exponent* η *in the order* $1/N^3$ *by the conformal bootstrap method*, Theor. Math. Phys. **50** (1982), 127–134.

[27] Broadhurst, D. J., Gracey, J. A., & Kreimer, D. *Beyond the triangle and uniqueness relations: non-zeta counterterms at large N from positive knots*, Z. Phys. **C75** (1997), 559–574.

[28] Kreimer, D. *Knots and divergences*, Phys. Lett. **B354** (1995), 117–124.

[29] Gracey, J. A. *Quark, gluon and ghost anomalous dimensions at* $O(1/N_f)$ *in quantum chromodynamics*, Phys. Lett. **B318** (1993), 177–183.

[30] Gracey, J. A. QCD *The QCD* β*-function at* $O(1/N_f)$, Phys. Lett. **B373** (1996), 178–184.

[31] Hasenfratz, A. & Hasenfratz, P. *The equivalence of the* $SU(N)$ *Yang-Mills theory with a purely fermionic model*, Phys. Lett. **B297** (1992), 166–170.

[32] Vasil'ev, A. N. & Nalimov, M. Y. *Analog of dimensional regularization for calculation of the renormalization group functions in the* $1/N$ *expansion for arbitrary dimension of space*, Theor. Math. Phys. **55** (1983), 423–431.

[33] Ciuchini, M., Derkachov, S. E., Gracey, J. A. & Manashov, A. N. *Computation of quark mass anomalous dimension at* $O(1/N_f^2)$ *in quantum chromodynamics*, Nucl. Phys. **B579** (2000), 56–100.

[34] Dudal, D., Gracey, J. A., Lemes, V. E. R., Sobreiro, R. F., Sorella, S. P. & Verschelde, H. *Renormalization properties of the mass operator* $A^a_\mu A^a_\mu$ *in three dimensional Yang-Mills theories in the Landau gauge*, Annals Phys. **317** (2005), 203–219.

[35] Dudal, D., Verschelde, H. & Sorella, S. P. *The anomalous dimension of the composite operator* A^2 *in the Landau gauge*, Phys. Lett. **B555** (2003), 126–131

[36] Gracey, J. A. *Three loop* $\overline{MS}$ *renormalization of the Curci-Ferrari model and the dimension two BRST invariant composite operator in QCD*, Phys. Lett. **B552** (2003), 101–110.

[37] Chetyrkin, K. G. *Four loop renormalization of QCD: full set of renormalization constants and anomalous dimensions*, Nucl. Phys. **B710** (2005), 499–510.

[38] Kondo, K. I. *A physical meaning of mixed gluon ghost condensate of mass dimension two*, Phys. Lett. **B572** (2003), 210-215.

[39] Dudal, D., Gracey, J. A., Lemes, V. E. R., Sarandy, M. S., Sobreiro, R. F., Sorella, S. P. & Verschelde, H. *An analytic study of the off-diagonal mass generation for Yang-Mills theories in the maximal abelian gauge*, Phys. Rev. **D70** (2004), 114038 1–18.

[40] Min, H., Lee, T. & Pac, P. Y. *Renormalization of Yang-Mills in the abelian gauge*, Phys. Rev. **D32** (1985), 440–449; Fazio, A. R., Lemes, V. E. R., Sarandy, M. S. & Sorella, S. P. *The diagonal ghost equation Ward identity for Yang-Mills theories in the maximal abelian gauge*, Phys. Rev. **D64** (2001), 085003 1–7.

[41] Curci, G. & Ferrari, R. *On a class of Lagrangian models for massive and massless Yang-Mills fields*, Nuovo Cim. **A32** (1976), 151–168.

[42] Abbott, L. F. *The background field method beyond one loop*, Nucl. Phys. **B185** (1981), 189–203; Capper, D. M. & MacLean, A. *The background field method at two loops: a general gauge Yang-Mills calculation*, Nucl. Phys. **B203** (1982), 413–422.

[43] Browne, R. E. & Gracey, J. A. *The Curci-Ferrari model with massive quarks at two loops*, Phys. Lett. **B540** (2002), 68–74.

[44] Collins, J. C. *Renormalization*, Cambridge Monographs on Mathematical Physics, Cambridge University Press, Cambridge, 1984.

[45] Gribov, V. N. *Quantization of non-abelian gauge theories*, Nucl. Phys. **B139** (1978), 1–19.

[46] Zwanziger, D. *Action from the Gribov horizon*, Nucl. Phys. **B321** (1989), 591–604.

[47] Zwanziger, D. *Local and renormalizable action from the Gribov horizon*, Nucl. Phys. **B323** (1989), 513–544.

[48] Zwanziger, D. *Renormalizability of the critical limit of lattice gauge theory by BRS invariance*, Nucl. Phys. **B399** (1993), 477–513.

[49] Dudal, D., Sobreiro, R. F., Sorella, S. P. & Verschelde, H. *The Gribov parameter and the dimension two condensate in Euclidean Yang-Mills theories in the Landau gauge*, Phys. Rev. **D72** (2005), 014016 1–25.

[50] Gracey, J. A. *Two loop correction to the Gribov mass gap equation in Landau gauge QCD*, Phys. Lett. **B632** (2006), 282–286.

[51] Kugo, T. & Ojima, I. *Local covariant operator formalism of nonabelian gauge theories and quark confinement problem*, Prog. Theor. Phys. Suppl. **66** (1979), 1–233,

[52] Capri, M. A. L., Dudal, D., Gracey, J. A., Lemes, V. E. R., Sobreiro, R. F., Sorella, S. P. & Verschelde, H. *A study of the gauge invariant, non-local mass operator* $F_{\mu\nu}\left(D^2\right)^{-1}F_{\mu\nu}$ *in Yang-Mills theories*, Phys. Rev. **D72** (2005), 105016 1–21.

[53] Gracey, J. A. *Classification and one loop renormalization of dimension six and dimension eight operators in quantum gluodynamics*, Nucl. Phys. **B634** (2002), 192–208.

Fields Institute Communications
Volume **50**, 2007

Quantum Field Theory in Curved Spacetime

Stefan Hollands
Inst. f. Theor. Physik
Georg-August-Universität
D-37077 Göttingen, Germany
hollands@theorie.physik.uni-goe.de

Abstract. We review recent developments in quantum field theory on curved spacetimes, name what in our opinion are the major open problems, and make some remarks towards their possible resolution.

1 Introduction

During the past 6–7 years, there has been considerable progress describing the dynamics of quantum fields propagating on a—mostly rigidly fixed—curved spacetime background. The theory is interesting from the physical point of view because it predicts and describes important effects, such as the production of density fluctuations in the early universe via quantum fluctuations, the creation of particles in the expanding universe, and the radiance of black holes. It is also a very interesting mathematical structure, since it forces one to unify the principles of quantum theory and the principle of general covariance, both of which are cornerstones of modern theoretical physics. Finally, the theory might even provide a fundamental description of quantum gravity, via the AdS-CFT correspondence, which puts quantum gravity in a "bulk" into correspondence with a QFT on a (generally curved) "holographic boundary", see e.g. [24, 26, 23].

It is far from straightforward to generalize the well-studied structure of quantum field theory in flat Minkowski spacetime to arbitrary curved spacetimes. This is definitely because quantum field theory is traditionally formulated in terms of concepts and principles that are tied to the very rigid mathematical structures of flat Minkowski space. Indeed, most formulations are based upon one or more of the following ideas:

- The existence of an S-matrix.
- A preferred "vacuum state".
- The notion of "particles".
- Conserved quantities associated with spacetime symmetries, such as energy-momentum, and derived concepts such as particle masses etc.

2000 *Mathematics Subject Classification.* Primary 81T20; Secondary 81T05.

- Thermal equilibrium states.
- Momentum space techniques.
- Preferred Hilbert space representation.
- Euclidean techniques ("Wick rotation"), path integrals, dimensional regularization, etc.

All of these concepts involve the structures of Minkowski spacetime in a key way and cannot in general be generalized to quantum field theory on a generic curved spacetime. For example, the idea of an S-matrix is closely related to the existence of suitable asymptotic spacetime regions in the far past and future, in which the fields can be said to approximately behave like "free fields". A generic curved spacetime may simply not have such regions. For example, our very own universe is known to become more and more curved as one goes back into the past. Also, even a static spacetime of topology $\mathbb{R} \times \Sigma$, where Σ is compact, will not admit the definition of an S-matrix, because field excitations cannot get infinitely far away from each other and thus never behave like "free wave packets". Actually, even if a spacetime has asymptotically flat regions, an S-matrix still might not exist, because the spacetime might not become "flat" at a sufficiently fast rate so as to allow for the S-matrix to exist. The notion of a vacuum state is closely related to the translation invariance of Minkowski spacetime: It is defined as the unique quantum state invariant under the Poincaré group. Roughly, in order to say that a state is devoid of particles, we need to have a notion that it is "uniformly empty" in all directions, and this requires sufficiently many translational symmetries of the spacetime. Given that one cannot even define a vacuum state in a general curved spacetime, it is clear that similar difficulties apply to the characterization "particle states". Indeed, these are normally associated with eigenstates of the energy-momentum operator. However, in a spacetime without time-translation symmetry, there is no conserved energy, and hence no invariant definition of a quantities like energy-eigenstates. Finally, the difficulties with technical tools such as Euclidean techniques stem from the fact that a generic Lorentzian spacetime cannot be viewed as a real section of a complex spacetime that also has a real, Riemannian section.

Thus, even though the above concepts are suitable—and in fact highly successful—for describing and interpreting quantum field theory in flat Minkowski spacetime, they have problems in curved space. In particular, it is far from clear how one would formulate the notion of general covariance. A better approach is to formulate the theory in terms of objects that can equally well be defined in all spacetimes, independent of the particular circumstances, such as the accidental existence of spacetime symmetries. Such objects are provided by the local quantum fields of the theory, and it has indeed proven to be fruitful to formulate quantum field theory in curved spacetime entirely in terms of the interacting fields. Normally, quantum fields are thought of as operators acting on the quantum states in the Hilbert space of the system. In curved spacetime, there is no preferred Hilbert space representation, but we may think of quantum fields more abstractly as members of some *-algebra, which is defined via abstract relations not involving a particular representation. In this way, one is lead to the following basic framework:

A quantum field theory is an assignment which associates with each globally hyperbolic spacetime (M, g) a *-algebra[1] $\mathcal{A}(M, g)$ containing the observables of the the theory. The algebra is supposed to be generated—at least in a weak sense—by

[1]More specifically, one may consider C^*-algebras, topological algebras, W^*-algebras, etc.

"smeared" quantum field expressions $\mathcal{O}_i(f)$. Here, i is a label that distinguishes the various kinds of fields in the given theory. We allow the fields to have a tensor or spinor character, so the label i in particular incorporates the specification of a bundle corresponding to the tensor or spinor character (or gauge vector bundle character) of the field $\mathcal{O}_i$. The smeared fields should be thought of heuristically as weighted spacetime averages of informal pointlike fields, as indicated in the informal notation

$$\mathcal{O}_i(f) = \int_M \mathcal{O}_i(x) f(x)\, e(x)\,, \tag{1}$$

where $e = \sqrt{g}\, dx^0 \wedge \cdots \wedge dx^{n-1}$ is the natural integration element with respect to some fixed choice of orientation. States in the algebraic framework are just linear, continuous, positive, normalized expectation functionals $\omega : \mathcal{A}(M,g) \to \mathbb{C}$, where positive means that $\omega(F^*F) \geq 0$ for any element F, and where normalized means that $\omega(\mathbb{1}) = 1$.

To express the local covariance property of the theory, one considers a situation in which a "smaller" spacetime (M,g) is isometrically embedded into a "larger" spacetime (M',g'), [15, 5]. Let $\psi : M \hookrightarrow M'$ denote this embedding map. Then it is postulated that there exist a corresponding embedding of the algebras, that is, a *-homomorphism

$$\alpha_\psi : \mathcal{A}(M,g) \hookrightarrow \mathcal{A}(M',g')\,. \tag{2}$$

It is further assumed that this *-homomorphism behaves naturally if we have a chain of embeddings of spacetime manifolds, in the sense that it should satisfy the obvious composition law. The existence of the above homomorphism asserts that the relations between elements in $\mathcal{A}(M,g)$ are not changed if we imagine that these elements actually live on a bigger spacetime, i.e., are identified with elements in $\mathcal{A}(M',g')$. They give the assignment $(M,g) \longrightarrow \mathcal{A}(M,g)$ the structure of a covariant functor from the category of globally hyperbolic spacetime manifolds, whose arrows are the isometric embeddings, to the category of topological *-algebras, whose arrows are the injective *-homomorphisms. Actually, the above statements are somewhat imprecise because the construction of the algebras $\mathcal{A}(M,g)$ typically depends not only upon the spacetime M and metric g, but also on a choice of time orientation, a choice of orientation, a choice of spin-structure, as well as on coupling constants. For example, if spinor fields are present, a spin structure is already needed at the classical level in order to define spinors on a manifold. Moreover, in case several spinor structures exist, the specification of the algebra also will depend on the precise choice of the spin-structure. In all cases, if we consider embeddings between spacetimes, then the additional structures such as orientations etc., should also be mapped into each other. Thus, the functor is really an assignment of the objects

$$\mathcal{M} = (M, g, \text{time orientation}\, T, \text{orientation}\, e, \text{spin structure, coupling constants}) \tag{3}$$

to *-algebras, and the morphisms are embeddings preserving the causal structure, as well as the other additional structures (acting trivially on the coupling constants). To save ink, we shall often write $\mathcal{A}(\mathcal{M})$, where $\mathcal{M}$ is supposed to be given by the above tuple.

A local, covariant field can now be defined as a field satisfying a natural "transformation law" under spacetime embeddings [5, 13]. More precisely, a local covariant field is an assignment that associates to any spacetime $\mathcal{M}$ a distribution valued in the algebra $\mathcal{A}(\mathcal{M})$. If $\psi : \mathcal{M} \to \mathcal{M}'$ is an isometric embedding preserving the spin-structures and orientations and spin-structures, and if f is smooth and of compact support, then it is required that

$$\alpha_\psi[\mathcal{O}_\mathcal{M}(f)] = \mathcal{O}_{\mathcal{M}'}(\psi_* f) . \tag{4}$$

where ψ_* means the push-forward of a compactly supported section on M to a compactly supported section on M'. This requirement reflects the property of the field of being locally defined, and generally covariant. If $M = M'$ is Minkowski spacetime, and the isometry ψ is given by a Poincare-transformation (Λ, a), then the above requirement reduces to the usual relativistic transformation law for quantum fields (with $\alpha_\psi = U(\Lambda, a) \,.\, U(\Lambda, a)^{-1}$ defined by a unitary on Hilbert space.) Thus, by formulating the theory in terms of algebras, we can implement the covariance principle [13, 5]. On the other hand, quantities related to states such as amplitudes, correlation functions are always global in nature and not covariant.

Another important ingredient in the formulation of quantum field theory is the condition of Einstein causality. This requirement is essentially unchanged in curved space. It is

$$[\mathcal{O}_1(f), \mathcal{O}_2(h)] = 0 \quad \text{If } f \text{ and } h \text{ are have spacelike related support,} \tag{5}$$

with the only difference that "spacelike" now refers to the curved metric g. For fermionic fields, one should use the graded commutator, as usual. Thus, the Einstein causality principle may be viewed as an algebraic relation between the fields.

Given the above axiomatic framework, one naturally wonders whether general properties of quantum field theory in Minkowski spacetime, such as the spin-statistics theorem or the Parity-Charge-Time (PCT)-theorem can be proven in the above framework in a model-independent fashion. Concerning the PCT-theorem, one notes that it is a priori not even clear precisely what one means by this, because a generic spacetime does not have any discrete isometries corresponding to P (parity reversal) and T (time reversal). However, as noted in [19], the combination PT may be viewed as a change in the orientations

$$(e, T) \to (e, -T), \tag{6}$$

where $T : M \to \mathbb{R}$ is now a global time function, and where e is the n-form on M defining the orientation. Thus, we may ask how the algebra $\mathcal{A}(\mathcal{M})$ associated with the background data $\mathcal{M} = (M, g, e, T)$ is related to the algebra $\mathcal{A}(\overline{\mathcal{M}})$ associated with the background data $\overline{\mathcal{M}} = (M, g, e, -T)$. The PCT-theorem asserts that there exists an anti-linear PCT-isomorphism $\theta^{\rm PCT}_\mathcal{M}$ such that the diagram

$$\begin{array}{ccc} \mathcal{A}(\mathcal{M}) & \xrightarrow{\theta^{\rm PCT}_{\mathcal{M}}} & \mathcal{A}(\overline{\mathcal{M}}) \\ {\scriptstyle \alpha_\psi}\downarrow & & \downarrow{\scriptstyle \alpha_\psi} \\ \mathcal{A}(\mathcal{M}') & \xrightarrow{\theta^{\rm PCT}_{\mathcal{M}'}} & \mathcal{A}(\overline{\mathcal{M}}') \end{array} \tag{7}$$

commutes. We could state this more abstractly by saying that the functors $\mathcal{M} \to \mathcal{A}(\mathcal{M})$ and $\mathcal{M} \to \overline{\mathcal{A}}(\mathcal{M}) = \mathcal{A}(\overline{\mathcal{M}})$ are equivalent. The existence of such an anti-linear isomorphism is far from straightforward. It has been proved (in the sense of an invariance of the operator product expansion) by [19] for analytic spacetimes, under certain assumptions about the microlocal properties of the singularities in operator products. The proof heavily relies on the theory of microlocal analysis (in the analytic setting), and the theory of several complex variables.

A proof of the spin-statistics theorem in the kind of abstract setting described above has also been given by Verch [28].

2 Free fields and perturbation theory

The formulation of quantum field theory in curved space we have outlined so far is very general in nature, and the character of a particular model is supposed to be reflected in the algebraic relations. Thus, to characterize, or construct, a particular model more precisely, we must specify these relations. For a linear, scalar quantum field ϕ, classically described by the action

$$S_0 = \int_M [(\nabla\phi)^2 + m^2\phi^2]\, e \tag{8}$$

this is quite simple. We may define the CCR-algebra $\mathcal{C}(M,g)_{S_0}$ to be the quotient of the free *-algebra generated by the symbols $\{1\!\!1, \phi(f);\ f \in C_0^\infty(M)\}$, by the relations

1. Linearity: $\phi(af + bh) = a\phi(f) + b\phi(h)$
2. Hermitian: $\phi(f)^* = \phi(\bar{f})$
3. Klein-Gordon: $\phi((\Box - m^2)f) = 0$
4. Commutator: $[\phi(f), \phi(h)] = \mathrm{i}\Delta(f,h)1\!\!1$, where $\Delta = \Delta_A - \Delta_R$ is the difference between the fundamental advanced and the fundamental retarded solution to the Klein-Gordon equation, which is uniquely defined on any globally hyperbolic spacetime.

The properties postulated in the previous section can easily be established for the assignment $(M,g) \to \mathcal{C}(M,g)_{S_0}$ using the well-known properties of the advanced and retarded propagators in curved spacetime. The field ϕ is (by definition) a local and covariant field. Of course, we would like to consider within this theory also other composite fields, such as the stress-energy tensor $T_{\mu\nu}$. However, as it stands, none of them is contained in the algebra $\mathcal{C}(M,g)_{S_0}$. However, they are contained in a suitably enlarged algebra, $\mathcal{A}(M,g)_{S_0}$, which may be defined as follows [13, 6, 7, 8]:

Let $\omega_2 \in \mathcal{D}'(M \times M)$ be a bidistribution of Hadamard type, meaning (a) that ω_2 is a bisolution to the Klein-Gordon equation $\Box - m^2$ in each entry, (b) that the anti-symmetric part of ω_2 is given by

$$\omega_2(x_1, x_2) - \omega_2(x_2, x_1) = \mathrm{i}\Delta(x_1, x_2)\,, \tag{9}$$

and (c) that it has a wave front set [21] of Hadamard type [25]:

$$\mathrm{WF}(\omega_2) = \{(x_1, k_1; x_2, k_2) \in T^*(M \times M) \setminus 0;\ k_1 = p, k_2 = -p, p \in \bar{V}_+^*\}\,. \tag{10}$$

Here, it is understood that the set can only contain those x_1 and x_2 that can be joined by a null geodesic, γ, and that $p = p_\mu dx^\mu$ is a parallel co-vector field tangent to that null geodesic, meaning that $\nabla_{\dot\gamma} p = 0$, and $V_\pm^*$ is the dual of the future/past lightcone. Consider now the algebra generated by an identity $1\!\!1$, and the following symbols F:

$$F = \int_{M^n} \left(:\phi^{\otimes n}:_\omega\ u_n\right)(x_1, \dots, x_n) \bigwedge_{i=1}^n e(x_i)\,, \tag{11}$$

where u_n is a symmetric, compactly supported distribution on M^n subject to the wave front condition [6]

$$\mathrm{WF}(u_n) \cap [(\bar{V}_+^*)^n \cup (\bar{V}_-^*)^n] = \emptyset\,. \tag{12}$$

The relations in this *-algebra, $\mathcal{A}(M,g)_{S_0}$, are defined as follows: The *-operation is defined by letting F^* be given by the same expression as F, but with u_n replaced by its complex conjugate, and the product is defined by

$$\begin{aligned} &: \phi^{\otimes n} :_\omega (x_1, \dots, x_n) : \phi^{\otimes m} :_\omega (x_{n+1}, \dots, x_{n+m}) \\ &= \sum_k \frac{n!m!}{(n-k)!(m-k)!k!} \sum_{p_1,\dots,p_k \in P} \prod_i \omega_2(x_{p_i(1)}, x_{p_i(2)}) \\ &\times : \phi^{\otimes n+m-2k} :_\omega (\{x_j;\ j \notin |P|\}) \end{aligned} \tag{13}$$

where P is the set of all pairs $p_i \in \{1, \dots, n\} \times \{n+1, \dots, n+m\}$. This formula is identical in nature to the standard Wick theorem for normal ordered quantities (relative to a Gaussian state with 2-point function ω_2). We finally impose the relation that any element F such that u_n is in the image of the Klein-Gordon operator vanishes. The wave front conditions on u_n and ω_2 are needed in order guarantee that the product between the quantities in (13) exists, because the latter involves the pointwise products of distributions [21]. The definition of $\mathcal{A}(M,g)_{S_0}$ superficially seems to depend on the particular choice of ω_2, but this is in fact not so: A change of ω_2 merely corresponds to a relabeling of the generators, and does not change the algebra [13].

The relation between the algebra $\mathcal{A}(M,g)_{S_0}$ and the CCR-algebra $\mathcal{C}(M,g)_{S_0}$ is as follows. If f is a smooth test function, then the generator (11) with $n=1$ and $u_1 = f$ can be identified with the smeared field $\phi(f)$. Indeed, using the Wick formula (13), and the antisymmetric part of ω_2, eq. (9), one easily derives the above commutation relation for $\phi(f)$. Similarly, one derives the Klein-Gordon equation. Thus, $\mathcal{C}(M,g)_{S_0} \subset \mathcal{A}(M,g)_{S_0}$. The higher order generators (11) with u_n given by the n-fold tensor product $f^{\otimes n} = f \otimes \cdots \otimes f$ of a smooth testfunction correspond to a smeared normal ordered product, and are elements in the CCR-algebra. They are related to $\phi(f)$ by the generating functional type relation

$$\sum_n \frac{i^n}{n!} : \phi^{\otimes n}(f^{\otimes n}) :_\omega = \exp\left\{ i\phi(f) + \frac{1}{2}\omega_2(f,f) \right\}. \tag{14}$$

These elements, too, are in the CCR algebra $C(M,g)_{S_0}$ since they can be written in terms of $\phi(f)$. However, the elements F involving distributional u_n are not in the CCR algebra, and so $\mathcal{A}(M,g)_{S_0}$ is genuinely bigger.

It can be shown that $\mathcal{A}(M,g)_{S_0}$ has a local and covariant dependence upon the spacetime in the sense described above despite the apparently non-local and non-covariant choice of the arbitrary 2-point function ω_2. Thus, we have the notion of a local covariant field in the algebra, and we can now seek a definition of composite fields as certain local covariant elements. It would appear that normal ordering with respect to ω_2 would be a natural way to define composite fields such as powers of the field ϕ. For example, the normal ordered square of ϕ is

$$\begin{aligned} : \phi^2(y) :_\omega &= \lim_{\varepsilon \to 0} : \phi^{\otimes 2} :_\omega (x_1(\varepsilon), x_2(\varepsilon)) \\ &= \lim_{\varepsilon \to 0} \left\{ \phi(x_1(\varepsilon))\phi(x_2(\varepsilon)) - \omega_2(x_1(\varepsilon), x_2(\varepsilon))\mathbb{1} \right\}, \end{aligned} \tag{15}$$

where $\vec{x}_i(\varepsilon)$ are smooth curves such that $x_i(0) = y$. One can see that this defines an algebra valued distribution in $\mathcal{A}(M,g)_{S_0}$ but it is not a local covariant field, because ω_2 necessarily is a non-local quantity. The possible definitions of Wick powers giving rise to local and covariant fields (satisfying also various other natural conditions) were classified in [13]. It turns out that the definition of a given classical expression

$$\mathcal{O}_i = \nabla^{a_1}\phi\nabla^{a_2}\phi\cdots\nabla^{a_n}\phi, \quad i = \{a_1, a_2, \dots, a_n\} \in \mathcal{I} \equiv \bigoplus_n \mathbb{Z}^n_{\geq 0} \tag{16}$$

as a local, covariant field in $\mathcal{A}(M,g)_{S_0}$ is not unique, but contains certain ambiguities. As proven in [13], these ambiguities correspond to the possibility of adding to a given field lower order Wick power powers times certain polynomials of the Riemann tensor and its derivatives $\nabla_{(\mu_1}\cdots\nabla_{\mu_k)}R_{\nu_1\nu_2\nu_3\nu_4}$ of the same dimension as $\mathcal{O}_i$. Here, the dimension of a Wick power is the map $[\cdot]:\mathcal{I}\to\mathbb{N}$ from the index set labelling the various fields, into the natural numbers defined by

$$[i] = n + \sum_{i=1}^{n} a_i\,. \tag{17}$$

One definition which is local and covariant (and satisfies also the other natural conditions given in [13]) is the following "local normal ordering prescription". It is based upon the use of the local Hadamard parametrix H instead of ω_2 in (15). This is the bidistribution defined on a convex normal neighborhood of the diagonal $\{(x,x);\ x\in M\}$ of $M\times M$ by[2]

$$H = \frac{v_0}{\sigma + i0t} + \left(\sum_{n\geq 0}\frac{1}{2^n n!}v_{n+1}\sigma^n\right)\ln(\sigma + i0t)\,. \tag{18}$$

In this expression, $\sigma(x,y)$ is the signed squared geodesic distance between two points, we have defined $t(x,y) = T(x) - T(y)$, where $T: M\to\mathbb{R}$ is a time-function, and the $v_n(x,y)$ are smooth symmetric functions that are determined by requiring that H be a parametrix, i.e., a solution to the Klein-Gordon equation $\Box - m^2$ in each entry modulo a smooth remainder. Explicitly $v_0 = D^{1/2}/2\pi^2$ is given in terms of the so-called VanVleck determinant D, and the higher v_n are defined by certain transport equations. The key point is that these functions are locally and covariantly defined in terms of the metric near the diagonal (in the sense of germs). Where it is well-defined, H has a wave front set of Hadamard type (10). Next, fix a convex normal neighborhood of the diagonal in M^n, and in that neighborhood define locally normal ordered products $:\phi^{\otimes n}:_{\mathrm{H}}(x_1,\dots,x_n)$ by the same formula as (14), but with ω_2 in that formula replaced by H. As the expressions (14), their expectation value in any so-called "Hadamard state" is smooth. Following [13] we define the local covariant n-th Wick power of the field as the distribution valued in $\mathcal{A}(M,g)_{S_0}$ given by

$$\phi^n(y) = \lim_{\varepsilon\to 0}\ :\phi^{\otimes n}:_{\mathrm{H}}(\vec{x}(\varepsilon)) \quad \vec{x}(\varepsilon) = (\exp_x(\varepsilon\xi_1),\dots,\exp_x(\varepsilon\xi_n))\,, \tag{19}$$

where ξ_i are tangent vectors at y. More generally, fields containing derivatives are defined by first acting with the derivatives on the appropriate tensor factor before taking the above "coincidence limit".

Having described the algebra $\mathcal{A}(M,g)_{S_0}$ of field observables in the linear field theory associated with an action S_0 without the interaction terms, we now turn to

---

[2] The infinite sum is to be understood in the sense of an asymptotic expansion.

the interacting theory, classically described by an action $S = S_0 + I$, where

$$I = \int_M \sum_i \kappa_i \mathcal{O}_i \, e \tag{20}$$

is a local polynomial interaction. For this it is technically convenient at an intermediate step to assume that the couplings κ_i in I are not constants, but actually smooth functions of compact support in M, which we assume are locally constant,

$$\kappa_i(x) = \kappa_i \chi(x)\,, \tag{21}$$

where $\chi \in C_0^\infty(M)$, and $\chi(x) = 1$ in an open set in M with compact closure. The cutoff functions χ serve as an infra red cutoff and are removed at a later stage. With the introduction of cutoff functions understood, the interacting fields are defined by the so-called Bogoliubov formula [12, 22] in terms of the retarded products $\mathbf{R}_n$:

$$\mathcal{O}_I(x) = \sum_n \frac{\mathrm{i}^n}{n!} \mathbf{R}_n(\mathcal{O}(x); I^{\otimes n}) \tag{22}$$

The expressions $\mathbf{R}_n$ in this formula are the so-called "retarded products". Each retarded product is a continuous, bilinear map

$$\mathbf{R}_n : \mathcal{S} \times \left(\bigotimes^n \mathcal{S} \right) \to \mathcal{A}(M,g)_{S_0} \quad (I_0, I_1 \otimes \cdots \otimes I_n) \mapsto \mathbf{R}_n(I_0, I_1 \otimes \cdots \otimes I_n)\,, \tag{23}$$

from the tensor powers of the space $\mathcal{S}$ of all classical action functionals $I_j = \int \mathcal{L}_j \, e$ that are local and polynomial in the field ϕ and whose couplings are compactly supported functions on M. The map is taking values in the algebra $\mathcal{A}(M,g)_{S_0}$ associated with the linear field theory, and is symmetric in $I_1, \dots, I_n$. Note that the power series expression (22) for the interacting fields is only a formal series, and no statement is made about its convergence. Since each term in these series is an element in $\mathcal{A}(M,g)_{S_0}$, the interacting fields are elements of the algebra $\mathcal{P} \otimes \mathcal{A}(M,g)_{S_0}$, where $\mathcal{P}$ is the corresponding ring of formal series in the couplings. All operations, such as multiplication in this ring and in the algebra $\mathcal{P} \otimes \mathcal{A}(M,g)_{S_0}$, are defined by simply formally multiplying out the corresponding formal series term by term.

We want the zeroth order contribution $\mathbf{R}_0(\mathcal{O}_i(x))$ in the Bogoluibov formula (22) to be given by our definition of Wick powers $\mathcal{O}_i(x)$ in the linear field theory[3]. Thus, we define $\mathbf{R}_0(\mathcal{O}_i(x))$ to be equal to the locally normal ordered field $\mathcal{O}_i(x)$ given in eq. (19). The terms with $n \geq 1$ in the formal series (22) represent the perturbative corrections coming from the interaction, I. They involve the higher, non-trivial, retarded products. The construction of these retarded products can be reduced[4] to the construction of the so-called "time-ordered products," because there exists a well-known formula of combinatorial nature relating these two quantities, see e.g. the appendix of [8]. The construction of the time-ordered products in turn has been given in [13, 14, 16], which is based on work of [3]. The strategy in these papers is to first write down a number of *functional relations* for the time

[3]Note that the argument of the retarded product $\mathbf{R}_0(\mathcal{O}_i(x))$ is a *classical* action (or density), while the corresponding Wick power $\mathcal{O}_i(x)$ is a distribution valued in the quantum algebra $\mathcal{A}(M,g)_{S_0}$. We should strictly speaking distinguish these quantities by introducing a new notation for the Wick power, but we shall not do this for simplicity.

[4]Alternatively, it should also be possible to construct the retarded products directly along the lines of [6, 7, 8], by suitably generalizing the arguments of that paper from Minkowski spacetime to curved spacetime.

ordered (or retarded) products that are motivated by corresponding properties of the interacting fields defined by the Bogoliubov formula. These properties then dictate to a large extent the construction of the time-ordered (and hence the retarded) products. However, there remains an extension problem for the distributions r in (r4) below at each order in perturbation theory, which corresponds the renormalization step in Minkowski spacetime. The task is to perform the renormalization step in such a way that the functional relations (in particular covariance) are not destroyed. The key functional relations are as follows:

(r1) **Causality:** Let $F, G, S_i \in \mathcal{S}$. Suppose that there is a Cauchy surface such that $\operatorname{supp} G$ is in its future and $\operatorname{supp} F$ in its past. Then

$$\mathbf{R}_n\left(F, G \otimes \bigotimes_i S_i\right) = 0. \tag{24}$$

(r2) **Field equation:**

$$\mathbf{R}_n\left(\bigotimes_i S_i, \int_M f(\Box - m^2)\phi\, e\right) = \sum_k \mathbf{R}_{n-1}\left(\bigotimes_{i\neq k} S_i; \int_M f\frac{\delta S_k}{\delta\phi}\right). \tag{25}$$

The field equation implies the interacting field equation in perturbation theory, $(\Box - m^2)\phi_I = (\delta I/\delta\phi)_I$.

(r3) **GLZ factorization formula** [8]:

$$\mathbf{R}_n\left(G, \bigotimes_i S_i \otimes F\right) - \mathbf{R}_n\left(F, \bigotimes_i S_i \otimes G\right)$$
$$= \sum_{I\cup J=\{1,\dots,n-1\}} \left[\mathbf{R}_{|I|,1}\left(F, \bigotimes_{i\in I} S_i\right), \mathbf{R}_{|J|,1}\left(G, \bigotimes_{j\in J} S_j\right)\right]. \tag{26}$$

The causality relation and the GLZ formula together imply the local commutativity $[\mathcal{O}_I(f), \mathcal{O}_I(h)] = 0$ when the supports of f, h are spacelike related.

(r4) **Expansion:** There exist local covariant c-number distributions r near the total diagonal $\{(x, x, \dots, x);\ x \in M\}$ such that (with $l_j < i_j$)

$$\mathbf{R}_n(\mathcal{O}_{i_0}(x), \mathcal{O}_{i_1}(y_1)\cdots\mathcal{O}_{i_n}(y_n)) =$$
$$\sum_{l_0,l_1,\dots,l_n} r^{l_0 l_1\dots l_n}_{i_0 i_1\dots i_n}(x, y_1, \dots, y_n) :\mathcal{O}_{l_0}(x)\mathcal{O}_{l_1}(y_1)\dots\mathcal{O}_{l_n}(y_n):_{\mathrm{H}}. \tag{27}$$

(r5) **Scaling:** The distributions r have a local, covariant dependence upon the metric, and

$$\frac{\partial^s}{\partial \ln\lambda^s}\left\{\lambda^D r^{l_0\dots l_n}_{i_0\dots i_n}[\lambda^2 g, \lambda^2 m^2]\right\} = 0 \tag{28}$$

for some s, where

$$D = \sum_k [i_k] - [l_k]. \tag{29}$$

The scaling implies that the naive scaling behavior of the theory is “broken” by a polynomial of degree at most s in the logarithm. It implies the existence of a renormalization group.

(r6) **Locality and covariance:** Let $f : (M, g) \to (M', g')$ be a causality and orientation preserving isometric embedding. Then the retarded products satisfy

$$\alpha_\psi \left[\mathbf{R}_n \left(F, \bigotimes_i S_i \right) \right] = \mathbf{R}_n \left(\psi_* F, \bigotimes_i \psi_* S_i \right) , \tag{30}$$

where ψ_* denotes the natural push-forward of a local action functional on M to the corresponding action functional on M'. This condition implies that the interacting field algebras are local and covariant.

(r7) **Microlocal condition:** The distributions r in the expansion (27) have the following wave front set:

$$\begin{aligned} \mathrm{WF}(r) \quad \subset \quad & \Big\{ (x, k; y_1, l_1; \dots; y_m, l_m) \in T^* M^{m+1}; \text{there is graph in } \mathcal{G}_{1,n} \text{ such that} \\ & k = \sum_{e:s(e)=x} p_e - \sum_{e:t(e)=x} p_e, \quad l_i = \sum_{e:s(e)=y_i} p_e - \sum_{e:t(e)=y_i} p_e \\ & y_i \in J^-(x) \quad i = 1, \dots, m \Big\} . \end{aligned} \tag{31}$$

The valence of the vertices y_i in the graph is restricted to be less or equal than the maximum power of ϕ occurring an the operators $\mathcal{O}_{i_1}, \dots, \mathcal{O}_{i_n}$ in eq. (27). The microlocal condition implies that the correlation functions have a corresponding microlocal spectrum property, which expresses the conservation of energy-momentum in curved space.

In the formulation of the last condition, we are using a graph theoretical notation [2]. We consider the set $\mathcal{G}_{n,1}$ of embedded, oriented graphs in the spacetime M with $n+1$ vertices corresponding to spacetime points. These vertices are of arbitrary valence, and are joined by edges, e, which are null-geodesic curves $\gamma_e : (0,1) \to M$. It is assumed that an abstract ordering[5] $<$ of the vertices is defined. If e is an edge joining two vertices, then $s(e)$ (the source) and $t(e)$ (the target) are the two vertices $\gamma_e(0)$ and $\gamma_e(1)$, where the curve is oriented in such a way that it starts at the smaller vertex relative to the fixed vertex ordering. Each edge carries a future directed, tangent parallel covector field, p_e, meaning that $\nabla_{\dot\gamma_e} p_e = 0$, and $p_e \in \partial V_+^*$.

Similar to the case of local covariant Wick products (the case $n = 0$), the above functional relations (together with other functional relations described in detail in [13, 14]) do not uniquely fix the retarded products: There remains a number of real constants at each order n which parametrize the set of possible definitions of $\mathbf{R}_n$ that are compatible with (r1)–(r7). These correspond to the usual "renormalization ambiguities" in perturbative quantum field theory, see [14, 15, 16] for details. Further conditions may have to be imposed if one wants to guarantee the conservation of various currents in the theory, e.g., in our case we would like the interacting stress tensor associated with the action S to be conserved

$$\nabla^\mu T_{\mu\nu\, I} = 0 . \tag{32}$$

This requires one to establish a hierarchy of corresponding Ward identities for the $\mathbf{R}_n$-products. The solution to this complicated problem is given in [16].

[5]The ordering is not assumed to be related to the causal structure of the manifold at this stage.

One finally needs to remove the dependence of the interacting fields on the arbitrary cutoff function χ. For this, one investigates how the interacting field changes when the cutoff function is varied. Assume that χ and χ' are two different cutoff functions, both of which are equal to 1 in an open globally hyperbolic neighborhood $U \hookrightarrow M$. Let I and I' be the corresponding interactions. Note that, as a classical functional, the difference $I - I'$ is supported in a compact region, and vanishes in the neighborhood U where the cutoff functions coincide. The key fact [3], which follows from the above functional relations (r1) and (r3), is now that there exists a unitary operator $V \in \mathcal{P} \otimes \mathcal{A}(M,g)_{S_0}$, depending upon I, I', with the property that

$$\mathcal{O}_i(x)_I = V\,\mathcal{O}_i(x)_{I'}\,V^*, \quad \text{for all } x \in U,\ i \in \mathcal{I}. \tag{33}$$

This relation may be interpreted as saying that the algebraic relations between the interacting fields within the region U where the cutoff function is constant do not depend on how the cutoff function is chosen outside this region, and this observation may be used to construct an abstract interacting field algebra $\mathcal{A}(M,g)_S$ associated with the entire spacetime (M,g) that is independent of the choice of cutoff function [3].

3 Operator product expansion

In the previous section, we have outlined how to construct free fields and perturbative interacting fields in the sense of algebras. However, while the construction of the free field was more or less straightforward, the definition of the interacting field algebra was quite indirect. One naturally wonders whether one could not present the interacting field algebra in a somewhat more direct manner, such that the algebraic relations are more explicit, in closer parallel with the relations for a free hermitian field given above in 1)–4).

In a finite dimensional algebra, the relations are encoded in the structure constants of the algebra, so an algebra may be presented explicitly for example by giving its structure constants. Thus, by analogy, what are the "structure constants" of the interacting field algebras $\mathcal{A}(M,g)$? In a non-linear theory, the only relations of simple polynomial type that one can think of are local (anti-) commutativity, and the field equation(s), such as $(\Box - m^2)\phi - \kappa\phi^3 = 0$. Now, the local commutativity fields in any quantum field theory does not specify a specific theory. Also the field equation type relations are essentially empty, unless one specifies precisely how, say, the field ϕ^3 is supposed to be related to ϕ. As we saw above, this is a non-trivial problem in quantum field theory, due to the short distance singularities present in products of fields. However, one can take the point of view [17] that that Wilson's Operator Product Expansion (OPE) should be viewed as an (asymptotic) algebraic relation and hence in particular can be viewed as characterizing a particular model, which one would expect to at least partially characterize an algebra. The OPE is the property that

$$\mathcal{O}_{i_1}(x_1)_I\mathcal{O}_{i_2}(x_2)_I\cdots\mathcal{O}_{i_n}(x_n)_I \sim \sum_j C_{i_1 i_2\dots i_n}{}^j(x_1,x_2,\dots,x_n)\,\mathcal{O}_j(y)_I\,. \tag{34}$$

Here the symbol $\sim$ roughly means that the left side can be approximated to any desired precision by the right side as the points $x_1,\dots,x_n \in M$ approach $y = x_n$, provided sufficiently many terms are included in the right side. The sum is over a finite set fields, such that

$$[j] < \Delta([i_1],\dots,[i_n])\,, \tag{35}$$

where $[\,]: \mathcal{I} \to \mathbb{R}$ is a suitable "dimension function" which assigns to each field (identified with the label $i \in \mathcal{I}$ in the abstract index set $\mathcal{I}$ enumerating the fields) its dimension $[i] \in \mathbb{R}$, and where Δ is some function (depending on the model), also assumed to be specified. It is assumed that only a finite number of terms in the sum are needed. Thus, given an OPE with suitable properties, one might hope to be able to reconstruct the algebras $\mathcal{A}(M, g)_S$ of the model [17].

This hope has not yet been fullfilled in this form, but, starting from the formulation of perturbation theory given in the previous section, it has been possible to construct the operator product expansion [19], and an explicit algorithm for obtaining the coefficients C to all orders in perturbation theory is now known. The basic idea is as follows: Suppose we have linear functionals Ψ_y^k from the algebra into the complex numbers, that are labelled by an index k enumerating the different composite fields, and a reference spacetime point y, and which form a "dual basis" to the composite quantum fields in the sense that $\Psi_y^k(\mathcal{O}_j(y)) = \delta^k{}_j$. Now apply the functional with label k to the OPE. Then we immediately see that the operator product coefficient in front of the k-th term in the sum on the right side of the OPE ought to be given precisely by the c-number distribution obtained by applying that functional to the product of fields on the left side of the OPE. One can give a perturbative construction of such a dual basis of functionals in the context of a scalar, renormalizable field theory model in any 4 dimensional Lorentzian spacetime [19]. In this way, one obtains the desired perturbative formula for the OPE coefficients for renormalizable theories. Non-renormalizable theories do not have a meaningful OPE.

One can derive the following important general properties about the OPE coefficients [19]:

(a) They have a local and covariant dependence upon the spacetime metric.
(b) They satisfy an associativity property.
(c) They satisfy a renormalization group equation.
(d) They satisfy a certain microlocal condition on their so-called wave front set.
(e) They can be expanded in a "scaling expansion".

Let us explain these properties.

(a) The local covariance property of the OPE coefficients states that if (M, g) and (M', g') are globally hyperbolic spacetimes with corresponding OPE coefficients C resp. C', and if $\psi : M \to M'$ is a causality and orientation preserving isometric embedding, then $\psi^* C'$ is equivalent to C at short distances. Thus, in this sense, the OPE coefficients are local functionals of the metric (and the coupling constants), and in particular do not depend upon the large scale structure of spacetime, such as the topology of M. This property can be understood from the fact that, as shown in [15], the interacting fields may be constructed in a local and covariant fashion, in the sense described above. Since the OPE may be interpreted as an asymptotic algebraic relation, and since the local and covariance property as just stated means that algebraic relations only depend on the metric locally and transform covariantly under spacetime embedding, it is natural to expect that also the OPE coefficients depend locally and covariantly upon the metric (and the couplings).

(b) The associativity property arises when one studies the different ways in which a configuration of n points can approach the diagonal in M^n. For example, in the case $n = 3$, we may consider a situation in which all three points approach each other at the same rate, or we may alternatively consider a situation in which two points approach each other faster than the third one. The possible ways in which n points can approach each other may be described by corresponding merger trees $\mathcal{T}$ which characterize the subsequent mergers The associativity property states that scaling together the points in an operator product according to a given tree is equivalent to performing subsequently the OPE in the hierarchical order represented by the tree $\mathcal{T}$. It can be proven [19]—using a general theorem of [17]—that this type of "short distance factorization property" indeed holds in perturbation theory. The associativity may again be understood intuitively from the fact that the OPE coefficients are in some sense the structure "constants" of the abstract associative *-algebra of which the interacting fields are elements. More formally, to characterize the various ways in which a configuration $\vec{x} \in M^n$ may approach the diagonal, one considers the Fulton-MacPherson compactification [11, 1], $M[n]$ of the space $M^n \setminus \cup_S \Delta_S$, where for each $S \subset \{1, \ldots, n\}$ the set Δ_S denotes the associated partial diagonal consisting of all configurations such that $x_i = x_j$ for all S. That compactification is the so-called "blow up" of M^n along the partial diagonals. The blow up $\text{Bl}(X, Y)$ of a manifold X along a submanifold Y is a general construction which can be understood as a generalization of the characteristic example $X = \mathbb{R}^n, Y = \{0\}$. In that example one identifies points in $\mathbb{R}^n \setminus \{0\}$ with their polar coordinates $(r, \widehat{v}) \in \mathbb{R}_{>0} \times S^{n-1}$, and one defines $\text{Bl}(\mathbb{R}^n, \{0\})$ to be the manifold with boundary $\mathbb{R}_{\geq 0} \times S^{n-1}$. The "blow down" map is $b : (r, \widehat{v}) \mapsto r\widehat{v}$, which collapses the entire boundary $\partial \text{Bl}(\mathbb{R}^n, \{0\}) = S^{n-1}$ to the point 0. It is clear in this example that points in the boundary of the blow up correspond to the "directions" from which the origin 0 may be approached.

In our example, the blow up $M[n]$ may be characterized as the set obtained by attaching to $M^n \setminus \cup_S \Delta_S$ a new type of boundary, $\partial M[n]$. The points in this boundary may be thought of as the endpoints of certain curves $\vec{x}(\varepsilon)$, which meet the boundary at $\varepsilon = 0$, and which describe a subsequent merger of the points in $\vec{x}$ as $\varepsilon \to 0$, according to a merger tree, $\mathcal{T}$. The following figure illustrates this definition in an example:

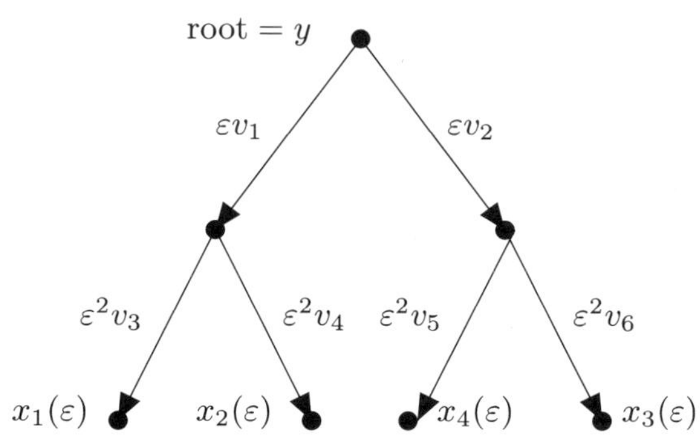

$$x_1(\varepsilon) = \varepsilon v_1 + \varepsilon^2 v_3, \; x_2(\varepsilon) = \varepsilon v_1 + \varepsilon^2 v_4, \; x_3(\varepsilon) = \varepsilon v_2 + \varepsilon^2 v_5, \; x_4(\varepsilon) = \varepsilon v_2 + \varepsilon^2 v_6$$

Here, v_i are vectors in the tangent space $T_y M$, which is identified with a neighborhood of y via the exponential map. If $\psi_{\mathcal{T}}(\varepsilon) : M^n \to M^n$ is the map $\vec{x}(1) \to \vec{x}(\varepsilon)$, then an equivalence relation $\sim_{\mathcal{T},\delta}$ may be introduced for distributions

u_1, u_2 on M^n by

$$u_1 \sim_{\mathcal{T},\delta} u_2 \quad :\Longleftrightarrow \quad \lim_{\varepsilon \to 0+} \varepsilon^{-\delta} (u_1 - u_2) \circ \psi_{\mathcal{T}}(\varepsilon) = 0 \,, \tag{36}$$

The OPE is supposed to hold in the sense of this equivalence relation, for all trees $\mathcal{T}$. The associativity property is the statement that

$$\begin{aligned} &C_{i_1 \dots i_n}{}^{j}(x_1, \dots, x_n)_I \\ &\sim_{\mathcal{T},\delta} \sum_{\vec{i}} \prod_{S \in \mathcal{T}} C_{\{i_{t(e)}; e \text{ such that } s(e) = S\}}{}^{i_S} \left(\{x_{t(e)}; e \text{ such that } s(e) = S\} \right)_I \,, \end{aligned} \tag{37}$$

where the sum is over all $\vec{i} : \mathcal{T} \to \mathcal{I}$, with the properties that

$$i_{\{k\}} = i_k, \quad k = 1, \dots, n, \quad i_{\{1,\dots,n\}} = j, \tag{38}$$

$$\sum_{e : s(e) = S} [i_{t(e)}] < \delta_S \quad \forall S \in \mathcal{T} \,, \tag{39}$$

where $\delta_S > 0$ are chosen sufficiently large. Here, a tree $\mathcal{T}$ is identified with a system of nested subsets $S_i \in \{1, \dots, n\}$ as in the following figure, and $s(e)$ is the subset which is the source of an edge of the tree, while $t(e)$ is the subset which is the target.

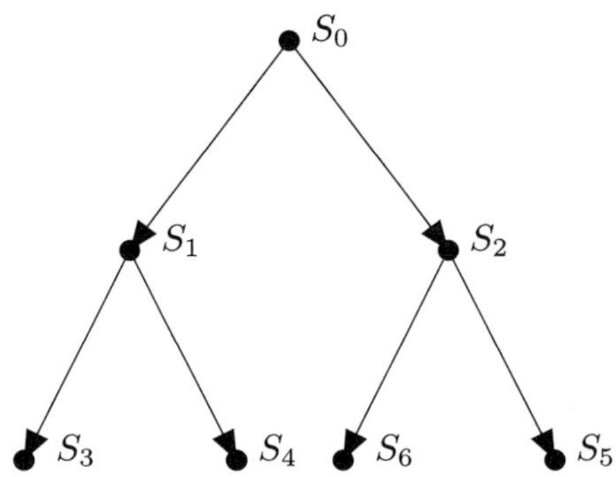

$$\mathcal{T} = \{S_0, S_1, \dots, S_6\}$$
$$S_0 = \{1,2,3,4\}, \ S_1 = \{1,2\}, \ S_2 = \{3,4\},$$
$$S_3 = \{1\}, \ S_4 = \{2\}, \ S_5 = \{3\}, \ S_6 = \{4\}$$

(c) The renormalization property of the OPE coefficients arises from the fact that the algebras of quantum fields satisfy a similar property [15]. Namely, if we rescale the metric by a constant conformal factor λ^2, then this is equivalent (in the sense of giving rise to isomorphic algebras) to redefining the field generators and at the same time letting the coupling constants of the interaction (which enter the structure of the algebra) flow in a particular way dictated by the "renormalization group flow" of the theory [15]. More precisely, there exists a flow of the couplings $\lambda \mapsto \kappa_i(\lambda) \in \mathcal{P}$ and $\lambda \mapsto Z^i{}_j(\lambda) \in \mathrm{Hom}(i,j) \otimes \mathcal{P}$, such that, if $I(\lambda)$ is the corresponding interaction, and $S(\lambda)$ is the full action (whose free part is defined with respect to the metric $\lambda^2 g$ and mass $\lambda^2 m^2$) then

$$R_\lambda : \mathcal{O}_i(x) \mapsto Z_i{}^j(\lambda) \mathcal{O}_j(x)_{I(\lambda)} \tag{40}$$

defines an isomorphism R_λ between the interacting field algebra $\mathcal{A}(M,\lambda^2 g)_S \to \mathcal{A}(M,g)_{S(\lambda)}$. Since the OPE coefficients are in a sense the structure constants of the algebra, they can be shown to have a corresponding invariance property.

(d) The microlocal property is a property describing the nature of the singularities in an OPE coefficient C. It states that the wave front set [21] $\mathrm{WF}(C)$ has a characteristic form that encodes the positivity of energy momentum in the tangent space. $\mathrm{WF}(C)$ is similar in nature to the so-called "microlocal spectrum condition" [2, 25] for the wave front set of correlation functions of linear field theories in curved space. However, our condition differs qualitatively from that proposed in [2, 25] in that the interactions may affect the form of the wave front set.

(e) The scaling expansion states that, if we scale n points $x_1, x_2, \dots, x_n$ in M together according to a merger tree, then a given OPE coefficient C can be approximated to any desired precision by a finite sum of terms each of which has the form of a polynomial in the mass and curvature tensors at point $y = x_n$, times a Lorentz invariant Minkowski distribution in the Riemannian normal coordinates of $x_1, \dots, x_{n-1}$ relative to $y = x_n$. These Minkowski distributions have, to each order in perturbation theory, a simple homogeneous scaling behavior modified by polynomials in the logarithm. They may be extracted from the given OPE coefficients by taking a certain "Mellin moment", which is an operation defined by first taking the Mellin-transform of a function and then extracting certain residue.

For example, the scaling expansion for the OPE coefficient in front of ϕ_I in the product $\phi_I\phi_I\phi_I$ for $I = \int \kappa\phi^4\, e$ is given to zeroth order in κ by [19]

$$C(x_1,x_2,x_3)_{(0)} = \frac{1}{2\pi^2}\sum_{i<j}\left\{\frac{1}{(\xi_i-\xi_j)^2} - \frac{1}{3}\frac{R_{\mu\nu\sigma\rho}\xi_i^\mu\xi_j^\nu\xi_i^\sigma\xi_j^\rho}{(\xi_i-\xi_j)^4} - \frac{1}{6}\frac{R_{\mu\nu}(\xi_i^\mu\xi_i^\nu+\xi_i^\mu\xi_j^\nu+\xi_j^\mu\xi_j^\nu)}{(\xi_i-\xi_j)^2} + \frac{1}{12}R\ln(\xi_i-\xi_j)^2\right\}, \tag{41}$$

and to first order in κ by

$$C(x_1,x_2,x_3)_{(1)} = \frac{\kappa}{2^6\pi^4}\frac{1}{a(\xi_1,\xi_2,\xi_3)}\sum_{i=1}^{3}\mathrm{Cl}_2\left(2\arctan\frac{a(\xi_1,\xi_2,\xi_3)}{(\xi_i-\xi_{i+1})(\xi_{i+1}-\xi_{i+2})}\right) + (\text{curvature terms}) \tag{42}$$

where ξ_i are the Riemannian normal coordinates of x_i, where a is the area of the triangle spanned by ξ_i (supposed to be in a spacelike plane), and where Cl_2 is the Clausen function. Here, we are giving only the first terms in the scaling expansion at each order in κ, but one can in principle calculate arbitrary many higher terms (which have correspondingly higher curvature terms).

4 Outlook

As we have described, there is now available a clean conceptual framework for the discussion of quantum field theory in curved spacetimes, and there are powerful techniques to construct concrete models in pertrubation theory. However, there remain many unanswered questions. Let us name some of them.

Quantum gauge theories: One obvious challenge is to try to generalize the perturbative constructions to quantum gauge theories in 4 dimensions. Here, the principle of gauge invariance raises a new issue, which is not so easy to solve even in flat space and requires the machinery of BRST cohomolgy. In the algebraic framework of perturbation theory sketched above, one would first construct algebras $\mathcal{B}(M,g)_S$ that contain besides the physical interacting field observables also a set of auxilary ghost fields and possibly anti-fields, and that depend upon background gauge fields and a choice of gauge fixing procedure. The BRST transformation would then be defined as a graded derivation $\delta_{\rm B} : \mathcal{B}(M,g)_S \to \mathcal{B}(M,g)_S$, i.e., a linear map satisfying

$$\delta_{\rm B}(F_1 F_2) = \delta_{\rm B}(F_1)F_2 \pm F_1 \delta_{\rm B}(F_2)\,, \tag{43}$$

where the "+" sign is taken according to the Grassmann parity of F_1 and F_2. Given such a structure, the true algebra of physical gauge invariant observables $\mathcal{A}(M,g)_S$ would then be defined as the factor algebra $\text{Ker}\delta_{\rm B}/\text{Im}\delta_{\rm B}$, i.e., the cohomology of the BRST derivation. The idea is that this factor algebra would be manifestly independent of the background gauge fields and gauge fixing procedures, but of course one would need to prove that.

These ideas have been developed to some extend by [6, 7, 8, 9] in Minkowski space, and we believe that they should also apply in curved space. But there still remain unsolved issues how to construct the BRST-transformations (as graded derivations) at the quantum level in a consistent way. We think that carrying out this program in detail is a highly interesting problem.

Physical states: Another set of questions is related to the issue of choosing physical states in curved spacetimes. Namely, given a spacetime (M,g) and the algebra $\mathcal{A}(M,g)$, how to find the state(s) describing a physical situation? The answer to this question obviously depends on precisely what the physical situation is, but we may consider various "standard situations". For example given a region $U \subset M$ much "smaller" than the typical curvature radius of U. Is there a class of states that approximately describe the vacuum within the region U? We feel that the answer to this question should be yes, but we have as yet not been able to come up with a simple, unambiguous algorithm for constructing such a state. Similarly, one may ask for states that are locally in thermal equilibrium. Finding states that are in local thermal equilibrium would be an important step towards finding notions of local phase space densities, and deriving suitable "transport equations" for such quantities analogous to the Bolzmann equation.

Spacetime position measurements: In quantum field theory, spacetime plays the role of a parameter space, which serves to label the field operators that are the true quantum observables. However, one may ask how to construct the true physical quantum operators for measurements of space and time position. This question is analogous to the question how to define a "time-operator" in quantum mechanics (which is to be distinguished from the parameter t appearing e.g. as labels of the Heisenberg operators $\hat{q}(t)$). It is well known that there are problems with a naive definition of such an observable, but recently a new proposal has been made in [4], which has now also been generalized to quantum field theory on Minkowski spacetime [10], and interesting commutation relations for the spacetime position operators are obtained. Unfortunately, these prescriptions still make heavy use of the special structures of Minkowski spacetime, and it appears that new ideas would be required in order to generalize these constructions to curved spacetime.

Beyond perturbation theory: One may finally ask about ways to construct interacting field models non-perturbatively. Of course, this problem is essentially open even for Minkowski quantum field theory in 4 dimensions, and it seems obvious that going to curved space will still complicate matters. However, we feel that the insights into the formulation of field theory in curved spacetime gained so far suggest new ways and directions how this ambitious goal may be achieved. Traditionally, one typically tries to define the theory via its correlation functions in the vacuum, mostly using a Euclidean formulation of the theory in which those correlation functions are given in terms of a path integral. The task is then to make sense of the ill-defined Lebesgue measure in the path integral, and this is a very challenging problem that has so far not been carried out successfully in 4 dimensions. Maybe one reason for the failure of that strategy so far is that one is trying to construct too many things at the same time: The vacuum state *and* the interacting quantum fields. Quantum field theory in curved spacetime teaches us that it is fruitful to disentangle these issues, by first constructing the quantum fields as members of an abstract algebra, and then finding interesting quantum states in a second step.

How would one try to define the algebras $\mathcal{A}(M, g)$ non-perturbatively for a given model, say, an asymptotically free gauge theory? Of course, we do not have an answer to this question, but we suggest that it may be a good approach to first try to construct the algebra $\mathcal{A}(M, g)$ via its Operator Product Expansion, and leave the analysis of physical states for a second step. Of course, the OPE is not an algebraic relation in the usual sense but only asymptotically, but one may still hope that it characterizes an algebra. We believe that the OPE coefficients might be easier to construct because, unlike quantum states, they are local in nature and independent of the (in general complicated) long distance properties of the theory which are reflected in the quantum states and corresponding correlation functions.

We even propose (provocatively) that the OPE coefficients might be given in terms of a convergent expansion in the coupling constant. Of course, it is "well-known" that the perturbation series in interacting quantum field theories in 4 dimensions do not converge, so let us try to clarify our claim. The arguments usually brought up against the convergence of the perturbation series (or the convergence of a suitably "resummed" series) fall into the following categories:

- "Dyson's argument:" If the perturbation series would converge, then it would define an analytic function in the coupling(s) κ, around $\kappa = 0$. In particular, in $\kappa\phi^4$-theory, we would have convergence for either sign of κ. But one would expect the theory to be unstable for $\kappa < 0$ and therefore the perturbation series for the ground state should break down, thus leading to a contradiction.
- "Number of Feynman diagrams too large:" The typical number of terms contributing to an n-th order perturbative expression (roughly, the number Feynman diagrams) is of order $n!$. Thus, unless there are cancellations, the maginitude of the n-th order perturbative contribution to a quantity should be of order $n!$. This cancels out the $1/n!$ in the denominator of (22) and similar other perturbative series, thus raising serious doubts about the convergence of the expansion. In fact, using this type of argument, it has been shown rigorously that the series for the ground state energy does not converge in certain quantum mechanical systems [27].

- "Instantons:" In a path integral formulation, one sees that so-called "instanton contributions" will typically contribute a factor $\exp(-S) \sim \exp(-const./g^2)$ in non-abelian gauge theories. Such a contribution is not analytic in the gauge coupling g at $g = 0$ (in fact, all Taylor coefficients vanish), so it cannot arise from a perturbation series in g. Also, an instanton is a topologically non-trivial configuration, which one cannot obtain by perturbing the trivial field configuration.
- "Renormalons:" In gauge theories, certain *individual* Feynman diagrams at order $2n$ give contributions of order $n!$. This indicates again that the terms in the perturbation grow too rapidly with the order.

Our basic remark here is that all the above arguments convincingly indicate that the perturbation series for global quantities (such as effective actions, ground states, energies etc.) have no chance to converge, but they do not indicate that also local quantities like the OPE coefficients would have those problems. For example, instantons always refer to the entire spacetime, M (in fact, even the dimension of the instanton spaces are dependent upon the topology of M), while the OPE coefficients are completely insensitive to global properties of M. Renormalons arise typically from the "soft momenta", i.e., the low energy contributions to the mentioned graphs. Again, this is a long distance property and should not be an issue for the OPE coefficients. Of course these arguments do not show that the series for the OPE coefficients converge, but maybe they are an indication that the non-perturbative construction of the OPE coefficients is not impossible.

Acknowledgements: The remarks in section 4 explained in the paragraphs following "Beyond perturbation theory" go back to conversations with R.M. Wald.

References

[1] S. Axelrod and I. M. Singer, "Chern-Simons perturbation theory. 2," J. Diff. Geom. **39**, 173 (1994)

[2] R. Brunetti, K. Fredenhagen and M. Köhler: "The microlocal spectrum condition and Wick polynomials on curved spacetimes," Commun. Math. Phys. **180**, 633-652 (1996)

[3] R. Brunetti and K. Fredenhagen: "Microlocal Analysis and Interacting Quantum Field Theories: Renormalization on physical backgrounds," Commun. Math. Phys. **208**, 623-661 (2000)

[4] R. Brunetti and K. Fredenhagen, "When does a detector click?," arXiv:quant-ph/0103144.

[5] R. Brunetti, K. Fredenhagen and R. Verch, "The generally covariant locality principle: A new paradigm for local quantum physics," Commun. Math. Phys. **237**, 31 (2003), [math-ph/0112041]; see also K. Fredenhagen, "Locally covariant quantum field theory," [arXiv:hep-th/0403007].

[6] M. Dütsch and K. Fredenhagen: "Algebraic quantum field theory, perturbation theory, and the loop expansion," Commun. Math. Phys. **219**, 5 (2002) [arXiv: hep-th/0001129]; "Perturbative algebraic field theory, and deformation quantization," [arXiv: hep-th/0101079]

[7] M. Dütsch and K. Fredenhagen: "The Master Ward Identity and Generalized Schwinger-Dyson Equation in Classical Field Theory", Commun. Math. Phys. **243**, 275 (2003), [hep-th/0211242]

[8] M. Dütsch and K. Fredenhagen: "Causal Perturbation Theory in Terms of Retarded Products, and a Proof of the Action Ward Identity," [hep-th/0403213]

[9] M. Dutsch and K. Fredenhagen, "Perturbative renormalization and BRST," arXiv:hep-th/0411196.

[10] K. Fredenhagen, to appear

[11] W. Fulton and R. MacPherson: "A compactification of configuration spaces," Ann. Math. 139 183 (1994)

[12] R. Haag, "On Quantum Field Theories," Z. Phys. **141**, 217 (1955) [Phil. Mag. **46**, 376 (1955)].

[13] S. Hollands and R. M. Wald: "Local Wick Polynomials and Time Ordered Products of Quantum Fields in Curved Space," Commun. Math. Phys. **223**, 289-326 (2001), [gr-qc/0103074]

[14] S. Hollands and R. M. Wald: "Existence of local covariant time-ordered-products of quantum fields in curved spacetime," Commun. Math. Phys. **231**, 309-345 (2002), [gr-qc/0111108]
[15] S. Hollands and R. M. Wald: "On the Renormalization Group in Curved Spacetime," Commun. Math. Phys. **237**, 123-160 (2003), [gr-qc/0209029]
[16] S. Hollands and R. M. Wald, "Conservation of the stress tensor in interacting quantum field theory in curved spacetimes," Rev. Math. Phys. **17**, 227 (2005)
[17] S. Hollands, R. M. Wald: "On the Operator Product Expansion in Curved Spacetime," in preparation
[18] S. Hollands, "A general PCT theorem for the operator product expansion in curved spacetime," Commun. Math. Phys. **244**, 209 (2004)
[19] S. Hollands, "The operator product expansion for perturbative quantum field theory in curved spacetime," arXiv:gr-qc/0605072.
[20] S. Hollands: in preparation
[21] L. Hörmander: *The Analysis of Linear Partial Differential Operators I,* Berlin, Springer-Verlag 1983
[22] G. Källen, "Formal integration of the equations of quantum theory in the Heisenberg representation," Ark. Fysik **2** (1950) 371
[23] V. Balasubramanian and P. Kraus, "Spacetime and the holographic renormalization group," Phys. Rev. Lett. **83**, 3605 (1999) [arXiv:hep-th/9903190].
[24] J. M. Maldacena, "The large N limit of superconformal field theories and supergravity," Adv. Theor. Math. Phys. **2**, 231 (1998) [Int. J. Theor. Phys. **38**, 1113 (1999)] [arXiv:hep-th/9711200], O. Aharony, S. S. Gubser, J. M. Maldacena, H. Ooguri and Y. Oz, "Large N field theories, string theory and gravity," Phys. Rept. **323**, 183 (2000) [arXiv:hep-th/9905111].
[25] M. J. Radzikowski, "Micro-local approach to the Hadamard condition in quantum field theory on curved space-time," Commun. Math. Phys. **179**, 529 (1996).
[26] K. Skenderis, "Lecture notes on holographic renormalization," Class. Quant. Grav. **19**, 5849 (2002) [arXiv:hep-th/0209067].
[27] B. Simon: "$P(\phi)_2$ Euclidean," Princeton University Press, Princeton (1974)
[28] R. Verch, "A spin-statistics theorem for quantum fields on curved spacetime manifolds in a generally covariant framework," Commun. Math. Phys. **223**, 261 (2001)

Fields Institute Communications
Volume **50**, 2007

Entropy of XY Spin Chain and Block Toeplitz Determinants

A. R. Its
Department of Mathematical Sciences
Indiana University-Purdue University Indianapolis
Indianapolis, IN 46202-3216 USA
itsa@math.iupui.edu

B.-Q. Jin
College of Physics and Electronic Information
Wenzhou University
Wenzhou, Zhejiang, P.R. China
jinbq@wzu.edu.cn

V. E. Korepin
C.N. Yang Institute for Theoretical Physics
State University of New York at Stony Brook
Stony Brook, NY 11794-3840 USA
korepin@insti.physics.sunysb.edu

Abstract. We consider entanglement in the ground state of the XY model on infinite chain. We use von Neumann entropy of a sub-system as a measure of entanglement. The entropy of a large block of neighboring spins approaches a constant as the size of the block increases. We evaluate this limiting entropy as a function of anisotropy and transverse magnetic field. We use integrable Fredholm operators and the Riemann-Hilbert approach. The entropy reaches minimum at highly ordered states but increases boundlessly at phase transitions.

1 Introduction

Entanglement is a primary resource for quantum computation and information processing [22, 23, 1, 8, 9, 10]. It is necessary for quantum control. It shows how much quantum effects we can use to control one system by another. Stable and large scale entanglement is necessary for scalability of quantum computation [27, 26, 24]. Entropy of a subsystem as a measure of entanglement was discovered in [1]. Essential progress is achieved in understanding of entanglement in various quantum systems [5, 6, 7, 4, 11, 12, 13, 14, 15, 27, 26, 28, 16, 17, 21, 30, 51, 25, 18, 29, 52, 20, 19, 32, 51].

XY model in a transverse magnetic field was studied from the point of view of quantum information in [3, 5, 40, 66]. In this paper we evaluate the entropy of

2000 *Mathematics Subject Classification.* Primary 82B23; Secondary 47B35.

a block of L neighboring spins in the ground state of the XY model in the limit $L \to \infty$ analytically. Our approach uses representation of [3] and the Riemann-Hilbert method of the theory of integrable Fredholm operators. The final answer is given in terms of the elliptic functions and is presented in Eq. (1.26) and (1.29) below. The Hamiltonian of the XY model can be written as

$$\mathcal{H} = - \sum_{n=-\infty}^{\infty} (1+\gamma)\sigma_n^x \sigma_{n+1}^x + (1-\gamma)\sigma_n^y \sigma_{n+1}^y + h\sigma_n^z. \tag{1.1}$$

Here $0 < \gamma < 1$ is the anisotropy parameter; σ_n^x, σ_n^y and σ_n^z are the Pauli matrices and h is the magnetic field. The model was solved in [41, 43, 44]. The methods of Toeplitz determinants, as well as the techniques based on integrable Fredholm operators, were used for the evaluation of some correlation functions, see [44, 49, 45, 46, 47, 48].

We consider the ground state $|GS\rangle$ of the model. We evaluate the entropy of a sub-system of the ground state. We shall calculate the entropy of a block of L neighboring spins, it measures the entanglement between the block and the rest of the chain [1]. We treat the whole chain as a binary system $|GS\rangle = |A\&B\rangle$. We denote this block of L neighboring spins by subsystem A and the rest of the chain by subsystem B. The density matrix of the ground state can be denoted by $\rho_{AB} = |GS\rangle\langle GS|$. The density matrix of subsystem A is $\rho_A = Tr_B(\rho_{AB})$. The von Neumann entropy $S(\rho_A)$ of subsystem A can be represented as follows:

$$S(\rho_A) = -Tr_A(\rho_A \ln \rho_A) \tag{1.2}$$

This entropy also defines the dimension of the Hilbert space of states of subsystem A.

A set of Majorana operators were used in [3] with self-correlations described by the following matrix:

$$\mathbf{B}_L = \begin{pmatrix} \Pi_0 & \Pi_{-1} & \dots & \Pi_{1-L} \\ \Pi_1 & \Pi_0 & & \vdots \\ \vdots & & \ddots & \vdots \\ \Pi_{L-1} & \dots & \dots & \Pi_0 \end{pmatrix}$$

Here

$$\Pi_l = \frac{1}{2\pi}\int_0^{2\pi} d\theta\, e^{-\mathrm{i}l\theta}\mathcal{G}(\theta), \quad \mathcal{G}(\theta) = \begin{pmatrix} 0 & g(\theta) \\ -g^{-1}(\theta) & 0 \end{pmatrix}$$

$$\text{and} \qquad g(\theta) = \frac{\cos\theta - \mathrm{i}\gamma\sin\theta - h/2}{|\cos\theta - \mathrm{i}\gamma\sin\theta - h/2|}. \tag{1.3}$$

One can use an orthogonal matrix V to transform $\mathbf{B}_L$ to a canonical form:

$$V\mathbf{B}_L V^T = \oplus_{m=1}^{L} \nu_m \begin{pmatrix} 0 & 1 \\ -1 & 0 \end{pmatrix}, \tag{1.4}$$

The real numbers $-1 < \nu_m < 1$ play an important role. We shall call them eigenvalues. The entropy of a block of L neighboring spins was represented in [3] as

$$S(\rho_A) = \sum_{m=1}^{L} H(\nu_m) \tag{1.5}$$

with

$$H(\nu) = -\frac{1+\nu}{2}\ln\frac{1+\nu}{2} - \frac{1-\nu}{2}\ln\frac{1-\nu}{2}. \tag{1.6}$$

In order to calculate the asymptotic form of the entropy it is not convenient to use formula (1.4) and (1.5) directly. Following the idea we have already used in [4], let us introduce:

$$\widetilde{\mathbf{B}}_L(\lambda) = \mathrm{i}\lambda I_L - \mathbf{B}_L, \quad D_L(\lambda) = \det \widetilde{\mathbf{B}}_L(\lambda) \tag{1.7}$$

and

$$e(x,\nu) = -\frac{x+\nu}{2}\ln\frac{x+\nu}{2} - \frac{x-\nu}{2}\ln\frac{x-\nu}{2}. \tag{1.8}$$

Here I_L is the identity matrix of the size $2L$. By definition, we have $H(\nu) = e(1,\nu)$ and

$$D_L(\lambda) = (-1)^L \prod_{m=1}^{L} (\lambda^2 - \nu_m^2). \tag{1.9}$$

With the help of the Cauchy residue theorem we rewrite formula (1.5) in the following form:

$$S(\rho_A) = \lim_{\epsilon\to 0^+} \frac{1}{4\pi \mathrm{i}} \oint_{\Gamma'} \mathrm{d}\lambda\, e(1+\epsilon,\lambda) \frac{\mathrm{d}}{\mathrm{d}\lambda} \ln D_L(\lambda) \qquad \star \tag{1.10}$$

Here the contour Γ' is depicted in Fig 1; it encircles all zeros of $D_L(\lambda)$. We also

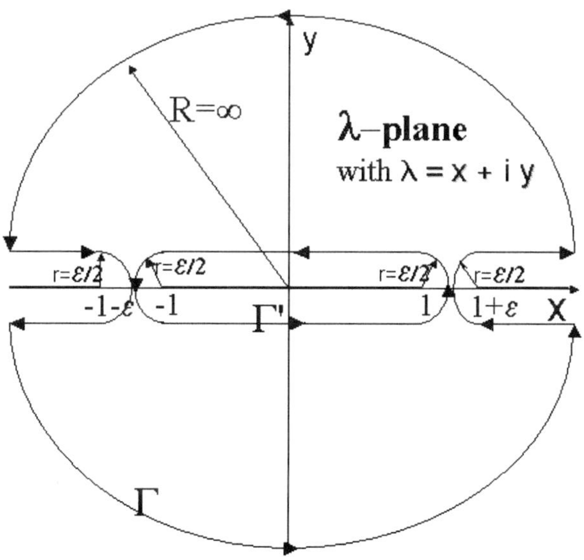

Figure 1 *Contours Γ' (smaller one) and Γ (larger one). Bold lines $(-\infty,-1-\epsilon)$ and $(1+\epsilon,\infty)$ are the cuts of the integrand $e(1+\epsilon,\lambda)$. Zeros of $D_L(\lambda)$ (Eq. 1.9) are located on the bold line $(-1,1)$. The arrows indicate the directions of integrations, and* r *and* R *are the radius of the circles.* ¶

notice that $\widetilde{\mathbf{B}}_L(\lambda)$ is the block Toeplitz matrix,

$$\widetilde{\mathbf{B}}_L(\lambda) = \begin{pmatrix} \widetilde{\Pi}_0 & \widetilde{\Pi}_{-1} & \dots & \widetilde{\Pi}_{1-L} \\ \widetilde{\Pi}_1 & \widetilde{\Pi}_0 & & \vdots \\ \vdots & & \ddots & \vdots \\ \widetilde{\Pi}_{L-1} & \dots & \dots & \widetilde{\Pi}_0 \end{pmatrix} \quad \text{with}$$

$$\widetilde{\Pi}_l = \frac{1}{2\pi \mathrm{i}} \oint_{\Xi} \mathrm{d}z\, z^{-l-1} \Phi(z), \tag{1.11}$$

where the matrix generator $\Phi(z)$ is defined by the equations,

$$\Phi(z) = \begin{pmatrix} \mathrm{i}\lambda & \phi(z) \\ -\phi^{-1}(z) & \mathrm{i}\lambda \end{pmatrix} \tag{1.12}$$

and

$$\phi(z) = \left(\frac{\lambda_1^*}{\lambda_1} \frac{(1-\lambda_1 z)(1-\lambda_2 z^{-1})}{(1-\lambda_1^* z^{-1})(1-\lambda_2^* z)} \right)^{1/2}. \tag{1.13}$$

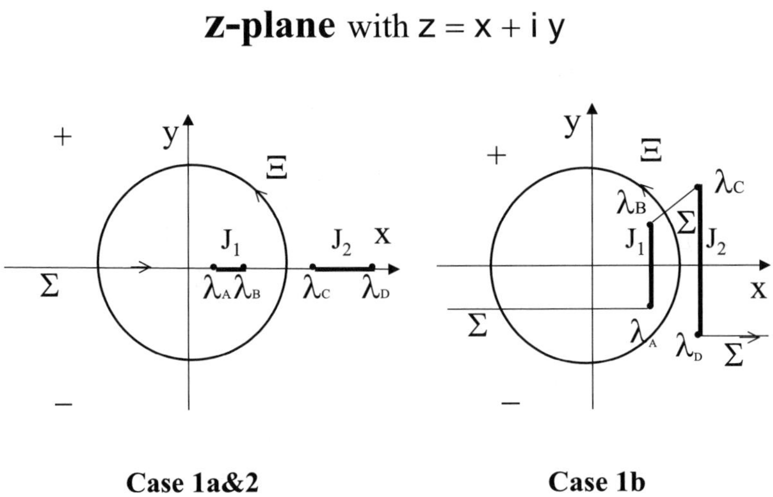

Figure 2 *The polygonal line Σ (oriented as it is indicated) separates the complex z plane into two parts: the part Ω_+ which lies to the left of Σ, and the part Ω_- which lies to the right of Σ. Curve Ξ is the unit circle in anti-clockwise direction. Cuts J_1, J_2 for the functions $\phi(z), w(z)$ are labeled by bold on the line Σ. The definition of the end points of the cuts $\lambda_{\dots}$ depends on the case:* **Case** *1a: $\lambda_A = \lambda_1$, $\lambda_B = \lambda_2^{-1}$, $\lambda_C = \lambda_2$, and $\lambda_D = \lambda_1^{-1}$.* **Case** *1b: $\lambda_A = \lambda_1$, $\lambda_B = \lambda_2^{-1}$, $\lambda_C = \lambda_1^{-1}$, and $\lambda_D = \lambda_2$.* **Case** *2: $\lambda_A = \lambda_1$, $\lambda_B = \lambda_2$, $\lambda_C = \lambda_2^{-1}$, and $\lambda_D = \lambda_1^{-1}$.* ¶

We fix the branch of $\phi(z)$ by requiring that $\phi(\infty) > 0$. We use $*$ to denote complex conjugation, and Ξ is the unit circle shown in Fig. 2. The points λ_1 and λ_2 are defined differently for the different values of γ and h.

There are three different cases:
Case 1a **is defined by inequality** $2\sqrt{1-\gamma^2} < h < 2$.
It describes moderate magnetic field.
Case 2 **is defined by** $h > 2$. This is strong magnetic field.
In both cases λ_1 and λ_2 are real and given by the formulae

$$\lambda_1 = \frac{h - \sqrt{h^2 - 4(1-\gamma^2)}}{2(1+\gamma)}, \quad \lambda_2 = \frac{1+\gamma}{1-\gamma}\lambda_1. \tag{1.14}$$

Case 1b **is defined by** $h^2 < 4(1-\gamma^2)$.
It is weak magnetic field, including zero magnetic field.
Both λ_1 and λ_2 are complex and given by the equations

$$\lambda_1 = \frac{h - \mathrm{i}\sqrt{4(1-\gamma^2) - h^2}}{2(1+\gamma)}, \quad \lambda_2 = 1/\lambda_1^*. \tag{1.15}$$

Note that in Case 1 the poles of the function $\phi(z)$ (Eq. 1.13) coincide with the points λ_A and λ_B, while in Case 2 they coincide with the points λ_A and λ_C.

We are going now to formulate our main results. To this end we need to introduce the Jacobi theta-function,

$$\theta_3(s) = \sum_{n=-\infty}^{\infty} e^{\pi i \tau n^2 + 2\pi i s n}. \tag{1.16}$$

We remind the following characteristic properties of this function (see e.g. [39]):

$$\theta_3(-s) = \theta_3(s), \quad \theta_3(s+1) = \theta_3(s) \tag{1.17}$$

$$\theta_3(s+\tau) = e^{-\pi i \tau - 2\pi i s}\theta_3(s) \tag{1.18}$$

$$\theta_3\left(n + m\tau + \frac{1}{2} + \frac{\tau}{2}\right) = 0, \quad n, m \in \mathbb{Z}. \tag{1.19}$$

The modulus parameter τ, in our case, is determined by the physical quantities h and γ according to the following equations.

$$\tau = i\tau_0, \quad \tau_0 = I(k')/I(k) \tag{1.20}$$

where $I(k)$ denotes the complete elliptic integral of the first kind,

$$I(k) = \int_0^1 \frac{dx}{\sqrt{(1-x^2)(1-k^2x^2)}} \qquad \diamond$$

also $k' = \sqrt{1-k^2}$, and

$$k = \begin{cases} \sqrt{(h/2)^2 + \gamma^2 - 1} \,/\, \gamma, & \text{Case 1a} \\ \sqrt{1 - \gamma^2 - (h/2)^2} \,/\, \sqrt{1 - (h/2)^2}, & \text{Case 1b} \\ \gamma \,/\, \sqrt{(h/2)^2 + \gamma^2 - 1}, & \text{Case 2} \end{cases} \tag{1.21}$$

Define also the function

$$\beta(\lambda) = \frac{1}{2\pi i} \ln \frac{\lambda+1}{\lambda-1}, \tag{1.22}$$

and the infinite sequence of real numbers,

$$\lambda_m = \tanh\left(m + \frac{1-\sigma}{2}\right)\pi\tau_0, \quad m \geq 0, \tag{1.23}$$

where $\sigma = 1$ in Case 1 and $\sigma = 0$ in Case 2. Observe that

- $0 < \lambda_m < 1, \quad \lambda_m \to 1, \quad m \to \infty$
- in view of Eq. (1.19), the points $\pm\lambda_m$ are zeros of $\theta_3\left(\beta(\lambda) + \frac{\sigma\tau}{2}\right)$.

Theorem 1. *Let Ω be the complex λ - plane without arbitrary fixed neighborhoods of the points $\lambda = \infty$, $\lambda = \pm 1$ and $\lambda = \pm\lambda_m$, $m = 0, 1, \dots$. Then the Toeplitz determinant $D_L(\lambda)$ admits the following asymptotic representation, which is uniform in $\lambda \in \Omega$.*

$$\begin{aligned} \frac{d}{d\lambda} \ln D_L(\lambda) &= -\frac{2\lambda}{1-\lambda^2} L + \frac{d}{d\lambda} \ln \left[\theta_3\left(\beta(\lambda) + \frac{\sigma\tau}{2}\right)\theta_3\left(\beta(\lambda) - \frac{\sigma\tau}{2}\right)\right] \\ &\quad + O\left(\rho^{-L}\right), \quad L \to \infty. \end{aligned} \tag{1.24}$$

Here ρ is any positive number satisfying the inequality,

$$1 < \rho < |\lambda_C|.$$

This theorem shows that in the large L limit, the points $\pm\lambda_m$ (1.23) are double zeros of the $D_L(\lambda)$. More precisely, we see that in the large L limit the eigenvalues ν_{2m} and ν_{2m+1} from (1.5), (1.4) merge to λ_m:

$$\nu_{2m}, \nu_{2m+1} \to \lambda_m, \tag{1.25}$$

which in turn implies our **main result**:

The limiting entropy, $S(\rho_A)$, of the subsystem can be identified with the infinite convergent series,

$$\diamondsuit \qquad S(\rho_A) = \sum_{m=-\infty}^{\infty} H(\lambda_m) = \sum_{m=-\infty}^{\infty} (1+\lambda_m) \ln \frac{2}{1+\lambda_m} \qquad \diamondsuit \tag{1.26}$$

It is worth mentioning that relation (1.25) also indicates the degeneracy of the spectrum of the matrix $\mathbf{B}_L$ and an appearance of an **extra symmetry** in the large L limit.

Observe that equation (1.10) can be also rewritten as

$$S_L(\rho_A) = \lim_{\epsilon \to 0+} \frac{1}{4\pi \mathrm{i}} \oint_{\Gamma'} \mathrm{d}\lambda\, e(1+\epsilon, \lambda) \frac{\mathrm{d}}{\mathrm{d}\lambda} \ln\left(D_L(\lambda)(\lambda^2-1)^{-L}\right), \tag{1.27}$$

because of the following identity

$$\begin{aligned} \lim_{\epsilon \to 0+} \oint_{\Gamma'} \mathrm{d}\lambda\, e(1+\epsilon, \lambda) \frac{\mathrm{d}}{\mathrm{d}\lambda} \ln(\lambda^2-1)^{-L} &= L \lim_{\epsilon \to 0+} \oint_{\Gamma'} \mathrm{d}\lambda\, e(1+\epsilon, \lambda) \frac{2\lambda}{1-\lambda^2} \\ &= 2\pi i L \lim_{\epsilon \to 0+} \left((2+\epsilon)\ln\frac{2+\epsilon}{2} + \epsilon \ln\frac{\epsilon}{2}\right) = 0. \end{aligned}$$

In this paper, we will show that series (1.26) coincides with the result of the following double limit procedure

$$S(\rho_A) = \lim_{\epsilon \to 0+} \left[\lim_{L \to \infty} \frac{1}{4\pi \mathrm{i}} \oint_{\Gamma'} \mathrm{d}\lambda\, e(1+\epsilon, \lambda) \frac{\mathrm{d}}{\mathrm{d}\lambda} \ln\left(D_L(\lambda)(\lambda^2-1)^{-L}\right)\right], \tag{1.28}$$

here the contour Γ' is depicted in Fig 1. This result can be alternatively written down as the following integral,

$$\clubsuit \qquad S(\rho_A) = \frac{1}{2} \int_1^{\infty} \ln\left(\frac{\theta_3\left(\beta(\lambda) + \frac{\sigma\tau}{2}\right)\theta_3\left(\beta(\lambda) - \frac{\sigma\tau}{2}\right)}{\theta_3^2\left(\frac{\sigma\tau}{2}\right)}\right) d\lambda. \qquad \spadesuit \tag{1.29}$$

We conjecture that in fact the limits in (1.28) can be interchanged. After the shorten version of this paper appeared in quant-ph, I.Peschel [64] simplified our expression for the entropy for non-vanishing magnetic field [Cases 1a and 2]. He used the approach of [13]. He showed that in these cases our formula (1.26) is equivalent to formula (4.33) of [13]. Moreover, I. Peschel was able to sum it up into the following expressions for the entropy.

$$S(\rho_A) = \frac{1}{6}\left[\ln\left(\frac{k^2}{16k'}\right) + \left(1 - \frac{k^2}{2}\right)\frac{4I(k)I(k')}{\pi}\right] + \ln 2, \tag{1.30}$$

in **Case 1a**, and

$$S(\rho_A) = \frac{1}{12}\left[\ln\frac{16}{(k^2k'^2)} + (k^2 - k'^2)\frac{4I(k)I(k')}{\pi}\right], \tag{1.31}$$

in **Case 2** . Here, $I(k)$ denotes the complete elliptic integral of the first kind, $k' = \sqrt{1-k^2}$, and

$$k = \begin{cases} \sqrt{(h/2)^2 + \gamma^2 - 1}\ /\ \gamma, & \text{Case 1a} \\ \gamma\ /\ \sqrt{(h/2)^2 + \gamma^2 - 1}, & \text{Case 2} \end{cases} \tag{1.32}$$

These describes cases of moderate and strong magnetic field.

We used the equation (1.26) to calculate limiting entropy for weak magnetic field.

In **Case 1b** we derived:

$$\begin{aligned} S(\rho_A) &= \tfrac{1}{6}\left[\ln\left(\tfrac{k^2}{16k'}\right) + \left(1 - \tfrac{k^2}{2}\right)\tfrac{4I(k)I(k')}{\pi}\right] + \ln 2, \\ & \qquad k = \sqrt{\tfrac{1-h^2/4-\gamma^2}{1-h^2/4}} \qquad \heartsuit \\ \text{iff} & \qquad 0 \le h < 2\sqrt{1-\gamma^2} \end{aligned} \tag{1.33}$$

Note that this case includes zero magnetic field.

These expressions helped us to study the range of variation of limting entropy. Together with Dr. Franchini we found that the entropy has a local minimum at the boundary of **Cases 1a and 1b**:

$$S(\rho) = \ln 2 \qquad \text{at} \qquad \left(\frac{h}{2}\right)^2 + \gamma^2 = 1 \qquad \star$$

At this boundary the ground state is doubly degenerated, but the rest of energy levels are separated by a gap.

Note that the absolute minimum of asymptotic entropy $\min S(\rho) = 0$ is achieved at infinite magnetic field corresponding to $k = 0$ in Case 2 [the ground state is ferromagnetic].

2 The asymptotic of block Toeplitz determinants, Widom's theorems

Our objective is the asymptotic calculation of the block Toeplitz determinant $D_L(\lambda)$ or, rather, its λ-derivative $\frac{d}{d\lambda}\ln D_L(\lambda)$. A general asymptotic representation of the determinant of a block Toeplitz matrix, which generalizes the classical strong Szegö theorem to the block matrix case, was obtained by H. Widom in [53] (see also more recent work [62] and references therein). Here is Widom's result.

Let

$$\varphi(z) = \sum_{k=-\infty}^{\infty} \varphi_k z^k, \quad |z| = 1,$$

be an $p \times p$ matrix - valued function defined on the unite circle, $|z| = 1$, and satisfying the following conditions.

1. $||\varphi|| \equiv \sum_{k=-\infty}^{\infty} ||\varphi_k|| + \{\sum_{k=-\infty}^{\infty} |k| ||\varphi_k||^2\}^{1/2} < \infty$
2. $\det \varphi(z) \neq 0, \quad \Delta|_{|z|=1} \arg \det \varphi(z) = 0.$

Consider the block Toeplitz determinant generated by $\varphi(z)$, i.e. $D_L[\varphi] = \det T_L[\varphi]$ where $T_L[\varphi] = (\varphi_{j-k})$, $j, k = 0, ..., L-1$.

Theorem 2. ([54])*Define*[1]

$$G[\varphi] = \exp\left\{\frac{1}{2\pi i}\int_{|z|=1} \ln\det\varphi(z)\frac{dz}{z}\right\}. \tag{2.1}$$

Then the limit

$$E[\varphi] = \lim_{L\to\infty} \frac{D_L[\varphi]}{(G[\varphi])^L}, \tag{2.2}$$

exists. Moreover, the following general formula can be written for the quantity $E[\varphi]$,

$$E[\varphi] = \det\left(T_\infty[\varphi]T_\infty[\varphi^{-1}]\right). \tag{2.3}$$

This quite beautiful theorem is not very efficient in concrete applications. The remarkable fact though is that the proof of theorem 2 is based on an auxiliary fact which was established in the preceding work of Widom and which can be made an efficient tool for a large class of symbols φ, including the matrix function $\Phi(z)$ from (1.12) which we are concerned with in this paper.

Theorem 3 (theorem 4.1 of [53])*Suppose that, in addition to the conditions of theorem 2, the matrix symbol* φ^{-1} *admits the Weiner-Hopf factorization,*

$$\varphi^{-1}(z) = u_+(z)u_-(z) = v_-(z)v_+(z), \tag{2.4}$$

where the subscribes "+" and "-" indicate the analyticity inside and outside of the unite circle, respectively. Suppose also that $\varphi(z)$ *can be included into a differentiable family,* $\lambda \to \varphi(z,\lambda)$. *Then,* $\ln E[\varphi]$ *is a differentiable function of* λ *and in fact,*

$$\frac{d}{d\lambda}\ln E[\varphi] = \frac{i}{2\pi}\int_{|z|=1} trace\left[\left(u'_+(z)u_-(z) - v'_-(z)v_+(z)\right)\frac{\partial\varphi(z,\lambda)}{\partial\lambda}\right]dz, \tag{2.5}$$

where ($'$) *means the derivative with respect to* z.

Let Ω be the set on the λ - plane introduced in theorem 1. In the next section we will show that for every $\lambda \in \Omega$, the matrix-valued function $\Phi(z)$ admits the *explicit* Weiner-Hopf factorization, and hence the second Widom theorem - theorem 3 above, is applicable. Indeed, for the function $\Phi(z)$ this theorem can be specified as follows.

[1]The integration along the unit circle is always assumed to be done in the positive, i.e. anti-clockwise direction.

Theorem 4. *Let $\lambda \in \Omega$ then the following asymptotic representation for the logarithmic derivative of the determinant $D_L(\lambda) = \det T_L[\Phi]$ takes place.*

$$\begin{aligned} \frac{d}{d\lambda}\ln D_L(\lambda) &= -\frac{2\lambda}{1-\lambda^2}L \\ &+\frac{1}{2\pi}\int_{|z|=1} trace\left[\left(U_+'(z)U_+^{-1}(z)+V_+^{-1}(z)V_+'(z)\right)\Phi^{-1}(z)\right]dz \\ &+r_L(\lambda), \end{aligned} \tag{2.6}$$

where the error term $r_L(\lambda)$ satisfies the uniform estimate,

$$|r_L(\lambda)| \le C\rho^{-L}, \qquad \lambda\in\Omega, \quad L\ge 1, \tag{2.7}$$

and ρ is any positive number such that $1<\rho<|\lambda_C|$. In (2.6) the 2×2 matrix-valued functions $U_\pm(z)$ and $V_\pm(z)$ solve the following Weiner-Hopf factorization problem:

(i) $\Phi(z) = U_+(z)U_-(z) = V_-(z)V_+(z) \quad z\in\Xi$

(ii) *$U_-(z)$ and $V_-(z)$ ($U_+(z)$ and $V_+(z)$) are analytic outside (inside) the unit circle Ξ.*

(iii) $U_-(\infty) = V_-(\infty) = I$.

Remark. The replacement of the factorization of the inverse symbol by the factorization of the symbol itself has been made for a purely technical reason.

Asymptotic representation (2.6) plays an important role in our analysis as, in fact, its starting point. The truth of the matter is that we had derived formula (2.6) before we became aware of Widom's second theorem. In our original derivation of (2.6) and proof of theorem 4 we used an alternative approach to Toeplitz determinants suggested by P. Deift in [31]. It is based on the Riemann-Hilbert technique of the theory of "integrable integral operators", which was developed in [33], [63] for evaluation of correlation functions of quantum completely integrable (exactly solvable) models [2]. It turns out that, using the block matrix version of [33] suggested in [37], one can generalize Deift's scheme to the block Toeplitz matrices, which provides a rather simple derivation of equation (2.6) together with the estimation of the error term indicated in theorem 4.

The Riemann-Hilbert approach has already proved its usefulness in the theory of Toeplitz and Hankel determinants with scalar symbols (see e.g. [55], [69], [56], [57], [58]). Therefore, we believe that the block-version of the Riemann-Hilbert scheme is worthwhile to present. Indeed, although in the relatively simple case of smooth symbols theorem 4, as it turned out, is a direct (up to the error term estimation) corollary of one of the old Widom's results, it is conceivable that in more challenging situations of the matrix symbols with singularities the Riemann-Hilbert scheme might become very useful, as it has been already the case for the scalar symbols.

Having all the above reasons in mind, we have decided to include in the paper our original proof of theorem 4. We will do this in the next section providing all the

[2]In its turn, the approach of [33] is based on the ideas of [34]. Several principal aspects of the integrable operator theory, especially the ones concerning with the integrable differential systems appearing in random matrix theory, have been developed in [35]. Some of the important elements of modern theory of integrable operators were already implicitly present in the earlier work [36].

necessary facts concerning integrable Fredholm operators and the Riemann-Hilbert method of their analysis.

The explicit factorization of the matrix $\Phi(z)$ will be performed in section 4 and the final proof of our main result - theorem 1, will be done in section 5.

3 The Riemann-Hilbert approach to the block Toeplitz determinants. An alternative proof of theorem 4

3.1 The Fredholm determinant representation. Let $f_j(z)$ and $h_j(z)$, $j = 1, 2$, be 2×2 matrix functions. We introduce the class of **integrable operators** K defined on $L_2(\Xi, \mathbb{C}^2)$ by the following equations (cf. [37]),

$$(K\,X)(z) = \oint_{\Xi} K(z, z')\,X(z')dz' \quad \text{for} \quad X \in L_2, \tag{3.1}$$

where

$$K(z, z') = \frac{f^T(z)h(z')}{z - z'}, \quad f(z) = \begin{pmatrix} f_1^T(z) \\ f_2^T(z) \end{pmatrix}$$
$$h(z) = \begin{pmatrix} h_1(z) \\ h_2(z) \end{pmatrix}. \tag{3.2}$$

Let I_2 denote the 2×2 identity matrix. Put

$$f_1(z) = z^L I_2, \quad f_2(z) = I_2 \tag{3.3}$$

$$h_1(z) = z^{-L}\frac{I_2 - \Phi(z)}{2\pi i}, \quad h_2(z) = -\frac{I_2 - \Phi(z)}{2\pi i}. \tag{3.4}$$

Then, essentially repeating the arguments of [31], we have the following relation

$$D_L(\lambda) = \det(I - K), \tag{3.5}$$

where I is the identity operator in $L_2(\Xi, \mathbb{C}^2)$. The determinant in the l.h.s. is the $2L\times 2L$ matrix determinant, while the (Fredholm) determinant in the r.h.s. is taken in $L_2(\Xi, \mathbb{C}^2)$. This relation can be proved by writing down the matrix representation of the integral operator $I - K$ in the basis $\{z^k e_j\}_{-\infty<k<\infty, j=1,2}$, where $\{e_1, e_2\}$ is the canonical basic in $\mathbb{C}^2$ (cf. [31]). The matrix elements of operator $I - K$ can be defined by following relations:

$$(I - K)z^k\,e_\alpha = \sum_{j=0}^{L-1}\sum_{\beta=1}^{2} \Phi_{\beta,\alpha}^{j-k} z^j\,e_\beta, \qquad 0 \le k < L, \quad \alpha = \{1, 2\}, \tag{3.6}$$

$$(I - K)z^k\,e_\alpha = z^k\,e_\alpha + \sum_{j=0}^{L-1}\sum_{\beta=1}^{2} \Phi_{\beta,\alpha}^{j-k} z^j\,e_\beta, \quad k < 0 \text{ or } k \ge L, \quad \alpha = \{1, 2\}. \tag{3.7}$$

Here

$$\Phi_{\beta,\alpha}^j = \frac{1}{2\pi i}\int_{|z|=1}\Big(\Phi(z)\Big)_{\beta,\alpha}\frac{dz}{z^{j+1}} \equiv (\widetilde{\Pi}_j)_{\beta,\alpha}.$$

With $\widetilde{\mathbf{B}}_L(\lambda)$ defined as $(\Phi_{\beta,\alpha}^{j-k})$, $0 \le j, k \le L-1$; α, β taking 1 or 2; (j, β) numerating the rows and (k, α) numerating the columns , equation (3.5) becomes obvious.

So we represented $D_L(\lambda)$ as a Fredholm determinant of the integral operator K. Define the resolvent operator R by

$$(I - K)(I + R) = I \quad \Leftrightarrow \quad R = (I - K)^{-1}K.$$

Then, we have the general equation [33], [37],

$$R(z,z') = \frac{F^T(z)H(z')}{z-z'}, \tag{3.8}$$

where

$$F^T = (I-K)^{-1}f^T, \quad \text{and} \quad H = h(I-K)^{-1}$$

where in the first equation $(I-K)^{-1}$ is understood as acting to the right, while in the second equation it acts to the left.

3.2 The Riemann-Hilbert problem. One of the main ingredients of the theory of integrable operators is the following Riemann-Hilbert representation for the functions $F(z)$ and $H(z)$.

$$F(z) = Y_+(z)f(z), \quad z \in \Xi, \tag{3.9}$$

$$H(z) = (Y_+^T)^{-1}(z)h(z), \quad z \in \Xi, \tag{3.10}$$

where the 4×4 matrix function $Y(z)$ is the (unique) solution of the following Riemann-Hilbert problem:

1. $Y(z)$ is analytic outside of the circle Ξ.
2. $Y(\infty) = I_4$, where I_4 denote the 4×4 identity matrix.
3. $Y_-(z) = Y_+(z)J(z), \quad z \in \Xi$ where $Y_+(z)$ ($Y_-(z)$) denote the left (right) boundary value of $Y(z)$ on Ξ (note, "+" means: from inside of the unite circle!). The 4×4 jump matrix $J(z)$ is defined by the equations,

$$\begin{aligned} J(z) &= I_4 + 2\pi i f(z)h^T(z) \\ &= \begin{pmatrix} 2I_2 - \Phi^T(z) & -z^L(I_2-\Phi^T(z)) \\ z^{-L}(I_2-\Phi^T(z)) & \Phi^T(z) \end{pmatrix}. \end{aligned} \tag{3.11}$$

It is also worth noticing the integral formulae,

$$Y(z) = I_4 - \int_\Xi \frac{F(z')h^T(z')}{z'-z}dz', \tag{3.12}$$

and

$$Y^{-1}(z) = I_4 + \int_\Xi \frac{f(z')H^T(z')}{z'-z}dz'. \tag{3.13}$$

We are going now to relate the λ - derivative of $\det(I-K)$ and the resolvent. To this end we first notice that

$$\frac{d}{d\lambda}\Phi(z) = iI_2.$$

Therefore, we have that,

$$\frac{d}{d\lambda}K(z,z') = -\frac{1}{2\pi}\frac{z^L(z')^{-L}-1}{z-z'}I_2 \equiv -iK(z,z')(I_2-\Phi(z'))^{-1}.$$

From this it follows that

$$\left[(I-K)^{-1}\frac{d}{d\lambda}K\right](z,z') = -iR(z,z')(I_2-\Phi(z'))^{-1},$$

and hence

$$\begin{aligned} \frac{d}{d\lambda}\ln\det T_L\{\phi\} &= -\mathrm{Trace}\,(I-K)^{-1}\frac{d}{d\lambda}K \\ &= i\int_\Xi \mathrm{trace}\,\left(R(z,z)(I_2-\Phi(z))^{-1}\right)dz, \end{aligned} \tag{3.14}$$

where "Trace" means the trace taking in the space $L_2(\Xi, \mathbb{C}^2)$, while "trace" is the 2×2 matrix trace.

From (3.8) we conclude that

$$R(z,z) = \frac{dF^T(z)}{dz} H(z),$$

which in turn implies the equation

$$\frac{d}{d\lambda} \ln \det T_L\{\phi\} = i \int_{\Xi} \text{trace} \left(\frac{dF^T(z)}{dz} H(z)(I_2 - \Phi(z))^{-1} \right) dz. \tag{3.15}$$

3.3 The asymptotic solution of the Riemann-Hilbert problem. Equation (3.15), together with (3.9) and (3.10) it reduces the question to the asymptotic analysis of the solution $Y(z)$ of the Riemann-Hilbert problem (1 - 3). For the latter we shall follow [31].

The basic observation is that the jump matrix $J(z)$ admits the following algebraic factorization,

$$J(z) = M(z) J_0(z) N(z), \tag{3.16}$$

. where

$$M(z) = \begin{pmatrix} I_2 & z^L \left(I_2 - (\Phi^T)^{-1}(z)\right) \\ 0_2 & I_2 \end{pmatrix}, \tag{3.17}$$

$$N(z) = \begin{pmatrix} I_2 & 0_2 \\ -z^{-L} \left(I_2 - (\Phi^T)^{-1}(z)\right) & I_2 \end{pmatrix}, \tag{3.18}$$

and

$$J_0(z) = \begin{pmatrix} (\Phi^T)^{-1}(z) & 0_2 \\ 0_2 & \Phi^T(z) \end{pmatrix}. \tag{3.19}$$

Choose now a small ϵ and define the matrix function $X(z)$ according to the equations:

$$X(z) = Y(z) \quad \text{if} \quad |z| > 1 + \epsilon, \quad \text{or} \quad |z| < 1 - \epsilon, \tag{3.20}$$

$$X(z) = Y(z) M(z) \quad \text{if} \quad 1 - \epsilon < |z| < 1, \tag{3.21}$$

$$X(z) = Y(z) N^{-1}(z) \quad \text{if} \quad 1 < |z| < 1 + \epsilon. \tag{3.22}$$

The new function has a jump accross the unit circle Ξ with the jump matrix $J_0(z)$ and two more jumps - accross the circles,

$$\Xi_1 : |z| = 1 - \epsilon, \quad \text{jump matrix } M(z)$$

and

$$\Xi_2 : |z| = 1 + \epsilon, \quad \text{jump matrix } N(z).$$

In other words, the original Rimeann-Hilbert problem (1 - 3) is equivalent to the problem

1^0. $X(z)$ is analytic outside of the contour $\Gamma \equiv \Xi \cup \Xi_1 \cup \Xi_2$.

2^0. $X(\infty) = I_4$, where I_4 denote the 4×4 identity matrix.

3^0. The jumps of the function $X(z)$ across the contour Γ are given by the equations

- $X_-(z) = X_+(z) M(z), \quad z \in \Xi_1$
- $X_-(z) = X_+(z) N(z), \quad z \in \Xi_2$
- $X_-(z) = X_+(z) J_0(z), \quad z \in \Xi$

where the jump matrices $M(z)$, $N(z)$, and $J_0(z)$ are defined in (3.17), (3.18), and (3.19), respectively and each circle is oriented counterclockwise.

Observe that the differences of the jump matrices on Ξ_1 and Ξ_2 from the identity are exponentially small as $n \to \infty$. This means one can expect the following asymptotic relation for $X(z)$,

$$X(z) \sim X^0(z), \tag{3.23}$$

where the independent on L function $X^0(z)$ solves the Riemann-Hilbert problem which is the same as the Y - problem but with the jump matrix $J_0(z)$ instead of $J(z)$.

We notice that the function $X^0(z)$ can be found explicitly in terms of the 2×2 matrix-valued functions $U_\pm(z)$ and $V_\pm(z)$ solving the following Weiner-Hopf factorization problem :

(i) $\Phi(z) = U_+(z)U_-(z) = V_-(z)V_+(z), \quad z \in \Xi$

(ii) $U_-(z)$ and $V_-(z)$ ($U_+(z)$ and $V_+(z)$) are analytic outside (inside) the unit circle Ξ.

(iii) $U_-(\infty) = V_-(\infty) = I$.

Indeed we have that

$$X^0(z) = \begin{pmatrix} U_+^T(z) & 0_2 \\ 0_2 & (V_+^T)^{-1}(z) \end{pmatrix}, \quad \text{if} \quad |z| < 1, \tag{3.24}$$

and

$$X^0(z) = \begin{pmatrix} (U_-^T)^{-1}(z) & 0_2 \\ 0_2 & V_-^T(z) \end{pmatrix}, \quad \text{if} \quad |z| > 1. \tag{3.25}$$

In the next section, we will show that for all $\lambda \in \Omega$ the above Weiner-Hopf factorization can be found explicitely, in terms of the elliptic theta-functions. In this section though we won't need the explicit formulae for $U_\pm(z)$ and $V_\pm(z)$. The only information we will need in this section is the existence of $U_\pm(z)$ and $V_\pm(z)$, and the uniform estimate

$$|U_\pm(z)|, \quad |V_\pm(z)| < C, \quad \forall z, \quad \forall \lambda \in \Omega, \tag{3.26}$$

which is the direct consequence of the explicit formulae (4.43), (4.44), and (4.2), (4.3) of section 4.

We are now ready to formulate and prove the rigorous version of the formal asymptotic relation (3.23).

Theorem 5. *Let Ω be the set introduced in theorem 1, i.e. the complex λ-plane without arbitrary fixed neighborhoods of the points $\lambda = \infty$, $\lambda = \pm 1$, and $\lambda = \pm\lambda_m$, $m = 0, 1, ...,$ where λ_m are defined in (4.45) . Then, for sufficiently large L and all $\lambda \in \Omega$, the Riemann-Hilbert problem (1^0 - 3^0) has the unique solution $X(z)$ which satisfies the following uniform estimate,*

$$\left| X(z) \left(X^0(z) \right)^{-1} - I_4 \right| \le C \frac{\rho^{-L}}{1 + |z|}, \quad z \in \mathbb{C}, \quad \lambda \in \Omega, \quad L \ge L_0, \tag{3.27}$$

where ρ is any positive number such that $1 < \rho < |\lambda_C|$, and the block diagonal matrix-valued function $X^0(z)$ is defined by the equations (3.24) and (3.25).

Proof. Put

$$R(z) := X(z) \left(X^0(z) \right)^{-1}.$$

Observe that since $\det J_0(z) \equiv 1$, the function $d(z) = \det X^0(z)$ solves the following scalar Riemann-Hilbert problem:

- $d(z)$ is analytic outside the unit circle Ξ

- $d_-(z) = d_+(z)$, $z \in \Xi$
- $d(\infty) = 1$

By virtue of the Liouville theorem, this implies that $d(z) = \det X^0(z) \equiv 1$, and hence the matrix ratio $R(z)$ is well defined. In fact, in terms of the matrix function $R(z)$ the Riemann-Hilbert problem (1^0 - 3^0) can be rewritten as follows .

1^{00}. $R(z)$ is analytic outside of the contour $\Gamma_0 \equiv \Xi_1 \cup \Xi_2$ (no jump across the unite circle Ξ!).

2^{00}. $R(\infty) = I_4$, where I_4 denote the 4×4 identity matrix.

3^{00}. The jumps of the function $R(z)$ across the contour Γ_0 are given by the equations
- $R_-(z) = R_+(z)\hat{M}(z), \quad z \in \Xi_1$
- $R_-(z) = R_+(z)\hat{N}(z), \quad z \in \Xi_2$

where the jump matrices $\hat{M}(z)$ and $\hat{N}(z)$ are defined by the formulae

$$\hat{M}(z) = X^0(z)M(z)[X^0(z)]^{-1}, \quad \hat{N}(z) = X^0(z)N(z)[X^0(z)]^{-1}. \tag{3.28}$$

Let us denote $G_0(z)$ the jump matrix of the function $R(z)$, i.e.

$$G_0(z) = \begin{cases} \hat{M}(z), & z \in \Xi_1 \\ \hat{N}(z), & z \in \Xi_2 \end{cases}. \tag{3.29}$$

Then, taking into account the estimate (3.26) and the expressions for matrices $M(z)$ and $N(z)$ we immediately arrive at the estimate,

$$|I_4 - G_0(z)| \le C\rho^{-L}, \quad 1 < \rho < |\lambda_C|, \quad z \in \Gamma_0, \quad \lambda \in \Omega, \quad L \ge 1. \tag{3.30}$$

This in turn implies the $L_2 \cap L_\infty$ - norm estimate:

$$\|I_4 - G_0\|_{L_2(\Gamma_0)\cap L_\infty(\Gamma_0)} \le C\rho^{-L}, \tag{3.31}$$

which, by standard arguments based on the analysis of the relevant singular integral equation (see e.g. [56] or Appendix D of [69]), yields the estimate (3.27). For the reader's convenience we will now present those arguments.

According to the general theory of Riemann-Hilbert problems (see e.g. [59] or [60]) the solution $R(z)$ of the Riemann-Hilbert problem (1^{00} - 3^{00}) is given by the integral representation,

$$R(z) = I + \frac{1}{2\pi i}\int_{\Gamma_0} r(z')\,(I_4 - G_0(z'))\,\frac{dz'}{z' - z}, \quad z \notin \Gamma_0, \tag{3.32}$$

where $r(z) \equiv R_+(z)$ solves the singular integral equation

$$r(z) = I + C_+[r\,(I_4 - G_0)](z), \quad z \in \Gamma_0, \tag{3.33}$$

and C_+ is the corresponding Cauchy operator:

$$(C_+h)(z) = \lim_{z'\to z,\, z' \in (+)-\text{side}} \int_{\Gamma_0} \frac{h(s)}{s - z'}\,\frac{ds}{2\pi i}.$$

We note that in the case under consideration this statement is an (almost) immediate consequence of the Cauchy theorem applied to the Cauchy integral,

$$\frac{1}{2\pi i}\int_{\Gamma_0} R_+(z')\,(I_4 - G_0(z'))\,\frac{dz'}{z' - z}$$

in conjunction with the jump relations 3^{00}.

Put

$$r_0(z) = r(z) - I_4.$$

Then equation (3.33) can be rewritten as an equation in $L_2(\Gamma_0)$,

$$(1 - \mathbb{K})r_0 = f, \tag{3.34}$$

where the function $f(z)$ and the operator $\mathbb{K}$ are defined by the formulae,

$$f = C_+(I - G_0),$$

$$\mathbb{K} : h \mapsto C_+[h\,(I_4 - G_0)], \quad h \in L_2(\Gamma_0).$$

The L_2-boundness of the operator C_+ (see e.g. [61]) and the estimate (3.31) imply that

$$||\mathbb{K}||_{L_2(\Gamma_0)\to L_2(\Gamma_0)} \leq ||C_+||_{L_2(\Gamma_0)\to L_2(\Gamma_0)}||I_4 - G_0||_{L_\infty(\Gamma_0)} = O(\rho^{-L}),$$

and

$$||f||_{L_2(\Gamma_0)} \leq ||C_+||_{L_2(\Gamma_0)\to L_2(\Gamma_0)}||I_4 - G_0||_{L_2(\Gamma_0)} = O(\rho^{-L}).$$

Therefore, equation (3.34) is uniquely solvable in $L_2(\Gamma_0)$ for sufficiently large L, and its solution satisfies the estimate,

$$||r_0||_{L_2(\Gamma_0)} = O(\rho^{-L}),$$

or, in terms of $r(z)$,

$$||I_4 - r||_{L_2(\Gamma_0)} = O(\rho^{-L}).$$

This equation together with (3.32) allows us to estimate $R(z)$:

$$\begin{aligned}
|R(z) - I_4| \;&\leq\; C||I_4 - G_0||_{L_\infty(\Gamma_0)}\frac{1}{\text{dist }(z;\Gamma_0)} \\
&\quad + C||I_4 - G_0||_{L_2(\Gamma_0)}||I_4 - r||_{L_2(\Gamma_0)}\frac{1}{\text{dist }(z;\Gamma_0)} \\
&\leq\; C\frac{\rho^{-L}}{\text{dist }(z;\Gamma_0)}.
\end{aligned}$$

This leads to the announced estimate (3.27) for all z outside of some neighborhood of the contour Γ_0. Since of the flexibility in the choice of the circles Ξ_1 and Ξ_2 (i.e. the flexibility in the choice of ϵ in (3.20 - 3.22)), the estimate is extended to the neighborhood of the contour Γ_0. This completes the proof of the theorem.

3.4 The proof of theorem 4. Let us observe that in terms of the function $X(z)$ the equations (3.9) and (3.10), which determine the functions $F(z)$ and $H(z)$, can be transformed as follows.

$$\begin{aligned}
F(z) &= X_+(z)M^{-1}(z)f(z) \\
&= X_+(z)\begin{pmatrix} I_2 & -z^L\left(I_2 - (\Phi^T)^{-1}(z)\right) \\ 0_2 & I_2 \end{pmatrix}\begin{pmatrix} z^L I_2 \\ I_2 \end{pmatrix} \\
&= X_+(z)\begin{pmatrix} z^L(\Phi^T)^{-1}(z) \\ I_2 \end{pmatrix},
\end{aligned}$$

and, similarly,

$$\begin{aligned} & H(z)\left(I_2-\Phi(z)\right)^{-1} \\ = & \left(X_{+}^T\right)^{-1}(z) M^T(z) h(z)\left(I_2-\Phi(z)\right)^{-1} \\ = & \frac{1}{2 \pi i}\left(X_{+}^T\right)^{-1}(z)\left(\begin{array}{cc} I_2 & 0_2 \\ z^L\left(I_2-\Phi^{-1}(z)\right) & I_2 \end{array}\right)\left(\begin{array}{c} z^{-L} I_2 \\ -I_2 \end{array}\right) \\ = & \frac{1}{2 \pi i}\left(X_{+}^T\right)^{-1}(z)\left(\begin{array}{c} z^{-L} I_2 \\ -\Phi^{-1}(z) \end{array}\right) . \end{aligned}$$

This implies that

$$\begin{aligned} \operatorname{tr}\left(\frac{d F^T(z)}{d z} H(z)\left(I_2-\Phi(z)\right)^{-1}\right) & =\frac{1}{2 \pi i} \frac{L}{z} \operatorname{tr} \Phi^{-1}(z)+\frac{1}{2 \pi i} \frac{d}{d z} \operatorname{tr} \Phi^{-1}(z) \\ & +\frac{1}{2 \pi i} \operatorname{tr}\left[\frac{d X_{+}^T(z)}{d z}\left(X_{+}^T\right)^{-1}(z)\left(\begin{array}{cc} \Phi^{-1}(z) & z^{-L} I_2 \\ -z^L \Phi^{-2}(z) & -\Phi^{-1}(z) \end{array}\right)\right], \end{aligned}$$

and hence, our basic relation for the Toeplitz determinant (3.15) can be re-written as

$$\frac{d}{d \lambda} \ln D_L(\lambda)=\frac{L}{2 \pi} \int_{\Xi} \operatorname{tr} \Phi^{-1}(z) \frac{d z}{z}$$

$$+\frac{1}{2 \pi} \int_{\Xi} \operatorname{tr}\left[\frac{d X_{+}^T(z)}{d z}\left(X_{+}^T\right)^{-1}(z)\left(\begin{array}{cc} \Phi^{-1}(z) & z^{-L} I_2 \\ -z^L \Phi^{-2}(z) & -\Phi^{-1}(z) \end{array}\right)\right] d z . \tag{3.35}$$

The second term in the r.h.s. of equation (3.35) can be split into the following three integrals,

$$\begin{aligned} & \frac{1}{2 \pi} \int_{\Xi} \operatorname{tr}\left[\frac{d X_{+}^T(z)}{d z}\left(X_{+}^T\right)^{-1}(z)\left(\begin{array}{cc} \Phi^{-1}(z) & z^{-L} I_2 \\ -z^L \Phi^{-2}(z) & -\Phi^{-1}(z) \end{array}\right)\right] d z \\ & =\frac{1}{2 \pi} \int_{\Xi} \operatorname{tr}\left[\frac{d X_{+}^T(z)}{d z}\left(X_{+}^T\right)^{-1}(z)\left(\begin{array}{cc} \Phi^{-1}(z) & 0_2 \\ 0_2 & -\Phi^{-1}(z) \end{array}\right)\right] d z \\ & +\frac{1}{2 \pi} \int_{\Xi} \operatorname{tr}\left[\frac{d X_{+}^T(z)}{d z}\left(X_{+}^T\right)^{-1}(z)\left(\begin{array}{cc} 0_2 & 0_2 \\ -z^L \Phi^{-2}(z) & 0_2 \end{array}\right)\right] d z \\ & +\frac{1}{2 \pi} \int_{\Xi} \operatorname{tr}\left[\frac{d X_{+}^T(z)}{d z}\left(X_{+}^T\right)^{-1}(z)\left(\begin{array}{cc} 0_2 & z^{-L} I_2 \\ 0_2 & 0_2 \end{array}\right)\right] d z \\ & \equiv \mathbb{T}_1+\mathbb{T}_2+\mathbb{T}_3 . \end{aligned} \tag{3.36}$$

The integral $\mathbb{T}_2$ can be replaced by the integral over the contour Ξ_1,

$$\mathbb{T}_2=\frac{1}{2 \pi} \int_{\Xi_1} \operatorname{tr}\left[\frac{d X_{-}^T(z)}{d z}\left(X_{-}^T\right)^{-1}(z)\left(\begin{array}{cc} 0_2 & 0_2 \\ -z^L \Phi^{-2}(z) & 0_2 \end{array}\right)\right] d z, \tag{3.37}$$

and hence estimated as

$$\left|\mathbb{T}_2\right| \leq C \rho^{-L}, \quad 1<\rho<\left|\lambda_C\right|, \quad \lambda \in \Omega, \quad L \geq L_0 . \tag{3.38}$$

(We notice that by virtue of theorem 5, the matrix function $X(z)$ is uniformly bounded) Using the jump relation, $X_{-}(z)=X_{+}(z) J_0(z)$, the integral $\mathbb{T}_3$ can be rewritten as

$$\mathbb{T}_3=\frac{1}{2 \pi} \int_{\Xi} \operatorname{tr}\left[\frac{d X_{-}^T(z)}{d z}\left(X_{-}^T\right)^{-1}(z)\left(\begin{array}{cc} 0_2 & z^{-L} \Phi^{-2}(z) \\ 0_2 & 0_2 \end{array}\right)\right] d z,$$

and, similarly to the integral $\mathbb{T}_2$, further replaced by the integral over the contour Ξ_2,

$$\mathbb{T}_3 = \frac{1}{2\pi}\int_{\Xi_2} \operatorname{tr}\left[\frac{dX_+^T(z)}{dz}(X_+^T)^{-1}(z)\begin{pmatrix} 0_2 & z^{-L}\Phi^{-2}(z) \\ 0_2 & 0_2 \end{pmatrix}\right] dz, \tag{3.39}$$

which in turn yields the same estimate (3.38) as in the case of integral $\mathbb{T}_2$,

$$|\mathbb{T}_3| \le C\rho^{-L}, \quad 1 < \rho < |\lambda_C|, \quad \lambda \in \Omega, \quad L \ge 1. \tag{3.40}$$

Finally we notice that, by virtue of theorem 5, the matrix function $X_+(z)$ in the integral $\mathbb{T}_1$ can be replaced, within the same error (3.40), by the block diagonal matrix function $X^0(z)$ from equation (3.24). Therefore, we derive from (3.35) the following asymptotic representation for the logarithmic derivative of $D_L(\lambda)$.

$$\begin{aligned}\frac{d}{d\lambda}&\ln D_L(\lambda)\\ &= \frac{L}{2\pi}\int_{\Xi} \operatorname{tr}\Phi^{-1}(z)\frac{dz}{z}\\ &+\frac{1}{2\pi}\int_{\Xi} \operatorname{tr}\left[\frac{d(X_+^0)^T(z)}{dz}((X_+^0)^T)^{-1}(z)\begin{pmatrix} \Phi^{-1}(z) & 0_2 \\ 0_2 & -\Phi^{-1}(z) \end{pmatrix}\right] dz + r_L(\lambda)\end{aligned} \tag{3.41}$$

where the error term $r_L(\lambda)$ satisfies the estimates,

$$|r_L(\lambda)| \le C\rho^{-L}, \quad 1 < \rho < |\lambda_C|, \quad \lambda \in \Omega, \quad L \ge 1 \tag{3.42}$$

Taking into account formula (3.24) for $X^0(z)$ we can simplify (3.41) as follows,

$$\begin{aligned}\frac{d}{d\lambda}\ln D_L(\lambda) &= \frac{L}{2\pi}\int_{\Xi} \operatorname{tr}\Phi^{-1}(z)\frac{dz}{z}\\ &+\frac{1}{2\pi}\int_{|z|=1} \operatorname{trace}\left[\left(U_+'(z)U_+^{-1}(z) + V_+^{-1}(z)V_+'(z)\right)\Phi^{-1}(z)\right] dz\\ &+r_L(\lambda),\end{aligned} \tag{3.43}$$

or

$$\begin{aligned}\frac{d}{d\lambda}\ln D_L(\lambda) &= -\frac{2\lambda}{1-\lambda^2}L\\ &+\frac{1}{2\pi}\int_{|z|=1} \operatorname{trace}\left[\left(U_+'(z)U_+^{-1}(z) + V_+^{-1}(z)V_+'(z)\right)\Phi^{-1}(z)\right] dz\\ &+r_L(\lambda),\end{aligned} \tag{3.44}$$

where we have used the relation

$$\operatorname{tr}\Phi^{-1}(z) = \frac{2i\lambda}{1-\lambda^2}\,.$$

In the last formulae, the symbol ($'$) means the derivative with respect to z. This completes the proof of theorem 4.

4 Wiener-Hopf factorization of matrix $\Phi(z)$

By explicit calculation, one can find that

$$(1-\lambda^2)\sigma_3\Phi^{-1}(z)\sigma_3 = \Phi(z), \quad \sigma_3 = \begin{pmatrix} 1 & 0 \\ 0 & -1 \end{pmatrix}. \tag{4.1}$$

Hence,

$$V_-(z) = \sigma_3 U_-^{-1}(z)\sigma_3 \tag{4.2}$$
$$V_+(z) = \sigma_3 U_+^{-1}(z)\sigma_3(1-\lambda^2), \quad \lambda \neq \pm 1, \tag{4.3}$$

and one only needs the explicit expressions for $U_\pm(z)$.

Our observation is that, for all λ outside of a certain discrete subset of the interval $[-1,1]$, the solution to the auxiliary Riemann-Hilbert problem (i-iii) exists; moreover, the functions $U_\pm(z)$ can be expressed in terms of the Jacobi theta-functions. Indeed, the auxiliary Riemann-Hilbert problem (i-iii) can be easily reduced to a type of the "finite-gap" Riemann-Hilbert problems which have already appeared in the analysis of the integrable statistical mechanics models (see [38]). Before we give detail expressions, let us first define some basic objects:

$$w(z) = \sqrt{(z-\lambda_1)(z-\lambda_2)(z-\lambda_2^{-1})(z-\lambda_1^{-1})}, \tag{4.4}$$
$$\beta(\lambda) = \frac{1}{2\pi i}\ln\frac{\lambda+1}{\lambda-1}, \tag{4.5}$$

where $w(z)$ is analytic on the domain $\mathbb{C}\setminus\{J_1 \cup J_2\}$ shown in Fig. 2 and fixed by the condition: $w(z) \to z^2$ as
$z \to \infty$. Next we define

$$\tau = \frac{2}{c}\int_{\lambda_B}^{\lambda_C}\frac{dz}{w(z)}, \quad c = 2\int_{\lambda_A}^{\lambda_B}\frac{dz}{w(z)}, \tag{4.6}$$

$$\delta = \frac{2}{c}\left(-\pi i - \int_{\lambda_A}^{\lambda_B}\frac{z dz}{w(z)}\right), \quad \omega(z) = \frac{1}{c}\int_{\lambda_A}^{z}\frac{dz}{w(z)}, \tag{4.7}$$

$$\Delta(z) = \frac{1}{2}\int_{\lambda_A}^{z}\frac{z+\delta}{w(z)}dz, \quad \kappa = \int_{\lambda_A}^{\infty}d\omega(z), \tag{4.8}$$

Points $\lambda_A, \lambda_B, \lambda_C, \lambda_D$ and cuts J_1, J_2 and curves Σ and Ξ are shown in Fig. 2. We shall also need:

$$\Delta_0 = \lim_{z\to\infty}\left[\Delta(z) - \frac{1}{2}\ln(z-\lambda_1)\right]. \tag{4.9}$$

Here, the contours of integration for c and δ are taken along the left side of the cut J_1. The contour of integration for τ is the segment $[\lambda_B, \lambda_C]$. The contours of integration for κ and in (4.9) are taken along the line Σ to the left from λ_A; also in (4.9), $\arg(z-\lambda_1) = \pi$. The contours of integration in the integrals $\Delta(z)$ and $\omega(z)$ are taking according to the rule: The contour lies entirely in the domain Ω_+ (Ω_-) for z belonging to Ω_+ (Ω_-). It also worth noticing that, for the general reasons (see e.g. [67]; see also Chapter 2 of [68]), we always have that $\mathrm{Re}\, i\tau < 0$. Moreover, as we will see later on, the following inequality takes place

$$i\tau < 0. \tag{4.10}$$

It also should be noted that $\omega(z)$ is an elliptic integral of the first kind, normalized by the period condition

$$2\int_{\lambda_A}^{\lambda_B} \mathrm{d}\omega(z) = 1. \tag{4.11}$$

and $\Delta(z)$ is an elliptic integral of the third kind, normalized by the conditions

$$2\int_{\lambda_A}^{\lambda_B} \mathrm{d}\Delta(z) = -\pi i, \tag{4.12}$$

and

$$\Delta(z) \sim \frac{1}{2}\ln z \quad \text{for } z \to \infty. \tag{4.13}$$

The rules formulated above define the integrals $\Delta(z)$ and $\omega(z)$ as the single-valued analytic functions in the domains Ω_+ and Ω_-. Let us denote Σ_A and Σ_D the parts of the contour Σ which lie to the left of $\lambda_A \equiv \lambda_1$ and to the right of $\lambda_D \equiv \lambda_1^{-1}$, respectively[3]. Then, on the polygonal line Σ the integrals $\Delta(z)$ and $\omega(z)$ have jumps which can be described by the equations,

$$\omega_+(z) - \omega_-(z) = 0 \text{ and } \Delta_+(z) - \Delta_-(z) = 0 \quad \text{for } z \in \Sigma_A, \tag{4.14}$$

$$\omega_+(z) + \omega_-(z) = 0 \text{ and } \Delta_+(z) + \Delta_-(z) = 0 \quad \text{for } z \in (\lambda_A, \lambda_B) \tag{4.15}$$

$$\omega_+(z) - \omega_-(z) = 1 \text{ and } \Delta_+(z) - \Delta_-(z) = -\pi i \quad \text{for } z \in (\lambda_B, \lambda_C) \tag{4.16}$$

$$\omega_+(z) + \omega_-(z) = \tau \text{ and } \Delta_+(z) + \Delta_-(z) = 2\int_{\lambda_B}^{\lambda_C} \mathrm{d}\Delta(z) = -2\pi i \int_{\lambda_1}^{\infty} \mathrm{d}\omega(z)$$

for $z \in (\lambda_C, \lambda_D)$ and integration is taken over the ray Σ_A (4.17)

$$\omega_+(z) - \omega_-(z) = 0 \text{ and } \Delta_+(z) - \Delta_-(z) = -\pi i \quad \text{for } z \in \Sigma_D \tag{4.18}$$

The only relation above which is not straightforward is the last equation in (4.17). It easy follows, however, from the application of the classical Riemann bilinear relations (see e.g. [67] or [68]) to the periods abelian (elliptic) integrals $\omega(z)$ and $\Delta(z)$.

Let

$$\theta_3(s) = \sum_{n=-\infty}^{\infty} e^{\pi i \tau n^2 + 2\pi i s n} \tag{4.19}$$

be the third Jacobi theta-function. We remind again the basic properties of the function$\theta_3(s)$ (see e.g. [39]):

$$\theta_3(-s) = \theta_3(s), \quad \theta_3(s+1) = \theta_3(s) \tag{4.20}$$

$$\theta_3(s+\tau) = e^{-\pi i \tau - 2\pi i s}\theta_3(s) \tag{4.21}$$

$$\theta_3\left(n + m\tau + \frac{1}{2} + \frac{\tau}{2}\right) = 0, \quad n, m \in \mathbb{Z} \tag{4.22}$$

[3]In particular, for the Cases 1a and 2, we have that Σ_A and Σ_D are the half lines $(-\infty, \lambda_A)$ and $(\lambda_D, +\infty)$, respectively.

We also introduce the 2×2 matrix valued function $\Theta(z)$ with the entries,

$$\begin{aligned}
\Theta_{11}(z) &= (z-\lambda_1)^{-\frac{1}{2}} e^{\Delta(z)} \\
&\quad \times \frac{\theta_3\left(\omega(z)+\beta(\lambda)-\kappa+\frac{\sigma\tau}{2}\right)}{\theta_3\left(\omega(z)+\frac{\sigma\tau}{2}\right)} \\
\Theta_{12}(z) &= -(z-\lambda_1)^{-\frac{1}{2}} e^{-\Delta(z)} \\
&\quad \times \frac{\theta_3\left(\omega(z)-\beta(\lambda)+\kappa-\frac{\sigma\tau}{2}\right)}{\theta_3\left(\omega(z)-\frac{\sigma\tau}{2}\right)} \\
\Theta_{21}(z) &= -(z-\lambda_1)^{-\frac{1}{2}} e^{-\Delta(z)} \\
&\quad \times \frac{\theta_3\left(\omega(z)+\beta(\lambda)+\kappa-\frac{\sigma\tau}{2}\right)}{\theta_3\left(\omega(z)-\frac{\sigma\tau}{2}\right)} \\
\Theta_{22}(z) &= (z-\lambda_1)^{-\frac{1}{2}} e^{\Delta(z)} \\
&\quad \times \frac{\theta_3\left(\omega(z)-\beta(\lambda)-\kappa+\frac{\sigma\tau}{2}\right)}{\theta_3\left(\omega(z)+\frac{\sigma\tau}{2}\right)}, \qquad (4.23)
\end{aligned}$$

where $\sigma = 1$ in Case 1 and $\sigma = 0$ in Case 2, and $\beta(\lambda)$, $\omega(z)$ and κ are defined in Eqs. (4.4-4.9). The branch of $(z-\lambda_1)^{-\frac{1}{2}}$ is defined on the z-plane cut along the part of the line Σ which is to the right of $\lambda_1 \equiv \lambda_A$, and it is fixed by the condition $\arg(z-\lambda_1) = \pi, \quad \text{if} \quad z-\lambda_1 < 0$.

The matrix function $\Theta(z)$ is defined on $\mathbb{C}\setminus\Sigma$. However, taking into account equations (4.14 - 4.18) describing the jumps of the integrals $\omega(z)$ and $\Delta(z)$ over the line Σ and the properties (4.20) and (4.21) of the theta function, one can see that $\Theta(z)$ is actually extended to the analytic function defined on $\mathbb{C}\setminus\{J_1 \cup J_2\}$ which satisfies the jump relations

$$\Theta_+(z) = \Theta_-(z)\sigma_1 \qquad z \in J_1 \qquad (4.24)$$

$$\Theta_+(z) = \Theta_-(z)\Lambda\sigma_1\Lambda^{-1} \qquad z \in J_2. \qquad (4.25)$$

$$\Lambda = i\begin{pmatrix} \lambda+1 & 0 \\ 0 & \lambda-1 \end{pmatrix}, \quad \sigma_1 = \begin{pmatrix} 0 & 1 \\ 1 & 0 \end{pmatrix}. \qquad (4.26)$$

The end points of the intervals $J_{1,2}$ are singular points of the function $\Theta(z)$. Specifically, one can notice that

$$\Theta(z) = O\left(\frac{1}{\sqrt{z-\lambda_1}}\right), \quad \text{as} \quad z \sim \lambda_1,$$

and

$$\Theta(z) = O\left(\frac{1}{\sqrt{z-\lambda_2^{-1}}}\right), \quad \text{as} \quad z \sim \lambda_2^{-1}.$$

The last estimate follows from the fact that, according to (4.11) and (4.22), the only zero of the denominators of the right hand sides in (4.23) is the point $z = \lambda_2^{-1}$ (which is λ_B in the Case 1, i.e. when $\sigma = 1$, and λ_C in the Case 2, i.e. when $\sigma = 0$). At the end points $z = \lambda_1^{-1}$ and $z = \lambda_2$ the function $\Theta(z)$ is holomrphic with respect to the variables $\sqrt{z-\lambda_1^{-1}}$ and $\sqrt{z-\lambda_2}$, respectively, although $\det\Theta(z)$ vanishes at these points. In more detail, the following representations take place at the points $z = \lambda_A$, λ_B, λ_C, and λ_D :

- in a neighborhood of $z = \lambda_A$, cut along J_1,

$$\Theta(z) = \Theta_A(z)(z-\lambda_A)^{\begin{pmatrix} -1/2 & 0 \\ 0 & 0 \end{pmatrix}} \begin{pmatrix} 1 & -1 \\ 1 & 1 \end{pmatrix} ; \tag{4.27}$$

- in a neighborhood of $z = \lambda_B$, cut along J_1,

$$\Theta(z) = \Theta_B(z)(z-\lambda_B)^{\begin{pmatrix} 1/2-\sigma & 0 \\ 0 & 0 \end{pmatrix}} \begin{pmatrix} 1 & -1 \\ 1 & 1 \end{pmatrix} ; \tag{4.28}$$

- in a neighborhood of $z = \lambda_C$, cut along J_2,

$$\Theta(z) = \Theta_C(z)(z-\lambda_C)^{\begin{pmatrix} \sigma-1/2 & 0 \\ 0 & 0 \end{pmatrix}} \Lambda \begin{pmatrix} 1 & -1 \\ 1 & 1 \end{pmatrix} \Lambda^{-1} ; \tag{4.29}$$

- in a neighborhood of $z = \lambda_D$, cut along J_2,

$$\Theta(z) = \Theta_D(z)(z-\lambda_D)^{\begin{pmatrix} 1/2 & 0 \\ 0 & 0 \end{pmatrix}} \Lambda \begin{pmatrix} 1 & -1 \\ 1 & 1 \end{pmatrix} \Lambda^{-1}. \tag{4.30}$$

In these formulae, the left matrix multipliers $\Theta_A(z)$, $\Theta_B(z)$, $\Theta_C(z)$, and $\Theta_D(z)$ are matrix - valued functions holomorphic (with respect to the variable z itself) and invertible as matrices at the points $z = \lambda_A$, $z = \lambda_B$, $z = \lambda_C$, and $z = \lambda_D$, respectively. To prove, say, representation (4.27) let us introduce the model function,

$$\Theta_0(z) := (z-\lambda_A)^{\begin{pmatrix} -1/2 & 0 \\ 0 & 0 \end{pmatrix}} \begin{pmatrix} 1 & -1 \\ 1 & 1 \end{pmatrix}.$$

This function is well defined in a small disk centered at λ_A and cut along J_1. By a simple direct computation we observe that the function $\Theta_0(z)$ satisfies accross the portion of J_1 lying in the disk *exactly* the same jump condition as the one indicated in (4.24). Therefore, the matrix ratio,

$$\Theta_A(z) := \Theta(z)\Theta_0^{-1}(z),$$

does not have any jumps at all in the neighborhood of λ_A. Hence, $z = \lambda_A$ is its isolated singular point. From the explicit formulae (4.23) it follows that, at worst,

$$|\Theta_A(z)| \le \frac{C}{|z-\lambda_A|^{1/2}},$$

which implies that $z = \lambda_A$ is in fact a removable singularity. Equation (4.27) follows. The other equations from the list (4.27 - 4.30) can be proven in the similar way.

Strictly speaking, we have not completed the proof of equations (4.27 - 4.30). We need to show that all the left multiplies are invertible matrices. To see this we note that

$$\Theta_{11}(\infty) = e^{\Delta_0}\frac{\theta_3\left(\beta(\lambda)+\frac{\sigma\tau}{2}\right)}{\theta_3\left(\kappa+\frac{\sigma\tau}{2}\right)} \tag{4.31}$$

$$\Theta_{22}(\infty) = e^{\Delta_0}\frac{\theta_3\left(\beta(\lambda)-\frac{\sigma\tau}{2}\right)}{\theta_3\left(\kappa+\frac{\sigma\tau}{2}\right)} \tag{4.32}$$

$$\Theta_{12}(\infty) = \Theta_{21}(\infty) = 0 , \tag{4.33}$$

and

$$\det\Theta(z) \equiv \phi(z)\det\Theta(\infty)\sqrt{\frac{\lambda_2}{\lambda_1}}. \tag{4.34}$$

To prove the latter equation, consider the ratio

$$\psi(z) := \det\Theta(z)/\phi(z).$$

The jump conditions (4.24) and (4.25) imply that the function $\psi(z)$ has no jumps while the representations (4.27 - 4.30) show that neither it has poles. Hence, this function is a constant and we have that

$$\psi(z) \equiv \psi(\infty) = \det\Theta(\infty)\sqrt{\frac{\lambda_2}{\lambda_1}}.$$

This proves (4.34). In its turn, equation (4.34) implies the invertability of $\Theta_k(\lambda_k)$ for all $k = A, B, C, D$. In addition we also have that

$$\det\Theta(z) \neq 0, \quad \text{for all} \quad z \notin \{\lambda_1^{-1}, \lambda_2\}.$$

In addition to matrix-valued function $\Theta(z)$, we introduce the matrix-valued function

$$Q(z) = \begin{pmatrix} \phi(z) & -\phi(z) \\ i & i \end{pmatrix} \tag{4.35}$$

Note that $Q(z)$ diagonalizes original jump matrix $\Phi(z)$:

$$\Phi(z) = Q(z)\Lambda Q^{-1}(z) \tag{4.36}$$

and $Q(z)$ is analytic on $\mathbb{C}\setminus\{J_1 \cup J_2\}$ and

$$Q_+(z) = Q_-(z)\sigma_1, \quad z \in J_1 \cup J_2. \tag{4.37}$$

Moreover, at the end points of the intervals $J_{1,2}$, which are the only singularities of the matrix function $Q(z)$, it behaves in the way similar to the behavior of the function $\Theta(z)$, i.e. we have (cf. (4.27 - 4.30)),

- in a neighborhood of $z = \lambda_A$, cut along J_1,

$$Q(z) = Q_A(z)(z-\lambda_A)^{\begin{pmatrix} -1/2 & 0 \\ 0 & 0 \end{pmatrix}} \begin{pmatrix} 1 & -1 \\ 1 & 1 \end{pmatrix}; \tag{4.38}$$

- in a neighborhood of $z = \lambda_B$, cut along J_1,

$$Q(z) = Q_B(z)(z-\lambda_B)^{\begin{pmatrix} 1/2-\sigma & 0 \\ 0 & 0 \end{pmatrix}} \begin{pmatrix} 1 & -1 \\ 1 & 1 \end{pmatrix}; \tag{4.39}$$

- in a neighborhood of $z = \lambda_C$, cut along J_2,

$$Q(z) = Q_C(z)(z-\lambda_C)^{\begin{pmatrix} \sigma-1/2 & 0 \\ 0 & 0 \end{pmatrix}} \begin{pmatrix} 1 & -1 \\ 1 & 1 \end{pmatrix}; \tag{4.40}$$

- in a neighborhood of $z = \lambda_D$, cut along J_2,

$$Q(z) = Q_D(z)(z-\lambda_D)^{\begin{pmatrix} 1/2 & 0 \\ 0 & 0 \end{pmatrix}} \begin{pmatrix} 1 & -1 \\ 1 & 1 \end{pmatrix}, \tag{4.41}$$

with some holomorphic and invertible at the respective points matrices $Q_A(z)$, $Q_B(z)$, $Q_C(z)$, and $Q_D(z)$. We are now ready to present the solution $U_\pm(z)$ of the Riemann-Hilbert problem (i-iii). Put

$$A = Q(\infty)\Lambda^{-1}\Theta^{-1}(\infty), \tag{4.42}$$

and define

$$U_-(z) := A\Theta(z)\Lambda Q^{-1}(z), \quad |z| \geq 1 \tag{4.43}$$
$$U_+(z) := Q(z)\Theta^{-1}(z)A^{-1}, \quad |z| \leq 1. \tag{4.44}$$

By virtue of Eq. (4.36), we only need to be sure that $U_-(z)$ and $U_+(z)$ are analytic and matrix invertible for $|z| > 1$ and $|z| < 1$ respectively. From the jump properties (4.24 - 4.25) of $\Theta(z)$ and (4.37) of $Q(z)$ it follows that $U_\pm$ have no jumps across $J_{1,2}$, and hence they might have only possible isolated singularities at $\lambda_{1,2}, \lambda_{1,2}^{-1}$. The analyticity at these points follows immediately from the representations (4.27 - 4.30) and (4.38 - 4.41) which show that the singularities, which the functions $\Theta(z)$ and $Q(z)$ do have at the end points of the segments $J_{1,2}$, are canceled out in the products (4.43)-(4.44).

The excluded values of λ for which the above construction fails are $\lambda = \pm 1$ and, in view of Eq. (4.34), the zeros of $\theta_3\left(\beta(\lambda) + \frac{\sigma\tau}{2}\right)$, i.e. (see (4.22)),

$$\pm\lambda_m, \quad \lambda_m = \tanh\left(m + \frac{1-\sigma}{2}\right)\pi\tau_0, \quad m \geq 0, \tag{4.45}$$

where,

$$\tau_0 = -i\tau = -i\frac{\int_{\lambda_B}^{\lambda_C}\frac{dz}{w(z)}}{\int_{\lambda_A}^{\lambda_B}\frac{dz}{w(z)}} > 0.$$

We conclude this section by noticing that, using the standard reduction (see e.g. [39]) of an arbitrary elliptic integral to the canonical elliptic integrals, we can rewrite the above expression for τ_0 as

$$\tau_0 = I(k')/I(k). \tag{4.46}$$

Here $I(k)$ denotes the complete elliptic integral of the first kind,

$$I(k) = \int_0^1 \frac{dx}{\sqrt{(1-x^2)(1-k^2x^2)}},$$

$k' = \sqrt{1-k^2}$, and

$$k = \begin{cases} \sqrt{(h/2)^2 + \gamma^2 - 1}\ /\ \gamma, & \text{Case 1a} \\ \sqrt{1-\gamma^2-(h/2)^2}\ /\ \sqrt{1-(h/2)^2}, & \text{Case 1b} \\ \gamma\ /\ \sqrt{(h/2)^2+\gamma^2-1}, & \text{Case 2} \end{cases} \tag{4.47}$$

5 The proof of theorem 1. Evaluation of the entropy

Denote

$$s(\lambda) := \frac{1}{2\pi}\int_{|z|=1} \text{trace}\ \left[\left(U_+'(z)U_+^{-1}(z) + V_+^{-1}(z)V_+'(z)\right)\Phi^{-1}(z)\right] dz \tag{5.1}$$

The proof of the theorem 1 will be achieved by showing that

$$s(\lambda) = \frac{d}{d\lambda}\ln\left[\theta_3\left(\beta(\lambda) + \frac{\sigma\tau}{2}\right)\theta_3\left(\beta(\lambda) - \frac{\sigma\tau}{2}\right)\right]. \tag{5.2}$$

From equation (4.3) we obtain

$$s(\lambda) = \frac{1}{2\pi}\int_{|z|=1} \text{trace}\,\left[U_+'(z)U_+^{-1}(z)\left(\Phi^{-1}(z) - \sigma_3\Phi^{-1}(z)\sigma_3\right)\right] dz. \tag{5.3}$$

Denote

$$\Psi(z) := \Phi^{-1}(z) - \sigma_3\Phi^{-1}(z)\sigma_3 = \frac{2}{1-\lambda^2}\begin{pmatrix} 0 & -\phi(z) \\ \phi^{-1}(z) & 0 \end{pmatrix}.$$

From equation (4.44), we have

$$U_+^{-1}(z) = A\Theta(z)Q^{-1}(z), \quad U_+'(z) = Q'(z)\Theta^{-1}(z)A^{-1} + Q(z)(\Theta^{-1})'(z)A^{-1}.$$

and from equation (4.35), we have

$$Q^{-1}(z) = \frac{1}{2}\begin{pmatrix} \phi^{-1}(z) & -i \\ -\phi^{-1}(z) & -i \end{pmatrix}.$$

Then

$$\begin{aligned} s(\lambda) &= \frac{1}{2\pi}\int_{|z|=1} \text{trace}\,\left[U_+'(z)U_+^{-1}(z)\Psi(z)\right] dz \\ &= \frac{i}{\pi(1-\lambda^2)}\int_{|z|=1} \text{trace}\,\left[\Theta^{-1}(z)\frac{d}{dz}\Theta(z)\sigma_3\right] dz. \end{aligned} \tag{5.4}$$

Let

$$\alpha(z) := \text{trace}\,\left[\Theta^{-1}(z)\frac{d}{dz}\Theta(z)\sigma_3\right]. \tag{5.5}$$

Then, one can easily see that the jumps of $\alpha(z)$ across $J_1 \cup J_2$ satisfy the equation,

$$\alpha_+(z) = -\alpha_-(z). \tag{5.6}$$

From the explicit formulae (4.23) for the Θ-matrix we derive the following asymptotic representation (actually, the Laurent series) at $z = \infty$

$$\Theta(z) = \Theta(\infty)\left(I + \frac{1}{z}\mathbb{M} + \ldots\right), \tag{5.7}$$

where $\Theta(\infty)$ is the diagonal matrix defined in (4.31)-(4.31) and the matrix coefficient $\mathbb{M}$ is given by the equations,

$$\mathbb{M}_{11} = \Delta_1 + \frac{\lambda_1}{2} + \frac{1}{c}\frac{\theta_3'\left(\kappa + \frac{\sigma\tau}{2}\right)}{\theta_3\left(\kappa + \frac{\sigma\tau}{2}\right)} - \frac{1}{c}\frac{\theta_3'\left(\beta(\lambda) + \frac{\sigma\tau}{2}\right)}{\theta_3\left(\beta(\lambda) + \frac{\sigma\tau}{2}\right)}, \tag{5.8}$$

$$\mathbb{M}_{22} = \Delta_1 + \frac{\lambda_1}{2} + \frac{1}{c}\frac{\theta_3'\left(\kappa + \frac{\sigma\tau}{2}\right)}{\theta_3\left(\kappa + \frac{\sigma\tau}{2}\right)} + \frac{1}{c}\frac{\theta_3'\left(\beta(\lambda) - \frac{\sigma\tau}{2}\right)}{\theta_3\left(\beta(\lambda) - \frac{\sigma\tau}{2}\right)}, \tag{5.9}$$

$$\mathbb{M}_{12} = -e^{-2\Delta_0}\frac{\theta_3\left(\beta(\lambda) - 2\kappa + \frac{\sigma\tau}{2}\right)}{\theta_3\left(\beta(\lambda) + \frac{\sigma\tau}{2}\right)}\frac{\theta_3\left(\kappa + \frac{\sigma\tau}{2}\right)}{\theta_3\left(\kappa - \frac{\sigma\tau}{2}\right)} \tag{5.10}$$

$$\mathbb{M}_{21} = -e^{-2\Delta_0}\frac{\theta_3\left(\beta(\lambda) + 2\kappa - \frac{\sigma\tau}{2}\right)}{\theta_3\left(\beta(\lambda) - \frac{\sigma\tau}{2}\right)}\frac{\theta_3\left(\kappa + \frac{\sigma\tau}{2}\right)}{\theta_3\left(\kappa - \frac{\sigma\tau}{2}\right)}. \tag{5.11}$$

In the above equations, Δ_1 is the coefficient of z^{-1} term of the expansion of the integral $\Delta(z)$ at $z = \infty$ and θ_3' means the derivative of the theta function with respect to its own argument (not the derivative with respect to λ !), i.e.

$$\theta_3'(s) = \frac{d}{ds}\theta_3(s).$$

In its turn, representation (5.7) immediately implies the following asymptotic behavior of $\alpha(z)$ as $z \to \infty$,

$$\alpha(z) = \frac{m(\lambda)}{z^2} + O\left(\frac{1}{z^3}\right), \quad z \to \infty. \tag{5.12}$$

where

$$\begin{aligned} m(\lambda) &= -\text{trace}\,[\mathbb{M}\sigma_3] \\ &= \frac{1}{c}\left(\frac{\theta_3'\left(\beta(\lambda)+\frac{\sigma\tau}{2}\right)}{\theta_3\left(\beta(\lambda)+\frac{\sigma\tau}{2}\right)} + \frac{\theta_3'\left(\beta(\lambda)-\frac{\sigma\tau}{2}\right)}{\theta_3\left(\beta(\lambda)-\frac{\sigma\tau}{2}\right)}\right). \end{aligned} \tag{5.13}$$

The representations (4.27) - (4.30) of the function $\Theta(z)$ at the end points show that at the end points, i.e. at the points λ_1, λ_2, λ_1^{-1}, λ_2^{-2}, the function $\alpha(z)$ has at most square root singularities. This, together with (5.6) and (5.12) yield the final explicit formula for $\alpha(z)$ (cf. the derivation of equation (4.34) given above):

$$\alpha(z) = \frac{m(\lambda)}{w(z)}. \tag{5.14}$$

Equation (5.14) means that equation (5.4) can be rewritten as

$$s(\lambda) = -\frac{2i}{\pi}\frac{m(\lambda)}{1-\lambda^2}\int_{\lambda_A}^{\lambda_B}\frac{dz}{w(z)}, \tag{5.15}$$

where the integration is taken on the left side of the cut $[\lambda_A, \lambda_B]$. By virtue of (5.13) and formula (4.6) for the constant c, the last equation can be transformed into

$$\begin{aligned} s(\lambda) &= -\frac{i}{\pi}\frac{1}{1-\lambda^2} \\ &\times \left(\frac{\theta_3'\left(\beta(\lambda)+\frac{\sigma\tau}{2}\right)}{\theta_3\left(\beta(\lambda)+\frac{\sigma\tau}{2}\right)} + \frac{\theta_3'\left(\beta(\lambda)-\frac{\sigma\tau}{2}\right)}{\theta_3\left(\beta(\lambda)-\frac{\sigma\tau}{2}\right)}\right). \end{aligned} \tag{5.16}$$

Taking into account that

$$\frac{d}{d\lambda}\beta(\lambda) = -\frac{i}{\pi}\frac{1}{1-\lambda^2},$$

we arrive to the formula (5.2). Theorem 1 is proven.

We are now going to show that the two expressions for the entropy $S(\rho_A)$, i.e. equations (1.26) and (1.28) coincide. We will proceed by defining $S(\rho_A)$ by equation (1.28) and showing that formula (1.26) holds. To this end we rewrite (1.28) as

$$S(\rho_A) = \lim_{\epsilon\to 0^+} S_\epsilon(\rho_A), \tag{5.17}$$

$$S_\epsilon(\rho_A) = \lim_{L\to\infty}\frac{1}{4\pi \mathrm{i}}\oint_{\Gamma'} \mathrm{d}\lambda\, e(1+\epsilon,\lambda)\frac{\mathrm{d}}{\mathrm{d}\lambda}\ln\left(D_L(\lambda)(\lambda^2-1)^{-L}\right),$$

and observe that, by virtue of theorem 1,

$$\frac{1}{4\pi \mathrm{i}}\oint_{\Gamma'} \mathrm{d}\lambda\, e(1+\epsilon,\lambda)\frac{\mathrm{d}}{\mathrm{d}\lambda}\ln\left(D_L(\lambda)(\lambda^2-1)^{-L}\right) = \frac{1}{4\pi \mathrm{i}}\oint_{\Gamma'} \mathrm{d}\lambda\, e(1+\epsilon,\lambda)s(\lambda) + O(\rho^{-L}),$$

where $s(\lambda)$ is given by (5.2). Therefore,

$$S_\epsilon(\rho_A) = \frac{1}{4\pi \mathrm{i}}\oint_{\Gamma'} \mathrm{d}\lambda\, e(1+\epsilon,\lambda)s(\lambda). \tag{5.18}$$

The function $s(\lambda)$ is analytic on the λ - plane minus the points ± 1 and $\pm\lambda_m$. Moreover, it is an odd function with the zero at $\lambda = \infty$. Therefore, its Laurent series at infinity is of the form,

$$s(\lambda) = \frac{c_3}{z^3} + \frac{c_4}{z^4} + \ldots$$

This implies that (5.18) can be transformed as follows (see Fig. 1).

$$\begin{aligned}
S_\epsilon(\rho_A) &= \frac{1}{4\pi \mathrm{i}} \oint_{\Gamma'} \mathrm{d}\lambda\, e(1+\epsilon,\lambda)s(\lambda) \\
&= \lim_{R\to\infty} \frac{1}{4\pi \mathrm{i}} \oint_{\Gamma} \mathrm{d}\lambda\, e(1+\epsilon,\lambda)s(\lambda) \\
&= \frac{1}{4\pi i} \int_{-\infty}^{-1-\epsilon} \left[-\frac{1+\epsilon+\lambda}{2} \left(\ln\left|\frac{1+\epsilon+\lambda}{2}\right| + i\pi \right) \right. \\
&\quad \left. + \frac{1+\epsilon+\lambda}{2} \left(\ln\left|\frac{1+\epsilon+\lambda}{2}\right| - i\pi \right) \right] s(\lambda)d\lambda \\
&\quad + \frac{1}{4\pi i} \int_{1+\epsilon}^{\infty} \left[-\frac{1+\epsilon-\lambda}{2} \left(\ln\left|\frac{1+\epsilon-\lambda}{2}\right| - i\pi \right) \right. \\
&\quad \left. + \frac{1+\epsilon-\lambda}{2} \left(\ln\left|\frac{1+\epsilon-\lambda}{2}\right| + i\pi \right) \right] s(\lambda)d\lambda \\
&= -\frac{1}{4} \int_{-\infty}^{-1-\epsilon} (1+\epsilon+\lambda)s(\lambda)d\lambda + \frac{1}{4} \int_{1+\epsilon}^{\infty} (1+\epsilon-\lambda)s(\lambda)d\lambda \\
&= \frac{1}{2} \int_{1+\epsilon}^{\infty} (1+\epsilon-\lambda)s(\lambda)d\lambda, \qquad (5.19)
\end{aligned}$$

where in the last equation we have again used the oddness of the function $s(\lambda)$. Representing the function $s(\lambda)$ as

$$s(\lambda) = \frac{d}{d\lambda} \ln \frac{\theta_3\left(\beta(\lambda) + \frac{\sigma\tau}{2}\right)\theta_3\left(\beta(\lambda) - \frac{\sigma\tau}{2}\right)}{\theta_3^2\left(\frac{\sigma\tau}{2}\right)}, \qquad (5.20)$$

we can perform in (5.19) integration by parts. Indeed, the function

$$t(\lambda) \equiv \ln \frac{\theta_3\left(\beta(\lambda) + \frac{\sigma\tau}{2}\right)\theta_3\left(\beta(\lambda) - \frac{\sigma\tau}{2}\right)}{\theta_3^2\left(\frac{\sigma\tau}{2}\right)}$$

is smooth for all $\lambda \in [1+\epsilon, +\infty)$, and

$$t(\lambda) = O\left(\frac{1}{\lambda^2}\right), \quad \lambda \to \infty.$$

This yields the following, alternative to (5.18), integral representation to the entropy $S_\epsilon(\rho_A)$.

$$S_\epsilon(\rho_A) = \frac{1}{2} \int_{1+\epsilon}^{\infty} \ln \frac{\theta_3\left(\beta(\lambda) + \frac{\sigma\tau}{2}\right)\theta_3\left(\beta(\lambda) - \frac{\sigma\tau}{2}\right)}{\theta_3^2\left(\frac{\sigma\tau}{2}\right)} d\lambda, \qquad (5.21)$$

It is an easy corollary of the second periodicity property (1.18) of the theta function that

$$\ln \theta_3(ia+b) = \frac{\pi}{\tau_0} a^2 \left(1 + O\left(\frac{1}{a}\right)\right), \quad a \to \pm\infty, \qquad (5.22)$$

$$\mathrm{dist}\left(ia+b; \left\{\frac{1}{2} + \frac{\tau}{2} + n + m\tau\right\}\right) \geq \delta_0 > 0.$$

Observe that

$$\beta(\lambda) \in i\mathbb{R} \pmod{\mathbb{Z}}, \quad \forall \lambda \in (-\infty, -1) \cup (1, \infty),$$

and hence

$$\operatorname{dist}\left(\beta(\lambda) \pm \frac{\sigma\tau}{2}; \left\{\frac{1}{2} + \frac{\tau}{2} + n + m\tau\right\}\right) \geq \frac{1}{2} > 0.$$

Also,

$$i\beta(\lambda) \to \pm\infty,$$

as $\lambda \to \pm 1 \pm 0$. Therefore, (5.22) is applicable and we see that the function $t(\lambda)$ satisfies the estimate,

$$t(\lambda) \sim \frac{1}{2\pi\tau_0} \ln^2 \left|\frac{\lambda - 1}{\lambda + 1}\right|, \quad \lambda \to \pm 1 \pm 0, \tag{5.23}$$

and hence is integrable at $\lambda = \pm 1$. This means, we can take limit $\epsilon \to 0$ in (5.21) at arrive to the final integral formula for the entropy $S(\rho_A)$ (cf. equation (1.29))

$$S(\rho_A) = \frac{1}{2}\int_1^\infty \ln \frac{\theta_3\left(\beta(\lambda) + \frac{\sigma\tau}{2}\right)\theta_3\left(\beta(\lambda) - \frac{\sigma\tau}{2}\right)}{\theta_3^2\left(\frac{\sigma\tau}{2}\right)}\, d\lambda. \tag{5.24}$$

To obtain the infinite sum representation (1.26) we introduce the positively oriented contour $\Gamma_{m,R}$ in Fig 3.

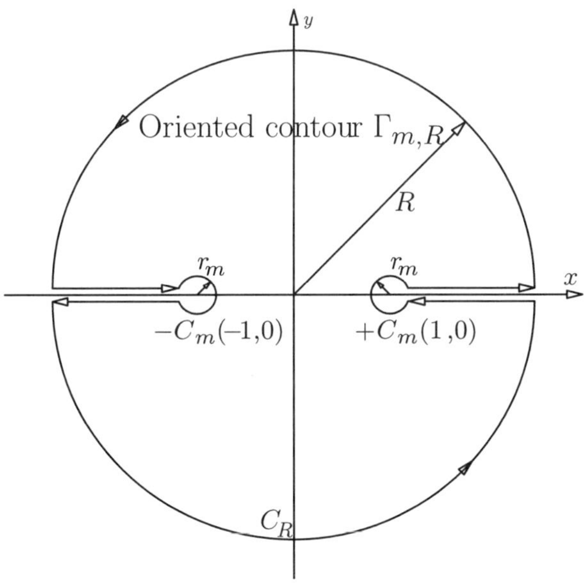

Figure 3 *Contours $\Gamma_{m,R}$ with small circles $\pm C_m$ around the points ± 1 with the radius $r_m = 1 - \left(\frac{1}{1+e^{\pi\tau_0}}\lambda_m + \frac{e^{\pi\tau_0}}{1+e^{\pi\tau_0}}\lambda_{m+1}\right)$ and the big circle C_R centered at zero with radius R, the two copies of the interval $[1 + r_m, R]$ - each lying on one of the sides of the cut from 1 to $+\infty$, and two similar copies of the symmetric interval, $[-R, -1 - r_m]$*

Then, by residue theorem,

$$\sum_{k=-m-(1-\sigma)}^{m} e(1, \lambda_k) = \frac{1}{4\pi \mathrm{i}} \oint_{\Gamma_{m,R}} d\lambda\, e(1, \lambda) s(\lambda)$$

$$= \lim_{R\to\infty} \frac{1}{4\pi \mathrm{i}} \oint_{\Gamma_{m,R}} \mathrm{d}\lambda\, e(1,\lambda)s(\lambda).$$

The last integral can be transformed (cf. derivation of (5.19)) to the expression

$$\begin{aligned}
&\frac{1}{4\pi \mathrm{i}} \oint_{C_m} \mathrm{d}\lambda\, e(1,\lambda)s(\lambda) + \frac{1}{4\pi \mathrm{i}} \oint_{-C_m} \mathrm{d}\lambda\, e(1,\lambda)s(\lambda) \\
&\quad + \frac{1}{4\pi i} \int_{-\infty}^{-1-r_m} \left[-\frac{1+\lambda}{2}\left(\ln\left|\frac{1+\lambda}{2}\right| + i\pi\right)\right. \\
&\quad \left. + \frac{1+\lambda}{2}\left(\ln\left|\frac{1+\lambda}{2}\right| - i\pi\right)\right] s(\lambda)d\lambda \\
&\quad + \frac{1}{4\pi i} \int_{1+r_m}^{\infty} \left[-\frac{1-\lambda}{2}\left(\ln\left|\frac{1-\lambda}{2}\right| - i\pi\right) + \frac{1-\lambda}{2}\left(\ln\left|\frac{1-\lambda}{2}\right| + i\pi\right)\right] s(\lambda)d\lambda \\
&= \frac{1}{4\pi \mathrm{i}} \oint_{C_m} \mathrm{d}\lambda\, e(1,\lambda)s(\lambda) + \frac{1}{4\pi \mathrm{i}} \oint_{-C_m} \mathrm{d}\lambda\, e(1,\lambda)s(\lambda) \\
&\quad - \frac{1}{4} \int_{-\infty}^{-1-r_m} (1+\lambda)s(\lambda)d\lambda + \frac{1}{4} \int_{1+r_m}^{\infty} (1-\lambda)s(\lambda)d\lambda \\
&= \frac{1}{4\pi \mathrm{i}} \oint_{C_m} \mathrm{d}\lambda\, e(1,\lambda)s(\lambda) + \frac{1}{4\pi \mathrm{i}} \oint_{-C_m} \mathrm{d}\lambda\, e(1,\lambda)s(\lambda) + \frac{1}{2} \int_{1+r_m}^{\infty} (1-\lambda)s(\lambda)d\lambda.
\end{aligned} \tag{5.25}$$

By a straightforward calculation we find that uniformly in $\lambda \in C_m$ the following estimates hold,

$$\beta(\lambda) = -\frac{\vartheta}{2\pi} - \tau\left(m + 1 - \frac{\sigma}{2}\right) + O\left(e^{-2\pi\tau_0}\right), \quad \lambda = 1 + r_m e^{i\vartheta}, \quad m \to \infty,$$

and hence

$$\mathrm{dist}\left(\beta(\lambda) \pm \frac{\sigma\tau}{2}; \left\{\frac{1}{2} + \frac{\tau}{2} + n + m'\tau\right\}\right) \geq \frac{\tau_0}{4} > 0,$$

for all $\lambda \in \pm C_m$ and sufficiently large m. This again implies the applicability of (5.22) and, as a consequence, the estimate

$$|s(\lambda)| \leq \frac{C}{r_m} \ln\frac{1}{r_m}, \quad \lambda \in \pm C_m.$$

The last estimate in its turn means that the integrals over $\pm C_m$ in (5.25) vanish as $m \to \infty$ while the integral over $[1 + r_m, \infty]$ becomes the integral over $[1, \infty]$. Therefore we arrive at the relation,

$$\sum_{k=-\infty}^{\infty} e(1,\lambda_k) \equiv \sum_{m=-\infty}^{\infty} H(\lambda_m) \equiv \sum_{m=-\infty}^{\infty} (1+\lambda_m)\ln\frac{2}{1+\lambda_m}$$

$$= \frac{1}{2}\int_1^{\infty} (1-\lambda)s(\lambda)d\lambda = \frac{1}{2}\int_1^{\infty} \ln\frac{\theta_3\left(\beta(\lambda) + \frac{\sigma\tau}{2}\right)\theta_3\left(\beta(\lambda) - \frac{\sigma\tau}{2}\right)}{\theta_3^2\left(\frac{\sigma\tau}{2}\right)}\, d\lambda,$$

which completes our evaluation of the entropy.

6 Some critical cases

The entropy has singularities at *phase transitions*. When $\tau \to 0$ we can use Landen transform (see [39]) to get the following estimate of the theta-function for small τ and pure imaginary s:

$$\ln \frac{\theta_3\left(s \pm \frac{\sigma\tau}{2}\right)}{\theta_3\left(\frac{\sigma\tau}{2}\right)} = \frac{\pi}{i\tau} s^2 \mp \pi i \sigma s + O\left(\frac{e^{-i\pi/\tau}}{\tau^2} s^2\right), \quad \text{as } \tau \to 0.$$

Now the leading term in the expression for the entropy (5.24) can be replaced by

$$S(\rho_A) = \frac{\mathrm{i}\pi}{6\tau} + O\left(\frac{e^{-i\pi/\tau}}{\tau^2}\right) \qquad \text{for } \tau \to 0. \tag{6.1}$$

Let us consider two physical situations corresponding to small τ depending on the case defined on the page 2:

1. *Critical magnetic field*: $\gamma \neq 0$ and $h \to 2$.
   This is included in our Case 1a and Case 2, when $h > 2\sqrt{1-\gamma^2}$. As $h \to 2$ the end points of the cuts $\lambda_B \to \lambda_C$, so τ given by Eq. (4.6) simplifies and we obtain from Eq. (6.1) that the entropy is very large:
   $$S(\rho_A) = -\tfrac{1}{6}\ln|2-h| + \tfrac{1}{3}\ln 4\gamma \Longrightarrow +\infty \qquad h \longrightarrow 2 \quad \text{and} \quad \gamma \neq 0. \tag{6.2}$$
   Next correction is $O(|2-h|\ln^2|2-h|)$. This limit agrees with predictions of conformal approach [13]. The first term in the right hand side of (6.2) can be represented as $(1/6)\ln\xi$, this confirms a conjecture of [13]. The correlation length ξ was evaluated in [42, 43].
2. *An approach to XX model*: $\gamma \to 0$ and $h < 2$: It is included in Case 1b, when $0 < h < 2\sqrt{1-\gamma^2}$. Now $\lambda_B \to \lambda_C$ and $\lambda_A \to \lambda_D$, we can calculate τ explicitly. The entropy increases without a bound:
   $$S^0(\rho_A) = -\tfrac{1}{3}\ln\gamma + \tfrac{1}{6}\ln(4-h^2) + \tfrac{1}{3}\ln 2 \Longrightarrow +\infty, \qquad \gamma \longrightarrow 0 \quad \text{at} \quad h < 2 \tag{6.3}$$
   correction is $O(\gamma\ln^2\gamma)$. This agrees with [4].

It is interesting to compare this critical behavior to Lipkin-Meshkov-Glick model. It is similar to XY model but each pair of spins interact with equal force, one can say that it is a model on a complete graph. The critical behavior in Lipkin-Meshkov-Glick was described in [52], it is similar to XY, but actual critical exponents are different.

Remark. The zeros λ_m satisfy an estimate:

$$|\lambda_{m+1} - \lambda_m| \leq 4\pi\tau_0 \quad \text{with} \quad \tau_0 = -i\tau.$$

This means that $(\lambda_{m+1} - \lambda_m) \to 0$ as $\tau \to 0$ for every m. This is useful for understanding of large L limit of the XX case corresponding to $\gamma \to 0$, as considered in [4]. The estimate explains why in the XX case the singularities of the logarithmic derivative of the Toeplitz determinant $d\ln D_L(\lambda)/d\lambda$ form a cut along the interval $[-1,1]$, while in the XY case it has a discrete set of poles at points $\pm\lambda_m$ of Eq. (4.45).

It would be interesting to generalize our approach to the a new class of quantum spin chains introduced recently by J. Keating and F. Mezzadri, while study matrix models [40].

Acknowledgments. We would like to thank P. Deift, B. McCoy, I. Peschel, F. Franchini and H. Widom for useful discussions. This work was supported by NSF Grants DMR-0302758, DMS-0099812 and DMS-0503712. The first co-author thanks B. Conrey, F. Mezzardi, P. Sarnak, and N. Snaith - the organizers of the 2004 program at the Isaac Newton Institute for Mathematical Sciences on Random Matrices, where part of this work was done, for an extremely stimulating research environment and hospitality during his visit.

References

[1] C.H. Bennett, H.J. Bernstein, S. Popescu, and B. Schumacher, *Concentrating partial entanglement by local operations*, Phys. Rev. **A 53**, 2046, (1996)

[2] C. Holzhey, F. Larsen and F. Wilczek,*Geometric and renormalized entropy in conformal field theory*, Nucl. Phys. B **424**, 443, (1994)

[3] G. Vidal, J.I. Latorre, E. Rico, and A. Kitaev, *Entanglement in Quantum Critical Phenomena*, Phys. Rev. Lett. **90**, 227902, (2003);

[4] B.-Q. Jin and V. E. Korepin,*Quantum Spin Chain, Toeplitz Determinants and Fisher-Hartwig Conjecture*, J. Stat. Phys. **116**, 79, (2004)

[5] A. Osterloh, L. Amico, G. Falci and R. Fazio ,*Scaling of entanglement close to a quantum phase transition*, Nature **416** 608, (2002)

[6] T.J.Osborne,M.A.Nielsen, *Entanglement in a simple quantum phase transition*, Phys.Rev.A **66**,032110(2002).

[7] V.Vedral, *High-temperature macroscopic entanglement*, New J.Phys.**6**,102 (2004).

[8] S. Popescu, A. J. Short, A. Winter, *Entanglement and the Foundations of Statistical Mechanics*, quant-ph/0511225

[9] M-H Hsieh, I. Devetak, A. Winter, *Entanglement-Assisted Capacity of Quantum Multiple Access Channels*, quant-ph/0511228

[10] P. Hayden, D. W. Leung, A. Winter, *Aspects of generic entanglement*, quant-ph/0407049

[11] J.I.Latorre, E.Rico, and G.Vidal,*Ground state entanglement in quantum spin chains*, QIC.**4**,048 (2004).

[12] V.E.Korepin, *Universality of Entropy Scaling in One Dimensional Gapless Models*,Phys.Rev. Lett.**92**, 096402 (2004).

[13] P. Calabrese, J. Cardy, *Entanglement entropy and quantum field theory*, JSTAT 0406 (2004) P002

[14] M.C.Arnesen, S.Bose, and V.Vedral,*Natural Thermal and Magnetic Entanglement in the 1D Heisenberg Model* Phys.Rev.Lett.**87**, 017901 (2001).

[15] F.Verstraete, M.A.Martín-Delgado, J.I.Cirac, *Diverging Entanglement Length in Gapped Quantum Spin Systems*, Phys. Rev. Lett. **92**, 087201 (2004).

[16] R.Orus, J.I.Latorre, *Universality of entanglement and quantum-computation complexity*, Phys. Rev. A **69**, 052308 (2004).

[17] J.K.Pachos,M.B.Plenio, *Three-Spin Interactions in Optical Lattices and Criticality in Cluster Hamiltonians*,Phys.Rev.Lett.**93**,056402(2004).

[18] J. Eisert, T. J. Osborne, *General entanglement scaling laws from time evolution*, quant-ph/0603114

[19] S. Anders, M. B. Plenio, W. Dür, F. Verstraete, H.-J. Briegel,*Ground state approximation for strongly interacting systems in arbitrary dimension*, quant-ph/0602230

[20] M. Cramer, J. Eisert, M.B. Plenio, J. Dreissig, *An entanglement-area law for general bosonic harmonic lattice systems*, quant-ph/0505092

[21] H.Fan and S.Lloyd,*Entanglement of eta-pairing state with off-diagonal long-range order*, quant-ph/0405130.

[22] C.H.Bennett, D.P.DiVincenzo,*Quantum information and computation*, Nature **404**,247 (2000).

[23] S. Lloyd,*A Potentially Realizable Quantum Computer*, Science **261**,1569(1993); *Envisioning a Quantum Supercomputer*, *ibid* **263**,695(1994).

[24] S.Ghosh, T.F.Rosenbaum, G.Aeppli, S.N.Coppersmith, *Entangled quantum state of magnetic dipoles*, Nature **425**, 48 (2003); V.Vedral, *Quantum physics: Entanglement hits the big time*, Nature **425**, 28 (2003).

[25] J. Wang, S, Kais, F. Remacle, R. D. Levine, *Size effects in the electronic properties of finite arrays of exchange coupled quantum dots: A renormalization group approach*, quant-ph/0405088; J. Wang, S, Kais, *Scaling of entanglement at quantum phase transition for two-dimensional array of quantum dots*, quant-ph/0405087; J. Wang, S, Kais,*Scaling of entanglement in finite arrays of exchange-coupled quantum dots*, quant-ph/0405085

[26] Y. Chen, P. Zanardi, Z. D. Wang, F. C. Zhang, *Entanglement and Quantum Phase Transition in Low Dimensional Spin Systems*, quant-ph/0407228; Y. Zhao, P. Zanardi, G. Chen, *Quantum Entanglement and the Self-Trapping Transition in Polaronic Systems*, quant-ph/0407080; A.Hamma, R.Ionicioiu, P.Zanardi, *Ground state entanglement and geometric entropy in the Kitaev's model*, quant-ph/0406202; P. Giorda, P. Zanardi, *Ground-State Entanglement in Interacting Bosonic Graphs*, quant-ph/0311058

[27] P. Zanardi, M. Rasetti, *Holonomic quantum computation*, Phys.Lett. **A264** 94-99 (1999); A. Marzuoli, M. Rasetti, *Spin network quantum simulator* , Phys.Lett. **A306** 79-87(2002); M. Rasetti, *A consistent Lie algebraic representation of quantum phase and number operators*, cond-mat/0211081

[28] Shi-Jian Gu, Hai-Qing Lin, and You-Quan Li, *Entanglement, quantum phase transition, and scaling in the XXZ chain*, Phys. Rev. A 68, 042330(2003); Shi-Jian Gu, Haibin Li,You-Quan Li, and Hai-Qing Lin, *The entanglement of Heisenberg chain with next-nearest-neighbor interaction*, quant-ph/0403026

[29] M.B. Plenio, J. Eisert, J. Dreissig, M. Cramer, *Entropy, entanglement, and area: analytical results for harmonic lattice systems*, quant-ph/0405142

[30] V. Popkov, M. Salerno, *Logarithmic divergence of the block entanglement entropy for the ferromagnetic Heisenberg model*, quant-ph/0404026

[31] P. A. Deift, *Integrable operators,* in Differential Operators and Spectral Theory, Amer. Math. Soc. Transl., Ser. 2, vol. 189. pp. 69-84, 1999.

[32] S. Michalakis, B. Nachtergaele, *Entanglement in Finitely Correlated Spin States*, math-ph/0606018

[33] A. R. Its, A. G. Izergin, V. E. Korepin, N. A. Slavnov, *Differential Equations for Quantum Correlation Functions*, J. Mod. Phys. B, **1003**, (1990); Proc. of a Conf. "Yang-Baxter Equations, Conformal Invariance and Integrability in Statistical Mechanics and Field Theory", CMA, ANU, Canberra, Australia, July 10-14, 1989, eds. M.N. Barber, P.A. Pearce, World Scientific, 1990, p. 303-338.

[34] M. Jimbo, T. Miwa, Y. Mori, and M. Sato, *Density matrix of an impenetrable Bose gas and the fifth Painlevé transcendent*, Physica **1D**, 80–158, (1980)

[35] C. A. Tracy and H. Widom, *Fredholm determinants, differential equations and matrix models*, Commun. Math. Phys. **163**, 33–72, (1994)

[36] L. A. Sakhnovich , *Operators similar to unitary operators*, Functional Anal.and Appl. **2** No 1, 48–60, (1968)

[37] J. Harnad, A. R. Its, *Integrable Fredholm Operators and Dual Isomonodromic Deformations*, Commun. Math. Phys. 226 (2002), 497-530

[38] P. A. Deift, A. R. Its, X. Zhou, *A Riemann-Hilbert Approach to Asymptotic Problems Arising in the Theory of Random Matrix Models, and also in the Theory of Integrable Statistical Mechanics*, Annals of mathematics, **146**, 149–235 (1997)

[39] E. T. Whittaker, G. N. Watson, *A Course of Modern Analysis*, Cambridge, University Press, 1927

[40] J.P. Keating, F. Mezzadri, *Random Matrix Theory and Entanglement in Quantum Spin Chains*, Commun. Math. Phys., **252**, 543-579(2004)

[41] E. Lieb, T. Schultz and D. Mattis,*Two soluble models of an antiferromagnetic chain*, Ann. Phys. **16**, 407, (1961)

[42] E. Barouch and B.M. McCoy,*Statistical Mechanics of the X Y Model. III. Spin-Correlation Functions*, Phys. Rev. **A 3**, 2137, (1971)

[43] E. Barouch and B.M. McCoy, *Statistical Mechanics of the X Y Model. II. Spin-Correlation Functions*, Phys. Rev. **A 3**, 786, (1971)

[44] E. Barouch, B.M. McCoy and M. Dresden,*Statistical Mechanics of the XY Model. I*, Phys. Rev. **A 2**, 1075, (1970)

[45] A.R. Its, A.G. Izergin, V.E. Korepin, and N.A. Slavnov, *Temperature correlations of quantum spins*, Phys. Rev. Lett. **70** , 1704, 1993

[46] P.A. Deift, X. Zhou, *Long-time asymptotic for the autocorrelation function of the transverse Ising chain at the critical magnetic field*, in the book: Singular limits of dispersive waves, N. M. Ercolani, I. R. Gabitov, C. D. Levermore and D. Serre (eds.), NATO ASI series B, Physics **320** Plenum Press, New York and London, 1994.

[47] Zvonarev, M. B.; Izergin, A. G.; Pronko, A. G. *Long-wave asymptotic of the correlation function of the third spin components in the XXO model with interaction across two nodes.* (Russian) Zap. Nauchn. Sem. S.-Peterburg. Otdel. Mat. Inst. Steklov. (POMI) 269 (2000), Vopr. Kvant. Teor. Polya i Stat. Fiz. 16, 207–218, 369; English translation in J. Math. Sci. (N. Y.) 115 (2003), no. 1, 2002–2008 82B20

[48] Kapitonov, V. S.; Pronko, A. G. *Time-dependent correlators of local spins in the one-dimensional Heisenberg XY-chain.* (Russian) Zap. Nauchn. Sem. S.-Peterburg. Otdel. Mat. Inst. Steklov. (POMI) 269 (2000), Vopr. Kvant. Teor. Polya i Stat. Fiz. 16, 219–261, 369–370; English translation in J. Math. Sci. (N. Y.) 115 (2003), no. 1, 2009–2032 (Reviewer: Anatoliy Yu. Zakharov) 82B20 (82B23)

[49] A.G. Abanov, F. Franchini,*Emptiness formation probability for the anisotropic XY spin chain in a magnetic field*, Phys. Lett. A **316**, 342, 2003

[50] M. Shiroishi, M. Takahahsi and Y.Nishiyama, *Emptiness Formation Probability for the One-Dimensional Isotropic XY Model*, J. Phys. Soc. Jpn **70** 3535, 2001

[51] H. Fan, V. Korepin, V. Roychowdhury, *Entanglement in a Valence-Bond-Solid State*, Phys. Rev. Lett. **93**, 227203, (2004)

[52] J. I. Latorre, R. Orus, E. Rico, and J. Vidal, *Entanglement entropy in the Lipkin-Meshkov-Glick model*, Phys. Rev. A**71**, 064101 (2005); J.Vidal, R. Mosseri, and J. Dukelsky, *Entanglement in a first-order quantum phase transition*, Phys. Rev. A **69**, 054101 (2004); J.Vidal, G. Palacios and R. Mosseri, *Entanglement in a second-order quantum phase transition*, Phys. Rev. A **69**, 022107 (2004)

[53] H. Widom, Asymptotic behavior of block Toeplitz matrices and determinants, Advances in Math. **13**, 284, (1974)

[54] H. Widom, On the limit of block Toeplitz determinants, Proceedings of the American Mathematical Society, **50**, 167, (1975)

[55] J. Baik, P. Deift, K. Johansson, *On the distribution of the length of the longest increasing subsequence of random permutations,* J. Amer. Math. Soc., **12** (1999) No. 4, 1119–1178.

[56] P. Deift, T. Kriecherbauer, K. T-R. McLaughlin, S. Venakides, and X. Zhou, *Uniform asymptotics for polynomials orthogonal with respect to varying exponential weights and applications to universality questions in random matrix theory*, Commun. Pure Appl. Math., **52**, 1335-1425 (1999).

[57] I. V. Krasovsky, *Gap probability in the spectrum of random matrices and asymptotics of polynomials orthogonal on an arc of the unit circle*, Int. Math. Res. Not. **2004**, 1249–1272 (2004)

[58] P. Deift, A. Its, I. Krasovsky, X. Zhou, *The Widom-Dyson constant and related questions of the asymptotic analysis of the Toeplitz determinants*, accepted to the Journal of Computational and Applied Mathematics (2006)

[59] K. Clancey and I. Gohberg,*Factorization of matrix functions and singular integral operators*, Operator Theory. **3**, Birkhäuser Verlag Basel (1981)

[60] R. Beals, P. A. Deift, and C. Tomei, *Direct and inverse scattering on the line.* Mathematical Surveys and Monographs **28**. AMS Providence, Rhode Island (1988).

[61] G. Litvinchuk and T. Spitkovskii, *Factorization of measurable matrix functions.* Birkhäuser Verlag, Basel, 51 (1987).

[62] A. Böttcher, *On the Determinant Formulas by Borodin, Okounkov, Baik, Deift and Rains*, Operator Theory: Advances and Applications, **135**, 91–99 (2002)

[63] N.M. Bogoliubov, A.G. Izergin, and V.E. Korepin, *Quantum Inverse Scattering Method and Correlation Functions*, Cambridge Univ. Press, Cambridge, 1993

[64] I. Peschel, *On the entanglement entropy for a XY spin chain*, cond-mat/0410416

[65] Z. Song, P. Zhang, T. Shi, C. P. Sun, *Effective boson-spin model for nuclei ensemble based universal quantum memory*, quant-ph/0409185

[66] M.-F. Yang, *Entanglement and Quantum Phase Transition Revisited*, Phys. Rev. **A 71**, 030302(R) (2005)

[67] H. M. Farkas, I. Kra, *Riemann Surfaces*, Springer-Verlag, Graduate Texts in Mathematics, 1980

[68] E. D. Belokolos, A. I. Bobenko, V. Z. Enol'skii, A. R. Its, V. B. Matveev, *Algebro-Geometric Approach to Nonlinear Integrable Equations*, Springer-Verlag, Springer Series in Nonlinear Dynamics, 1994

[69] P. Bleher and A. Its, *Semiclassical asymptotics of orthogonal polynomials, Riemann-Hilbert problem, and universality in the matrix model*, Annals of mathematics, **150** (1999), 185 - 266.

Fields Institute Communications
Volume **50**, 2007

On the Quantitative Boundary Behavior of SLE

Nam-Gyu Kang
Department of Mathematics
Massachusetts Institute of Technology
Cambridge, Massachusetts 02139 USA
kang@math.mit.edu

Abstract. We survey the results on the conjectured sharp Hölder exponent of SLE_κ ($\kappa \neq 4$), L^2 expectation of (pre-)Schwarzian derivatives of SLE, and their exponential decay of correlations. We estimate the exceptional set of SLE_4 in terms of logarithmic capacity. We also survey the result on (one-sided) Makarov's law of iterated logarithms for SLE.

1 Introduction

Based on the pioneering work of O. Schramm, a scaling limit of a number of discrete models in statistical physics at criticality can be described by a random conformal map which satisfies the Loewner differential equation. It is conjectured that the examples of the discrete models include the self-avoiding walk and the Ising model. On the other hand, it has been shown that the critical site percolation on the triangular grid, the loop-erased random walk, the uniform spanning tree, and the harmonic explorer have conformally invariant scaling limits and their scaling limits correspond to the various Schramm-Loewner evolutions.

For each $\kappa \geq 0$ and each $z \in \mathbb{H}$, let $g_t(z)$ be the solution of the chordal Loewner equation

$$\partial_t g_t(z) = \frac{2}{g_t(z) - \sqrt{\kappa} B_t}, \quad g_0(z) = z, \tag{1.1}$$

where B_t is a one-dimensional standard Brownian motion on the real line, starting from 0. The solution exists whenever $g_t(z) - \sqrt{\kappa}B_t$ is bounded away from zero. This implies $g_t(z)$ is well-defined up to the first time $\tau(z)$ such $\lim_{t\uparrow\tau(z)}(g_t(z) - \sqrt{\kappa}B_t) = 0$. For each $t > 0$, the map g_t is a conformal mapping from the domain $H_t := \{z \in \mathbb{H} : \tau(z) > t\}$ onto $\mathbb{H}$. The process $t \mapsto g_t$ is called *chordal Schramm-Loewner evolution* in $\mathbb{H}$ with parameter κ, or SLE_κ. The sets $K_t := \{z \in \overline{\mathbb{H}} : \tau(z) \leq t\}$ are called the *hulls* of the SLE. There is (almost surely) a uniquely defined continuous path $\gamma : [0, \infty) \to \overline{\mathbb{H}}$ such that H_t is the unbounded component of $\mathbb{H} \setminus \gamma[0, t]$ for all

2000 *Mathematics Subject Classification.* Primary 60K35; Secondary 30C99.
The author is supported in part by NSF Grant DMS 05-05751.

$t \geq 0$. This path is called the *SLE_κ trace* from 0 to ∞ in $\mathbb{H}$ and is given by

$$\gamma(t) = \lim_{z \to 0} g_t^{-1}(z + \sqrt{\kappa} B_t).$$

Define the backward flow $f_t = g_{-t}$ of g_t for nonnegative t. Then f_t is a conformal map from $\mathbb{H}$ into a subset of $\mathbb{H}$ satisfying

$$\partial_t f_t(z) = \frac{-2}{f_t(z) + \sqrt{\kappa} B_t}, \qquad f_0(z) = z. \tag{1.2}$$

For each fixed $t \in \mathbb{R}$, the map $f_t(z)$ has the same distribution as the map $z \mapsto g_t^{-1}(z + \sqrt{\kappa} B_t) - \sqrt{\kappa} B_t$.

From the dimension estimate for the trace and outer boundary of the hull, B. Duplantier conjectured that $\mathrm{SLE}_{\kappa'}$ ($\kappa' = 16/\kappa$) should describe the boundary of the hull of SLE_κ when $\kappa > 4$ (see [5]). Duplantier duality has been shown to hold for $\kappa = 8$ and $\kappa = 6$. In the case $\kappa = 8$, the frontier of the uniform spanning tree Peano curve consists of two loop-erased random walks. When $\kappa = 6$, the restriction property makes it possible to describe the outer boundary of conditioned SLE_6 in terms of $\mathrm{SLE}_{8/3}$. Using arguments of conformal field theory, B. Duplantier and I. Binder derived the mixed multifractal spectrum $f_{\mathrm{mixed}}(\alpha, \lambda)$ for the scaling and winding (with respect to harmonic measure) in terms of the central charge:

$$f_{\mathrm{mixed}}(\alpha, \lambda) = \alpha + b - \frac{b\alpha^2}{2\alpha - 1 - \lambda^2}, \tag{1.3}$$

where $b = (25 - c)/12$ and c is the central charge (see [6] for the definition)

$$c = 1 - 6\Big(\sqrt{\frac{\kappa}{4}} - \sqrt{\frac{4}{\kappa}}\Big)^2.$$

See [7] for definitions and more details of the results.

2 Hölder continuity

G. F. Lawler, O. Schramm, and W. Werner computed the derivative expectation $\mathbb{E}[|g_t'(1)|^p]$ to obtain the crossing exponent for SLE_κ. This computation led them to prove the Mandelbrot conjecture. To obtain more information about the regularity of the backward flows, the derivative expectation $\mathbb{E}[|g_t'(z)|^p]$ for arbitrary complex number z has been computed in [23]. Using this, S. Rohde and O. Schramm proved that g_t^{-1} is almost surely Hölder continuous when $\kappa \neq 4$.

In [14], we reexamined their derivative expectation to derive the conjectured sharp estimate for the Hölder exponent. In this section, we survey this result.

Theorem 2.1 *Suppose $\kappa \neq 4$, a sufficiently small $c > 0$, and a bounded set $D \subset \mathbb{H}$ are given. Then almost surely f_1 is h-Hölder continuous in D on the event that* $\operatorname{Im} f_1(z) \geq c$ *in D, provided*

$$h < h(\kappa) := 1 - \frac{1}{\mu} - \sqrt{\frac{1}{\mu^2} + \frac{2}{\mu}},$$

where $\mu = \kappa/4 + 2 + 4/\kappa$.

Note that $\lim_{\kappa \to 0} h(\kappa) = 1$, $\lim_{\kappa \to \infty} h(\kappa) = 1$, and $\lim_{\kappa \to 4} h(\kappa) = 0$. Also, the exponent $h(\kappa)$ satisfies the Duplantier duality. After oral communication with us, I. Binder and B. Duplantier derived the same formula from the multifractal spectrum of SLE_κ independently. Indeed, the exponent $h(\kappa)$ satisfies $f(1/h(\kappa)) = 0$ for the multifractal spectrum f of SLE_κ. Here, $f(\alpha) = f_{\mathrm{mixed}}(\alpha, \lambda = 0)$ for the mixed multifractal spectrum f_{mixed} in (1.3). We note that J. Lind has also obtained Theorem 2.1 independently. She has shown that $(f_t(\sqrt{z})+\sqrt{\kappa}B_t)^2$ is almost surely h-Hölder continuous provided $h < h(\kappa)$. See [17] for this.

We will present the heuristic way to derive Theorem 2.1. For a rigorous proof, see [14]. For $z \in \mathbb{H}$ and $u \in \mathbb{R}$, set

$$T_u = T_u(z) := \sup\{t \in \mathbb{R} : \operatorname{Im}(g_t(z)) \geq e^u\} \tag{2.1}$$

and

$$X_u = \operatorname{Re}(g_{T_u(z)}(z) - \sqrt{\kappa}B_{T_u}).$$

The stopping time T_u is well-defined since $\operatorname{Im} g_t(z)$ is monotone decreasing in t for each $z \in \mathbb{H}$. It is well known that for all $z \in \mathbb{H}$, almost surely $T_u \neq \pm\infty$. Recall S. Rohde and O. Schramm's derivative expectation [23].

Theorem 2.2 (S. Rohde, O. Schramm) *Suppose $z = x+iy \in \mathbb{H}$ with $y < 1$ given. For each $b \in \mathbb{R}$, define p and q by*

$$p := 2b + \kappa b(1-b)/2, \quad q := 4b + \kappa b(1-2b)/2. \tag{2.2}$$

Then

$$y^p \mathbb{E}\big[(1+X_0^2)^b |g'_{T_0(z)}(z)|^p\big] = (1+x^2/y^2)^b y^q. \tag{2.3}$$

From (2.3), we have the following estimate for the derivative expectation: For each bounded set $D \subset \mathbb{H}$,

$$\mathbb{E}|g'_{T_0}(z)|^p \leq C y^{p(1-\frac{1-\sqrt{1-2p/\mu}}{p/\mu})}, \tag{2.4}$$

where $\mu = \kappa/4 + 2 + 4/\kappa$ and C is a constant depending on D. Set the distribution $G(\lambda) = \mathbb{P}[|g'_{T_0}(z)/y| > \lambda]$. Then estimate (2.4) says

$$\int_0^\infty pG(\lambda)\lambda^{p-1}\, d\lambda \leq Cy^{-\mu+\mu\sqrt{1-2p/\mu}}.$$

Use the substitution $r = 1 - 2p/\mu$ $(0 < r < 1)$ and $\nu = \lambda^{-\mu/2}$ to obtain

$$\int_0^\infty y^\mu \frac{G(\lambda)}{\nu} \nu^{r-1}\, d\nu \leq C\frac{e^{-\sqrt{ar}}}{1-r}, \tag{2.5}$$

where $\sqrt{a} = \mu \log y^{-1}$. Fix $\kappa \neq 4$. By scaling, we may assume $D = [-1,1] \times (0,1]$. With $\nu = \lambda^{-\mu/2}$, $\sqrt{a} = \mu \log y^{-1}$, and $e^{-x} = \nu$, we take the inverse Mellin transform on both sides of (2.5) to formally derive the following estimate without rigorous proof:

$$\mathbb{P}\Big[\Big|\frac{g'_{T_0}(z)}{y}\Big| > \lambda\Big] \leq C\Big(\operatorname{erfc}\big(\sqrt{\frac{\mu}{2}}\frac{\log(\lambda y)}{\sqrt{\log \lambda}}\big) - \frac{1}{y^{2\mu}}\operatorname{erfc}\big(\sqrt{\frac{\mu}{2}}\frac{\log(\lambda/y)}{\sqrt{\log\lambda}}\big)\Big),$$

where erfc is the complementary error function. From the above estimate, it is not hard to obtain the following estimate. Let $c > 0$. Then

$$\begin{aligned} &\mathbb{P}\Big[\Big|\frac{f_1'(z)}{y}\Big| > \lambda \,\Big|\, \inf_{z\in D} \operatorname{Im} f_1(z) \geq c\Big] \\ &\quad\leq C\Big(\operatorname{erfc}\big(\sqrt{\frac{\mu}{2}}\frac{\log(\lambda y)}{\sqrt{\log \lambda}}\big) - \frac{1}{y^{2\mu}}\operatorname{erfc}\big(\sqrt{\frac{\mu}{2}}\frac{\log(\lambda/y)}{\sqrt{\log\lambda}}\big)\Big). \end{aligned} \tag{2.6}$$

With $\lambda = y^{h-2}$ and $y = 2^{-n}$, the above estimate gives

$$\mathbb{P}\Big[\Big|\frac{f_1'(z)}{y}\Big| > \lambda \,\Big|\, \inf_{z\in D} \operatorname{Im} f_1(z) \geq c\Big] \leq C 2^{-n\frac{\mu}{2}\frac{(1-h)^2}{2-h}}. \tag{2.7}$$

Consider the Whitney decomposition $\{Q_{j,n}\}(n \geq 0, 1 \leq j \leq 2^n)$ of D. We denote the center of the Whitney square $Q_{j,n}$ by $z_{j,n}$. By the Borel-Cantelli lemma and the distortion theorem, f_1 is almost surely h-Hölder in D on the event $\operatorname{Im} f_1(z) \geq c$ in D, if

$$\sum_{n=0}^{\infty} \sum_{j=1}^{2^n} \mathbb{P}\big[|f_1'(z_{j,n})| > 2^{n(1-h)} \mid \inf_{z\in D} \operatorname{Im} f_1(z) \geq c\big] < \infty.$$

This is summable if

$$1 - \frac{\mu}{2}\frac{(1-h)^2}{2-h} < 0, \tag{2.8}$$

or

$$h < h(\kappa) = 1 - \frac{1}{\mu} - \sqrt{\frac{1}{\mu^2} + \frac{2}{\mu}},$$

where $\mu = \kappa/4 + 2 + 4/\kappa$. This motivates the formulation of Theorem 2.1.

3 Exceptional set for SLE$_4$

In this section, we measure the exceptional set of points at which the derivative of SLE$_4$ fails to satisfy a certain growth condition of logarithmic type. Consider the Whitney decomposition of $D := (1/2, 1] \times (0, 1)$. Then for $x \in (1/2, 1]$ there is a unique Whitney square $Q_n(x)$ of side length 2^{-n} such that a projection of $Q_n(x)$ onto x-axis contains x. For $\alpha > 0$, define a collection $\mathcal{E}_n(\alpha)$ of exceptional Whitney squares Q of side length 2^{-n} such that $|f_1'(z_Q)| > 2^n/n^\alpha$ for a Whitney center z_Q of Q. We also define an exceptional set $\mathcal{E}_\alpha$ by a set of points in $(1/2, 1]$ such that

$$\mathcal{E}_\alpha = \{x : Q_n(x) \in \mathcal{E}_n(\alpha) \text{ i.o. }\}.$$

One naturally asks how small this exceptional set $\mathcal{E}_\alpha$ is. In this section, we prove that $\mathcal{E}_\alpha$ is polar almost surely for sufficiently small α. Before doing it, we recall the definitions of potentials, energies, and the logarithmic capacities. See [22] or [9].

Definition 3.1 For a finite Borel measure μ on $\mathbb{C}$ with a compact support, its *potential* is the function $p_\mu : \mathbb{C} \to [-\infty, \infty)$ defined by

$$p_\mu(z) := \int \log|z - w| \, d\mu(w)$$

and its *energy* $I(\mu)$ is given by

$$I(\mu) := \iint \log|z - w| \, d\mu(z) \, d\mu(w) = \int p_\mu(z) \, d\mu(z).$$

Definition 3.2 The *logarithmic capacity* of a subset E of $\mathbb{C}$ is given by

$$\operatorname{cap}(E) := \sup_\mu e^{I(\mu)},$$

where the supremum is taken over all Borel probability measures μ on $\mathbb{C}$ such that $\operatorname{supp}\mu$ is a compact subset of E. E is called *polar* if E has logarithmic capacity zero.

The logarithmic capacity is not sub-additive. However, there is a relation between logarithmic capacity and unions which we will use to prove our main result of this section. See pp. 130–131 in [22] for the next lemma.

Lemma 3.3 ([22]) *Let $\{B_n\}$ be a sequence of Borel subsets of $\mathbb{C}$, let $B = \bigcup_n B_n$, and let $d > 0$. If $\operatorname{diam}(B) \le d$, then $\operatorname{cap}(B) \le d$ and*

$$\frac{1}{\log(d/\operatorname{cap}(B))} \le \sum_n \frac{1}{\log(d/\operatorname{cap}(B_n))}. \tag{3.1}$$

Theorem 3.4 *For given $\epsilon > 0$ and $\alpha < 4/15$, the exceptional set $\mathcal{E}_\alpha$ is polar almost surely on the event $E := \{\inf_{z\in D} \operatorname{Im} f_1(z) \ge \epsilon\}$.*

Proof Let $z = x + iy \in \mathbb{H}$ with $y < 1$ be given. For $p \in [0, \infty)$, define the derivative expectation $F = F_p$ by

$$F(x, y) = y^p \mathbb{E}\big[|g'_{T_0(z)}(z)|^p\big]. \tag{3.2}$$

Then F satisfies the parabolic PDE

$$\frac{2y}{x^2+y^2} F_y = -\frac{\kappa}{2} F_{xx} + \frac{2x}{x^2+y^2} F_x + p\frac{4y^2}{(x^2+y^2)^2} F$$

with an initial condition $F(\cdot, 1) = 1$ (Refer to [23]). With a change of variable $\theta = \sinh^{-1}(x/y)$ and $F(x, y) = \widetilde{F}(\theta, y)$, $\widetilde{F}$ satisfies

$$y\widetilde{F}_y = -\frac{\kappa}{4}\widetilde{F}_{\theta\theta} + (2 + \frac{\kappa}{4})\frac{\sinh\theta}{\cosh\theta}\widetilde{F}_\theta + \frac{2p}{\cosh^2\theta}\widetilde{F}.$$

Substituting $y = e^{-\frac{4}{\kappa}t}$ and $\widetilde{F}(\theta, y) = u(\theta, t)e^{-(\frac{4}{\kappa}+\frac{1}{2})^2 t}\cosh^{\frac{4}{\kappa}+\frac{1}{2}}\theta$, u satisfies a heat equation with a potential

$$u_t = u_{\theta\theta} - \Big(\frac{8p}{\kappa} - (\frac{4}{\kappa} + \frac{1}{2})(\frac{4}{\kappa} + \frac{3}{2})\Big)\frac{1}{\cosh^2\theta} u. \tag{3.3}$$

Also note that the initial condition is $u(0, \theta) = \cosh^{-\frac{4}{\kappa}-\frac{1}{2}}\theta$ when $t = 0$.

The parabolic PDE (3.3) is reduced to a heat equation without a potential when $p = \frac{\kappa}{8}(\frac{4}{\kappa} + \frac{1}{2})(\frac{4}{\kappa} + \frac{3}{2}) = 15/8$. In this case, we use a heat kernel to get an estimate:

$$u(\theta, t) \le C_1 e^{\gamma^2 t}\Big(e^{-\gamma\theta}\operatorname{erfc}\big(\gamma\sqrt{t} - \frac{1}{2}\frac{\theta}{\sqrt{t}}\big) + e^{\gamma\theta}\operatorname{erfc}\big(\gamma\sqrt{t} + \frac{1}{2}\frac{\theta}{\sqrt{t}}\big)\Big), \tag{3.4}$$

where $\gamma = \frac{4}{\kappa} + \frac{1}{2} = 3/2$. By (3.4) and the tail estimate of the complementary error function, there is a uniform constant C such that

$$F(x, y) \le C n^{-\frac{1}{2}} 2^{-n} \tag{3.5}$$

for all (x, y) in a given Whitney square Q of side length 2^{-n} in D.

Take $m \in \mathbb{Z}$ such that $e^m \le \epsilon < e^{m+1}$. Recall the stopping time (2.1) and the fact that f'_1 has the same distribution as g'_{-1}. As in the proof of Corollary 3.5 in [23], we may assume $T_0(z_{j,n}) \ge 1$ with probability one. For fixed $z = z_{j,n}$ and $k = m, \cdots, 0$, let E_k be the event $E_k := \{T_k(z) \le 1 < T_{k+1}(z)\}$. Then

$$\begin{aligned}
&\mathbb{P}\big[y|f'_1(z)| > \lambda \,\big|\, E\big] \\
&\le C_1 \sum_{k=m}^{0} \mathbb{P}\big[y|g'_{T_k(z)}(z)| > C_2\lambda \,\big|\, EE_k\big]\mathbb{P}\big[E_k \,\big|\, E\big] \\
&\le C \sum_{k=m}^{0} \lambda^{-p} e^{kp} F_p(e^{-k} z),
\end{aligned}$$

where F_p is given by (3.2) and $p = 15/8$.

It follows from the above estimate and the estimate (3.6) that

$$\mathbb{P}[Q_n(x) \in \mathcal{E}_n(\alpha)] \leq C_\epsilon n^{\frac{15}{8}\alpha - \frac{1}{2}} 2^{-n} \tag{3.6}$$

on the event E. By definition, note that

$$\mathcal{E}_\alpha = \bigcap_{k=1}^{\infty} \bigcup_{n=k}^{\infty} \bigcup_{Q_n \in \mathcal{E}_n(\alpha)} \Pi_x(Q_n),$$

where Π_x denotes the projection onto the x-axis. By Lemma 3.3,

$$\mathbb{E}\Big[\frac{1}{\log(1/\operatorname{cap}(\mathcal{E}_\alpha))}\Big] \leq C_\epsilon \lim_{k\to\infty} \sum_{n=k}^{\infty} \frac{n^{\frac{15}{8}\alpha - \frac{1}{2}}}{\log 2^n}$$

on the event E. If $\alpha < 4/15$, then the series on the right hand side above is convergent. This implies

$$\mathbb{E}[1/\log(\operatorname{cap}(\mathcal{E}_\alpha))] = 0$$

or $\operatorname{cap}(\mathcal{E}_\alpha) = 0$ almost surely on the event E. □

4 (Pre-)Schwarzian derivatives

The estimate for the upper bound on the Hausdorff dimension of SLE boundary is already established by S. Rohde and O. Schramm [23]. While the Hausdorff dimension of the SLE_κ trace was proved by V. Beffara [1], it remains an open conjecture for the boundary of the hull in the case of $\kappa > 4$. A missing element in the study of SLE is a theory for decay of correlations for the map f_t'. In [14], we computed the correlation estimates for the normalized (pre-)Schwarzian derivatives of SLE. It seems possible that this approach could lead to a determination of the boundary dimension when $\kappa > 4$. In this section, we survey the results on the (pre-)Schwarzian derivatives of SLE in [14].

The *logarithmic derivative* or *pre-Schwarzian derivative* Lf of a locally univalent function f is defined by

$$Lf(z) = \frac{f''(z)}{f'(z)}.$$

The *Schwarzian derivative* Sf of a locally univalent function f is defined by

$$Sf(z) = \Big(\frac{f''(z)}{f'(z)}\Big)' - \frac{1}{2}\Big(\frac{f''(z)}{f'(z)}\Big)^2.$$

Suppose f maps $\mathbb{D}$ conformally into $\mathbb{C}$. Then, by the distortion theorem,

$$(1-|z|^2)|Lf(z)| \leq 6 \text{ and } (1-|z|^2)^2|Sf(z)| \leq 6. \tag{4.1}$$

The asymptotic behavior of the (pre-)Schwarzian derivative measures the degree of non-conformality of the boundary [20].

In order to study the subsets of rectifiable curves, given a set E in the complex plane and a square Q, P. W. Jones introduced an L^∞ version of the *beta number*

$$\beta(Q) = \beta_E(Q) := \frac{1}{\ell(Q)} \inf_{L\in\mathcal{L}} \sup_{z\in E\cap 3Q} \operatorname{dist}(z, L),$$

where $\mathcal{L}$ is the set of all lines L intersecting Q and an L^∞ *Jones content*

$$\mathcal{J}_\infty(E) = \sum_{Q\in\mathcal{D}} \beta_\infty^2(Q)\ell(Q),$$

where the summation extends over all dyadic squares.

Theorem 4.1 (Jones [13]) *Suppose E is a subset of $\mathbb{R}^2$. Then E is contained in a rectifiable curve if and only if* $\operatorname{diam}(E)$ *and the L^∞ Jones content $\mathcal{J}_\infty(E)$ are finite.*

A connected set E is called *uniformly wiggly* with constant β_0 if $\beta_E(Q) > \beta_0$ for every Q such that $3^{-1}Q$ intersects E and $\ell(Q) \le \operatorname{diam}(E)$.

Theorem 4.2 (Bishop, Jones [3]) *Suppose $E \subset \mathbb{R}^2$ is a closed, connected, uniformly wiggly set with constant β_0. Then* $\dim(E) \ge 1 + C\beta_0^2$*, where C is an absolute constant.*

C. J. Bishop, P. W. Jones, R. Pemantle, and Y. Peres used a stochastic version of this theorem to prove that the dimension of the Brownian frontier is greater than 1 [4]. Also, C. J. Bishop and P. W. Jones showed that large Schwarzian implies large beta numbers. For the precise statement, see [2] or [9]. Here, we state the quasiconformal version. Using this, J. Graczyk and P. W. Jones proved that every subarc of the boundary of the Siegel disk (for rotation numbers of "constant type") has the Hausdorff dimension strictly larger than 1 [11]. A Jordan curve Γ is *K-quasiconformal* (or a *K-quasicircle*) if it is the image of the unit circle by a K-quasiconformal homeomorphism $h : \mathbb{C} \to \mathbb{C}$.

Theorem 4.3 (Graczyk, Jones [11]) *Suppose that Γ is a K-quasicircle. Assume also that there exist $\Delta > 0$ and $\epsilon > 0$ so that for every $z_0 \in \mathbb{D}$,*

$$\sup_{\rho(z,z_0)\le\Delta} \left|S(h)\right|\left|1-|z|\right|^2 \ge \epsilon.$$

Then there exists $\epsilon_0 > 0$ which depends solely on K, Δ, and ϵ so that

$$\beta_\Gamma > \epsilon_0.$$

Just as the Hardy space H^1 is an appropriate substitute for L^1 in many results concerning singular integrals, the BMO space, or the space of functions of bounded mean oscillation, is a natural substitute for L^∞. The BMO space on $\mathbb{R}^n$ is the set of equivalence classes of locally integrable functions f (modulo additive constants) for which the inequality

$$\frac{1}{|Q|}\int_Q |f(x) - f_Q|\,dx \le C \tag{4.2}$$

holds for all cubes Q. Here, f_Q denotes the mean value of f over the cube Q. The smallest such C is taken to be BMO norm and is denoted by $\|f\|_{\mathrm{BMO}}$ (see [24]).

Theorem 4.4 (John and Nirenberg) *There exist positive constants c and C such that for each cube Q, each $f \in$ BMO, and each λ,*

$$|\{x \in Q : |f(x) - f_Q| > \lambda\}| \le C|Q|\exp(-c\lambda/\|f\|_{\mathrm{BMO}}).$$

This notion can be modified in the setting of continuous martingales. A continuous martingale M belongs to the space $\mathcal{BMO}$ of martingales of bounded mean oscillation if there exists C such that for all stopping times τ

$$\mathbb{E}\big(|M_\infty - M_\tau|^2 \big| \mathcal{F}_\tau\big) \le C^2, \quad \text{almost surely.}$$

The smallest such C is defined as the $\mathcal{BMO}$ norm of M and is denoted by $\|M\|_{\mathcal{BMO}}$. Let $M^* = \sup_t |M_t|$. For the following results, see pp. 208–211 in [8].

Theorem 4.5 (Probabilistic analogue of John and Nirenberg inequality) *There exists a positive constant C such that for each $M \in \mathcal{BMO}$,*

$$\mathbb{P}[M^* > \lambda \|M\|_{\mathcal{BMO}}] \leq Ce^{-\lambda/e}.$$

For the SLE related maps, it is well known that $\operatorname{Im} f_t(z)$ is monotone increasing in t for every $z \in \mathbb{H}$. For $z \in \mathbb{H}$ and $u \in \mathbb{R}$, set a stopping time

$$T_u = T_u(z) := \inf\{t \in \mathbb{R} : \operatorname{Im}(f_t(z)) \geq e^u\}. \tag{4.3}$$

It is also well known that for all $z \in \mathbb{H}$ almost surely $T_u \neq \infty$. See [23].

Theorem 4.6 *For any $u \in \mathbb{R}$ and $x \neq 0$, the normalized pre-Schwarzian $yLf_t(x+iy)$ at $t = T_u$ has the asymptotic second moment*

$$\lim_{y\to 0} \mathbb{E}\big[|yLf_{T_u}(x+iy)|^2\big] = \frac{1}{2(\sqrt{\kappa/4}+\sqrt{4/\kappa})^2}. \tag{4.4}$$

Furthermore, for fixed $z = x + iy (x \neq 0) \in \mathbb{H}$,

$$\lim_{t\to\infty} \mathbb{E}\big[|yLf_t(z)|^2\big] = \frac{1}{2(\sqrt{\kappa/4}+\sqrt{4/\kappa})^2}. \tag{4.5}$$

The normalized pre-Schwarzian $yLf_t(z)$, after we subtract a negligible term, is a complex martingale of $\mathcal{BMO}$. To see this, define a random conformal map F_t from $\mathbb{H}$ into a subset of $\mathbb{C}\setminus\mathbb{R}_+$ by $F_t(z) = (f_t(z)+\sqrt{\kappa}B_t)^2$ and set

$$\begin{aligned} L_t = L_t(z) &:= yLf_t(z) + \frac{4}{4+\kappa}\Big(\frac{yf_t'(z)}{f_t(z)+\sqrt{\kappa}B_t} - \frac{y}{z}\Big) \\ &= \frac{-2}{\sqrt{\kappa/4}+\sqrt{4/\kappa}} \int_0^t \frac{yf_s'(z)}{(f_s(z)+\sqrt{\kappa}B_s)^2}\, dB_s. \end{aligned} \tag{4.6}$$

Theorem 4.7 *The process L is a $\mathcal{BMO}$ martingale in t and*

$$\|L\|^2_{\mathcal{BMO}} = \frac{1}{2(\sqrt{\kappa/4}+\sqrt{4/\kappa})^2}.$$

Theorem 4.8 *For any $u \in \mathbb{R}$ and $x \neq 0$, the normalized Schwarzian $y^2Sf_t(x+iy)$ at $t = T_u$ has the asymptotic second moment*

$$\lim_{y\to 0} \mathbb{E}\big[|y^2Sf_{T_u}(x+iy)|^2\big] = \frac{9\kappa}{(\kappa+6)(3\kappa+8)}. \tag{4.7}$$

Furthermore, for $x \neq 0$, we have

$$\lim_{y\to 0}\Big(\lim_{t\to\infty} \mathbb{E}\big[|y^2Sf_t(x+iy)|^2\big]\Big) = \frac{9\kappa}{(\kappa+6)(3\kappa+8)}. \tag{4.8}$$

The normalized Schwarzian $y^2Sf_t(z)$, after subtracting a negligible term, is a complex martingale of $\mathcal{BMO}$. To see this, set

$$\begin{aligned} S_t = S_t(z) &:= y^2Sf_t(z) + \frac{12}{3\kappa+8}\Big(\frac{y^2}{z^2} - \frac{y^2f_t'(z)^2}{(f_t(z)+\sqrt{\kappa}B_t)^2}\Big) \\ &= \frac{24\sqrt{\kappa}}{3\kappa+8}\int_0^t \frac{y^2f_s'(z)^2}{(f_s(z)+\sqrt{\kappa}B_s)^3}\, dB_s. \end{aligned} \tag{4.9}$$

Theorem 4.9 *The process S is a $\mathcal{BMO}$ martingale in t. Furthermore, for $x \neq 0$,*

$$\lim_{y\to 0} \|S(x+iy)\|^2_{\mathcal{BMO}} = \frac{9\kappa}{(3\kappa+8)(\kappa+6)}.$$

Recall the definition of the hyperbolic distance $d_{\mathbb{H}}$:

$$\cosh d_{\mathbb{H}}(z_1, z_2) = 1 + \frac{|z_1 - z_2|^2}{2y_1y_2} = -1 + \frac{|z_1 - \overline{z_2}|^2}{2y_1y_2}. \tag{4.10}$$

The normalized (pre-)Schwarzian has correlations that decay exponentially in the hyperbolic distance from $z_1 = x_1 + iy_1$ to $z_2 = x_2 + iy_2$. Thus, the normalized (pre-)Schwarzian derivatives are nearly independent if z_1 and z_2 are far away from each other.

Theorem 4.10 *The $\mathcal{BMO}$ martingale L has exponential decay of correlations:*

$$\left|\mathbb{E}L_t(z_1)\overline{L_t(z_2)}\right| \le \frac{1}{(\sqrt{\kappa/4} + \sqrt{4/\kappa})^2} \cosh^{-2} \frac{d_{\mathbb{H}}(z_1, z_2)}{2} \; (t \in (0, \infty)), \tag{4.11}$$

and

$$\left|\mathbb{E}L_\infty(z_1)\overline{L_\infty(z_2)}\right| = \frac{1}{2(\sqrt{\kappa/4} + \sqrt{4/\kappa})^2} \cosh^{-2} \frac{d_{\mathbb{H}}(z_1, z_2)}{2}. \tag{4.12}$$

Furthermore, given positive u and a compact subset K of the real line, L_t at $T_u^* := \max\left(T_u(z_1), T_u(z_2)\right)$ has exponential decay of correlations:

$$\limsup_{y_1, y_2 \to 0} y_1 y_2 \left|\mathbb{E}Lf_{T_u^*}(z_1)\overline{Lf_{T_u^*}(z_2)}\right| \le \frac{1}{2(\sqrt{\kappa/4} + \sqrt{4/\kappa})^2} \cosh^{-2} \frac{\rho}{2}, \tag{4.13}$$

where the limsup is taken over all $(z_1, z_2) \in \left(K \times (0,1)\right)^2$ such that $d_{\mathbb{H}}(z_1, z_2) \ge \rho$.

Theorem 4.11 *The $\mathcal{BMO}$ martingale S has exponential decay of correlations: There exists an absolute constant C such that for all $t \in [0, \infty]$, $z_1, z_2 \in \mathbb{H}^2$,*

$$\left|\mathbb{E}S_t(z_1)\overline{S_t(z_2)}\right| \le C \cosh^{-2} \frac{d_{\mathbb{H}}(z_1, z_2)}{2}. \tag{4.14}$$

Furthermore, given positive u and a compact subset $K (0 \notin K)$ of the real line, S_t at T_u^* has exponential decay of correlations:

$$\limsup_{y_1, y_2 \to 0} \left|\mathbb{E}S_{T_u^*}(z_1)\overline{S_{T_u^*}(z_2)}\right| \le \frac{9\kappa}{(\kappa + 6)(3\kappa + 8)} \cosh^{-4} \frac{\rho}{2}, \tag{4.15}$$

where the limsup is taken over all $(z_1, z_2) \in \left(K \times (0,1)\right)^2$ such that $d_{\mathbb{H}}(z_1, z_2) \ge \rho$.

For the proofs of Theorems 4.6–4.11, see [14]. We used a chordal version of Goluzin's identities to obtain the estimates in Theorems 4.6–4.11. See p.118 in [10] for a radial version of the identities. It is likely that this result leads to an estimate on the lower bound for the Hausdorff dimension of the SLE boundary. This should be true because Theorems 4.1, 4.2, 4.3 mean that "most often" the boundary is wiggly near $f_t(z)$. Furthermore, the decay of correlations from Theorem 4.11 means that a statistical version of Theorem 4.3 should hold, as it does in [4].

5 Makarov's law of the iterated logarithms for SLE

In this section, we derive (one-sided) Makarov's law of iterated logarithms (LIL) for SLE. As a consequence, we compare the harmonic measure of the SLE boundary to the Hausdorff measure associated with a logarithmicoexponential function. We will sketch a non-rigorous, but heuristic way to derive LIL for SLE. For more details and the rigorous proof, see [15].

An analytic function g on $\mathbb{D}$ is called a *Bloch function* if

$$\|g\|_{\mathcal{B}} = \sup_{z \in \mathbb{D}} (1 - |z|^2)|g'(z)| < \infty.$$

This defines a semi-norm and the Bloch functions form a complex Banach space $\mathcal{B}$ with the norm $|g(0)| + \|g\|_{\mathcal{B}}$. See [9] or [20]. There is a close connection between Bloch functions and univalent functions, in particular their derivatives: For any conformal map $f : \mathbb{D} \to \mathbb{C}$, $\log f'$ is a Bloch function and $\|\log f'\|_{\mathcal{B}} \leq 6$. Conversely, if $\|g\|_{\mathcal{B}} \leq 1$, then $g = \log f'$ for some conformal map f.

Theorem 5.1 (Makarov) *If $g \in \mathcal{B}$, then for almost every $\zeta \in \partial\mathbb{D}$,*

$$\limsup_{r\to 1} \frac{|g(r\zeta)|}{\sqrt{\log\frac{1}{1-r}\log\log\log\frac{1}{1-r}}} \leq \|g\|_{\mathcal{B}}.$$

This inequality is called *Makarov's LIL* because it was first proved by N. G. Makarov in [18] and because it looks like the classical Khinchin's LIL for Brownian motion.

Theorem 5.1 has a converse, proved by P. W. Jones in [12]. Let $\rho(\cdot,\cdot)$ denote the standard hyperbolic metric on $\mathbb{D}$.

Theorem 5.2 (Jones) *If $\|g\|_{\mathcal{B}} \leq 1$ and if there exists β and $M < \infty$ such that for all $z_0 \in \mathbb{D}$,*

$$\sup_{\{z:\rho(z,z_0)<M\}} (1-|z|^2)|g'(z)| \geq \beta,$$

then there exists $c = c(\beta, M) > 0$ such that almost every $\zeta \in \partial\mathbb{D}$,

$$\limsup_{r\to 1} \frac{\operatorname{Re} g(r\zeta)}{\sqrt{\log\frac{1}{1-r}\log\log\log\frac{1}{1-r}}} \geq c.$$

A real valued function φ on an interval $(0, r_0)$ is called a *logarithmicoexponential function* or an *L-function* if φ is defined by a finite algebraic combination of exponential functions and logarithmic functions. A *measure function* is a positive, increasing and continuous function. For a measure function φ, define

$$\Lambda_\varphi^\delta(E) = \inf\Big\{\sum \varphi(r_j) : E \subseteq \bigcup B(z_j, r_j), r_j \leq \delta\Big\}.$$

Then the limit $\Lambda_\varphi(E) = \lim_{\delta\to 0}\Lambda_\varphi^\delta(E)$ exists in $[0,\infty]$, and $\Lambda_\varphi(E)$ is called the *φ-Hausdorff measure* of E. For any measure function φ, Λ_φ is a metric outer measure in the sense of Carathéodory, and therefore its restriction to the Borel sets is a measure.

Theorem 5.3 (Makarov) *Suppose φ is a positive continuous increasing L-function satisfying*

$$\lim_{r\to 0}\frac{\varphi(r)}{r} = \infty$$

and f is a conformal mapping from $\mathbb{D}$ onto a simply connected domain Ω. Let ω denote the harmonic measure for some point $w_0 \in \Omega$. Then

a. $\omega \ll \Lambda_\varphi$ *if and only if* $\liminf_{r\to 1}\dfrac{(1-r^2)|f'(r\zeta)|}{\varphi^{-1}(1-r)} > 0$ *for a.e.* $\zeta \in \partial\mathbb{D}$,

b. $\omega \perp \Lambda_\varphi$ *if and only if* $\liminf_{r\to 1}\dfrac{(1-r^2)|f'(r\zeta)|}{\varphi^{-1}(1-r)} = 0$ *for a.e.* $\zeta \in \partial\mathbb{D}$,

c. *there is a set $E \subseteq \partial\Omega$ of σ-finite Λ_φ measure such that $\omega(E) = 1$ if and only if*

$$\liminf_{r\to 1}\frac{(1-r^2)|f'(r\zeta)|}{\varphi^{-1}(1-r)} < \infty$$

for almost all $\zeta \in \partial\mathbb{D}$.

N. G. Makarov proved that the harmonic measure for simply connected planar domain is supported on a set of Hausdorff dimension 1. Moreover, he showed that the associated Hausdorff measure with the function $\varphi_C(r) = re^{C\sqrt{\log\frac{1}{r}\log\log\log\frac{1}{r}}}$ $(C > 0)$ is the proper measure for the size of support of the harmonic measure. See [19] or [9].

Theorem 5.4 (Makarov) *There is an absolute constant $C > 0$ such that*

$$\omega \ll \Lambda_{\varphi_C}$$

for every simply connected domain Ω. Conversely, there is $c < C$ such that there exists a Jordan domain Ω for which

$$\omega \perp \Lambda_{\varphi_c}.$$

Theorem 5.5 (Makarov) *Suppose ∂ is a Jordan repeller on $\partial\Omega$. Then ∂ is rectifiable, or there exists $c_0 > 0$ such that*

$$\begin{aligned} \omega &\ll \Lambda_{\varphi_c} \quad \text{for } c > c_0, \\ \omega &\perp \Lambda_{\varphi_c} \quad \text{for } c \leq c_0. \end{aligned}$$

Let $D = [-1/2, 1/2] \times (0, 1)$. For a fixed time t and a sufficiently small $\epsilon > 0$, consider the event $E = \{\inf_{z\in D} \operatorname{Im} f_t(z) \geq \epsilon\}$. Observe that SLE trace in $\mathbb{H}$ with $\partial\mathbb{H}$ has a sharp corner at the base point if κ is sufficiently small. Thus, the local geometry at this point is different from that at any other points. However, on the event E, the geometry of the base point will not be taken into consideration. Or, we consider the SLE trace in the slit domain $\mathbb{C}\setminus\mathbb{R}_+$. More precisely, we define a random conformal map from $\mathbb{H}$ into a subset of $\mathbb{C}\setminus\mathbb{R}_+$ by $F_t(z) = (f_t(z) + \sqrt{\kappa}B_t)^2$. Take the Itô derivative of $\log F_t'(z)$:

$$d\log F_t'(z) = d\log f_t'(z) + d\log(f_t(z) + \sqrt{\kappa}B_t),$$

where

$$d\log f_t'(z) = \frac{2}{(f_t(z) + \sqrt{\kappa}B_t)^2}\,dt \tag{5.1}$$

and

$$\begin{aligned} d\log(f_t(z) + \sqrt{\kappa}B_t) = &\frac{-2-\kappa/2}{(f_t(z) + \sqrt{\kappa}B_t)^2}\,dt \\ &+ \frac{\sqrt{\kappa}\,dB_t}{f_t(z) + \sqrt{\kappa}B_t}. \end{aligned}$$

This Itô calculus on $\log(f_t(z) + \sqrt{\kappa}B_t)$ gives us the following key identity:

$$\begin{aligned} d\log f_t'(z) = &\frac{2}{\sqrt{\kappa/4} + \sqrt{4/\kappa}}\frac{dB_t}{f_t(z) + \sqrt{\kappa}B_t} \\ &- \frac{4}{4+\kappa} d\log(f_t(z) + \sqrt{\kappa}B_t). \end{aligned} \tag{5.2}$$

The derivative of (5.2) with respect to the complex variable was derived to compute the L^2 expectation and the correlation of the pre-Schwarzian derivative of SLE backward flow in [14]. Also, see the equation (4.6). The other version of (5.2) for the SLE map g_t has been studied by S. Sheffield:

$$d\log g_t'(z) = \frac{2}{\sqrt{\kappa/4}-\sqrt{4/\kappa}}\frac{dB_t}{g_t(z)-\sqrt{\kappa}B_t} - \frac{4}{4-\kappa}d\log(g_t(z)-\sqrt{\kappa}B_t). \tag{5.3}$$

The value of $-2\,\mathrm{Im}(g_t(z)-\sqrt{\kappa}B_t)^{-1}$ represents the harmonic measure of the tip of $\gamma(t)$ as seen from the point z.

Note that the SLE$_\kappa$ trace is almost surely a continuous path. See [23] for $\kappa \neq 8$. In the special case $\kappa = 8$, this was shown as a consequence of the theorem that the scaling limit of the uniform spanning tree Peano curve is chordal SLE$_8$ [16]. Due to Carathéodory, the backward flow f_t extends continuously to $\partial\mathbb{H}$. It follows from the boundedness of $|f_t(z)+\sqrt{\kappa}B_t|$ in $\overline{D}$ on the event E that there is a random constant $C = C(t,\omega)$ such that

$$|\,\mathrm{Re}\log(f_t(z)+\sqrt{\kappa}B_t)| \le C$$

in D. Hence, we have almost surely

$$\limsup_{y\to 0}\frac{\mathrm{Re}\log(f_t(z)+\sqrt{\kappa}B_t)}{\sqrt{\log\frac{1}{y}\log\log\log\frac{1}{y}}} = 0 \tag{5.4}$$

on the event E. Hence, $\log f_t'(z)$, after we subtract a negligible term, is a complex martingale M_t, where

$$dM_t = \frac{2}{\sqrt{\kappa/4}+\sqrt{4/\kappa}}\frac{dB_t}{f_t(z)+\sqrt{\kappa}B_t}.$$

Now, it is easy to compute $\mathbb{E}[|M_{T_0}|^2]$:

$$\begin{aligned}\mathbb{E}[|M_{T_0}|^2] &= \Big(\frac{2}{\sqrt{\kappa/4}+\sqrt{4/\kappa}}\Big)^2\mathbb{E}\int_0^{T_0}\frac{1}{|f_t(z)+\sqrt{\kappa}B_t|^2}\,dt\\ &= \Big(\frac{2}{\sqrt{\kappa/4}+\sqrt{4/\kappa}}\Big)^2\frac{\mathbb{E}[\log\mathrm{Im}\,f_{T_0}(z)-\log\mathrm{Im}\,f_0(z)]}{2}\\ &= \frac{2}{(\sqrt{\kappa/4}+\sqrt{4/\kappa})^2}\log\frac{1}{y}.\end{aligned}$$

In [15], we compute $\mathbb{E}[|\,\mathrm{Re}\,M_{T_0}|^2]$ to derive the following theorem.

Theorem 5.6 *For given $x \in [-1/2,1/2]\setminus\{0\}$,*

$$\mathbb{E}|\,\mathrm{Re}\log f_{T_0}'(x+iy)|^2 = \frac{1}{(\sqrt{\kappa/4}+\sqrt{4/\kappa})^2}\log\frac{1}{y}+o(\log\frac{1}{y}). \tag{5.5}$$

Based on this theorem, we formulate the following conjecture.

Conjecture *Let $D = [-1/2,1/2]\times(0,1)$. For a fixed time t and a sufficiently small $\epsilon > 0$, almost surely, for almost all $x \in [-1/2,1/2]$,*

$$\limsup_{y\to 0}\frac{\mathrm{Re}\log f_t'(x+iy)}{\sqrt{\log\frac{1}{y}\log\log\log\frac{1}{y}}} = \frac{1}{\sqrt{\frac{\kappa}{8}}+\sqrt{\frac{2}{\kappa}}} \tag{5.6}$$

on the event $E = \{\inf_{z\in D}\mathrm{Im}\,f_t(z)\ge\epsilon\}$. Let $c(\kappa) := 1/(\sqrt{\kappa/8}+\sqrt{2/\kappa})$. Suppose ω is a harmonic measure restricted to $\{f_t(x) : x\in[-1/2,1/2]\}$. For a sufficiently *small $\epsilon > 0$, almost surely,*

$$\omega \ll \Lambda_{\varphi_c} \quad \textit{for } c > c(\kappa),$$
$$\omega \perp \Lambda_{\varphi_c} \quad \textit{for } c \le c(\kappa)$$

on the event E.

Compare this conjecture with Theorem 5.5. In [15], we prove the conjectured sharp upper bound.

Theorem 5.7 *Let $D = [-1/2, 1/2] \times (0,1)$. For a fixed time t and a sufficiently small $\epsilon > 0$, almost surely, for almost all $x \in [-1/2, 1/2]$,*

$$\limsup_{y\to 0} \frac{\operatorname{Re}\log f_t'(x+iy)}{\sqrt{\log \frac{1}{y} \log\log\log \frac{1}{y}}} \le \frac{1}{\sqrt{\frac{\kappa}{8}} + \sqrt{\frac{2}{\kappa}}} \tag{5.7}$$

on the event $E = \{\inf_{z\in D} \operatorname{Im} f_t(z) \ge \epsilon\}$.

Corollary *Let $c(\kappa) := 1/(\sqrt{\kappa/8} + \sqrt{2/\kappa})$. Suppose ω is a harmonic measure restricted to $\{f_t(x) : x \in [-1/2, 1/2]\}$. For a sufficiently small $\epsilon > 0$, almost surely,*

$$\omega \ll \Lambda_{\varphi_c} \quad \textit{for } c > c(\kappa)$$

on the event E.

Here, we sketch a non-rigorous, but heuristic way to derive Theorem 5.7. For given $r > 1$ and $\alpha > 0$, set $y_n = \exp(-r^n)$ and

$$\mathbb{P}_n = \mathbb{P}\big[\log|f_1'(x+iy_n)| > \alpha\sqrt{\log\frac{1}{y_n}\log\log\log\frac{1}{y_n}} \mid E\big].$$

It follows from the estimate (2.6) that

$$\sum_{n=1}^{\infty} \mathbb{P}_n \le C \sum_{n=1}^{\infty} (n\log r)^{-\frac{\mu}{2}\alpha^2}.$$

Note that $\sum_{n=1}^{\infty} \mathbb{P}_n < \infty$ if $\alpha > c(\kappa)$. By the Borel-Cantelli lemma and Fubini theorem, we derive the conjectured sharp upper bound. For more details and the rigorous proof, see [15].

Remark 5.8 Due to N. G. Makarov, there is a positive number C, a domain Ω, and $E \subset \partial\Omega$ such that $\Lambda_{\varphi_C}(E) = 0$ but $\omega(E) = 1$. In [12], P. W. Jones called such a domain a Makarov domain and proved that if $\partial\Omega$ is wiggly enough on all scales, then Ω is a Makarov domain. See Theorem 5.2. Based on the experimental evidence from computer simulations, he conjectured that the exterior of the Diffusion Limited Aggregation (DLA) is essentially one of the Makarov domains. The question raised by P. W. Jones whether Makarov's LIL is related to DLA, as in the case of SLE, seems to have an affirmative answer.

Remark 5.9 The quantity $\log|f_t'(z)|$ can be approximated by the partial sum of the normalized pre-Schwarzian derivatives. We showed that the linear dependence correlation coefficients of the normalized pre-Schwarzian derivatives have the exponential decay. See Theorem 4.10. However, one can expect that the ρ-mixing coefficients or the maximal correlation coefficients of $\operatorname{Re}\big[yLf_t(z)\big]$ have the exponential decay. This computation may lead to the almost sure invariance property or the strong approximation of $\log|f_t'(z)|$. As a consequence, it may result in the lower bound of the LIL for SLE. See [21] for the ρ-mixing condition and the almost sure invariance principles for partial sums of weakly dependent random variables.

Acknowledgements

The author would like to express his gratitude to the organizers and the staffs of Workshop on Percolation, SLE, and Related Topics of September 2005 at Fields Institute for their gracious hospitality.

References

[1] Beffara, V. *The dimension of the SLE curves*, preprint, math.PR/0211322, 2002.

[2] Bishop, C. J., and Jones, P. W. *Harmonic measure, L^2 estimates and the Schwarzian derivative.* J. Anal. Math. **62** (1994), 77–113.

[3] Bishop, C. J., and Jones, P. W. *Wiggly sets and limit sets.* Ark. Mat. **35**, 2 (1997), 201–224.

[4] Bishop, C. J., Jones, P. W., Pemantle, R., and Peres, Y. *The dimension of the Brownian frontier is greater than 1.* J. Funct. Anal. **143**, 2 (1997), 309–336.

[5] Duplantier, B. *Conformally invariant fractals and potential theory.* Phys. Rev. Lett. **84**, 7 (2000), 1363–1367.

[6] Duplantier, B. *Conformal Fractal Geometry and Boundary Quantum Gravity.* In *Fractal geometry and applications: a jubilee of Benoît Mandelbrot, Part 2: Multifractals, Probability and Statistical Mechanics, Applications*, Proc. Sympos. Pure Math. vol. 72, Amer. Math. Soc., Providence, RI, 2004, pp. 365–482.

[7] Duplantier, B., and Binder, I. *Harmonic measure and winding of conformally invariant curves.* Phys. Rev. Lett. **89** (2002), 264101.

[8] Durrett, R. *Brownian motion and martingales in analysis.* Wadsworth, Belmont, 1984.

[9] Garnett, J., and Marshall, D. E. *Harmonic measure.* Cambridge University Press, Cambridge, 2005.

[10] Goluzin, G. M. *Geometric Theory of Functions of a Complex Variable*, Translations of Mathematical Monographs. vol. 26, Amer. Math. Soc., Providence, 1969.

[11] Graczyk, J., and Jones, P. W. *Dimension of the boundary of quasiconformal Siegel disks.* Invent. Math. **148**, 3 (2002), 465–493.

[12] Jones, P. W. *Square functions, Cauchy integrals, analytic capacity, and harmonic measure.* In *Harmonic analysis and partial differential equations (El Escorial, 1987)*, Lecture Notes in Math. vol. 1384, Springer, Berlin, 1989, pp. 24–68.

[13] Jones, P. W. *Rectifiable sets and the traveling salesman problem.* Invent. Math. **102**, 1 (1990), 1–15.

[14] Kang, N. *Boundary behavior of SLE.* J. Amer. Math. Soc. 20, 1 (2007), 185–210.

[15] Kang, N. *The law of the iterated logarithm for SLE*, preprint.

[16] Lawler, G. F., Schramm, O., and Werner, W. *Conformal invariance of planar loop-erased random walks and uniform spanning trees.* Ann. Probab. **32**, 1B (2004), 939–995.

[17] Lind, J. *Hölder regularity for the SLE trace*, preprint, 2005.

[18] Makarov, N. G. *On the distortion of boundary set under conformal mappings.* Proc. London Math. Soc., **51** (1985), 369–384.

[19] Makarov, N. G. *Probability methods in the theory of conformal mappings.* Algebra i Analiz, **1**, 1 (1989), 3–59, English translation in Leningrad Math. J. **1**, 1 (1990), 1–56.

[20] Pommerenke, C. *Boundary behaviour of conformal maps.* Springer-Verlag, Berlin, 1992.

[21] Philipp, W., and Stout, W. *Almost Sure Invariance Principles for Partial Sums of Weakly Dependent Random Variables.* Mem. Amer. Math. Soc., vol. 161, Amer. Math. Soc., Providence, 1975.

[22] Ransford, T. *Potential theory in the complex plane.* Cambridge Univesity Press, Cambridge, 1995.

[23] Rohde, S., and Schramm, O. *Basic properties of SLE.* Ann. of Math. **161**, 2 (2005), 883–924.

[24] Stein, E. M. *Harmonic analysis: real-variable methods, orthogonality, and oscillatory integrals.* Princeton University Press, Princeton, 1993.

Fields Institute Communications
Volume **50**, 2007

The Configurational Measure on Mutually Avoiding SLE Paths

Michael J. Kozdron
Department of Mathematics & Statistics
University of Regina
Regina, SK S4S 0A2 Canada
kozdron@math.uregina.ca

Gregory F. Lawler
Department of Mathematics
University of Chicago
Chicago, IL 60637 USA
lawler@math.uchicago.edu

Abstract. We define multiple chordal SLEs in a simply connected domain by considering a natural configurational measure on paths. We show how to construct these measures so that they are conformally covariant and satisfy certain boundary perturbation and Markov properties, as well as a cascade relation. As an example of our construction, we derive the scaling limit of Fomin's identity in the case of two paths directly; that is, we prove that the probability that an SLE_2 and a Brownian excursion do not intersect can be given in terms of the determinant of the excursion hitting matrix. Finally, we define the λ-SAW, a one-parameter family of measures on self-avoiding walks on $\mathbb{Z}^2$.

1 Introduction

The Schramm-Loewner evolution (SLE) is a powerful tool for describing continuum limits that arise in critical two-dimensional models in statistical physics. The starting point [16] is to write down some properties that the continuum limit is expected to satisfy and then to show that this leaves only a one-parameter family of possibilities. This works beautifully in the case of one path in a simply connected domain.

For multiple paths or non-simply connected domains, the approach in [16] is not sufficient to determine what the natural definition(s) should be. Essentially

2000 *Mathematics Subject Classification.* Primary 60G50; Secondary 60J45, 60J65.

Research of the first author is supported in part by the Natural Sciences and Engineering Research Council of Canada.

Research of the second author is supported by the National Science Foundation grant DMS-0405021.

the conformal Markov property, which determines the distribution of a single SLE path in simply connected domains, is not sufficient to describe the more complicated situations. A number of authors have tried to consider these processes using infinitesimal descriptions. In this paper, we would like to try to suggest that this is not the most natural approach, and that instead one should look at configurational measures on paths. What we do here is describe the configurational measure associated to multiple chordal SLEs in a simply connected domain. This construction is based on the work in [10] and is also discussed in [3]. For a statistical mechanics argument describing multiple SLEs, consult [1]. Nonetheless, it still seems useful to describe the configurational measure in some detail.

We will start by writing down a number of properties that we expect the measure to have. We will not worry at the moment whether some of these properties are redundant or consistent. Let D be a Jordan domain, and let $z_1, \ldots, z_n, w_n, \ldots, w_1$ be $2n$ distinct points on ∂D in counterclockwise order. We write $\mathbf{z} = (z_1, \ldots, z_n)$, $\mathbf{w} = (w_1, \ldots, w_n)$, and say that D is (locally) analytic at $\mathbf{z}$, $\mathbf{w}$ if conformal maps sending D to the unit disk $\mathbb{D}$ can be extended analytically to neighborhoods of $z_1, \ldots, z_n, w_n, \ldots, w_1$. We fix $b \in \mathbb{R}$, a parameter that is called the *boundary scaling exponent* or *boundary conformal weight.* We want to define a measure $Q_{D,b,n}(\mathbf{z}, \mathbf{w})$ on mutually avoiding n-tuples of simple paths $(\gamma^1, \ldots, \gamma^n)$ in D. More precisely, γ^j is an equivalence class of curves such that there is a representative $\gamma^j : [0,1] \to \mathbb{C}$ which is simple, $\gamma^j(0) = z_j$, $\gamma^j(1) = w_j$, and $\gamma^j(0,1) \subset D$. We will write just γ^j for $\gamma^j(0,1)$. When we refer to measures on curves in this paper, we will implicitly mean measures on equivalence classes of curves under equivalence by reparametrization; consult, for example, [5] or [8].

- **Conformal Covariance.** If D is analytic at $\mathbf{z}$, $\mathbf{w}$, then $Q_{D,b,n}(\mathbf{z}, \mathbf{w})$ is a non-zero, finite measure supported on n-tuples $(\gamma^1, \ldots, \gamma^n)$ where γ^j is a simple curve in D connecting z_j and w_j and
$$\gamma^j \cap \gamma^k = \emptyset, \quad 1 \le j < k \le n.$$
Moreover, if $f : D \to f(D)$ is a conformal transformation and $f(D)$ is analytic at $f(\mathbf{z})$, $f(\mathbf{w})$, then
$$f \circ Q_{D,b,n}(\mathbf{z}, \mathbf{w}) = |f'(\mathbf{z})|^b \, |f'(\mathbf{w})|^b \, Q_{f(D),b,n}(f(\mathbf{z}), f(\mathbf{w})) \tag{1.1}$$
where $f(\mathbf{z}) = (f(z_1), \ldots, f(z_n))$ and $f'(\mathbf{z}) = f'(z_1) \cdots f'(z_n)$; see Figure 1.

If we write
$$Q_{D,b,n}(\mathbf{z}, \mathbf{w}) = H_{D,b,n}(\mathbf{z}, \mathbf{w}) \, \mu^{\#}_{D,b,n}(\mathbf{z}, \mathbf{w}),$$
where $H_{D,b,n}(\mathbf{z}, \mathbf{w}) = |Q_{D,b,n}(\mathbf{z}, \mathbf{w})|$ and $\mu^{\#}_{D,b,n}(\mathbf{z}, \mathbf{w})$ is a probability measure, then the conformal covariance condition (1.1) becomes the scaling rule for H,
$$H_{D,b,n}(\mathbf{z}, \mathbf{w}) = |f'(\mathbf{z})|^b \, |f'(\mathbf{w})|^b \, H_{f(D),b,n}(f(\mathbf{z}), f(\mathbf{w})), \tag{1.2}$$
and the conformal *invariance* rule for $\mu^{\#}$, namely
$$f \circ \mu^{\#}_{D,b,n}(\mathbf{z}, \mathbf{w}) = \mu^{\#}_{f(D),b,n}(f(\mathbf{z}), f(\mathbf{w})). \tag{1.3}$$
Since $\mu^{\#}$ is a conformal invariant, we can define $\mu^{\#}_{D,b,n}(\mathbf{z}, \mathbf{w})$ even if the boundaries are not smooth at $\mathbf{z}$, $\mathbf{w}$. (Compare this with the construction of the Brownian excursion measure in [5].) We will write $Q_{D,b}(z,w) = Q_{D,b,1}(z,w)$ and similarly for $H_{D,b}(z,w)$ and $\mu^{\#}_{D,b}(z,w)$.

The next condition describes what happens when we remove part of the domain.

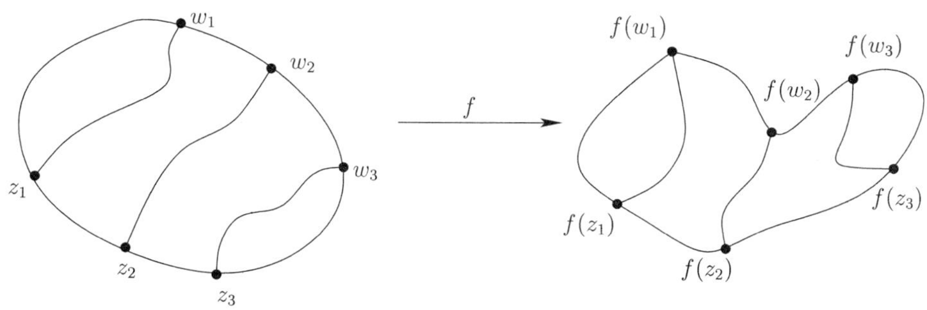

Figure 1 Conformal covariance

- **Boundary Perturbation.** Suppose $D \subset D'$ are Jordan domains and ∂D, $\partial D'$ agree and are analytic in neighborhoods of $\mathbf{z}$, $\mathbf{w}$. Then $Q_{D,b,n}(\mathbf{z},\mathbf{w})$ is absolutely continuous with respect to $Q_{D',b,n}(\mathbf{z},\mathbf{w})$. Moreover, the Radon-Nikodym derivative

$$Y_{D,D',b,n}(\mathbf{z},\mathbf{w}) = \frac{dQ_{D,b,n}(\mathbf{z},\mathbf{w})}{dQ_{D',b,n}(\mathbf{z},\mathbf{w})}$$

is a conformal invariant. In other words, if $f : D' \to f(D')$ is a conformal map that extends analytically in neighborhoods of $\mathbf{z}, \mathbf{w}$, then

$$Y_{f(D),f(D'),b,n}(f(\mathbf{z}), f(\mathbf{w}))(f \circ \bar{\gamma}) = Y_{D,D',b,n}(\mathbf{z},\mathbf{w})(\bar{\gamma}), \tag{1.4}$$

where $\bar{\gamma} = (\gamma^1, \ldots, \gamma^n)$ and $f \circ \bar{\gamma} = (f \circ \gamma^1, \ldots, f \circ \gamma^n)$; see Figure 2.

As with $\mu^{\#}_{D,b,n}(\mathbf{z},\mathbf{w})$, the last condition (1.4) implies that $Y_{D,D',b,n}(\mathbf{z},\mathbf{w})$ is well-defined even if the boundaries are not smooth at $\mathbf{z},\mathbf{w}$.

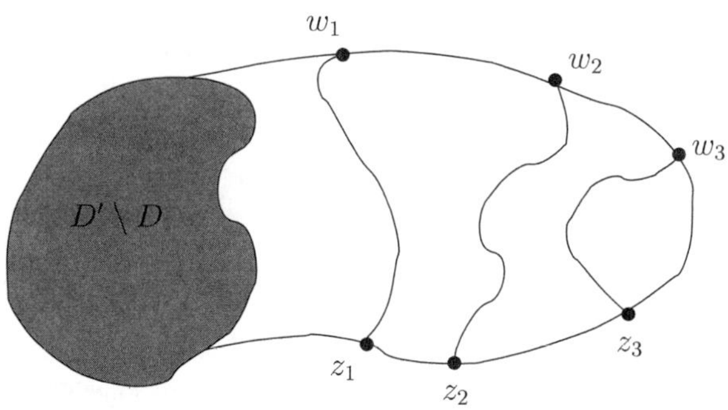

Figure 2 Boundary perturbation

The third relation relates the measures for different n.

- **Cascade Relation.** Let

$$\hat{\mathbf{z}} = (z_1, \ldots, z_{j-1}, z_{j+1}, \ldots, z_n), \quad \hat{\mathbf{w}} = (w_1, \ldots, w_{j-1}, w_{j+1}, \ldots, w_n),$$
$$\hat{\gamma} = (\gamma^1, \ldots, \gamma^{j-1}, \gamma^{j+1}, \ldots, \gamma^n).$$

The marginal distribution on $\hat{\gamma}$ induced by $Q_{D,b,n}(\mathbf{z},\mathbf{w})$ is absolutely continuous with respect to $Q_{D,b,n-1}(\hat{\mathbf{z}},\hat{\mathbf{w}})$ with Radon-Nikodym derivative

$H_{\hat{D},b}(z_j, w_j)$. Here $\hat{D}$ is the subdomain of $D \setminus \hat{\gamma}$ whose boundary includes z_j, w_j; see Figure 3.

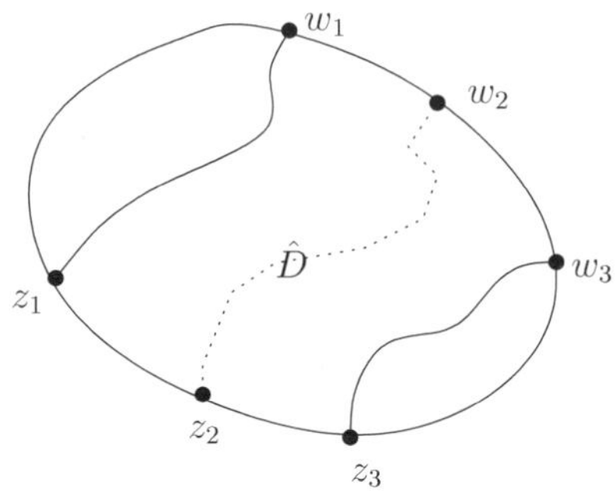

Figure 3 Cascade relation. $\hat{z} = (z_1, z_3)$, $\hat{w} = (w_1, w_3)$.

In order to construct such a collection of measures it is useful to include a fourth condition. This condition is stated in terms of $\mu^{\#}_{D,b}$ rather than $Q_{D,b}$.

- **Markov Property.** In the measure $\mu^{\#}_{D,b}(z, w)$, the conditional distribution on γ given an initial segment $\gamma[0, t]$ is $\mu^{\#}_{D \setminus \gamma[0,t],b}(\gamma(t), w)$; see Figure 4.

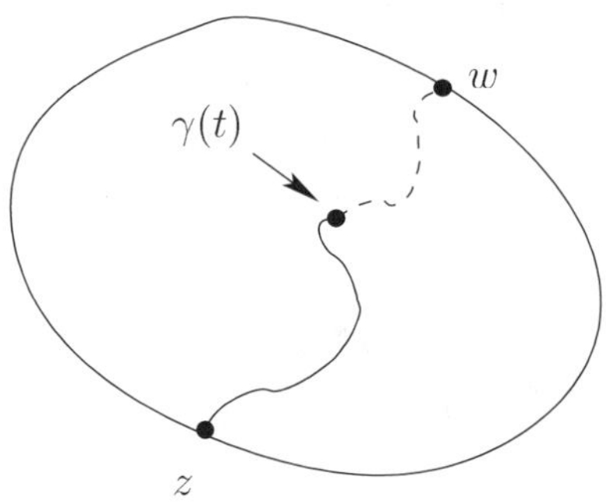

Figure 4 Markov property

We have stated this condition in a way that does not use two dimensions and conformal invariance. The *conformal Markov property* is the combination of the Markov property and (1.3). Schramm [16] showed that there is a one-parameter family of measures, which he parametrized by κ, satisfying the conformal Markov property. While these measures are well-defined for $\kappa > 0$, they are supported on simple curves only for $0 < \kappa \leq 4$; see [8, Proposition 6.9]. Therefore, we will restrict our consideration only to $\kappa \leq 4$. The scaling rule (1.2) implies that, up to an arbitrary multiplicative constant,

$$H_b(x, y) = H_{\mathbb{H},b}(x, y) = H_{\mathbb{H},b,1}(x, y) = |y - x|^{-2b}. \tag{1.5}$$

We will continue to write H_b for $H_{\mathbb{H},b}$ $(= H_{\mathbb{H},b,1})$ where $\mathbb{H} = \{z \in \mathbb{C} : \Im(z) > 0\}$ is the upper half plane. We will choose the constant to be 1 so that (1.5) holds. Given this, the measures $Q_{D,b,n}(\mathbf{z}, \mathbf{w})$ can be determined for all simply connected

D and $\mathbf{z}$, $\mathbf{w}$ at which ∂D is analytic by a combination of conformal mapping and the cascade relation. The question is whether or not we can find b, κ such that all four of the properties above hold. The answer is yes, assuming we choose $\kappa = \kappa(b)$ satisfying

$$\kappa = \frac{6}{2b+1} \quad \text{so that} \quad b = \frac{6-\kappa}{2\kappa}.$$

Notice that b decreases from ∞ to $1/4$ as κ increases from 0 to 4 which puts the restriction $b \geq 1/4$ on the scaling exponent. There are four other parameters that can be used to parametrize this family, namely

$$a = \frac{2}{\kappa} = \frac{2b+1}{3}, \quad \lambda = \frac{(3a-1)(4a-3)}{2a}, \quad c = -2\lambda, \quad \text{and} \quad d = 1 + \frac{1}{4a}. \tag{1.6}$$

These parameters have interpretations which can be summarized as follows.

- $b \in [1/4, \infty)$ is the *boundary scaling exponent* or *boundary conformal weight.*
- $\kappa \in (0, 4]$ is the variance of the driving Brownian motion in SLE_κ if the half-plane capacity at time t is $2t$.
- $a \in [1/2, \infty)$ and at is the half-plane capacity at time t for SLE_κ if the driving Brownian motion is chosen to have variance 1.
- $\lambda \in [-1/2, \infty)$ is the exponent of the Brownian loop measure that arises in the Radon-Nikodym derivatives.
- $c \in (-\infty, 1]$ is the *central charge.* We will not use this parameter in this paper, but c comes up naturally in representations of conformal field theories.
- $d \in (1, 3/2]$ is the Hausdorff dimension of the paths. This is also a parameter that will not be discussed in this paper.

We will tend to use b as our parameter, but any of the quantities above could equally well be used.

We have defined the measure $Q_{D,b,n}(\mathbf{z}, \mathbf{w})$ for fixed $\mathbf{z}, \mathbf{w}$. However, if D has a piecewise analytic boundary, we can define

$$Q_{D,b,n} = \int_{(\partial D)^n \times (\partial D)^n} Q_{D,b,n}(\mathbf{z}, \mathbf{w})\, d\mathbf{z}\, d\mathbf{w}.$$

In the case $b = 1$, the scaling rule implies that $Q_{D,1,n}$ is conformally invariant.

2 Preliminaries

2.1 SLE_κ. We review the definition of chordal SLE_κ in simply connected domains. We will only consider the case $\kappa \leq 4$ in this paper, and let $a = 2/\kappa$ as in (1.6). Then SLE_κ in $\mathbb{H}$ from 0 to ∞ is the random curve $\gamma : (0, \infty) \to \mathbb{H}$ with $\gamma(0+) = 0$ satisfying the following. Let $g_t : \mathbb{H} \setminus \gamma(0, t] \to \mathbb{H}$ be the unique conformal transformation with $g_t(z) - z = o(1)$ as $z \to \infty$. Then g_t satisfies the chordal Loewner differential equation

$$\dot{g}_t(z) = \frac{a}{g_t(z) - U_t}, \quad g_0(z) = z, \tag{2.1}$$

where the driving function U_t is a standard one-dimensional Brownian motion. The curve has been parametrized so that $\gamma(0, t]$ has half-plane capacity at, i.e., so that g_t has the expansion

$$g_t(z) = z + \frac{at}{z} + O(|z|^{-2}), \qquad z \to \infty. \tag{2.2}$$

If D is a simply connected domain and z, w are distinct boundary points of D, then chordal SLE_κ from z to w is defined to be the conformal image of SLE_κ

in $\mathbb{H}$ under a conformal transformation taking 0 to z and ∞ to w. We consider this as a probability measure on curves modulo reparametrization, and denote this probability measure by $\mu^{\#}_{D,b}(z,w)$.

There are a number of special values for κ. In this paper we will look at special properties of $\kappa = 2$. As we now show, the stochastic process $\dot{g}_t(z)$ is a martingale iff $\kappa = 2$ (equivalently, $a = 1$).

Proposition 2.1 *Suppose that $g_t(z)$ satisfies the chordal Loewner eq. (2.1). If $M_t = M_t^{z,a} = \dot{g}_t(z)$, then M_t is a martingale if and only if $a = 1$.*

Proof If $Z_t = g_t(z) - U_t$, then a simple calculation using Itô's formula shows that

$$dZ_t = \frac{a}{Z_t}dt - dU_t,$$

so that Z_t is a Bessel process with $d\langle Z\rangle_t = dt$. Using Itô's formula again therefore gives

$$dM_t = d\left(\frac{a}{Z_t}\right) = -\frac{a}{Z_t^2}dZ_t + \frac{a}{Z_t^3}d\langle Z\rangle_t = \frac{a}{Z_t^2}dU_t + \frac{a(1-a)}{Z_t^3}dt.$$

Hence, M_t is a martingale iff $a = 1$, in which case $dM_t = M_t^2 dU_t$. □

2.2 Brownian loop measure. We review the Brownian loop measure as introduced in [14]; see also [8, Section 5.7]. (That paper considered the loop soup, which is a Poissonian realization of the loop measure. This loop soup is very useful heuristically but is not needed for the definition of the measure.) A *rooted loop* of time duration t is a continuous function $\eta : [0,t] \to \mathbb{C}$ with $\eta(0) = \eta(t)$. It can also be considered as an ordered pair (t, η) where $\eta : (-\infty, \infty) \to \mathbb{C}$ is a continuous function with period t. An *unrooted loop* is an equivalence class of rooted loops under the equivalence $(t, \eta) \sim (t, \eta_s)$ for all s where $\eta_s(r) = \eta(s + r)$. Each rooted loop can be written uniquely as $(t, z, \bar{\eta})$ where $t > 0$ is the time duration, z is the root, and $\bar{\eta}$ is a loop of time duration 1 rooted at the origin. More specifically, η is given by translation and Brownian scaling, $\eta(s) = z + t^{1/2}\,\bar{\eta}(s/t)$. The *rooted Brownian loop measure* is given by choosing $(t, z, \bar{\eta})$ from the measure

$$\frac{1}{2\pi t}\,dt \times \text{ area } \times \nu_{BB}.$$

where ν_{BB} denotes the probability measure associated to Brownian bridges. The *Brownian loop measure* is the measure on unrooted loops obtained from this measure by "forgetting the loops." The loop measure in a domain D is exactly this measure restricted to loops that stay in the domain. An important fact is that the Brownian loop measure is conformally invariant: if $f : D \to D'$ is a conformal transformation, then the image of the loop measure on D by f is the loop measure on D'. (If we are worrying about the parametrization of the loops, we have to change the parametrization by the usual rule for conformal invariance of Brownian motion. In this paper, the parametrization will not be important, only the traces of the loops.) The Brownian loop measure can also be obtained as the scaling limit of a simple measure on random walk loops; see Section 5.1.

If $D \subsetneq \mathbb{C}$ is simply connected and V_1, V_2 are closed sets, we let $m(D; V_1, V_2) < \infty$ denote the Brownian loop measure of loops in D that intersect both V_1 and V_2. If $f : D \to D'$ is a conformal transformation, then $m(f(D); f(V_1), f(V_2)) = m(D; V_1, V_2)$. The two main properties of the Brownian loop measure that we will use are conformal invariance, and the fact that if D is simply connected, then

$0 < m(D; V_1, V_2) < \infty$, provided that V_1, V_2 are disjoint closed sets with at least one V_i compact; $\mathrm{dist}(V_1, V_2) > 0$; and non-polar, i.e., such that with positive probability a Brownian motion starting at $z \in D$ hits both of these sets.

2.3 Boundary perturbation of SLE. Suppose $D \subset \mathbb{H}$ is a simply connected domain such that $\mathbb{H} \setminus D$ is bounded, and $x \in \mathbb{R}$ with $\mathrm{dist}(x, \mathbb{H} \setminus D) > 0$. For any such domain there is a unique conformal transformation $\Phi = \Phi_{D,x} : D \to \mathbb{H}$ such that $\Phi(x) = 0$, $\Phi(\infty) = \infty$, $\Phi'(\infty) = 1$. Moreover, Schwarz reflection tells us how to extend Φ analytically in a neighborhood of x. Schramm [16] defined SLE_κ (modulo time parametrization) in D from x to ∞ to be the image of SLE_κ in $\mathbb{H}$ from 0 to ∞ under the conformal map Φ^{-1}. By using properties of conformal mappings and Itô's formula (see, e.g., [8, Section 4.6]) one can give an equivalent formulation (see, e.g., [9, Section 4.3]). Let γ and g_t be as above, and let $D_t = g_t(D)$, $\Phi_t = \Phi_{D_t, U_t}$. Consider the solution to (2.1) where the driving function U_t satisfies the SDE

$$dU_t = b \, \frac{\Phi_t''(U_t)}{\Phi_t'(U_t)} \, dt + dB_t, \tag{2.3}$$

where B_t is a standard one-dimensional Brownian motion. Then the distribution of the path γ is exactly that of SLE_κ in D. In this formulation we have parametrized the paths so that g_t satisfies (2.2). This is not the same parametrization as one would get if one started with an SLE_κ in $\mathbb{H}$ from 0 to ∞, say $\tilde{\gamma}$, and let $\gamma(t) = \Phi^{-1}\tilde{\gamma}(t)$. However, we only define SLE_κ in domains up to time reparametrization, so this is not an issue. In fact, we could have used (2.3) as the *definition* of SLE_κ in D, and then conformal invariance of the process would be a *result*.

There is a another equivalent way of defining SLE_κ in D (for $\kappa \le 4$) that comes from a martingale introduced in [10]. Let γ be an SLE_κ in $\mathbb{H}$ from x to ∞ parametrized as in (2.2). Let $T = T_D = \inf\{t > 0 : \gamma(t) \in \mathbb{H} \setminus D\}$. If $t < T$, then we can let D_t and Φ_t be as in the previous paragraph. Define Y_t by

$$Y_t = 1\{\gamma[0,t] \cap (\mathbb{H} \setminus D) = \emptyset\} \exp\left\{ \frac{a\lambda}{12} \int_0^t S\Phi_s(U_s)\, ds \right\},$$

where, as before,

$$\lambda = \lambda(\kappa) = \frac{(3a-1)(4a-3)}{2a} = \frac{(6-\kappa)(8-3\kappa)}{4\kappa},$$

and S denotes the Schwarzian derivative. In fact (see [10] and [14]),

$$-\frac{a}{12} \int_0^t S\Phi_s(U_s)\, ds = m(\mathbb{H}; \mathbb{H} \setminus D, \gamma(0,t]).$$

(The factor a comes from the fact that γ is parametrized as in (2.2).) Hence, we can write

$$Y_t = 1\{\gamma[0,t] \cap (\mathbb{H} \setminus D) = \emptyset\} \exp\{-\lambda \, m(\mathbb{H}; \mathbb{H} \setminus D, \gamma(0,t])\}. \tag{2.4}$$

In [10] it was shown that $M_t = \Phi_t'(U_t)^b \, Y_t$ is a local martingale. It is easy to check that $\Phi_t'(U_t) \le 1$ which implies that M_t is a bounded martingale for $\lambda \ge 0$ ($\kappa \le 8/3$). In this case, Girsanov's theorem can be used to find the effective SDE for paths weighted by the random variable M_∞. The resulting equation is (2.3). Since this is the same as SLE_κ in D, and we know that (for $\kappa \le 4$) SLE_κ in D never leaves D, we conclude that with respect to the weighted measure, $T = \infty$. What this says is that for $\kappa \le 8/3$, the probability measure on paths given by starting with SLE_κ from x to ∞ in $\mathbb{H}$, and then taking the Radon-Nikodym derivative M_∞ / M_0, is the same as SLE_κ in D. Since $M_0 = \Phi'(x)^b$ and $M_\infty = Y_\infty$, we conclude that $\mathbf{E}[Y_\infty] = \Phi'(x)^b$.

For $8/3 < \kappa \leq 4$, we have $\lambda < 0$ and hence Y_t is not bounded. However, the same conclusions hold. The reason is as follows. For a fixed $\gamma : (0,\infty) \to \mathbb{H}$ parametrized as in (2.2) with $\gamma(t) \to \infty$ and $\mathrm{dist}(\gamma(0,\infty), \mathbb{H} \setminus D) > 0$, standard conformal mapping estimates show that $Y_t \to Y_\infty < \infty$. In other words, the only problem that one can have is that Y_t blows up as $\gamma(t) \to \mathbb{H} \setminus D$. If we consider a stopping time of the form $\tau_n = \inf\{t : Y_t \geq n\}$, then we know that $Y_{t\wedge\tau_n}$ is a martingale. If we use Girsanov to weight the paths by this martingale, the paths satisfy (2.3). We know (for $\kappa \leq 4$) that paths that are given by this SDE never reach $\mathbb{H} \setminus D$ and go to infinity. Hence, M_t is actually a uniformly integrable martingale with $\mathbf{E}[M_\infty] = \mathbf{E}[M_0] = \Phi'(x)^b \leq 1$.

As a prelude to the next section, we can see that we could define the SLE_κ measure in D to be the measure on paths whose Radon-Nikodym derivative with respect to SLE_κ in D is Y_∞. This is a finite measure with total measure $\Phi'(x)^b$. If we normalize to make this a probability measure, then the corresponding measure is that of SLE_κ in D from x to ∞.

3 SLE measure

3.1 One path ($n = 1$). The configurational approach is to view chordal SLE_κ as not just a probability measure on paths connecting two specified points on the boundary, but rather as a finite measure on paths that when normalized gives chordal SLE_κ as defined by Schramm. Recall that we restrict our consideration to $\kappa \leq 4$. Suppose D is a simply connected domain and z, w are distinct boundary points at which ∂D is analytic. Then the SLE_κ measure on paths in D from z to w is defined to be

$$Q_{D,b}(z,w) = H_{D,b}(z,w)\; \mu^{\#}_{D,b}(z,w)$$

where $\mu^{\#}_{D,b}(z,w)$ is the SLE_κ probability measure, and $H_{D,b}(z,w) = |Q_{D,b}(z,w)|$ is determined by the scaling rule (1.2) from the kernel

$$H_b(x,y) = H_{\mathbb{H},b}(x,y) = |y-x|^{-2b}$$

as in (1.5). The conformal covariance rule

$$f \circ Q_{D,b}(z,w) = |f'(z)|^b\, |f'(w)|^b\; Q_{f(D),b}(f(z),f(w))$$

follows immediately from the conformal invariance of $\mu^{\#}_{D,b}(z,w)$ and the scaling rule (1.2) for $H_{D,b}(x,y)$. An important property of the measure is the boundary perturbation rule.

Proposition 3.1 (Boundary Perturbation) *Suppose $D \subset D' \subsetneq \mathbb{C}$ are simply connected domains. Suppose that z, w are distinct points of ∂D; ∂D is analytic in neighborhoods of z, w; and ∂D, $\partial D'$ agree in neighborhoods of z, w. Then $Q_{D,b}(z,w)$ is absolutely continuous with respect to $Q_{D',b}(z,w)$ with Radon-Nikodym derivative*

$$Y_{D,D',b}(z,w)(\gamma) = 1\{\gamma \subset D\}\, \exp\big\{ -\lambda\, m(D';\gamma, D'\setminus D)\big\}.$$

In particular, the Radon-Nikodym derivative is a conformal invariant, and moreover

$$H_{D,b}(z,w) \leq H_{D',b}(z,w). \tag{3.1}$$

Proof It suffices to prove this for $D' = \mathbb{H}$ and this is discussed in Section 2.3. □

Remark If $f : D' \to f(D')$ is a conformal transformation, then

$$Y_{f(D),f(D'),b}(f(z), f(w))(f \circ \gamma) = Y_{D,D',b}(z,w)(\gamma).$$

It follows that the Radon-Nikodym derivative

$$Y_{D,D',b}(z,w) = \frac{dQ_{D,b}(z,w)}{dQ_{D',b}(z,w)}$$

makes sense even for non-smooth boundaries provided that $D \subset D'$ and the boundaries agree near z and w. If $f : D' \to \mathbb{H}$ is a conformal transformation with $f(z) = 0, f(w) = \infty$, then $f(D' \setminus D)$ is a bounded subset of $\mathbb{H}$ bounded away from 0.

3.2 Multiple paths. We will now define the measures $Q_{D,b,n}$ for positive integers n. Let $\mathbf{z} = (z_1, \ldots, z_n)$, $\mathbf{w} = (w_1, \ldots, w_n)$ denote n-tuples of distinct points in ∂D ordered counterclockwise as in the introduction. Then $Q_{D,b,n}(\mathbf{z}, \mathbf{w})$, the n-path SLE$_\kappa$ measure in D, is defined to be the measure that is absolutely continuous with respect to the product measure

$$Q_{D,b}(z_1, w_1) \times \cdots \times Q_{D,b}(z_n, w_n)$$

with Radon-Nikodym derivative $Y(\bar\gamma) = Y_{D,b,\mathbf{z},\mathbf{w}}(\gamma^1, \ldots, \gamma^n)$ given by

$$Y(\bar\gamma) = 1\{\gamma^k \cap \gamma^l = \emptyset,\ 1 \le k < l \le n\} \exp\left\{-\lambda \sum_{k=1}^{n-1} m(D; \gamma^k, \gamma^{k+1})\right\}. \tag{3.2}$$

If $\lambda \ge 0$, it is immediate that $Q_{D,b,n}(\mathbf{z}, \mathbf{w})$ is a finite measure. As we will see after we prove the next proposition, this is also true for $\lambda \ge -1/2$; see (3.3).

Note that Y is conformally invariant. Consequently, the measures $Q_{D,b,n}$ satisfy the conformal covariance relation (1.1). Other important properties are given in the next two propositions.

Proposition 3.2 (Cascade Relation) *For $1 \le j \le n$, if*

$$\mathbf{z} = (z_1, \ldots, z_n), \quad \mathbf{w} = (w_1, \ldots, w_n), \quad \hat\gamma = (\gamma^1, \ldots, \gamma^{j-1}, \gamma^{j+1}, \ldots, \gamma^n),$$
$$\hat{\mathbf{z}} = (z_1, \ldots, z_{j-1}, z_{j+1}, \ldots, z_n), \quad \hat{\mathbf{w}} = (w_1, \ldots, w_{j-1}, w_{j+1}, \ldots, w_n),$$

then the marginal measure on $\hat\gamma$ in $Q_{D,b,n}(\mathbf{z}, \mathbf{w})$ is absolutely continuous with respect to $Q_{D,b,n-1}(\hat{\mathbf{z}}, \hat{\mathbf{w}})$ with Radon-Nikodym derivative $H_{\hat D,b}(z_j, w_j)$. Here $\hat D$ is the subdomain of $D \setminus \hat\gamma$ whose boundary includes z_j, w_j. (If z_j, w_j are contained in the boundary of different components, then $H_{\hat D,b}(z_j, w_j) = 0$.) Moreover, the conditional distribution of γ^j given $\hat\gamma$ is that of SLE$_\kappa$ from z_j to w_j in $\hat D$.

Proof Let $\hat Y(\hat\gamma) = Y_{D,b,\hat{\mathbf{z}},\hat{\mathbf{w}}}(\hat\gamma)$ denote the density of $Q_{D,b,n-1}(\hat{\mathbf{z}}, \hat{\mathbf{w}})$ with respect to the appropriate product measure as defined in (3.2). Then

$$Y = \hat Y \, 1\{\gamma^l \cap \gamma^k = \emptyset,\ 1 \le l < k \le n\} \exp\{-\lambda \hat m\}$$

where

$$\hat m = m(D; \gamma^{j-1}, \gamma^j) + m(D; \gamma^j, \gamma^{j+1}) - m(D; \gamma^{j-1}, \gamma^{j+1})$$

Simple inclusion-exclusion shows that $\hat m$ is the measure of loops in D that intersect both γ^j and $D \setminus \hat D$. The result now follows from (3.2) and the discussion in Section 2.3. □

Remark Let $H_{D,b,n}(\mathbf{z},\mathbf{w}) = |Q_{D,b,n}(\mathbf{z},\mathbf{w})|$. It follows from the proposition and (3.1), that

$$H_{D,b,n}(\mathbf{z},\mathbf{w}) \le H_{D,b,n-1}(\hat{\mathbf{z}},\hat{\mathbf{w}})\, H_{D,b}(z_j,w_j).$$

By iterating, we get

$$H_{D,b,n}(\mathbf{z},\mathbf{w}) \le \prod_{j=1}^{n} H_{D,b}(z_j,w_j), \tag{3.3}$$

so that we conclude $Q_{D,b,n}(\mathbf{z},\mathbf{w})$ is a finite measure for all $b \ge 1/4$, or equivalently, all $\lambda \ge -1/2$.

Proposition 3.3 (Boundary Perturbation) *Suppose $D \subset D' \subsetneq \mathbb{C}$ are simply connected domains. Then $Q_{D,b,n}(\mathbf{z},\mathbf{w})$ is absolutely continuous with respect to $Q_{D',b,n}(\mathbf{z},\mathbf{w})$ with Radon-Nikodym derivative*

$$Y_{D,D',b,n}(\mathbf{z},\mathbf{w})(\bar{\gamma}) = 1\{\gamma^j \subset D,\ j=1,\ldots,n\}\ \exp\{-\lambda\, m(D';\gamma^1\cup\cdots\cup\gamma^n, D'\setminus D)\}.$$

In particular, the Radon-Nikodym derivative is a conformal invariant.

Proof It follows from Proposition 3.1 that the Radon-Nikodym derivative of

$$Q_{D,b}(z_1,w_1)\times\cdots\times Q_{D,b}(z_n,w_n)$$

with respect to

$$Q_{D',b}(z_1,w_1)\times\cdots\times Q_{D',b}(z_n,w_n)$$

is

$$1\{\gamma^j \subset D,\ j=1,\ldots,n\}\ \exp\left\{-\lambda\sum_{k=1}^{n} m(D';\gamma^k, D'\setminus D)\right\}.$$

Hence by (3.2), $Y_{D,D',b,n}(\mathbf{z},\mathbf{w})(\bar{\gamma})$ equals this quantity times

$$\exp\left\{-\lambda\sum_{k=1}^{n-1}[m(D;\gamma^k,\gamma^{k+1}) - m(D';\gamma^k,\gamma^{k+1})]\right\}.$$

But $m(D';\gamma^k,\gamma^{k+1}) - m(D;\gamma^k,\gamma^{k+1})$ denotes the measure of the set of loops in D' that intersect all three of γ^k, γ^{k+1}, and $D'\setminus D$. Therefore, for $k=1,\ldots,n-1$,

$$m(D';\gamma^k, D'\setminus D) - [m(D';\gamma^k,\gamma^{k+1}) - m(D;\gamma^k,\gamma^{k+1})]$$

denotes the measure of the set of loops in D that intersect γ^k and $D'\setminus D$ but do not intersect γ^{k+1}. For topological reasons, they cannot possibly intersect γ^l for $l \ge k+1$. For any loop in D' that intersects both $\gamma^1\cup\cdots\cup\gamma^n$, there is a largest k such that the loop intersects γ^k but not $\gamma^{k+1},\ldots,\gamma^n$. This is accounted for in the kth term of the sum and in no other. □

Recall that $H_{D,b,n}$ satisfies the scaling rule

$$H_{D,b,n}(\mathbf{z},\mathbf{w}) = |f'(\mathbf{z})|^b\, |f'(\mathbf{w})|^b\, H_{f(D),b,n}(f(\mathbf{z}),f(\mathbf{w})).$$

It also satisfies the following cascade relation. Suppose $\mathbf{z} = (z_1,\hat{\mathbf{z}})$, $\mathbf{w} = (w_1,\hat{\mathbf{w}})$ and all the points of $\hat{\mathbf{z}}$, $\hat{\mathbf{w}}$ lie on the same arc connecting z_1 and w_1 on ∂D. Then,

$$H_{D,b,n}(\mathbf{z},\mathbf{w}) = Q_{D,b}(z_1,w_1)\,[\,H_{D(\gamma),b,n-1}(\hat{\mathbf{z}},\hat{\mathbf{w}})\,]$$

where $D(\gamma)$ is the connected component of $D\setminus\gamma$ whose boundary contains $\hat{\mathbf{z}}$, $\hat{\mathbf{w}}$ and we use the shorthand notation $\mu\,[\,f\,] = \int f\,d\mu$. If $f: D\to\mathbb{H}$ is a conformal transformation with $f(z_1) = 0$, $f(w_1) = \infty$, and $\hat{\mathbf{z}}$, $\hat{\mathbf{w}}$ are sent to the positive real axis, then

$$H_{D,b,n}(\mathbf{z}, \mathbf{w}) = |Q_{D,b}(z_1, w_1)| \, |f'(\hat{\mathbf{z}})|^b \, |f'(\hat{\mathbf{w}})|^b \, \mathbf{E}[H_{(\mathbb{H}\setminus\gamma)_+,b,n-1}(f(\hat{\mathbf{z}}), f(\hat{\mathbf{w}}))], \quad (3.4)$$

where in this case the expectation is over SLE_κ from 0 to ∞ in $\mathbb{H}$ and $(\mathbb{H} \setminus \gamma)_+$ denotes the component of $\mathbb{H} \setminus \gamma$ to the right of γ. This shows that $Q_{D,b,n}$ for all n can be computed, at least in principle, from computations of SLE_κ in $\mathbb{H}$.

We will also define the *partition function*

$$\tilde{H}_{D,b,n}(\mathbf{z}, \mathbf{w}) = \frac{H_{D,b,n}(\mathbf{z}, \mathbf{w})}{H_{D,b}(z_1, w_1) \cdots H_{D,b}(z_n, w_n)}. \quad (3.5)$$

Note that this is a conformal *invariant*, i.e.,

$$\tilde{H}_{D,b,n}(\mathbf{z}, \mathbf{w}) = \tilde{H}_{f(D),b,n}(f(\mathbf{z}), f(\mathbf{w})).$$

In particular $H_{D,b,n}$,and $\tilde{H}_{D,b,n}$ are determined by the values of

$$H_{b,n}(\mathbf{x}, \mathbf{y}) = H_{\mathbb{H},b,n}(\mathbf{x}, \mathbf{y}),$$

for $\mathbf{x}$, $\mathbf{y} \in \mathbb{R}^n$. We also write $\tilde{H}_{b,n}$ for $\tilde{H}_{\mathbb{H},b,n}$.

Remark One case that is known is SLE_2, corresponding to *loop-erased walk* $(a = b = 1)$, in which case

$$H_{1,n}(\mathbf{x}, \mathbf{y}) = \det[(y_k - x_j)^{-2}]_{1 \le j,k \le n}.$$

This is a consequence of Fomin's identity and the $n = 1$ case. We describe this in more detail in Section 5.

If $0 < x_1 < \cdots < x_n < y_n < \cdots < y_1 < \infty$, we let

$$\tilde{H}^*_{b,n}(\mathbf{x}, \mathbf{y}) = \tilde{H}_{b,n+1}((0, \mathbf{x}), (\infty, \mathbf{y})) = \lim_{w \to \infty} \tilde{H}_{b,n+1}((0, \mathbf{x}), (w, \mathbf{y})),$$

$$H^*_{b,n}(\mathbf{x}, \mathbf{y}) = \left[\prod_{j=1}^{n} H_{b,1}(x_j, y_j)\right] \tilde{H}^*_{b,n}(\mathbf{x}, \mathbf{y}) = \lim_{w \to \infty} w^{2b} \, H_{b,n+1}((0, \mathbf{x}), (w, \mathbf{y})).$$

These functions satisfy the following scaling rules for $r > 0$:

$$\tilde{H}^*_{b,n}(r\mathbf{x}, r\mathbf{y}) = \tilde{H}^*_{b,n}(\mathbf{x}, \mathbf{y}), \quad H^*_{b,n}(r\mathbf{x}, r\mathbf{y}) = r^{-2bn} H^*_{b,n}(\mathbf{x}, \mathbf{y}). \quad (3.6)$$

3.3 The partition function for two paths. In this section we calculate $H^*_{b,1}(x, y) = (y - x)^{-2b} \, \tilde{H}^*_{b,1}(x, y) = (y - x)^{-2b} \tilde{H}_{b,2}((0, x), (\infty, y))$ for $0 < x < y < \infty$. The formula in the following proposition has appeared previously; see, for example, [3]. As usual, we will write just $H^*(x, y)$, $\tilde{H}^*(x, y)$ for $H^*_{b,1}(x, y)$, $\tilde{H}^*_{b,1}(x, y)$, respectively. We will use this result in Section 5.4 to derive the scaling limit of Fomin's identity.

Proposition 3.4 *If* $b \ge 1/4$, *then*

$$H^*_{b,1}(x, y) = (y - x)^{-2b} \, \frac{\Gamma(2a) \, \Gamma(6a - 1)}{\Gamma(4a) \, \Gamma(4a - 1)} \, (x/y)^a \, F(2a, 1 - 2a, 4a; x/y) = \frac{\phi_b(x/y)}{(y - x)^{2b}}, \quad (3.7)$$

where F *denotes the hypergeometric function and* $a = (2b + 1)/3$.

Proof By the scaling rule (3.6), we know that $\tilde{H}^*(x,y) = \phi(x/y)$ for some function $\phi = \phi_b$ of one variable. We will parametrize SLE_κ so that it satisfies (2.1) where the driving function $U_t = -B_t$ is a standard one-dimensional Brownian motion. Let $0 < x < y$, and let $X_t = g_t(x) + B_t$, $Y_t = g_t(y) + B_t$ so that

$$dX_t = \frac{a}{X_t}\,dt + dB_t, \qquad dY_t = \frac{a}{Y_t}\,dt + dB_t$$

as in the proof of Proposition 2.1. Let $(\mathbb{H}\setminus\gamma)_+$ denote the component of $\mathbb{H}\setminus\gamma$ whose boundary includes x, y. If $g : (\mathbb{H}\setminus\gamma)_+ \to \mathbb{H}$ is a conformal transformation with $g(\infty) = \infty$, then

$$H_{(\mathbb{H}\setminus\gamma)_+,b}(x,y) = \frac{g'(x)^b\, g'(y)^b}{[g(y)-g(x)]^{2b}}$$

as a consequence of (1.2) and (1.5). Using this we see that if

$$J_t = \frac{g_t'(x)^b\, g_t'(y)^b}{[g_t(y)-g_t(x)]^{2b}} \qquad \text{and} \qquad J_\infty = \lim_{t\to\infty} J_t$$

then from (3.4) and (3.5)

$$\tilde{H}^*(x,y) = (y-x)^{2b}\,\mathbf{E}^{x,y}[J_\infty] \tag{3.8}$$

so that $H^*(x,y) = \mathbf{E}^{x,y}[J_\infty]$. We will now give the differential equation for $\tilde{H}^*(x,y)$. Since

$$\partial_t\,[\log g_t'(x)] = -\frac{a}{X_t^2}, \qquad \partial_t\,[\log g_t'(y)] = -\frac{a}{Y_t^2}, \qquad \text{and}$$

$$\partial_t\,[\log(g_t(y)-g_t(x))] = \frac{1}{Y_t - X_t}\left[\frac{a}{Y_t} - \frac{a}{X_t}\right] = -\frac{a}{X_t\,Y_t},$$

we see that

$$\begin{aligned} J_t &= \exp\{\log J_t\} = J_0 \exp\left\{\int_0^t \partial_s[\log J_s]\,ds\right\} \\ &= \frac{1}{(y-x)^{2b}} \exp\left\{-ab\int_0^t \left(\frac{1}{X_s} - \frac{1}{Y_s}\right)^2 ds\right\}. \end{aligned}$$

Hence, (3.8) gives

$$\tilde{H}^*(x,y) = \mathbf{E}^{x,y}\left[\exp\left\{-ab\int_0^\infty \left(\frac{1}{X_s} - \frac{1}{Y_s}\right)^2 ds\right\}\right].$$

It now follows from the (usual) Markov property that $J_t\,\tilde{H}^*(X_t,Y_t)$ is a martingale. That is, if $M_t = \mathbf{E}^{x,y}[J_\infty|\mathcal{F}_t]$ so that M_t is a martingale, then

$$\begin{aligned} M_t &= \mathbf{E}^{x,y}\left[\exp\left\{-ab\int_0^\infty \left(\frac{1}{X_s} - \frac{1}{Y_s}\right)^2 ds\right\} \Bigm| \mathcal{F}_t\right] \\ &= \exp\left\{-ab\int_0^t \left(\frac{1}{X_s} - \frac{1}{Y_s}\right)^2 ds\right\} \\ &\qquad \times \mathbf{E}^{x,y}\left[\exp\left\{-ab\int_t^\infty \left(\frac{1}{X_s} - \frac{1}{Y_s}\right)^2 ds\right\} \Bigm| \mathcal{F}_t\right] \\ &= J_t\,\tilde{H}^*(X_t,Y_t). \end{aligned}$$

Itô's formula now implies that

$$-ab\left(\frac{1}{x}-\frac{1}{y}\right)^2 \tilde{H}^* + \frac{a}{x}\,\partial_x \tilde{H}^* + \frac{a}{y}\,\partial_y \tilde{H}^* + \frac{1}{2}\,\partial_{xx}\tilde{H}^* + \frac{1}{2}\,\partial_{yy}\tilde{H}^* + \partial_{xy}\tilde{H}^* = 0. \quad (3.9)$$

Since $\tilde{H}^*(x,y) = \phi(x/y)$, we find

$$\partial_x \tilde{H}^* = y^{-1}\,\phi'(x/y), \quad \partial_y \tilde{H}^* = -x\,y^{-2}\,\phi'(x/y), \quad \partial_{xx}\tilde{H}^* = y^{-2}\,\phi''(x/y),$$

$$\partial_{yy}\tilde{H}^* = 2\,x\,y^{-3}\,\phi'(x/y) + x^2\,y^{-4}\,\phi''(x/y), \quad \partial_{xy}\tilde{H}^* = -y^{-2}\,\phi'(x/y) - x\,y^{-3}\,\phi''(x/y),$$

so that after substituting into (3.9), multiplying by y^2, letting $u = x/y$, combining terms, and recalling that $b = (3a-1)/2$, we have

$$u^2\,(1-u)^2\,\phi''(u) + 2\,u\,(a-u+(1-a)\,u^2)\,\phi'(u) - a(3a-1)(1-u)^2\,\phi(u) = 0. \quad (3.10)$$

Note that $0 < u < 1$, and let $\psi(u) = u^{-a}(1-u)^{1-4a}\phi(u)$ so that (3.10) implies

$$u\,(1-u)\,\psi''(u) + (4a - 8au)\,\psi'(u) - 2a(6a-1)\,\psi(u) = 0. \quad (3.11)$$

We see that (3.11) is the well-known hypergeometric differential equation (see equation (9.5.5) of [15]) which has two linearly independent solution; hence

$$\psi(u) = C_1 F(2a, 6a-1, 4a; u) + C_2 u^{1-4a} F(1-2a, 2a, 2-4a; u)$$

where C_1 and C_2 are arbitrary constants. Equation (9.5.3) of [15] gives $F(2a, 6a-1, 4a; u) = (1-u)^{1-4a}\,F(2a, 1-2a, 4a; u)$ so that the general solution of (3.10) is therefore

$$\phi(u) = C_1 u^a\,F(2a, 1-2a, 4a; u) + C_2 u^{1-3a}\,(1-u)^{4a-1}\,F(2a, 1-2a, 2-4a; u).$$

Since $\phi(u) \to 0$ as $u \to 0+$, and $\phi(u) \to 1$ as $u \to 1-$, we find $C_2 = 0$ and

$$C_1^{-1} = F(2a, 1-2a, 4a; 1) = \lim_{u \to 1-} F(2a, 1-2a, 4a; u) = \frac{\Gamma(4a)\,\Gamma(4a-1)}{\Gamma(2a)\,\Gamma(6a-1)},$$

so that

$$\phi(u) = \phi_b(u) = \frac{\Gamma(2a)\,\Gamma(6a-1)}{\Gamma(4a)\,\Gamma(4a-1)}\,u^a\,F(2a, 1-2a, 4a; u).$$

□

Remark If $x > 0$ is fixed, then

$$\lim_{y \to \infty} y^{a+2b} H^*_{b,1}(x,y) = \frac{\Gamma(2a)\,\Gamma(6a-1)}{\Gamma(4a)\,\Gamma(4a-1)}\,x^a. \quad (3.12)$$

Remark If $\tilde{H}_{b,2} = \tilde{H}_{\mathbb{H},b,2}$ is as defined in (3.5), then this proposition implies for $x_1 < x_2 < y_2 < y_1$ that

$$\tilde{H}_{b,2}(x_1, x_2, y_1, y_2) = \phi_b(1-q), \quad (3.13)$$

where

$$q = q(x_1, x_2, y_2, y_1) = \frac{(y_1 - x_1)\,(y_2 - x_2)}{(y_1 - x_2)\,(y_2 - x_1)},$$

denotes the cross-ratio. As pointed out in [3], if the quantity on the left of (3.13) is to be a conformal invariant, it should be a function of q. If $x_1 = 0$, $x_2 = x$, $y_2 = y$, $y_1 = \infty$, then $q = 1 - (x/y)$ as expected.

Remark If $1-2a=-k$ where k is a non-negative integer, then $F(2a,1-2a,4a;x/y)$ is a kth degree polynomial. These values correspond to

$$b=\frac{3k+1}{4}.$$

In particular, $\phi_{1/4}(u)=\sqrt{u}$ and $\phi_1(u)=2u-u^2=1-(1-u)^2$. Several other special cases of (3.7) are listed in the following table.

$a=$	$b=$	$\kappa=$	$H^*_{b,1}(x,y)=$
2	5/2	1	$(y-x)^{-5}\,(x/y)^2\,[6-9(x/y)+5(x/y)^2-(x/y)^3]$
3/2	7/4	4/3	$(y-x)^{-7/2}\,(x/y)^{3/2}\,[(7/2)-(7/2)(x/y)+(x/y)^2]$
1	1	2	$(y-x)^{-2}\,(x/y)\,[2-(x/y)]$
1/2	1/4	4	$(y-x)^{-1/2}\,(x/y)^{1/2}$

Remark Our argument is essentially the same as the calculation of *Cardy's formula* for SLE_κ. An examination of the proof will show that the hypergeometric solution is valid for $a>1/4$, or, equivalently, $\kappa<8$. The only reason we restrict to $b\geq 1/4$ in the proposition is to have $\kappa\leq 4$. In fact, if $\kappa=6$ (corresponding to $b=0$, or $a=1/3$) then we recover Cardy's formula, namely

$$\phi_0(u)=\frac{\Gamma(2/3)}{\Gamma(4/3)\,\Gamma(1/3)}\,u^{1/3}\,F(2/3,1/3,4/3;u).$$

A proof for the $b=0$ case only may be found in either [2] or [8, Proposition 6.33].

4 Multiple SLEs

As was shown in Section 3.2, the definition of $Q_{D,b,n}(\mathbf{z},\mathbf{w})$ is fairly straightforward. Since $H_{D,b,n}(\mathbf{z},\mathbf{w})=|Q_{D,b,n}(\mathbf{z},\mathbf{w})|$, we can write

$$Q_{D,b,n}(\mathbf{z},\mathbf{w})=H_{D,b,n}(\mathbf{z},\mathbf{w})\,\mu^{\#}_{D,b,n}(\mathbf{z},\mathbf{w}),$$

for some probability measure $\mu^{\#}_{D,b,n}(\mathbf{z},\mathbf{w})$ on n-tuples of paths that can be called SLE_κ from $\mathbf{z}$ to $\mathbf{w}$ in D. This measure will be absolutely continuous with respect to the product measure given by taking n independent SLE_κ paths $\gamma^1,\ldots,\gamma^n$ where γ^j goes from z_j to w_j. In the present section, we use Girsanov's theorem to give the SDE.

Suppose $G(t,x_0,x_1,\ldots,x_n)$ is a smooth, non-negative function (at least on the set $\{x_j\neq x_k : 0\leq j<k\leq n\}$), and suppose $\gamma:(0,\infty)\to\mathbb{H}$ is an SLE_κ curve with corresponding maps g_t parametrized as in (2.1). We allow the function $G(t,x_0,x_1,\ldots,x_n)$ to be a random function adapted to the Brownian filtration, but we assume that with probability one G is C^1 in t. Let $X_t^j=g_t(x_j)$ where W_t is a standard Brownian motion, and note that

$$\dot{X}_t^j=\frac{a}{X_t^j-W_t}.$$

If $J_t = G(t, W_t, X_t^1, \ldots, X_t^n)$, then Itô's formula gives

$$dJ_t = J_t \left[R_t \, dt + \frac{\partial_0 G(t, W_t, \bar{X}_t)}{G(t, W_t, \bar{X}_t)} \, dW_t \right],$$

where

$$R_t = G(t, W_t, \bar{X}_t)^{-1} \left[\dot{G}(t, W_t, \bar{X}_t) + \frac{1}{2} \, \partial_{00} G(t, W_t, \bar{X}_t) + a \sum_{j=1}^{n} \frac{\partial_j G(t, W_t, \bar{X}_t)}{X_t^j - W_t} \right]$$

with $\bar{X}_t = (X_t^1, \ldots, X_t^n)$ and $\partial_j = \partial_{x_j}$, $\partial_{00} = \partial_{x_0 x_0}$. If

$$M_t = \exp\left\{ -\int_0^t R_s \, ds \right\} J_t,$$

then M_t is a local martingale satisfying

$$dM_t = M_t \, \frac{\partial_0 G(t, W_t, \bar{X}_t)}{G(t, W_t, \bar{X}_t)} \, dW_t.$$

If M_t is a martingale, Girsanov's theorem says that if we weight the paths by M_t then this is the same as if W_t satisfied

$$dW_t = \frac{\partial_0 G(t, W_t, \bar{X}_t)}{G(t, W_t, \bar{X}_t)} \, dt + dB_t, \tag{4.1}$$

for a standard Brownian motion B_t. Hence an SLE_κ weighted by G is the same as a solution to the chordal Loewner equation (2.1) where the driving function W_t satisfies (4.1). Even if M_t is not a martingale, we can get a similar expression for a stopped process $M_{t \wedge \tau}$ if the stopped process is a continuous martingale.

4.1 SLE in a domain D. Suppose that $D \subset \mathbb{H}$ is a simply connected domain with $\mathbb{H} \setminus D$ bounded and bounded away from zero. Let $D_t = g_t(D)$ and let Φ_t be a conformal transformation of D_t onto $\mathbb{H}$ with $\Phi_t(\infty) = \infty$, $\Phi_t'(\infty) = 1$. Let $G(t,x) = \Phi_t'(x)^b$. Then if we weight the paths of SLE_κ by this function, the paths satisfy

$$dW_t = b \, \frac{\Phi_t''(W_t)}{\Phi_t'(W_t)} \, dt + dB_t. \tag{4.2}$$

This is the same equation as given in Section 2.3 for SLE_κ in D when one takes the conformal image of SLE_κ from $\mathbb{H}$ to D. In other words, another way of viewing SLE_κ in D is as SLE_κ weighted by $\Phi_t(W_t)^b$.

A similar example is SLE_κ from 0 to y where $0 < y < \infty$. This measure on paths is absolutely continuous with respect to SLE_κ from 0 to ∞ provided that the paths are stopped at a stopping time smaller than the first visit to y. At that point the paths become singular. (Although this is somewhat imprecise, it is not too difficult to make it precise.) The equation (4.2) can be applied to see how the process evolves. If t is smaller than the hitting time of y, we define g_t as before and let $X_t = g_t(y)$. Then, $\Phi_t(z) = (z - X_t)^{-1}$ is a conformal map taking $g_t(\mathbb{H} \setminus \gamma(0,t])$ ($= \mathbb{H}$ in this case) to $\mathbb{H}$, mapping X_t to ∞. Using (4.2), we get

$$dW_t = \frac{2b}{X_t - W_t} \, dt + dB_t, \qquad dX_t = \frac{a}{X_t - W_t} \, dt. \tag{4.3}$$

This is an example of an $\mathrm{SLE}(\kappa, \rho)$ process as introduced in [10]. Note that

$$d(W_t - X_t) = \frac{(1-4b)/3}{W_t - X_t} \, dt + dB_t,$$

so comparison with a Bessel process shows that the path has finite lifetime. (Recall that we are parametrizing the curve by its half-plane capacity — if γ is a simple curve from 0 to y, then the path has finite capacity.) We can derive the same process by taking SLE$_\kappa$ from 0 to ∞ and weighting the paths by $G(t, W_t, X_t) = H_b(W_t, X_t) = (X_t - W_t)^{-2b}$. We can see this by comparing (4.1) and the first equation in (4.3).

4.2 Two SLE$_\kappa$ paths. Consider the case of the SLE$_\kappa$ measure corresponding to two mutually avoiding SLE$_\kappa$ paths; one from 0 to ∞ and the other from x_1 to x_2. This corresponds to choosing

$$G(t, x_0, x_1, x_2) = H^*(x_1 - x_0, x_2 - x_0) = (x_2 - x_1)^{-2b}\, \phi_b\left(1 - \frac{x_2 - x_1}{x_2 - x_0}\right) \tag{4.4}$$

where ϕ_b is as in Proposition 3.4. (Recall that $H^*(x, y) = H^*_{\mathbb{H},b,1}(x, y)$.) Then,

$$\frac{\partial_0 G(t, x_0, x_1, x_2)}{G(t, x_0, x_1, x_2)} = -\frac{1}{x_2 - x_0}\,\frac{q\,\phi_b'(1-q)}{\phi_b(1-q)} = -\frac{q}{x_2 - x_0}\,[\log \phi_b]'(1-q),$$

where $\partial_0 = \partial_{x_0}$ as before and $q = (x_2 - x_1)/(x_2 - x_0)$ denotes the appropriate cross-ratio. This is the marginal distribution of the SLE$_\kappa$ path from 0 to ∞ given by the measure $\mu^{\#}_{b,2}((0, x_1), (x_2, \infty))$. Two examples are as follows.

- If $b = 1$ (equivalently, $\kappa = 2$), we have $\phi_1(u) = 2u - u^2$ and hence
$$\frac{\partial_0 G(t, x_0, x_1, x_2)}{G(t, x_0, x_1, x_2)} = -\frac{2}{x_2 - x_0}\,\frac{q^2}{1 - q^2}.$$
- If $b = 1/4$ (equivalently, $\kappa = 4$), then $\phi_{1/4}(u) = \sqrt{u}$ and hence
$$\frac{\partial_0 G(t, x_0, x_1, x_2)}{G(t, x_0, x_1, x_2)} = -\frac{1}{2(x_2 - x_0)}\,\frac{q}{1 - q}.$$

If we let $x_2 \to \infty$, we can consider two mutually avoiding SLE$_\kappa$ paths both going to infinity. Here we use the function (see (3.12))

$$G(t, x_0, x_1) = (x_1 - x_0)^a.$$

In this case,

$$\frac{\partial_0 G(t, x_0, x_1)}{G(t, x_0, x_1)} = -\frac{a}{x_1 - x_0}.$$

From this we see that if we take a pair mutually avoiding SLE$_\kappa$ paths starting at x_0, x_1, respectively, then the measure on the path started at x_0 is obtained by solving the usual chordal Loewner equation with driving function U_t satisfying

$$dU_t = \frac{a}{U_t - V_t}\, dt + dB_t,$$

where $V_t = g_t(x_1)$ satisfies

$$dV_t = \frac{a}{V_t - U_t}\, dt.$$

This is an example of a SLE(κ, ρ) process in [10]. Note that $Z_t = U_t - V_t$ satisfies

$$dZ_t = \frac{2a}{Z_t}\, dt + dB_t. \tag{4.5}$$

4.2.1 *Growing both paths simultaneously.* It is possible to write down the equation for two mutually avoiding SLE paths growing simultaneously. We can write the corresponding conformal transformation g_t as the solution to the Loewner equation

$$\dot{g}_t(z) = \frac{a}{g_t(z) - U_t^1} + \frac{a}{g_t(z) - U_t^2},$$

for appropriate driving functions U_t^1, U_t^2. The parametrization of g_t is such that

$$g_t(z) = z + \frac{2a}{z} + O(|z|^{-2}), \qquad z \to \infty.$$

To find the equation of U_t^j we start by taking them to satisfy

$$dU_t^1 = \frac{a}{U_t^1 - U_t^2}\, dt + dW_t^1 \quad \text{and} \quad dU_t^2 = \frac{a}{U_t^2 - U_t^1}\, dt + dW_t^2,$$

where W_t^1, W_t^2 are independent standard Brownian motions.

Assume that $U_0^1 = x_0$ and $U_0^2 = x_1$. If $G(t, x_0, x_1) = (x_1 - x_0)^a$, then $G(t, U_t^1, U_t^2)$ satisfies

$$\begin{aligned}\frac{dG(t, U_t^1, U_t^2)}{G(t, U_t^1, U_t^2)} &= \frac{a(a-1)}{(U_t^2 - U_t^1)^2}\, dt + \frac{a}{U_t^1 - U_t^2}\, dU_t^1 + \frac{a}{U_t^2 - U_t^1}\, dU_t^2 \\ &= \frac{a(3a-1)}{(U_t^2 - U_t^1)^2}\, dt + \frac{a}{U_t^1 - U_t^2}\, dW_t^1 + \frac{a}{U_t^2 - U_t^1}\, dW_t^2\end{aligned}$$

Again,

$$M_t = \exp\left\{-a(3a-1)\int_0^t (U_s^2 - U_s^1)^{-2}\, ds\right\} G(t, U_t^1, U_t^2)$$

is a local martingale satisfying

$$dM_t = M_t \left[\frac{a}{U_t^1 - U_t^2}\, dW_t^1 + \frac{a}{U_t^2 - U_t^1}\, dW_t^2\right].$$

Compare this with [8, Example 1.18]. If we weight the paths by M_t, the effective equation on the W_t^j is

$$dW_t^j = \frac{a}{U_t^j - U_t^{3-j}}\, dt + dB_t^j,$$

where B_t^1, B_t^2 denote independent standard Brownian motions, and the effective equation on the U_t^j becomes

$$dU_t^j = \frac{2a}{U_t^j - U_t^{3-j}}\, dt + dB_t^j.$$

Note that if $Z_t = U_{t/2}^2 - U_{t/2}^1$, $B_t = B_{t/2}^2 - B_{t/2}^1$, then Z_t, B_t satisfy (4.5).

4.3 Multiple SLE$_\kappa$. Suppose $x_1 < \cdots < x_n < y_n < \cdots < y_1$. Then the measure $\mu^{\#}_{\mathbb{H},b,n}(\mathbf{x}, \mathbf{y})$ can be described by using the function

$$G(t, x_1, \ldots, x_n, y_1, \ldots, y_n) = H_{b,n}(\mathbf{x}, \mathbf{y}).$$

We can use this to describe the marginal distribution for a particular path. For example, suppose we are interested in the first path. Then it acts like SLE$_\kappa$ weighted locally by $G(t, W_t, X_t^2, \ldots, X_t^n, Y_t^1, \ldots, Y_t^n)$.

5 The scaling limit of Fomin's identity

The SLE measure is expected to be the limit of discrete critical processes. One example that is well understood at this point is the case of the loop-erased random walk (LERW) which is the prototypical model for the boundary scaling exponent $b = 1$. We will describe the discrete model in this section and some different ways in which one can take the limit to get the SLE measure with $b = 1$ (corresponding to $\kappa = 2$).

In [4], Fomin proved an identity showing that a certain probability for loop-erased random walk on $\mathbb{Z}^2$ can be given in terms of a determinant of hitting probabilities for simple random walk. He also conjectured that this identity holds for continuous processes. In fact, he wrote,

> "...we do not need the notion of loop-erased Brownian motion. Instead, we discretize the model, compute the probability, and then pass to the limit."

It is shown in [6] that the scaling limit of the determinant of hitting probabilities for simple random walk is, in fact, the determinant of the hitting densities of Brownian excursions, i.e., the so-called *excursion Poisson kernel determinant.*

We conclude this section by deriving the scaling limit of Fomin's identity directly in the case $n = 2$ using the ideas of the previous sections, and show how the probability that an SLE_2 avoids a Brownian excursion may be computed in terms of the excursion Poisson kernel determinant. This is the natural continuous analogue of the discrete $n = 2$ case, in which Fomin's identity gives the probability that a loop-erased walk avoids a simple random walk in terms of the determinant of the simple random walk hitting matrix.

5.1 Loop-erased random walk. We write the integer lattice $\mathbb{Z}^2$ in complex form $\mathbb{Z} + i\mathbb{Z}$. If $A \subsetneq \mathbb{Z}^2$, we let $\partial A = \{z \in \mathbb{Z}^2 : \mathrm{dist}(z, A) = 1\}$. We call $\omega = [\omega_0, \ldots, \omega_n]$ a *(nearest neighbor) random walk path* if $\omega_0, \ldots, \omega_n \in \mathbb{Z}^2$ with $|\omega_j - \omega_{j-1}| = 1$ for $j = 1, \ldots, n$. We write $|\omega| = n$ for the number of steps in the path. If z, $w \in \partial A$, we say that ω is an excursion in A from z to w if $\omega_0 = z$, $\omega_n = w$, and $\omega_1, \ldots, \omega_{n-1} \in A$. The *(simple random walk) excursion measure* $\nu(A; z, w)$ is the measure that assigns weight $4^{-|\omega|}$ to each excursion ω from z to w in A. We write $h_A(z, w) = |\nu(A; z, w)|$. The *loop-erased excursion* measure $\nu_{LE}(A; z, w)$ is the measure obtained from $\nu(A; z, w)$ by doing chronological loop-erasure on each excursion. Note that $|\nu_{LE}(A; z, w)| = |\nu(A; z, w)| = h_A(z, w)$. We also write $\nu^{\#}(A; z, w) = \nu(A; z, w)/h_A(z, w)$, and similarly for $\nu^{\#}_{LE}(A; z, w)$, to denote the measures normalized to be probability measures.

Let $\mathcal{H} = \{j + ik \in \mathbb{Z}^2 : k > 0\}$ denote the discrete upper half plane with $\partial\mathcal{H} = \mathbb{Z}$. It is not difficult to see (cf., equation (28) in [6]) that if $j_1 \neq j_2$, then

$$h_{\mathcal{H}}(j_1, j_2) = \frac{1}{16} G_{\mathcal{H}}(j_1 + i, j_2 + i)$$

where $G_{\mathcal{H}}$ denotes the random walk Green's function in $\mathcal{H}$. Indeed we get the expression above by dividing each excursion from j_1 to j_2 in $\mathcal{H}$ into its first step $(j_1, j_1 + i)$; its *middle*, i.e., any walk in A starting at $j_1 + i$ and ending at $j_2 + i$; and its final step $(j_2 + i, j_2)$. It is not hard to show (cf., equation (15.3) in [17]) that

$$G_{\mathcal{H}}(x, y) = a(x, \bar{y}) - a(x, y),$$

where $a(x,y) = a(y-x)$ denotes the potential kernel for random walk and $\bar{y}$ denotes the complex conjugate of y. The potential kernel has asymptotic behavior

$$a(z) = \frac{2}{\pi}\log|z| + C + O(|z|^{-2}),$$

for a known constant C. Also, this relation can be differentiated, i.e.,

$$a(z-2i) - a(z) = \frac{2}{\pi}\left[\log|z-2i| - \log|z|\right] + O(|z|^{-3}).$$

Using this, we get the relation

$$h_{\mathcal{H}}(j_1, j_2) = \frac{1}{4\pi\,(j_2 - j_1)^2} + O((j_2 - j_1)^{-3}). \tag{5.1}$$

Let us consider the scaling limit of this measure. Suppose N is a large integer. Let $\mathcal{H}_N = N^{-1}\mathcal{H}$ and let $\nu^{(N)}(j/N, k/N)$ be the measure that assigns measure $4\pi N^2 4^{-|\omega|}$ to each excursion ω in $\mathcal{H}$ from j to k. As $N \to \infty$, this measure approaches the *Brownian excursion measure* ν_B on $\mathbb{H}$ defined roughly as follows. The measure $\nu_B(x,y)$ has total mass $H_1(x,y) = (y-x)^{-2}$. If $\nu_B^{\#}(x,y)$ denotes the probability measure obtained by normalization, then this is the measure corresponding to "Brownian motion starting at x conditioned to leave $\mathbb{H}$ at y" which can be made precise using h-processes. (See, for example, [5, Section 3.4].)

In the limit above, we fixed an x, y and took the limit. A similar limit can be taken by considering the measure that gives weight $4\pi 4^{-|\omega|}$ to every excursion (with any possible endpoints). A similar argument shows that the limiting measure obtained is (a constant multiple of) Brownian excursion measure in $\mathbb{H}$ and can be written as

$$\int_{-\infty}^{\infty}\int_{-\infty}^{\infty} \frac{1}{(y-x)^2}\,\nu_B^{\#}(x,y)\,dx\,dy.$$

The advantage of this formulation is that it can be used to describe the scaling limit of excursions in Jordan domains without any smoothness assumptions on the boundary; see [5] and [6].

We can also take the scaling limit of the loop-erased measure

$$\nu_{LE}(\mathcal{H}; j,k) = h_{\mathcal{H}}(j,k)\,\nu_{LE}^{\#}(\mathcal{H}; j,k).$$

The scaling limit of the normalized measure $\nu_{LE}^{\#}(\mathcal{H}; j,k)$ was shown in [11] to be SLE_2 from x to y. (To be honest, the paper [11] considered a radial version, but the basic argument works in this case.)

A *rooted (random walk) loop* in A is a random walk path $\omega = [\omega_0, \ldots, \omega_{2n}]$ with $\omega_0 = \omega_{2n}$ and $\omega_1, \ldots, \omega_{2n} \in A$. *An unrooted (random walk) loop* is an equivalence class of loops under the equivalence

$$[\omega_0, \omega_1, \omega_2, \ldots, \omega_{2n}] \sim [\omega_1, \omega_2, \ldots, \omega_{2n}, \omega_1].$$

For each unrooted loop ω, let R_ω denote the number of distinct rooted loops that produce this unrooted loop. Note that R_ω is always a divisor of $|\omega|$ and the proportion of loops of length $2n$ with $R_\omega < 2n$ decays exponentially in n. The random walk loop measure in A is the measure that assigns weight $[R_\omega/|\omega|]\,4^{-|\omega|}$ to each unrooted loop in A. If we scale the loops by N^{-1} and take the limit, we get the Brownian loop measure [13] as mentioned in Section 2.2.

The *random walk loop soup* in $\mathcal{H}$ is a Poisson point process $\mathcal{C}_t$ from the loop measure. At any fixed time t, the set $\mathcal{C}_t$ is a multi-set of loops, where the number of times the unrooted loop ω appears has a Poisson distribution with parameter

$(R_\omega/|\omega|)4^{-|\omega|}t$. If $A \subset \mathbb{Z}^2$, then $\mathcal{C}_t(A)$ denotes the loop soup in A, i.e., $\mathcal{C}_t$ restricted to loops in A. Given a self-avoiding path $\eta = [\eta_0, \eta_1, \dots, \eta_k]$ and a realization of the loop soup $\mathcal{C}_t$, there is a systematic way of taking the loops in $\mathcal{C}_1(A)$ and adding them to η to produce a random walk path ω from η_0 to η_k whose chronological loop-erasure is η. It can be described as follows.

- Let $\omega^1, \dots, \omega^r$ denote the loops in $\mathcal{C}_1(A)$ that intersect η. It is not difficult to check that (with probability one) there are only a finite number of such loops (assuming A is a proper subset). Order the loops in the order that they appeared in the Poisson process.
- For each ω^j choose a rooted loop that represents it by letting η_l be the point with smallest index l such that η_l appears in ω^j and choosing the root of ω so that the rooted path starts at η_l. If ω visits η_l more than once, choose the root uniformly over all the possibilities.
- Attach a loop at η_l by adding all the loops rooted at η_l in the order they appear in the Poisson process.

Then straightforward combinatorial arguments show that if η is chosen according to the probability measure $\nu^{\#}_{LE}(A; z, w)$ and $\mathcal{C}_1$ is an independent realization of the loop soup in A and the loops are added as above, then the distribution of the resulting random walk is $\nu^{\#}(A; z, w)$.

Similarly, we can start with $\nu^{\#}(A; z, w)$ and get independent copies of the loop-erased measure $\nu^{\#}_{LE}(A; z, w)$ and the loop soup in A restricted to curves that hit the loop-erased path. We will explain this more now. Suppose $\eta = [\eta_0, \eta_1, \dots, \eta_k]$ is a self-avoiding excursion in A from z to w. The $\nu^{\#}_{LE}(A; z, w)$ measure of η is exactly the $\nu^{\#}(A; z, w)$ measure of all loops ω whose loop-erasure is η. It is straightforward to show (see, for example, Equation (12.2.2) of [7]) that

$$\nu^{\#}_{LE}(A; z, w)(\eta) = 4^{-|\eta|}\,\Theta_A(\eta) \quad \text{where} \quad \Theta_A(\eta)^{-1} = q_{A_1}(\eta_1) \cdots q_{A_{k-1}}(\eta_{k-1}).$$

Here $A_j = A \backslash \{\eta_1, \dots, \eta_{j-1}\}$ and $q_A(x)$ denotes the probability that a random walk starting at x leaves A before returning to x. We will show that

$$\Theta_A(\eta) = \mathbf{P}\{\mathcal{C}_1(A) \text{ contains no loop that intersects } \eta\}. \tag{5.2}$$

To verify (5.2), we first note that any path ω with loop-erasure η can be split into a unique fashion as

$$\omega = [\eta_0, L_1, L_2, \dots, L_{k-1}, \eta_k]$$

where L_j is a loop rooted at η_j in $A \setminus \{\eta_1, \dots, \eta_{j-1}\}$. Each of these loops can be split into a number of ways as

$$L_j = \omega^{j,1} \oplus \cdots \oplus \omega^{j,r_j}$$

where $\omega^{j,l}$ are loops at L_j. We give L_j weight $4^{-|L_j|}$. We let

$$\alpha(\omega^{j,l}) = 4^{-|\omega^{j,l}|}\theta(\omega^{j,l})^{-1}$$

where $\theta([\omega_0, \dots, \omega_l])$ is the number of $j \geq 1$ such that $\omega_j = \omega_0$. We give the l-tuple $(\omega^{j,1}, \dots, \omega^{j,l})$ weight

$$\alpha(\omega^{j,1}) \cdots \alpha(\omega^{j,l}) = \frac{1}{4^{|\omega^{j,1}| + \cdots + |\omega^{j,r}|}\,[\theta(\omega^{j,1}) \cdots \theta(\omega^{j,l})]}.$$

The following lemma yields (5.2).

Lemma 5.1 *Suppose A is a proper subset of $\mathbb{Z}^2$ and $z \in A$. Let $m = m_{z,A}$ denote the measure that assigns weight $4^{-|\omega|}\,\theta(\omega)^{-1}$ to each loop in A rooted at z. Let $\mathcal{A}_t$ denote a Poisson point process from this measure and let l_t denote the loop obtained by concatenating all the loops in $\mathcal{A}_t$ in chronological order. (If there are no loops, then l_t is the trivial loop of length 0.) Then*

$$\mathbf{P}\{l_1 = \omega\} = 4^{-|\omega|} q_A,$$

where q_A is the probability that simple random walk starting at z leaves A before returning to z.

Proof Let $\mathcal{L} = \mathcal{L}(A, z)$ denote the set of all loops in A rooted at z. Note that $1 - q_A = \sum_{\omega \in \mathcal{L}, \theta(\omega)=1} 4^{-|\omega|}$ so that

$$\sum_{k=0}^{\infty} \sum_{\omega \in \mathcal{L}, \theta(\omega)=k} 4^{-|\omega|} = \sum_{k=0}^{\infty} \left[\sum_{\omega \in \mathcal{L}, \theta(\omega)=1} 4^{-|\omega|} \right]^k = \sum_{k=0}^{\infty} (1 - q_A)^k = 1/q_A. \tag{5.3}$$

For the trivial loop $[z]$, $\mathbf{P}\{l_1 = [z]\}$ is the probability that there are no loops in the Poissonian realization, i.e.,

$$\mathbf{P}\{l_1 = [z]\} = \exp\left\{ -\sum_{\omega \in \mathcal{L}} 4^{-|\omega|} \theta(\omega)^{-1} \right\}.$$

It now follows from (5.3) that

$$\sum_{\omega \in \mathcal{L}} 4^{-|\omega|} \theta(\omega)^{-1} = \sum_{k=1}^{\infty} \frac{1}{k} \sum_{\omega \in \mathcal{L}, \theta(\omega)=k} 4^{-|\omega|} = \sum_{k=1}^{\infty} \frac{1}{k} (1 - q_A)^k = -\log q_A$$

so therefore

$$\mathbf{P}\{l_1 = [z]\} = q_A.$$

More generally, if ω is a loop that returns to the origin k times, and $j_1, j_2, \ldots, j_r$ are positive integers summing to k, we could get the loop ω by splitting the loop into r loops where the mth loop is from the $(j_1 + \cdots + j_{m-1})$th to the $(j_1 + \cdots + j_m)$th return to z. By considering all these possibilities, one can deduce that

$$\mathbf{P}\{l_1 = \omega\} = q_A\, 4^{-|\omega|} \sum_{j_1 + \cdots + j_r = k} \frac{1}{r!} \frac{1}{j_1 \cdots j_r}.$$

We leave it to the reader to show that the summation equals one and so $\mathbf{P}\{l_1 = \omega\} = q_A\, 4^{-|\omega|}$. (As a hint, write

$$\frac{1}{1-t} = \exp\{-\log(1-t)\}$$

and expand both sides in power series in t.) □

Computing the right side of (5.2) is very difficult; the local terms corresponding to small loops dominate the probability. However, we can immediately conclude a boundary perturbation rule. If $A \subset A'$ and $z, w \in \partial A \cap \partial A'$, then

$$\frac{\Theta_{A'}(\eta)}{\Theta_A(\eta)} = \mathbf{P}\{\mathcal{C}_1(A') \text{ contains no loop intersecting both } A' \setminus A \text{ and } \eta\}.$$

This should be compared to Proposition 3.1.

5.2 Fomin's identity. Suppose $A \subset \mathbb{Z}^2$ and $x_1, \ldots, x_n, y_1, \ldots, y_n \in \partial A$. Fomin [4] showed that the

$$\nu(A; x_1, y_1) \times \cdots \times \nu(A; x_n, y_n)$$

measure of the set of n-tuples of excursions $(\omega^1, \ldots, \omega^n)$ satisfying the condition

$$LE(\omega^{m-1}) \cap \omega^m = \emptyset, \qquad m = 2, \ldots, n,$$

where $LE(\omega^{m-1})$ denotes the loop erasure of ω^{m-1}, is given by

$$\det[h_A(x_l, y_m)]_{1 \le l,m \le n}.$$

In the case $A = \mathcal{H}$, topological considerations show that this can be non-zero only if the points appear in order, i.e., after relabeling, they satisfy

$$x_1 < \cdots < x_n < y_n < \cdots < y_1.$$

As Fomin points out, since the scaling limit of simple random walk is Brownian motion, which is conformally invariant, this formula in itself suggests the conformal invariance of loop-erased random walk. There are technical difficulties in taking the limit in domains with rough boundaries but these can be handled; see [6]. In the case $n = 2$, the determinant equals

$$h_{\mathcal{H}}(x_1, y_1)\, h_{\mathcal{H}}(x_2, y_2) \left[1 - \frac{h_{\mathcal{H}}(x_1, y_2)\, h_{\mathcal{H}}(x_2, y_1)}{h_{\mathcal{H}}(x_1, y_1)\, h_{\mathcal{H}}(x_2, y_2)}\right].$$

As the differences go to infinity, this is asymptotic to

$$h_{\mathcal{H}}(x_1, y_1)\, h_{\mathcal{H}}(x_2, y_2) \left[1 - \frac{(y_1 - x_1)^2\, (y_2 - x_2)^2}{(y_1 - x_2)^2\, (y_2 - x_1)^2}\right].$$

If $x_1 = 0$, $x_2 = x$, $y_2 = y$, then as $y_1 \to \infty$ the last expression is asymptotic to $h_{\mathcal{H}}(0, y_1)$ times

$$h_{\mathcal{H}}(x, y)\, \frac{x}{y} \left[2 - \frac{x}{y}\right],$$

and we have recovered the expression for $H^*_{1,1}(x, y)$ that was derived in Section 3.3.

5.3 Review of excursion Poisson kernel determinant. We now briefly review the necessary notation from [6] about the excursion Poisson kernel. Suppose that $D \subset \mathbb{C}$ is a simply connected Jordan domain and that ∂D is locally analytic at x and y. The *excursion Poisson kernel* is defined as

$$H_{\partial D}(x, y) = \lim_{\epsilon \to 0} \frac{1}{\epsilon}\, H_D(x + \epsilon \mathbf{n}_x, y)$$

where $H_D(z, y)$ for $z \in D$ is the usual Poisson kernel, and $\mathbf{n}_x$ is the unit normal at x pointing into D. Explicit formulæ are known when $D = \mathbb{D}$, the unit disk, or $D = \mathbb{H}$, namely

$$H_{\partial \mathbb{D}}(x, y) = \frac{1}{\pi\, |y - x|^2} = \frac{1}{2\pi(1 - \cos(\arg y - \arg x))} \quad \text{and} \quad H_{\partial \mathbb{H}}(x, y) = \frac{1}{\pi (y - x)^2}.$$

Notice that $\pi H_{\partial \mathbb{H}}(x, y) = H_{\mathbb{H},1,1}(x, y)$. Since the excursion Poisson kernel is known [6, Proposition 2.11] to be conformally covariant, we obtain (1.2) with $b = 1$, $n = 1$. Suppose now that $D' \subset \mathbb{C}$ is also a Jordan domain and that $f : D \to D'$ is a conformal transformation. Let $x_1, \ldots, x_n, y_1, \ldots, y_n$ be distinct boundary points at which ∂D is locally analytic, and assume that $\partial D'$ is locally analytic at $f(x_1), \dots, f(x_n), f(y_1), \dots, f(y_n)$. It follows [6, Proposition 2.16] that if $\mathbf{H}_{\partial D}(\mathbf{x},\mathbf{y}) = [H_{\partial D}(x_j, y_k)]_{1\le j,k\le n}$ denotes the $n \times n$ *hitting matrix*

$$\mathbf{H}_{\partial D}(\mathbf{x},\mathbf{y}) = \begin{bmatrix} H_{\partial D}(x_1,y_1) & \cdots & H_{\partial D}(x_1,y_n) \\ \vdots & \ddots & \vdots \\ H_{\partial D}(x_n,y_1) & \cdots & H_{\partial D}(x_n,y_n) \end{bmatrix}$$

then

$$\det \mathbf{H}_{\partial D}(\mathbf{x},\mathbf{y}) = \left(\prod_{\ell=1}^{n} |f'(x_\ell)|\,|f'(y_\ell)|\right) \det[H_{\partial D'}(f(x_j), f(y_k))]_{1\le j,k\le n}.$$

5.4 Direct derivation of the scaling limit. Suppose that γ is an SLE_2 in $\mathbb{H}$ from 0 to ∞ as in Section 2.1, and for every $0 < t < \infty$ let $\mathbb{H}_t$ denote the slit-plane $\mathbb{H}_t = \mathbb{H} \setminus \gamma[0,t]$. Fix two real numbers $0 < x < y < \infty$, and let $\beta : [0,1] \to \overline{\mathbb{H}}$ be a Brownian excursion from x to y in $\mathbb{H}$. It then follows that

$$\mathbf{P}\{\,\gamma[0,t] \cap \beta[0,1] = \emptyset \mid \gamma[0,t]\,\} = \frac{H_{\partial\mathbb{H}_t}(x,y)}{H_{\partial\mathbb{H}}(x,y)} = \frac{H_{\mathbb{H}_t,1,1}(x,y)}{H_{\mathbb{H},1,1}(x,y)} \tag{5.4}$$

so that taking expectations and limits of (5.4) gives

$$\mathbf{P}\{\,\gamma[0,\infty) \cap \beta[0,1] = \emptyset\,\} = \tilde{H}^*_{\mathbb{H},1,1}(x,y) = \frac{x}{y}\left(2 - \frac{x}{y}\right). \tag{5.5}$$

The first equality is similar to (3.4), and the second equality follows from Proposition 3.4.

Theorem 5.2 *If x, $y \in \mathbb{R}$ with $0 < x < y < \infty$, and that γ, β are as above, then*

$$\mathbf{P}\{\,\gamma[0,\infty) \cap \beta[0,1] = \emptyset\,\} = \frac{\det \mathbf{H}_{\partial\mathbb{D}}(f(\mathbf{x}), f(\mathbf{y}))}{H_{\partial\mathbb{D}}(f(0), f(\infty)) H_{\partial\mathbb{D}}(f(x), f(y))} \tag{5.6}$$

where $f : \mathbb{H} \to \mathbb{D}$ is a conformal transformation.

Remark By working in $\mathbb{H}$ and $\mathbb{D}$, it is possible to perform explicit calculations. Since the quantity on the right side of (5.6) is known to be a conformal invariant, we can show, with a combination of conformal transformations, that the probability that an SLE_2 avoids a Brownian excursion in D is given by the appropriate determinant of the matrix of excursion Poisson kernels.

Proof By the scaling rule (3.6), it suffices without loss of generality to assume that $0 < x < 1$ and $y = 1$. Furthermore, we may assume that the conformal transformation $f : \mathbb{H} \to \mathbb{D}$ is given by

$$f(z) = \frac{iz+1}{z+i},$$

so that $f(0) = -i$, $f(1) = 1$, $f(\infty) = i$, and

$$f(x) = \left(\frac{2x}{x^2+1}\right) + i\left(\frac{x^2-1}{x^2+1}\right) = \exp\left\{-i \arctan\left(\frac{1-x^2}{2x}\right)\right\}.$$

Notice that $(2x)^2 + (x^2-1)^2 = (x^2+1)^2$ so that $|f(x)| = 1$ as expected. Writing $f(x) = e^{i\theta}$, we find that

$$\begin{aligned}
&\frac{\det \mathbf{H}_{\partial\mathbb{D}}(f(\mathbf{x}), f(\mathbf{y}))}{H_{\partial\mathbb{D}}(f(0), f(\infty))H_{\partial\mathbb{D}}(f(x), f(y))} \\
&= \frac{\det \mathbf{H}_{\partial\mathbb{D}}(-i, i, f(x), 1)}{H_{\partial\mathbb{D}}(-i,i)H_{\partial\mathbb{D}}(f(x),1)} \\
&= \frac{H_{\partial\mathbb{D}}(-i,i)H_{\partial\mathbb{D}}(e^{i\theta},1) - H_{\partial\mathbb{D}}(-i,1)H_{\partial\mathbb{D}}(e^{i\theta},i)}{H_{\partial\mathbb{D}}(-i,i)H_{\partial\mathbb{D}}(e^{i\theta},1)} \\
&= \frac{\frac{1}{2\pi(1-\cos\pi)}\frac{1}{2\pi(1-\cos\theta)} - \frac{1}{2\pi(1-\cos(\frac{\pi}{2}))}\frac{1}{2\pi(1-\cos(\frac{\pi}{2}+\theta))}}{\frac{1}{2\pi(1-\cos\pi)}\frac{1}{2\pi(1-\cos\theta)}} \\
&= \frac{2\cos\theta + \sin\theta - 1}{1+\sin\theta}.
\end{aligned}$$

Since $\theta = -\arctan\left(\frac{1-x^2}{2x}\right)$ we see that $\cos\theta = \frac{2x}{x^2+1}$ and $\sin\theta = \frac{1-x^2}{x^2+1}$ which upon substitution gives

$$\frac{2\cos\theta + \sin\theta - 1}{1+\sin\theta} = \frac{\frac{4x}{x^2+1} + \frac{1-x^2}{x^2+1} - 1}{1 + \frac{1-x^2}{x^2+1}} = \frac{4x - 2x^2}{2} = x(2-x).$$

Comparison with (5.5) now yields the result. □

6 The λ-SAW

Here we will define a one-parameter family of measures on self-avoiding walks on the integer lattice $\mathbb{Z}^2$. We conjecture that the scaling limit of this measure will give the measure on SLE as described before.

A simple random walk path ω that does not visit any point more than once is called a self-avoiding walk. If ω is an excursion in A connecting boundary points z, w, we will call ω a self-avoiding excursion (SAE). Let r, $\lambda \in \mathbb{R}$ be parameters. If $A \subsetneq \mathbb{Z}^2$ and z, $w \in \partial A$, then the measure $W_{r,\lambda,z,w,A}$ is defined to be the measure that assigns to each SAE ω in A from z to w measure

$$\exp\{-r|\omega| + \lambda m^*(A;\omega)\},$$

where $m^*(A;\omega)$ denotes the random walk loop measure of the set of loops in A that intersect ω. In analogy with the discussion on SLE, we conjecture that for each $\lambda \geq -1/2$, there is a critical r_λ such that the total mass $|W_{r_\lambda,\lambda,0,N,\mathcal{H}}|$ neither decays nor grows exponentially as $N \to \infty$.

- If $\lambda = 1$ and $e^{-r} = 1/4$, then $W_{r,1,0,N,\mathcal{H}}$ is exactly the same as the loop-erased excursion measure $\nu_{LE}(\mathcal{H}; 0, N)$ which from (5.1) has total mass asymptotic to $(4\pi)^{-1}N^{-2}$. Therefore, $e^{-r_1} = 1/4$.
- If $\lambda = 0$, then this measure is the same as the usual self-avoiding walk measure that gives each walk of n steps the same measure. In this case e^{r_0} is the *connective constant* for self-avoiding walk which can be defined by saying that the number of SAWs of length n starting at the origin in $\mathbb{Z}^2$ grows like $e^{r_0 n}$.

We now make this conjecture stronger and say that there exist $C > 0$, $b \in \mathbb{R}$ (depending on λ) such that $|W_{r_\lambda,\lambda,0,N,\mathcal{H}}| \sim C N^{-2b}$. Consider the measure $N^{2b-2} C^{-1} W_{r_\lambda,\lambda,\cdot,\cdot,\mathcal{H}}$ where we use $\cdot,\cdot$ to mean that we take the union over all

x, $y \in \mathbb{R}$. We also scale the paths by N^{-1}. Then we conjecture that we have a measure on excursions in $\mathbb{H}$. If $x_1 < x_2 < x_3 < x_4$, then the total mass of the set of excursions connecting $[x_1, x_2]$ to $[x_3, x_4]$ should be

$$\int_{x_1}^{x_2} \int_{x_3}^{x_4} \frac{1}{(t-s)^{2b}} dt\, ds.$$

We conjecture that we can write this measure as

$$\int_{x_1}^{x_2} \int_{x_3}^{x_4} \frac{1}{(t-s)^{2b}}\, \mu_{\mathbb{H}}^{\#}(s,t)\, dt\, ds,$$

where $\mu_{\mathbb{H}}^{\#}(s,t)$ is a conformally invariant probability measure on paths.

We have not discussed the parametrization of the limiting measure. An even stronger conjecture would say that there is an exponent[1] ν defined roughly by saying that the typical number of steps in a walk from 0 to N in the measure above contains $N^{1/\nu}$ steps. Then we could let

$$\omega^{(N)}(t) = N^{-t}\omega(t\, N^{1/\nu}).$$

However, it is not necessary to include the parametrization to have a nontrivial result.

The measure on k-tuples of self-avoiding walks is defined similarly, by giving measure

$$\exp\{-r(|\omega_1| + \cdots + |\omega_k|) + \lambda m^*(A; \omega_1 \cup \cdots \cup \omega_k)\}.$$

We conjecture the following.

- If $\lambda \geq -1/2$, then the λ-SAW gives a limiting measure on paths that is the SLE_κ measure. In this case κ, λ, b are related as before.
- For $\lambda < -1/2$, this does not give a limiting measure on simple paths.

For $\lambda = 1$, this has been proved using the loop-erased random walk. For $\lambda = 0$, this is equivalent to very difficult conjectures about self-avoiding walks; see [12].

Acknowledgements

The authors would like to express their gratitude to the Fields Institute for their gracious hospitality during the *Percolation, SLE, and Related Topics Workshop* of September 2005.

References

[1] M. Bauer, D. Bernard, and K. Kytölä. Multiple Schramm-Loewner Evolutions and Statistical Mechanics Martingales. *J. Stat. Phys.*, 120:1125–1163, 2005.

[2] J. Cardy. Lectures on Conformal Invariance and Percolation. Preprint, 2001. Available online at `arXiv:math-ph/0103018`.

[3] J. Dubédat. Euler integrals for commuting SLEs. Preprint, 2005. Available online at `arXiv:math.PR/0507276`. To appear, *J. Stat. Phys.*

[4] S. Fomin. Loop-erased walks and total positivity. *Trans. Amer. Math. Soc.*, 353:3563–3583, 2001.

[5] M. J. Kozdron. On the scaling limit of simple random walk excursion measure in the plane. *Alea*, 2:125–155, 2006.

[6] M. J. Kozdron and G. F. Lawler. Estimates of random walk exit probabilities and application to loop-erased random walk. *Electron. J. Probab.*, 10:1442–1467, 2005.

[1] It is traditional to use ν for this critical exponent. Although we have also used ν with parameters to denote certain measures on paths, this should not cause confusion.

[7] G. F. Lawler. Loop-erased random walk, in *Perplexing Problems in Probability: Festschrift in Honor of Harry Kesten*, M. Bramson and R. Durrett, eds., pages 197–247. Birkhäuser, Boston, MA, 1999.

[8] G. F. Lawler. *Conformally Invariant Processes in the Plane*, American Mathematical Society, Providence, RI, 2005.

[9] G. F. Lawler. The Laplacian-b random walk and the Schramm-Loewner Evolution. Preprint, 2005. To appear, *Illinois J. Math.*

[10] G. F. Lawler, O. Schramm, and W. Werner. Conformal restriction: the chordal case. *J. Amer. Math. Soc.*, 16:917–955, 2003.

[11] G. F. Lawler, O. Schramm, and W. Werner. Conformal invariance of planar loop-erased random walks and uniform spanning trees. *Ann. Probab.*, 32:939–995, 2004.

[12] G. F. Lawler, O. Schramm, and W. Werner. On the scaling limit of planar self-avoiding walk. In *Fractal Geometry and Applications: A Jubilee of Benoît Mandelbrot: Part 2*, M. L. Lapidus and M. van Frankenhuijsen, eds., pages 339–364. American Mathematical Society, Providence, RI, 2004.

[13] G. F. Lawler and J. A. Trujillo Ferreras. Random walk loop soup. Preprint, 2004. Available online at `arXiv:math.PR/0409291`. To appear, *Trans. Amer. Math. Soc.*

[14] G. F. Lawler and W. Werner. The Brownian loop soup. *Probab. Theory Related Fields*, 128:565–588, 2004.

[15] N. N. Lebedev. *Special Functions & their Applications.* Dover, New York, NY, 1972.

[16] O. Schramm. Scaling limits of loop-erased random walks and uniform spanning trees. *Israel J. Math.*, 118:221–288, 2000.

[17] F. Spitzer. *Principles of Random Walk.* Springer-Verlag, New York, NY, second edition, 2001.

Fields Institute Communications
Volume **50**, 2007

Dyson–Schwinger Equations: From Hopf Algebras to Number Theory

Dirk Kreimer
Institut des Hautes Études Scientifiques
Le Bois Marie, 35, route de Chartres
F-91440 Bures-sur-Yvette, France
kreimer@ihes.fr

Abstract. We consider the structure of renormalizable quantum field theories from the viewpoint of their underlying Hopf algebra structure. We review how to use this Hopf algebra and the ensuing Hochschild cohomology to derive non-perturbative results for the short-distance singular sector of a renormalizable quantum field theory. We focus on the short-distance behaviour and thus discuss renormalized Green functions $G_R(\alpha, L)$ which depend on a single scale $L = \ln q^2/\mu^2$.

1 Introduction

The crucial notion of locality, the structure of Dyson–Schwinger equations and the appearance of mixed motives in the evaluation of Feynman graphs are intimately related. We want to exhibit how these notions come together in quantum field theory. We emphasize the role of Dyson–Schwinger equations in this interplay.

Renormalization theory is a time-tested subject put to daily use in many branches of physics. We have seen many of its facets illuminated here at the Fields Institute. In this paper, we focus on its applications in quantum field theory, where a standard perturbative approach is provided through an expansion in Feynman diagrams. In perturbation theory it is mainly a combinatorial problem: determine the needed correction to parameters in the Lagrangian such that the computation allows for finite results in the desired order of perturbation. Whilst the resulting combinatorics of the Bogoliubov recursion, solved by suitable forest formulas, has been known for a long time, the subject regained interest on the conceptual side with the discovery of an underlying Hopf algebra structure behind these recursions.

2000 *Mathematics Subject Classification.* Primary 81T16; Secondary 81T17.
Supported by CNRS and by grant NSF-DMS0603781 at Center for Mathematical Physics, Boston University.

Non-perturbatively, one faces the equations of motion which the full Green functions have to fulfill. These Dyson–Schwinger equations (DSEs) reflect the self-similarity of amplitudes in quantum field theory, upon studying their skeleton expansion: the computation of propagation or interaction of amplitudes proceeds by taking into account that the same propagation or interaction can happen in internal processes. Hence the notion of internal process demands objects which possess internal structure. These turn out to be the celebrated skeleton graphs of a theory. They are crucial in understanding at the same time the algebraic as well as the number-theoretic properties of a field theory.

The main goal of this review is to emphasize one crucial point: How the algebraic structure of the perturbative expansion can lead to non-perturbative solutions, by emphasizing the unifying role of algebraic structures in understanding local renormalization as well as the above self-similar structure of Green functions.

We will also point out along the way how internal symmetries are reflected in this set-up. This is a very recent insight [1], which we can only mention in passing. It goes a long way in establishing a quantum field theory which stands its own ground: instead of deriving quantum gauge invariance from the differential geometry of a classical gauge field theory, we obtain it as a consequence of the algebraic structures of a local quantum theory, and the classical Lagrangian is a derived quantity. It is indeed obtained from the residues of Feynman graphs which are primitive elements in the Hopf algebra of renormalization, an old result in field theory [2].

Finally, in the course of our discussion we will give an idea where one can see signs of universality from this set-up. It is at the time of writing an open research problem to connect the rather systematic account to DSEs exhibited in this talk to other areas where renormalization theory and universality are likewise fundamental notions, as these proceedings hopefully demonstrate.

We first have to review the algebraic structure of quantum field theory, and with it the Hopf algebra structure of a perturbative expansion. From a strictly perturbative viewpoint, this is summarized in this volume in the paper by Ebrahimi-Fard and Guo [3], with emphasis given to Rota–Baxter algebras. Our own summary below follows [4]. We hence will be short in our discussion of the perturbative viewpoint, and aim at a qualitative discussion of the above points, exemplified in the much stressed example of the propagator in massless Yukawa theory [5, 6]. Hochschild cohomology then leads the way from perturbative physics to non-perturbative results. We summarize here results detailed in previous works on the structure of DSEs [1, 5, 6, 7, 8, 9] and will end in discussing shortly the number-theory of skeleton graphs, which found a mathematical interpretation as periods of mixed motives recently [10]. We will be content in exhibiting how those primitive graphs mentioned above provide the connection to algebraic geometry and motives.

1.1 A splitting of amplitudes. We assume we work in a renormalizable quantum field theory which provides a finite set $\mathcal{R} \subset \mathcal{A}$ of amplitudes which need renormalization. Here, $\mathcal{A}$ is the set of all amplitudes in a given theory. We hence work with a theory which, from a Lagrangian perspective, provides a finite number of parameters which need renormalization. Note that our notion of an amplitude is such that each monomial in said Lagrangian corresponds to an amplitude. In comparison with the standard terminology, we distinguish the various form-factors provided by a given Green function and regard each corresponding structure function as a Green function. For example the vertex function for the photon decay into an electron-positron pair in quantum electrodynamics (QED) has a form-factor decomposition which provides twelve independent structure functions [11]. Only one of them needs renormalization and corresponds to the term $\bar{\psi}A\!\!\!/\psi$ in the QED Lagrangian. We denote the projection of the QED vertex function onto this form-factor as the amplitude $\in \mathcal{R}$ corresponding to this term in the Lagrangian.

A general amplitude $a \in \mathcal{A}$ is thus specified by a particular form-factor and the quantum numbers of particles participating in a scattering type experiment contributing to that form-factor. It suffices to consider one-particle irreducible (1PI) Green functions G^a for each amplitude a. The knowledge of these Green function suffices to determine all relevant physics by standard techniques.

We assume that the amplitudes a allow to specify an integer $n = n(a)$ which gives the number of external legs. We let $\mathcal{M}_a$ be the set of all 1PI graphs contributing to the amplitude a. We remind the reader that a 1PI graph is a graph which remains connected after removal of any one of its internal edges. By $|\Gamma|$ we denote the number of independent loops in Γ. By $\mathrm{sym}(\Gamma)$ we denote the rank of the automorphism group of the graph.

We then have in general

$$G^a = 1 \pm \sum_{\Gamma\in\mathcal{M}_a} \alpha^{|\Gamma|}\frac{\phi(\Gamma)}{\mathrm{sym}(\Gamma)} = 1 \pm \sum_{k\geq 1}\alpha^k\phi(c_k^a), \tag{1.1}$$

so that

$$c_k^a = \sum_{\Gamma\in\mathcal{M}_a,\ |\Gamma|=k}\frac{\Gamma}{\mathrm{sym}(\Gamma)}, \tag{1.2}$$

is the sum over all 1PI graphs of order k contributing to the amplitude a. In the above, we take the plus sign if $n(a) \geq 3$ and the minus sign for $n(a) = 2$. We discard tadpole amplitudes, $n(a) = 1$, and vacuum amplitudes, $n(a) = 0$. A main point of the subsequent discussion will concern the structure of the sum

$$\Gamma^a = \mathbb{I} \pm \sum_{\Gamma\in\mathcal{M}_a}\alpha^{|\Gamma|}\frac{\Gamma}{\mathrm{sym}(\Gamma)}, \tag{1.3}$$

in the above Green function G^a.

The set of amplitudes $\mathcal{A}$ decomposes for a renormalizable theory into two disjoint subsets

$$\mathcal{A} = \mathcal{A}_+ \cup \mathcal{R}. \tag{1.4}$$

Here, to stress it once more, $\mathcal{R}$ is the set of amplitudes for form-factors of one-particle irreducible graphs which need renormalization. On the other hand, $\mathcal{A}_+$ is the set of amplitudes which are overall finite.

Those amplitudes behave very differently. For an element $r \in \mathcal{R}$ we can write

$$\Gamma^r = \mathbb{I} \pm \sum_{\Gamma\in\mathcal{M}^r}\alpha^{|\Gamma|}\frac{\Gamma}{\mathrm{sym}(\Gamma)} = \mathbb{I} \pm \sum_{k\geq 1}\alpha^k B_+^{k;r}(\Gamma^r Q^{n_r k}), \tag{1.5}$$

where $B_+^{k;r}$ are Hochschild one-cocycles as explained below. The quantity Q is intimately related to the notion of an invariant charge [12]. For momentum space Feynman rules ϕ we have

$$\frac{\partial_L \ln\phi(Q^{n_r})}{\partial L}|_{L=0} = \beta(\alpha), \tag{1.6}$$

where β is the β-function of the theory which describes the running of the charge Q. Hence Q^{n_r} is a monomial in the Γ^r, $r \in \mathcal{R}$, or their inverses. n_r is an integer such that $Q^{n_r} - \mathbb{I}$ is of order one in α. Finally, $L = \ln q^2/\mu^2$ sets the scale for that running charge.

From (1.5), we have thus a self-similar recursive system determining the formal sums Γ^r, $r \in \mathcal{R}$ in terms of themselves and the action of suitable maps $B_+^{k;r}$. The study of the maps $B_+^{k;r}$ is crucial for a QFT. The Hopf algebra elements $B_+^{k;r}(\mathbb{I})$ provide the skeleton graphs underlying the DSEs and are the terms which drive the recursion which leads to the full theory, and which connects QFT to motives [10].

In contrast, for $a \in \mathcal{A}_+$, we have

$$\Gamma^a = \mathbb{I} \pm \sum_{\Gamma \in \mathcal{M}^a} \alpha^{|\Gamma|} \frac{\Gamma}{\mathrm{sym}(\Gamma)} = \mathbb{I} \pm \sum_k B_+^{k;a}(M_a Q^{n_r k}), \tag{1.7}$$

where M_a is another monomial in Γ^s, $s \in \mathcal{R}$, or their inverses. Hence, amplitudes from the set $\mathcal{A}_+$ are determined from the knowledge of the ones in $\mathcal{R}$.

Hence, we focus on amplitudes $r \in \mathcal{R}$ where it turns out that the self-similarity and the properties of the Hochschild one-cocycles $B_+^{k;r}$ are paramount to their understanding, as well with respect to perturbative renormalization as with respect to nonperturbative physics.

1.2 The example of massless QED. Massless quantum electrodynamics starts from the Lagrangian

$$\mathcal{L}_{\mathrm{QED}} = \bar{\psi}\,[\not\partial + \not A]\,\psi + \frac{1}{4} F \cdot F. \tag{1.8}$$

It has three monomials and as a renormalizable theory we hence can work with a set

$\mathcal{R}_{\mathrm{QED}} = \{$  , , $\}$, (1.9)

in standard notation.

The Green function for the vertex has an elaborate form-factor expansion [11], (with $n($  $) = 3$),

G_μ  $= \phi(\Gamma$  $) = \gamma_\mu F_1(p_1^2, p_2^2, p_3^3, \alpha, \mu) + \cdots ,$ (1.10)

where it is only F_1 which needs renormalization. We let P_1 be the projector onto this form factor.

The unrenormalized Green functions for these monomials are then given as

$G_{\mathrm{u}}^{\bar\psi \not A \psi} = 1 + \sum_{\Gamma \in \mathcal{M}} \alpha^{|\Gamma|} \mathrm{P}_1 \phi(\Gamma),$ (1.11) — subscript of $\mathcal{M}$: 

$G_{\mathrm{u}}^{\bar\psi \not\partial \psi} = 1 - \sum_{\Gamma \in \mathcal{M}} \alpha^{|\Gamma|} \mathrm{P}_2 \phi(\Gamma),$ (1.12) — subscript of $\mathcal{M}$: 

$G_{\mathrm{u}}^{F \cdot F} = 1 - \sum_{\Gamma \in \mathcal{M}} \alpha^{|\Gamma|} \mathrm{P}_3 \phi(\Gamma).$ (1.13) — subscript of $\mathcal{M}$:

The projectors P_2 onto the kinetic term of the inverse fermion propagator is redundant in a massless theory and only needed in a massive theory, and the projector P_3 onto the transversal part of the inverse photon propagator is likewise redundant in the Landau gauge where the tree level term remains transversal. We often write

$G^r = \phi(\Gamma^r)$ as a shorthand for such equations, with suitable projectors understood in the application of the Feynman rules ϕ.

Note that the Green function for the full vertex, containing eleven further amplitudes taken from the set $\mathcal{A}_+$, is obtained by omitting the projector P_1 in the above. We stick to the notion that each monomial in the Lagrangian has its own Green function, while all other amplitudes belong to the set $\mathcal{A}_+$.

The goal is to calculate the corresponding 1PI Green functions order by order in the fine-structure α of the theory, by applying Feynman rules to these 1PI graphs of a renormalizable theory under consideration. There are two problems here: each single graph is mapped by the Feynman rules to an ill-defined quantity, and furthermore, after labourosly eliminating these divergences, the resulting series is not of the convergent type.

Progress with both problems is possible thanks to the algebraic structures underlying Feynman graphs using the Lie and Hopf algebras discussed below.

As we said before, 1PI Green functions are parameterized by the quantum numbers, -masses, momenta, spin and such-, of the particles participating in the scattering process under consideration. Physicists denote propagating particles by lines, and the perturbative expansion in terms of graphs is organized such that external half-lines denote the particles parameterizing the amplitude a and the Green function G^a under consideration, while internal edges and vertices describe internal propagations and vertices.

Note that the Lagrangian L of massless quantum electrodynamics is obtained accordingly as

$$L = \hat{\phi}(\;\text{—}\;)^{-1} + \hat{\phi}(\;\text{⅄}\;) + \hat{\phi}(\;\text{∿∿}\;)^{-1} = \bar{\psi}\partial\!\!\!/\psi + \bar{\psi}A\!\!\!/\psi + \frac{1}{4}F \cdot F, \tag{1.14}$$

where $\hat{\phi}$ are coordinate space Feynman rules. Here, inversion like $\hat{\phi}(\;\text{—}\;)^{-1}$ takes account of the fact that monomials quadratic in the fields refer to inverse propagators, while — and ∿∿ refer to the free propagators of QED.

This is not to say that there are no other Green function in quantum electrodynamics. But we focus here on the Green functions which need renormalization, and this, for a renormalizable field theory, gives us a finite set of terms to be considered. It is indeed the recursive self-similar nature of amplitudes from the set $\mathcal{R}$ which drives the need for renormalization. That the unrenormalized amplitudes suffer from short-distance divergences is a mere accident of perturbation theory which disappears once the non-perturbative fixpoint equations of motion, the DSEs, have been taken into account.

The unrenormalized momentum space Feynman rules ϕ assign to a graph a function of the external momenta $\{p_f\}$, and other quantum numbers assiggned to external legs, ($\Gamma^{[0]}$ and $\Gamma^{[1]} = \Gamma^{[1]}_{\text{int}} \cup \Gamma^{[1]}_{\text{ext}}$ being the set of vertices v and internal and external edges e of Γ) of the form

$$\phi(\Gamma)(\{p_f\}) = \int \prod_{v \in \Gamma^{[0]}} \phi(v)\delta^{(4)}\left(\sum_{f \text{ incident } v} k_f\right) \prod_{e \in \Gamma^{[1]}_{\text{int}}} \text{Prop}(k_e)\frac{d^4 k_e}{4\pi^2}. \tag{1.15}$$

They are determined from the knowledge of the Feynman rules for interaction vertices v and the knowledge of free covariances $\text{Prop}(k_e)$ for each internal edge e.

In the above form (1.15) a Dirac mass

$$\delta^{(4)}\left(\sum_{f\in\Gamma^{[1]}_{\rm ext}} p_f\right), \tag{1.16}$$

will factor out of the expression for momentum conservation for $|\Gamma^{[1]}_{\rm ext}| = n(a)$ external momenta p_f, for $\Gamma \in \mathcal{M}_a$.

As a result, formally the unrenormalized Green function $G^a_{\rm u}$ is obtained as

$$G^r_{\rm u}(\alpha;\{p_f\};z) = \phi(\Gamma^r)\,(\alpha;\{p_f\};z)\,, \tag{1.17}$$

where we have introduced a suitably chosen regulator z, needed in perturbation theory but not non-perturbatively.

Note that in (1.15) the four-dimensional Dirac mass for each internal vertex guarantees momentum conservation at each such vertex and restricts the number of four-dimensional integrations to the number of independent cycles in the graph. These integrals suffer from UV singularities which render the integration over the momenta in internal cycles ill-defined. We remind the reader that the problem persists in coordinate space, where one confronts the continuation of products of distributions to regions of coinciding support. We restrict ourselves here to a discussion of the situation in momentum space and refer the reader to the literature for the situation in coordinate space [13].

The problem of perturbative renormalization is to make sense out of this situation term by term: We have to determine invertible series $Z^r(\alpha, z)$ for all $r \in \mathcal{R}$, hence in the parameters of the Lagrangian

$$L = \sum_{r\in\mathcal{R}} \hat{\phi}^{\pm1}(r), \tag{1.18}$$

such that the modified Lagrangian

$$\tilde{L} = \sum_{r\in\mathcal{R}} Z^r(\alpha,z)\,\hat{\phi}^{\pm1}(r), \tag{1.19}$$

produces a perturbation series in graphs which allows for the removal of the regulator z. Again, $\hat{\phi}$ are the Feynman rules in coordinate space, with the understanding that they evaluate an amplitude r to the corresponding tree level term.

Let us first describe how this transition is achieved using the Lie- and Hopf algebra structure of the perturbative expansion which we summarize below:

- Decide on the free fields and local interactions of the theory, appropriately specifying quantum numbers (spin, mass, flavor, color and such) of fields, restricting interactions so as to obtain a renormalizable theory.
- Determine the Feynman rules for free propagators from free field theory. The Feynman rules for interaction vertices then follow from locality: the ability to compensate by local counterterms actually fixes the structure of interaction vertices modulo the absolute values of masses and charges which parameterize the chosen theory.
- Consider the set of all 1PI graphs with edges corresponding to those free-field propagators. Together with the vertices this allows to construct a pre-Lie algebra of graph insertions. Anti-symmetrize this pre-Lie product to get a Lie algebra $\mathcal{L}$ of graph insertions and define the Hopf algebra $\mathcal{H}$ which is dual to the enveloping algebra $\mathcal{U}(\mathcal{L})$ of this Lie algebra.

- Realize that the coproduct and antipode of this Hopf algebra give rise to the forest formula which generates local counterterms upon introducing a suitable Rota–Baxter map, a renormalization scheme in physicists' parlance.
- Use the Hochschild cohomology of this Hopf algebra to show that you can absorb singularities in local counterterms, hence in the form described in (1.19) above.

This gives a rather satisfying account of renormalization theory. As an added bonus, if you work with an an intermediate complex regulator z these steps can be summarized as to construct the Birkhoff decomposition of the unrenormalized Feynman rules, regarded as an element in $\mathrm{Spec}(\mathcal{G})$, the character group of this Hopf algebra [15, 16, 17]. This settles perturbative renormalization.

The situation is even better when we look at the structure of the full non-perturbative Green functions and concentrate on the short-distance behaviour. Here, a regulator is not at all needed. Indeed, in recent years, we have learned how to make progress with non-perturbative physics [1, 5, 7, 8], and continue as follows:

- Show that the elements c_k^r form a sub Hopf algebra.
- Determine the Hochschild one-cocycles $B_+^{k;a}$ for this sub Hopf algebra from the primitive elements of the Hopf algebra.
- Choose a basis of amplitudes such that each Green function which needs renormalization only depends on a single scale L.
- Construct the DSEs as fixpoint equations with the elements $\phi(B_+^{k;a}(\mathbb{I}))$ as kernels corresponding to the Dyson skeleton expansion.
- For the set $\mathcal{R}$, determine the recursion relations which follow from the renormalization group upon rescaling L, applying the consequences of the representations of one-parameter groups of automorphisms of the above Hopf algebras [17] in this set-up.
- Solve the DSE for the remaining unknown anomalous dimensions.
- Aim to establish functional equations which connect an expansion in α to an expansion in $1/\alpha$.

Very recently we gained insight how to carry such a program through [6], and will exhibit below one example where it is carried to the end.

What is encouraging are the structural features one can establish for any renormalizable field theory which this program exhibits.

A first observation is that non-perturbatively, no regulator is needed. The anomalous dimensions of propagators and vertices at zero momentum transfer self-regulate the theory. Similar ideas, for the conformal case of a vanishing β-function only, have been employed in the past, and are reviewed in this volume [18].

A second feature is that upon organizing the DSEs in terms of Hochschild one-cocycles, the algebraic structures of the forest formulas remain invariant upon addition of more one-cocycles. This allows for a much more justified expansion, in terms of Hochschild cohomology, than the usual truncations done in DSEs. It goes hand in hand with a decomposition of field theory into periods of more and more complicated motives. In particular, the decomposition into the one-cocycles $B_+^{k;r}(\mathbb{I})$ and the restriction to a subset of these is a factorization of field theory reminiscent of a decomposition of a zeta function into its Euler factors. This rather far-fetched analogy [19] will be elaborated elsewhere.

From the viewpoint of DSEs, the Birkhoff decomposition becomes a decomposition into a homogenous part and an inhomogenity determined by the chosen renormalization condition. It thus survives the transition to non-perturbative physics.

If the charge $\phi(Q^{n_r})$ gives rise to a vanishing β function, the corresponding Hopf algebra governing the DSEs becomes cocommutative, the dual Lie algebra is abelian and we can solve non-perturbatively by a scaling solution and a simple Mellin transform of the one-cocycles $B_+^{k;r}$ [7, 9].

In the general case, starting from this Mellin transform, one realizes that the renormalization group allows to work with a very simplified form of the coproduct: we can project on both sides to terms in the span of linear generators. Hence we find two extremely useful simplifications: the restriction to sub Hopf algebras which are much easier to define than the full graph algebras, and the linearization in the use of the coproduct. Together with the fact that the primitives $B_+^{k;r}(\mathbb{I})$ exhibit conformal invariance at the renormalization point, which severely restricts the form of their Mellin transform, this goes a long way, I set my hopes for the future, in understanding better the phenomenon of universality, finding common anomalous dimensions in very different physical systems. But before we can exhibit these features in the promised example, we have to start with the structure of Feynman graphs.

2 Lie- and Hopf algebras of graphs

To capture the structure of a renormalizable quantum field theory, we organize it in terms of graphs. More formally, these graphs Γ will index generators δ_Γ of a Hopf algebra, so that as an algebra it is the free commutative algebra on generators indexed by the 1PI graphs of the theory. We will also consider pre-Lie and Lie algebras, with their generators Z_Γ indexed by the very same graphs. We will often simply write Γ for δ_Γ or Z_Γ when the context is clear.

All algebras are supposed to be over some field $\mathbb{K}$ of characteristic zero, associative and unital, and similarly for co-algebras. The unit (and by abuse of notation also the unit map) will be denoted by $\mathbb{I}$, the co-unit map by $\bar{e}$. Algebra homomorphisms are supposed to be unital. A bialgebra $(A = \bigoplus_{i=0}^\infty A_i, m, \mathbb{I}, \Delta, \bar{e})$ is called graded connected if $A_iA_j \subset A_{i+j}$ and $\Delta(A_i) \subset \bigoplus_{j+k=i} A_j \otimes A_k$, and if $\Delta(\mathbb{I}) = \mathbb{I}\otimes\mathbb{I}$ and $A_0 = k\mathbb{I}$, $\bar{e}(\mathbb{I}) = 1 \in \mathbb{K}$ and $\bar{e} = 0$ on $\bigoplus_{i=1}^\infty A_i$. We call $\ker \bar{e}$ the augmentation ideal of A and denote by P the projection $A \to \ker\bar{e}$ onto the augmentation ideal, $P = id - \mathbb{I}\bar{e}$.

Note that the augmentation ideal contains the quantum world: all graphs containing loops belong to the augmentation ideal. The classical world is captured by $A_0 = k\mathbb{I}$.

Furthermore, we use Sweedler's notation $\Delta(h) = \sum h' \otimes h''$ for the coproduct. We define

$$\mathrm{Aug}^{(k)} = \left(\underbrace{P \otimes \cdots \otimes P}_{k \text{ times}}\right) \Delta^{k-1},\ A \to \{\ker \bar{e}\}^{\otimes k}, \tag{2.1}$$

as a map into the k-fold tensorproduct of the augmentation ideal. We let $A^{(k)} = \ker \mathrm{Aug}^{(k+1)} / \ker \mathrm{Aug}^{(k)}$, $\forall k \geq 1$. All bialgebras considered here are bigraded in the sense that

$$A = \bigoplus_{i=0}^{\infty} A_i = \bigoplus_{k=0}^{\infty} A^{(k)}, \tag{2.2}$$

where $A_k \subset \oplus_{j=1}^k A^{(j)}$ for all $k \geq 1$. $A_0 \simeq A^{(0)} \simeq \mathbb{K}$.

The first construction we have to study is the pre-Lie algebra structure of 1PI graphs.

2.1 The Pre-Lie Structure. We are considering 1PI Feynman graphs. 1PI graphs are naturally graded by their number of independent loops, the rank of their first homology group $H_{[1]}(\Gamma, \mathbb{Z})$. We already wrote $|\Gamma|$ for this degree of a graph Γ.

As we stressed several times, a crucial notion is the external leg structure of a graph. It determines the relevant amplitude to which the graph contributes. The relevant contribution to the counterterm in the Lagrangian is then obtained by evaluating the Feynman rules on the tree level graph which corresponds to that amplitude. It is thus profitable to have a map which assigns to a given graph that tree level graph. This map $\mathbf{res}(\Gamma)$ is easily described: for any graph, we shrink all its internal edges to a point, so that a single vertex, with the external edges attached as half edges, remains. This is the desired tree level graph. For a graph with two external edges, the result is simply the two-point vertex of a form-factor of a single inverse edge, otherwise, if $n(r) > 2$, we get an interaction term in the Lagrangian.

When we evaluate a graph by the Feynman rules, we hence obtain a result in the form

$$\phi(\Gamma) = \phi(\mathbf{res}(\Gamma))X + Y, \tag{2.3}$$

where X is a superficially divergent Green function and Y contributes to amplitudes in $\mathcal{A}_+$.

We will see below how to construct a QFT from primitive graphs Γ. Primitive graphs have no divergent subgraphs which need renormalization. For such graphs Γ, the above decomposition reads

$$\phi(\Gamma) = \phi(\mathbf{res}(\Gamma)) \left[\frac{r}{z}\right] + \text{ finite terms} + Y, \tag{2.4}$$

for some regulator z. Here, r is the numerical residue of the graph. Thus,

$$\lim_{z \to 0} z\phi(\Gamma) = r\phi(\mathbf{res}(\Gamma)), \tag{2.5}$$

and hence the name residue also for the map which shrinks all internal edges. Note that $|\mathbf{res}(\Gamma)| = 0$. We emphasize that the set $\mathcal{M}_r$ contains only 1PI graphs Γ such that $|\Gamma| > 0$. The residue $\mathbf{res}(\Gamma)$ is not an element of this set, and hence is no generator in our Hopf algebra. The elements of degree zero are strictly given by the scalars so that we have a connected Hopf algebra, which is justified by the very fact that each Green function is a mere structure function corresponding to an amplitude in the Lagrangian.

Having specified free quantum fields and local interaction terms between them, one immediately obtains the set of 1PI graphs, and can consider for a given external leg structure r the set of graphs with that external leg structure. For a renormalizable theory, we can define a superficial degree of divergence using the first Betti number,

$$\omega = \sum_{r \in \Gamma_{\text{int}}^{[1]} \cup \Gamma^{[0]}} \omega_r - 4|H_{[1]}(\Gamma, \mathbb{Z})|, \tag{2.6}$$

for each such external leg structure: $\omega(\Gamma) = \omega(\Gamma')$ if $\mathbf{res}(\Gamma) = \mathbf{res}(\Gamma')$, all graphs with the same external leg structure have the same superficial degree of divergence

-the hallmark of a renormalizable theory-, and only for a finite number of distinct external leg structures $r \in \mathcal{R}$ will this degree indeed signify a divergence.

For the pre-Lie structure we define a bilinear operation

$$\Gamma_1 * \Gamma_2 = \sum_{\Gamma} n(\Gamma_1, \Gamma_2; \Gamma)\Gamma, \tag{2.7}$$

where the sum is over all 1PI graphs Γ. Here, $n(\Gamma_1, \Gamma_2; \Gamma)$ is a section coefficient which counts the number of ways a subgraph Γ_2 can be reduced (by shrinking it to its residue in the above sense) to a point in Γ such that Γ_1 is obtained. The above sum is evidently finite as long as Γ_1 and Γ_2 are finite graphs, and the graphs which contribute necessarily fulfill $|\Gamma| = |\Gamma_1| + |\Gamma_2|$ and $\mathbf{res}(\Gamma) = \mathbf{res}(\Gamma_1)$, as $\mathbf{res}(\Gamma_1 \star \Gamma_2) = \mathbf{res}(\Gamma_1)$ by construction.

One then has:

Theorem 2.1 *The operation $*$ is pre-Lie:*

$$[\Gamma_1 * \Gamma_2] * \Gamma_3 - \Gamma_1 * [\Gamma_2 * \Gamma_3] = [\Gamma_1 * \Gamma_3] * \Gamma_2 - \Gamma_1 * [\Gamma_3 * \Gamma_2]. \tag{2.8}$$

Together with the corresponding Lie algebra $\mathcal{L}$ one is led to consider the dual of its universal enveloping algebra $\mathcal{U}(\mathcal{L})$ using the theorem of Milnor and Moore and the above grading by the loop number. This graded dual, obtained from the usual Kronecker pairing, is a Hopf algebra $\mathcal{H}(m, \mathbb{I}, \Delta, \bar{e})$ which is commutative but not co-commutative. $\mathcal{H}$ is a graded commutative Hopf algebra which suffices to describe perturbative renormalization theory [14, 15, 16].

2.2 Multiplicative subtraction. The above algebra structures are available once one has decided on the set of 1PI graphs of interest, delivering the renormalization of any such chosen local quantum field theory. As to be expected, gauge theories provide particular properties with respect to the appearance of sub Hopf algebras which explain the Slavnov Taylor identities for the couplings [1], while the identities related to kinematics of Green functions expressing transversality of physical degrees of freedom for the gauge boson, are reflected in the presence of Hopf ideals [20] and a semi-direct product structure between radiation and matter [21].

From the above, one-particle irreducible graphs Γ provide the linear generators δ_Γ, with span $\mathcal{H}_{\rm lin} = \mathrm{span}(\delta_\Gamma)$, of the Hopf algebra $\mathcal{H} = \oplus_{i=0}^{\infty}\mathcal{H}_i$. Disjoint union of graphs provides the commutative product. We let P be the projector into the augmentation ideal $\mathcal{H}_{\rm aug} = \oplus_{i=1}^{\infty}\mathcal{H}_i$ and $P_{\rm lin}$ be the projector into $\mathcal{H}_{\rm lin} \subset \mathcal{H}_{\rm aug}$.

Let now Γ be a 1PI graph. We find the Hopf algebra $\mathcal{H}$ as described above to have a co-product explicitely given as $\Delta : \mathcal{H} \to \mathcal{H} \otimes \mathcal{H}$:

$$\Delta(\Gamma) = \Gamma \otimes \mathbb{I} + \mathbb{I} \otimes \Gamma + \sum_{\gamma \subset \Gamma} \gamma \otimes \Gamma/\gamma, \tag{2.9}$$

where the sum is over all unions of 1PI superficially divergent proper subgraphs (hence subgraphs with residue in $\mathcal{R}$ each), and we extend this definition to products of graphs so that we get a bi-algebra.

Having a co-product, two further structure maps of $\mathcal{H}$ are immediate, the co-unit and the antipode. The co-unit $\bar{e}$ vanishes on any non-trivial Hopf algebra element, $\bar{e}(\mathbb{I}) = 1$, $\bar{e}(X) = 0$. The antipode is

$$S(\Gamma) = -\Gamma - \sum_{\gamma \subset \Gamma} S(\gamma)\Gamma/\gamma. \tag{2.10}$$

Note that for each term in the restricted sum $[P \otimes P]\Delta(\Gamma) = \sum_i \Gamma'_{(i)} \otimes \Gamma''_{(i)}$ we have unique gluing data G_i such that

$$\Gamma = \Gamma''_{(i)} \leftarrow_{G_i} \Gamma'_{(i)}, \ \forall i. \tag{2.11}$$

These gluing date describe the necessary bijections to glue the components $\Gamma'_{(i)}$ back into $\Gamma''_{(i)}$ so as to obtain Γ: using them, we can reassemble the whole from its parts. Each possible gluing can be interpreted as a composition in the insertion operad of Feynman graphs [22, 23]. This gluing has a faithful representation as an iteration of Feynman integrals, upon replacing the δ-function for momentum conservation at a given insertion point by the expression for the inserted graph, and identifying the continuous quantum numbers at external legs of the inserted graphs with the momentum labels of the edges attached to the removed vertex, in accordance with the gluing data. It is one of the challenges with regard to analytic progress with Feynman graphs to understand this operation as a generalization of the notion of iterated integrals and their algebraic structure.

Disjoint scattering processes give rise to independent amplitudes, so one is led to the study of characters of the Hopf algebra, maps $\phi : \mathcal{H} \to V$ such that $\phi \circ m = m_V(\phi \otimes \phi)$.

Such maps assign to any element in the Hopf algebra an element in a suitable target space V. The study of tree-level amplitudes in lowest order perturbation theory justifies to assign to each edge a propagator and to each elementary scattering process a vertex which define the Feynman rules $\phi(\mathbf{res}(\Gamma))$ and the underlying Lagrangian, on the level of residues (in the graphical sense above) of these very graphs. As graphs themselves are constructed from edges and vertices, such residues indeed, one is led to assign to each Feynman graph an evaluation in terms of an integral over the continues quantum numbers assigned to edges or vertices, which leads to the familiar integrals over momenta in closed loops mentioned before, and hence leads to the Feynman rules (1.15), $\phi \in \mathrm{Spec}(\mathcal{G})$, as before.

Next, we choose a map $R : V \to V$, from which we obviously demand that is does not modify the UV-singular structure, and furthermore that it obeys

$$R(xy) + R(x)R(y) = R(R(x)y) + R(xR(y)), \tag{2.12}$$

an equation which guarantees the multiplicativity of renormalization and is at the heart of the Birkhoff decomposition which emerges below: it tells us that elements in V split into two parallel subalgebras given by the image and kernel of R. Such Rota–Baxter algebras play a role for associative algebras which is similar to the role Yang–Baxter algebras play for Lie algebras. The structure of these algebras allows to connect renormalization theory to integrable systems [3]. The situation is then remarkably similar to the factorization and Birkhoff decomposition in many studies of dynamical systems or problems in condensed matter theory, see for example Korepin et.al. in this volume [24]. Also, most of the results obtained initially for a specific renormalization scheme like minimal subtraction can be obtained in general upon a structural analysis of the corresponding Rota–Baxter algebras.

In renormalization theory we define a further character S_R^ϕ which deforms $\phi \circ S$ slightly and delivers the counterterm for Γ in the renormalization scheme R:

$$S_R^\phi(\Gamma) = -Rm_V(S_R^\phi \otimes \phi \circ P)\Delta = -R[\phi(\Gamma)] - R\left[\sum_{\gamma \subset \Gamma} S_R^\phi(\gamma)\phi(\Gamma/\gamma)\right], \tag{2.13}$$

a slight modification of the inverted Feynman rules

$$\phi \circ S = m_V(S \circ \phi \otimes \phi \circ P)\Delta = -\phi(\Gamma) - \sum_{\gamma \subset \Gamma} \phi \circ S(\gamma)\phi(\Gamma/\gamma). \tag{2.14}$$

Note that $S_R^\phi \in \mathrm{Spec}(\mathcal{G})$ thanks to (2.12).

The classical results of renormalization theory follow using this group structure: We obtain the renormalization of Γ by the application of a renormalized character

$$S_R^\phi \star \phi(\Gamma) = m_V(S_R^\phi \otimes \phi)\Delta, \tag{2.15}$$

and Bogoliubov's $\bar{R}$ operation as

$$\bar{R}(\Gamma) = m_V(S_R^\phi \otimes \phi)(\mathrm{id} \otimes P)\Delta(\Gamma) = \phi(\Gamma) + \sum_{\gamma \subset \Gamma} S_R^\phi(\gamma)\phi(\Gamma/\gamma), \tag{2.16}$$

so that we have

$$S_R^\phi \star \phi(\Gamma) = \bar{R}(\Gamma) + S_R^\phi(\Gamma). \tag{2.17}$$

Here, $S_R^\phi \star \phi$ is an element in the group of characters $\mathcal{G}$ of the Hopf algebra, with the group law given by the convolution

$$\phi_1 \star \phi_2 = m_V \circ (\phi_1 \otimes \phi_2) \circ \Delta, \tag{2.18}$$

so that the co-product, co-unit and co-inverse (the antipode) give the product, unit and inverse of this group, as befits a Hopf algebra. This Lie group has the previous Lie algebra $\mathcal{L}$ of graph insertions as its Lie algebra: $\mathcal{L}$ exponentiates to $\mathcal{G}$. This finishes perturbative renormalization theory. Further results in particular with regard to the vicinity of $\phi \in \mathrm{Spec}(\mathcal{G})$ as tested by one-parameter groups of automorphisms of the Hopf algebra were obtained in [17]. This illuminates in particular the renormalization group, whose ubiquitous applications are discussed, for example, in this volume [25].

3 The role of Hochschild cohomology and nonperturbative physics

The Hochschild cohomology of the combinatorial Hopf algebras which we discuss here plays three major roles in quantum field theory: it allows to prove locality from the accompanying filtration by the augmentation degree coming from the kernels $\ker \mathrm{Aug}^{(k)}$, it allows to write the quantum equations of motion in terms of the Hopf algebra primitives, elements in $\mathcal{H}_{\mathrm{lin}} \cap \{\ker \mathrm{Aug}^{(2)} / \ker \mathrm{Aug}^{(1)}\}$, and identifies the relevant sub-Hopf algebras formed by time-ordered products. Before we discuss these properties, let us first introduce the relevant Hochschild cohomology [15].

3.1 Hochschild cohomology of bialgebras. Let $(A, m, \mathbb{I}, \Delta, \epsilon)$ be a bialgebra, as before. We regard linear maps $L : A \to A^{\otimes n}$ as n-cochains and define a coboundary map b, $b^2 = 0$, by

$$bL := (\mathrm{id} \otimes L) \circ \Delta + \sum_{i=1}^{n} (-1)^i \Delta_i \circ L + (-1)^{n+1} L \otimes \mathbb{I}, \tag{3.1}$$

where Δ_i denotes the coproduct applied to the i-th factor in $A^{\otimes n}$, which defines the Hochschild cohomology of A.

For the case $n = 1$, (3.1) reduces to, for $L : A \to A$,

$$bL = (\mathrm{id} \otimes L) \circ \Delta - \Delta \circ L + L \otimes \mathbb{I}. \tag{3.2}$$

Note that for $h \in A^{(k)}$, we have $L(h) \in A^{k+1}$.

The category of objects (A, C) which consists of a commutative bialgebra A and a Hochschild one-cocycle C on A, and morphisms compatible with the cocycle actions, has an initial object $(\mathcal{H}_{\mathrm{rt}}, B_+)$, where $\mathcal{H}_{\mathrm{rt}}$ is the Hopf algebra of (non-planar) rooted trees and the closed but non-exact one-cocycle B_+ grafts a product of rooted trees together at a new root [15].

In Feynman graph Hopf algebras we will consider many one-cocycles $B_+^{k;r}$. The closedness of these $B_+^{k;r}$ in the context of Feynman graph Hopf algebras will turn out to be crucial for what follows. Also, we note that such one-cocycles transpose to the universal enveloping algebra on the dual side. With $B_+ : \mathcal{H} \to \mathcal{H}_{\mathrm{lin}}$ a one-cocycle, it turns out that by the standard Kronecker pairing [15, 16], the dual map $B_+^\vee : \mathcal{L} \to U(\mathcal{L})$ is a one-cocycle in Lie algebra cohomology.

3.2 The roles of Hochschild cohomology. The Hochschild cohomology of the Hopf algebras of 1PI graphs sheds light on the structure of 1PI Green functions. To determine the relevant Hochschild one-cocycles of a Feynman graph Hopf algebra $\mathcal{H}$, one determines first the primitives graphs γ of the Hopf algebra, which by definition fulfill

$$\Delta(\gamma) = \gamma \otimes \mathbb{I} + \mathbb{I} \otimes \gamma. \tag{3.3}$$

Using the pre-Lie product above, one then determines maps (possibly first on suitably chosen sub Hopf algebras based on graph which have γ as a cograph)

$$B_+^\gamma : \mathcal{H} \to \mathcal{H}_{\mathrm{lin}}, \tag{3.4}$$

such that

$$B_+^\gamma(h) = B_+^\gamma(h) \otimes \mathbb{I} + \left(\mathrm{id} \otimes B_+^\gamma\right) \Delta(h), \tag{3.5}$$

where $B_+^\gamma(h) = \sum_\Gamma n_+(\gamma, h, \Gamma)\Gamma$. The new section coefficients $n_+(\gamma, h, \Gamma)$ are closely related [1] to the section coefficients (2.7) we had before, but are normalized so that (3.5) holds.

Using the definition of the Bogoliubov map $\bar{R}$ this immediately shows that

$$\bar{R}(B_+^\gamma(h)) = \int \mathrm{D}_{\gamma \leftarrow G_i} \left[S_R^\phi * \phi(h) \right], \tag{3.6}$$

which proves locality of counterterms inductively upon recognizing that B_+^γ increases the augmentation degree. By D_γ we denote the measure assigned by the Feynman rules to the primitive γ,

$$\phi(B_+^\gamma(\mathbb{I})) =: \int D_\gamma. \tag{3.7}$$

The insertion of the functions for the subgraph is achieved using the relevant gluing data of (2.11), and the start of the induction on primitive graphs, which belong to $\mathcal{H}^{(1)}$, is immediate by Weinberg's theorem [29]. We hence obtain, by applying the Rota–Baxter map R to $\bar{R}$ in (3.6), the crucial statement of renormalization theory: the counterterm in perturbation theory is obtained by replacing all subgraphs by their renormalized expressions.

If we want to understand now why a field theory can be renormalized by local counterterms, we have to understand that it can be organized such that at each order all its contributions are in the image of a Hochschild closed one-cocycle.

It turns out that this can be achieved already at the combinatorial level. The sum over all graphs Γ^r contributing to a given amplitude can indeed be obtained as the solution to a fixpoint equation which puts these sums (for all $r \in \mathcal{R}$) into the image of those Hochschild one-cocycles $B_+^{k;r}$.

Indeed, to recover the quantum equation of motions, the DSEs, from Hochschild cohomology, one proves that

$$\Gamma^r = \mathbb{I} \pm \sum_{\gamma,\ \mathbf{res}(\gamma)=r} \frac{\alpha^{|\gamma|}}{\mathrm{sym}(\gamma)} B_+^\gamma(X_\gamma), \tag{3.8}$$

where

$$X_\gamma = \prod_{e \in \gamma_{\mathrm{int}}^{[1]}} \prod_{v \in \gamma^{[0]}} \Gamma^v / \Gamma^e = \Gamma^r Q^{n_r|\gamma|}, \tag{3.9}$$

in a renormalizable theory. Upon application of the Feynman rules the Hopf algebra elements $B_+^\gamma(\mathbb{I})$ turn to the integral kernels of the usual Dyson–Schwinger equations. We obtain the required recursive form (1.5) of the DSEs for $r \in \mathcal{R}$. This allows for new non-perturbative approaches as exhibited in a moment.

The one-cocycles introduced above allow to determine sub Hopf algebras of the form

$$\Delta(c_k^r) = \sum_{j=0}^{k} P_{k,j}^r \otimes c_{k-j}^r, \tag{3.10}$$

where the c_j^r are defined in Eq.(1.2) and the $P_{k,j}^r$ are polynomials in these variables given below. These sub Hopf algebras do not necessitate the considerations of single Feynman graphs any longer, but allow to establish renormalization directly for the sum of all graphs at a given loop order.

They hence establish a Hopf algebra structure on time-ordered products in momentum space. For theories with internal symmetries one expects and indeed finds that the existence of these sub-algebras establishes relations between graphs which are the Slavnov–Taylor identities between the couplings in the Lagrangian [1].

3.3 QED as an example. Let us present the DSEs for massless QED. Thanks to the Ward–Takahashi identity we only need to give them for two elements in $\mathcal{R}_{\mathrm{QED}}$, the vertex and the photon propagator.

We give them first in graphical form using the results of [26].

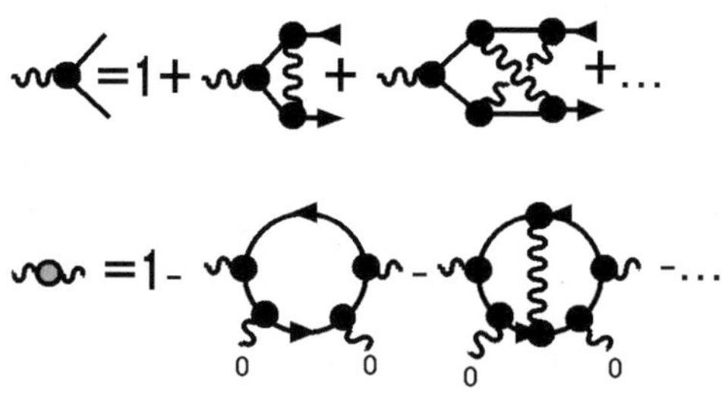

(3.11)

All internal edges are full propagators. Note that in the second equation we took a double derivative with respect to the external momentum at the inverse photon propagator. The corresponding Green function is indicated on the lhs. That makes it effectively a four-point function with two zero-momentum external photons. Our Feynman rules, incorporating our definition of an amplitude, are normalized to evaluate the tree-level term to unity, and we get an expansion in terms of 1PI four-photon amplitudes, organized in a skeleton expansion in this equation.

The combinatorial DSEs are thus simply

$$\Gamma^r = \mathbb{I} \pm \sum_{\gamma \in \mathcal{H}^{(1)} \cap \mathcal{H}_{\mathrm{lin}},\ \mathbf{res}(\gamma)=r} \alpha^{|\gamma|} B_+^{\gamma}(\Gamma^r Q^{2|\gamma|}) = \mathbb{I} \pm \sum_{k \geq 1} \alpha^k B_+^{k;r}(\Gamma^r Q^{2k}), \quad (3.12)$$

for $r \in \{\perp, \sim\}$. The sum is over all 1PI primitive graphs with the desired residue r. Also,

$$Q = \frac{\Gamma^{\perp}}{\Gamma^{\rightarrow} \sqrt{\Gamma^{\sim}}}. \quad (3.13)$$

Furthermore,

$$B_+^{k;r} = \sum_{\gamma \in \mathcal{H}^{(1)} \cap \mathcal{H}_{\mathrm{lin}}, \mathbf{res}(\gamma)=r, |\gamma|=k} \frac{1}{\mathrm{sym}(\gamma)} B_+^{\gamma}, \quad (3.14)$$

and B_+^{γ} glues vertex functions and propagator functions into edges of γ in accordance with Hochschild cohomology.

The sub Hopf algebra is given as follows [1]. Let us look at the set $\mathcal{R}_{\mathrm{QED}}$. Each of the series $\Gamma^r = \Gamma^r(\alpha)$ is a formal series in α, with coefficients c_k^r. The sought-after polynomial $P_{k,j}^r$ in the variables c_m^s is simply the Taylor coefficient of order j in the Taylor expansion in α of the expression

$$\Gamma^r Q^{2(k-j)}, \quad (3.15)$$

using (3.13). This is nothing but the argument of the one-cocycle in (3.12) and this is true in general. There is a wonderful and completely general phenomenon appearing here [1, 7]: if we restrict the sum over one-cocycles in (3.12) at some loop order, we obtain the same sub Hopf algebra. Those sub Hopf algebras are completely universal and independent of the order of the skeleton expansion we take into account. This is in my opinion a first clue as to why phenomena like universality can appear in recursive dynamical systems, to which quantum field theories evidently belong, at least in their short-distance sector.

To see this result in action, let us a look at a three loop example for the QED vertex.

$$\begin{aligned}
\Delta c_3^{\perp} = {} & \mathbb{I} \otimes c_3^{\perp} \\
& + [4c_1^{\rightarrow} + 5c_1^{\perp} + 2c_1^{\sim}] \otimes c_2^{\perp} \\
& + \Big[2c_2^{\rightarrow} + c_2^{\sim} + 3c_2^{v} + 3c_1^{\perp} c_1^{\perp} \\
& + 3c_1^{\rightarrow} c_1^{\rightarrow} + c_1^{\sim} c_1^{\sim} + 6c_1^{\rightarrow} c_1^{\perp} \\
& + 3c_1^{\sim} c_1^{\perp} + 2c_1^{\sim} c_1^{\rightarrow} \Big] \otimes c_1^{\perp} \\
& + c_3^{\perp} \otimes \mathbb{I}.
\end{aligned}$$

If we now impose the Hopf ideal generated by the Ward identities [20] the above simplifies to

$$\begin{aligned}
\Delta c_3^{\perp} = \;& \mathbb{I} \otimes c_3^{\perp}\\
&+\left[c_1^{\perp} + 2c_1^{\sim\!\sim\!\sim}\right] \otimes c_2^{\perp}\\
&+\left[c_2^{\perp} + c_2^{\sim\!\sim\!\sim} + c_1^{\sim\!\sim\!\sim}\, c_1^{\sim\!\sim\!\sim}\right] \otimes c_1^{\perp}\\
&+c_3^{\perp} \otimes \mathbb{I},
\end{aligned}$$

and the Hopf algebra structure were cocommutative were it not for the non-vanishing beta function, which ensures non-vanishing $c_k^{\sim\!\sim\!\sim}$.

Actually, taken into account the classical results of Baker, Johnson and Willey [27], the Hopf algebra would be trivial were the β-function to vanish. A thorough discussion of such structures in QED and their connection to the representation theory of the QED Hopf algebra will be given elsewhere [21].

Taking the Ward identity into account, the combinatorial DSEs are even simpler, as then

$$Q = \frac{1}{\sqrt{\Gamma^{\sim\!\sim\!\sim}}}. \tag{3.16}$$

In particular, for the photon we get

$$\Gamma^{\sim\!\sim\!\sim} = \mathbb{I} - \sum_{\gamma\in\mathcal{H}^{(1)}\cap\mathcal{H}_{\text{lin}},\ \mathbf{res}(\gamma)=\sim\!\sim\!\sim} \alpha^{|\gamma|} B_+^{\gamma}\left(\left[\Gamma^{\sim\!\sim\!\sim}\right]^{1-|\gamma|}\right). \tag{3.17}$$

Note that for the one-loop skeleton we have $|\gamma| = 1$, the argument of the one-cocycle is $\mathbb{I}$ trivially, and hence this term becomes an inhomogenous part in the DSE, and provides the fermion determinant for the QED Lagrangian.

4 The structure of Green functions

It is now time to summarize the structure of Green functions. We closely follow [1, 5, 6, 7].

4.1 Green functions for $\mathcal{R}$. For a given superficially divergent amplitude $r \in \mathcal{R}$ we let Γ^r be the sum

$$\Gamma^r = \mathbb{I} \pm \sum_{\mathbf{res}(\Gamma)=r} \alpha^{|\Gamma|} \frac{\Gamma}{\text{sym}(\Gamma)}, \tag{4.1}$$

over all 1PI graphs Γ contributing to that amplitude, with α a loop-counting small parameter. Projection onto suitable form factors $\phi(r)$ allows the sum to start with unity, so that by application of the Feynman rules, $\phi(\Gamma^r)$ is the corresponding structure function and the Lagrangian L is given by

$$L = \sum_{r\in\mathcal{R}} \hat{\phi}^{\pm 1}(r), \tag{4.2}$$

where we take the plus sign for $n(r) \geq 3$ and the minus sign for $n(r) = 2$.

We write

$$\Gamma^r = \mathbb{I} \pm B_+^r(\Gamma^r, Q(\{\Gamma^i\})), \tag{4.3}$$

where the Hochschild one-cocycle

$$B_+^r(\Gamma^r, Q) = \sum_{k\geq 1} \alpha^k B_+^{k;r}(\Gamma^r Q^{n_r k}), \tag{4.4}$$

is a sum of one-cocycles $B_+^{k,r}$ and Q^{n_r} is a monomial in the Γ^r:

$$Q^{n_r} = \frac{1}{\Gamma^{\mathbf{res}(\gamma)}} \frac{\prod_{v\in\gamma^{[0]}} \Gamma^v}{\prod_{e\in\gamma^{[1]}_{\text{int}}} \Gamma^e}, \tag{4.5}$$

for any $\gamma \in \mathcal{H}^{(1)}$ and $|\gamma| = 1$. In general, different such γ have different internal edges and vertices, and hence we obtain different $Q = Q(\gamma)$. For Q to be well-defined, we divide by the relations which equate them all. This is the origin of the Slavnov–Taylor identities for the couplings [1].

The $B_+^{k,r}$ themselves are obtained from the skeleton graphs γ of the theory:

$$B_+^{k,r} = \sum_{\gamma\in\mathcal{H}^{(1)}\cap\mathcal{H}_{\text{lin}},\mathbf{res}(\gamma)=r} \frac{1}{\text{sym}(\gamma)} B_+^\gamma, \tag{4.6}$$

where the sum is over all Hopf algebra primitives γ contributing to the amplitude r at k loops. These maps are defined to be closed Hochschild one-cocycles on the sub Hopf algebra generated by their concatentations and products [1, 7].

Effectively, (4.3) reduces the sum (4.1) over all graphs to a sum over primitive ones, making use of the recursive structure of this fixpoint equation which determines the sum of graphs which contribute to a chosen amplitude. The sums involved typically reflect the universal law of [28] and will be discussed in detail in upcoming work.

The c_j^r are the linear generators of a sub-Hopf algebra:

Theorem 4.1 *i) There exists a charge Q as above, maps $B_+^{k;r}$ and polynomials $P_{k,j}^r$ such that*

$$\Gamma^r = \mathbb{I} \pm \sum_k \alpha^k B_+^{k;r}\left(\Gamma^r Q^{n_r k}\right), \tag{4.7}$$

$$\Delta B_+^{k;r} = B_+^{k;r} \otimes \mathbb{I} + \left(\text{id} \otimes B_+^{k;r}\right) \Delta, \tag{4.8}$$

$$\Delta c_k^r = \sum_{j=0}^{k} P_{k,j}^r \otimes c_{k-j}^r, \tag{4.9}$$

which make the free commutative algebra in generators $\{c_k^r\}$ into a sub Hopf algebra $\mathcal{H}_c(\Delta, m, S, \epsilon)$ of the Feynman graph Hopf algebra.
ii) The polynomials $P_{k,j}^r$ are the Taylor coefficients of $\Gamma^r Q^{n_r(k-j)}$ (the expansion of the argument of the k-th one-cocycle) to order j in α.

We call the sub Hopf algebra $\mathcal{H}_c$ the Hopf algebra of time-ordered products, as the sum of all graphs with the same residue delivers precisely that. Feynman rules are then defined in accordance with the Hochschild cohomology as iterated integrals

$$\phi(\Gamma) = \int \mathrm{D}_{\gamma\leftarrow G_i}\phi(h), \tag{4.10}$$

on subgraphs h, using (2.11).

For a study of the short-distance behaviour of QFT we have to look at the amplitudes in $\mathcal{R}$. For such amplitudes, we can make use of the scaling properties of our Hopf algebras wrt one-parameter groups of automorphisms, and combine the sub Hopf algebra structure above with the general results on complex graded commutative Hopf algebras of [16, 17].

We have to make sure that any Green function which is superficially divergent and appears in the integrand generated by $B_+^{k;r}(\mathbb{I})$ only depends on a single scale. To achieve this in accordance with our approach is a non-trivial demand for amplitudes with $n(r) > 2$. The corresponding vertex functions $\phi_R(\Gamma^r)$, $\phi_R := S_R^\phi * \phi$, depend a priori on $n(r)-1$ linear independent external momenta p_f. Those external momenta will be internal momenta inside the DSEs, where they are to be integrated out thanks to the underlying iterated structure of those equations. Thanks to the fact that short-distance singularities are local we can isolate them into vertex functions at zero momentum transfer, and decompose a divergent amplitude further into an amplitude which depends on a single scale, and an amplitude in $\mathcal{A}_+$. The choice of such a decomposition corresponds to a choice of a basis of primitive elements in the Hopf algebra to work with, and an easy recursion argument [22] shows that we can choose an appropriate basis of such primitives so that we can set up the DSEs such that all divergent subgraphs depend only on a single scale.

Hence there exists a basis of graphs and external structures in the Hopf algebra such that

$$\phi_R(Q) = \phi_R(Q)(\alpha, L), \tag{4.11}$$

where $L = \ln q^2/\mu^2$ is the single scale which determines the running of the invariant charge. In this chosen basis, short distance singularities are captured by Green functions which are functions of two dimensionless variables α, L, with a remarkable duality between these two variables first observed in [5].

In perturbation theory the Feynman rules now allow us to write the renormalized Green functions (compare to the unrenormalized ones in (1.1)), in the momentum scheme so that $G_R^r(\alpha, 0) = 1$, as

$$G_R^r(\alpha, L) = \phi_R(\Gamma^r) = S_R^\phi * \phi(\Gamma^r) = 1 \pm \sum_k \alpha^k \phi_R(c_k^r)(L). \tag{4.12}$$

We can expand in a different manner

$$G_R^r(\alpha, L) = 1 \pm \sum_{k\geq 1} \gamma_k^r(\alpha) L^k, \tag{4.13}$$

and the renormalization group dictates relations between the γ_k^r. We work them out in a moment and note the renormalization group equations first:

$$\frac{\partial G_R^r(\alpha, L)}{\partial L} - \left[n_e(r)\gamma_1^e(\alpha) + \beta(\alpha)\frac{\partial}{\partial\alpha}\right] G_R^r(\alpha, L) = 0. \tag{4.14}$$

Here, n_e is the number of external legs of type e in the residue r, $e \in \mathcal{R}$, $n(e) = 2$ and γ_1^e is the corresponding anomalous dimension, obtained by taking a derivative in (4.13),

$$\gamma_1^e := \frac{\partial G^e(\alpha, L)}{\partial L}|_{L=0}. \tag{4.15}$$

$\beta(\alpha)$ is the β-function corresponding to the charge $\phi_R(Q)$. It is given through the anomalous dimensions in a simple form. If

$$Q^{n_r} = \frac{\prod_v [\Gamma^v]^{n_v}}{\prod_e [\Gamma^e]^{n_e/2}}, \tag{4.16}$$

then

$$\frac{1}{n_r}\beta(\alpha) = \sum_v n_v \gamma_1^v - \frac{1}{2}\sum_e n_e \gamma_1^e, \tag{4.17}$$

in accordance with (1.6). We let

$$\frac{1}{n_r}\beta_{\text{comb}} := \sum_v n_v \Gamma^v - \frac{1}{2}\sum_e n_e \Gamma^e, \tag{4.18}$$

be the corresponding series in α with coefficients in the Hopf algebra.

Next, we note that in the case of a linear DSE [7, 9], we get

$$\partial_L \phi(Q)(L) = 0. \tag{4.19}$$

As a result, scaling

$$G(\alpha, L) = e^{-\gamma(\alpha)L}, \tag{4.20}$$

solves the linear DSE so that

$$\gamma_k(\alpha) = \frac{\gamma_1(\alpha)^k}{k!}. \tag{4.21}$$

In such a case, the corresponding Hopf algebra structure of the c_k^r is cocommutative and the dual Lie algebra structure abelian. The violation of conformal invariance captured by a non-vanishing β-function reflects itself in the non-vanishing commutators of the Feynman graph Lie algebra.

A beautiful result of [17] is the scattering formula, which expresses higher poles in the regulator through iterated residues. This translates into a statement on the coefficients of higher logs in the leading log expansion of a quantum field theory. We are assured that to get to these higher log contributions from a Hopf algebra element, we simply have to apply the coproduct sufficiently often so as to decompose it into the components in $\mathcal{H}^{(1)}$. We then find the contribution to the k-th power of log by the product of the residues of all those primitives.

Hence, to proceed in general we consider the map

$$P_{\text{lin}}^{(n)} = \underbrace{P_{\text{lin}} \otimes \cdots \otimes P_{\text{lin}}}_{n \text{ times}} \Delta^{n-1} \tag{4.22}$$

where P_{lin} is the projector into the linear span of generators of the Hopf algebra. From [1, 7] we have:

Theorem 4.2 *The linearized coproduct is obtained as*

$$P_{\text{lin}}^{(2)}\Gamma^r = P_{\text{lin}}\Gamma^r \otimes P_{\text{lin}}\Gamma^r + P_{\text{lin}}Q^{n_r} \otimes \alpha\partial_\alpha \Gamma^r,$$

where

$$P_{\text{lin}}Q^{n_r} = \beta_{\text{comb}}. \tag{4.23}$$

This allows us to understand the iterative structure of the next-to... leading log expansion (4.13).

We define for $n > 1$

$$\sigma_n := \frac{1}{n!} m^{n-1} \underbrace{\sigma_1 \otimes \cdots \otimes \sigma_1}_{n \text{ times}} \Delta^{n-1}, \tag{4.24}$$

and σ_1 is the residue defined by

$$\sigma_1 = \partial_L \phi_R (S \star Y)(L)|_{L=0}. \tag{4.25}$$

Actually, σ_n evaluates to the coefficient of the L^n term in the evaluation of a Hopf algebra element by the renormalized Feynman rules, by the scattering type formula [17].

We have

$$h \notin H_{\text{lin}} \Rightarrow \sigma_1(h) = 0, \tag{4.26}$$

so we can use Theorem 2 and, by the above definition (4.13) of $\gamma_k^r(\alpha)$,

$$\gamma_k^r(\alpha) = \sigma_k(\Gamma^r(\alpha)). \tag{4.27}$$

Projection onto the linear generators delivers the desired formula for the expansion in L, $\forall k \geq 2$:

$$\gamma_k^r(\alpha) = \frac{1}{k}\left[\gamma_1^r(\alpha)\gamma_{k-1}^r(\alpha) + \sum_j s^j \gamma_1^j(\alpha)\alpha\partial_\alpha \gamma_{k-1}^r(\alpha)\right]. \tag{4.28}$$

This gives us a second clue towards universality in field theory: not only simplify the rather complicated graph Hopf algebras to the rather simple Hopf algebras of time-ordered products, but from these Hopf algebras we only need to remind ourselves of the simplest linear part in them: the underlying complications of diffeomorphisms of physical parameters only very mildly interfere with the short-distance sector of a theory.

With the the above choice of basis we can now introduce the Mellin transform by raising internal propagators $\text{Prop}(k_e)$ with momentum k_e to non-integer powers ρ. A derivative wrt ρ then amounts to the insertion of logarithmic corrections $\ln k_e^2$ into the Feynman integrals $\phi_R(B_+^{k;r}(\mathbb{I}))$, which is all what is needed to proceed in view of the expansion (4.13) for internal Green functions.

The DSEs turn into equations which determine γ_1^r as we will see in a moment, while the further terms in the L expansion are determined from (4.28) above. Green functions also have the usual expansion in α which is triangular wrt γ_k:

$$\gamma_k^r(\alpha) = \sum_{j \geq k} \gamma_{k,j}^r \alpha^j. \tag{4.29}$$

We can hence proceed to work out the recursion relations which express the functions γ_k^r through the functions γ_1^r for $k > 1$, and turn the Dyson–Schwinger equations into an implicit equation which allows to determine the sole unknown coefficients $\gamma_{1,j}^r$ from the knowledge of the above Mellin transforms.

Before we finally discuss an example, let us mention a third clue towards universality: the fact that by construction the above primitives are invariant under conformal inversion severely restricts the form of the Mellin transform and analytic differences can to a large extent be compensated by a redefinition of the relevant couplings. We hope to have some more concrete results along those lines in the future.

4.2 A simple example. For concreteness, we consider massless Yukawa theory and consider all self-iterations of the one-loop massless fermion propagator, with subtractions in the momentum scheme at $q^2 = \mu^2$. Our Green function is an inverse propagator with momentum q and a function of two variables a and $L = \ln q^2/\mu^2$. We ignore radiative corrections at the bosonic line and also at vertices, so the set $\mathcal{R}$ has a single element and the superscript r is suppressed henceforth. We restrict ourselves to a single element in $\mathcal{H}^{(1)} \cap \mathcal{H}_{\text{lin}}$ and hence to a single one-cocycle B_+. We rederive the results of [5] for this case.

We write the perturbative series for the Dyson–Schwinger equation as

$$X(a) = \mathbb{I} - aB_+\left(\frac{1}{X(a)}\right), \tag{4.30}$$

where $\phi(B_+(\mathbb{I}))$ provides the one-loop self-energy integral to be iterated. Note that upon setting $X(a) = \mathbb{I} - \underline{X}(a)$, this is the equation for the self-energy $\underline{X}(a) = -P_{\text{lin}}X(a)$ studied in [5].

We have

$$Q = 1/X^2 \to P_{\text{lin}}(Q) = -2\underline{X}(a), \tag{4.31}$$

and find the linearized coproduct

$$P^{(2)}_{\text{lin}}X(a) = P_{\text{lin}}X(a) \otimes (P_{\text{lin}} - 2a\partial_a)X(a). \tag{4.32}$$

This is Proposition 1 of [5] and we also get

Theorem 4.3 *The next-to next-to... leading log expansion in L is given through the anomalous dimension $\gamma_1(a)$ as*

$$\gamma_k(a) = \frac{1}{k}\gamma_1(a)(1 - 2a\partial_a)\gamma_{k-1}(a). \tag{4.33}$$

This is Proposition 2 of [5]. As a result, can work out with ease the recursions which express γ_k, $k > 1$ through the Taylor coefficients of γ_1.

Such recursions are obtained for any non-linear DSE by iterating Theorem 4.2. We observe that we only need the cocommutative part in the determination of the coproduct as is evident from the definition (4.24) of σ_k, $k > 1$. The non-cocommutative part is always of lower degree in L in the obvious filtration by L.

It remains to understand how to compute $\gamma_1(\alpha)$. We proceed here by the Mellin transform. It reads

$$F(\rho) = \frac{1}{q^2}\int d^4k \frac{k \cdot q}{[k^2]^{1+\rho}(k+q)^2}|_{q^2=1} = \frac{1}{\rho(\rho-2)} =: \frac{r}{\rho} + \sum_{i\geq 0} f_i\rho^i. \tag{4.34}$$

Let us introduce a short hand notation as

$$\gamma \cdot U := \sum_{k=1}^{\infty} \gamma_k(\alpha)U^k. \tag{4.35}$$

The Dyson–Schwinger equation becomes

$$\gamma \cdot L = \alpha(1 + \gamma \cdot \partial_{-\rho})^{-1}[e^{-L\rho} - 1]F(\rho)|_{\rho=0}, \tag{4.36}$$

where we evaluate the rhs at $\rho = 0$ after taking derivatives. The functional dependence of the non-linear DSE on $XQ = X^{-1}$ reflects itself on the rhs.

The only unknown quantities in this equation are the Taylor coefficients $\gamma_{1,j}$ which are implicitly defined through the Taylor coefficients of the Mellin transform (4.34) above.

Taking a derivative of (4.36) wrt L and setting L to zero allows us to read them off:

$$\gamma_1 = \alpha(1+\gamma\cdot\partial_{-\rho})^{-1}\rho F(\rho)|_{\rho=0} \quad (4.37)$$

$$= \alpha r + \alpha\left(\sum_{k\geq 1}[\gamma\cdot\partial_{-\rho}]^k\right)\left[\sum_{k=1}^{\infty}\rho^k f_{k-1}\right]|_{\rho=0}, \quad (4.38)$$

so $\gamma_{1,1} = r$ universally. The resulting recursions determine the non-perturbative solution given in [5].

As a final remark, we mention that a functional equation can be assigned to the solution of the DSE, relating a weak to a strong coupling expansion making use of the analytic structure behind the recursion above. It is reminiscent of a functional equation for a ζ-function in two variables for the function field case [6].

Furthermore, in the general case one needs multivariate Mellin transforms, which is an immediate source for the appearance of transcendentals in solutions to DSEs [6]. The same holds actually for the related study of higher one-cocycles [10]. A few short remarks finish our exploration of DSEs.

5 From primitives to motives

At long last we arrive in understanding the role of primitives $\gamma\in\mathcal{H}^{(1)}$ in field theory. Their Mellin transforms provide the basic constituents of a field theory, with each Taylor coefficient of such a transform an interesting period in its own right.

Knowledge of these Mellin transforms is at the time of writing sparse and restricted to low degrees in α, even if we are only to look at the the residue, the first Taylor coefficient in such a Mellin transform. Indeed, in accordance with (2.4) we find

$$\phi(B_+^\gamma)(\mathbb{I}) = \frac{r_\gamma}{\rho} + \text{finite terms}, \quad (5.1)$$

and our first task is to find the residue r_γ for primitive graphs γ.

Still, such numbers provide much to explore for a mathematician, and lead deep into the territory of algebraic geometry and motivic theory. Data on such numbers have been amply provided in collaboration with David Broadhurst [30, 31] and John Gracey [32, 33], and hide in many computations in particle physics [34].

While it is nice that those Feynman graphs could be given a motivic interpretation in [10], we are apparently (rather depressingly, the evaluation of the wheel with n-spokes studied by us in that paper is already a formidable task in this respect) a far cry from a full motivic classification of Feynman graphs, though it almost certainly exists. The results of [10] where a first step in this direction. Whilst incomplete, they give hope for the future. One proceeds by assigning to a graph Γ a graph polynomial and finds the relevant pairings which are responsible for the period assigned to the desired residue to be determined from the interplay between the zeros of the graph polynomial (the graph hypersurface) and the simplex of integration. The existence of subgraphs containing closed loops (albeit convergent) ensures that this interplay has non-trivial homology, and gives the resulting periods motivic interpretation in the simple cases we could study.

The message to be kept is that upon organizing the short-distance singular sector in a self-consistent manner, we find that we can solve the DSEs upon understanding the number-theoretic and motivic nature of the elements $\phi(B_+^{k;r}(\mathbb{I}))$.

Acknowledgment

Countless discussions with David Broadhurst and Spencer Bloch on these topics, and encouragement from my students Christoph Bergbauer, Kurusch Ebrahimi-Fard and Karen Yeats, and some proofreading by Abhijnan Rej, are gratefully acknowledged.

References

[1] D. Kreimer, *Anatomy of a gauge theory*, Annals of Physics (2006), online first, in press, hep-th/0509135.

[2] A. I. Vainshtein, V. I. Zakharov, V. A. Novikov, M. A. Shifman, *The ABC of Instantons*, Sov. Phys. Ushpekhi **25** (1982), 195-215.

[3] K. Ebrahimi-Fard, L. Guo, *Rota–Baxter algebras in renormalization of perturbative quantum field theory*, this volume; hep-th/0604116.

[4] D. Kreimer, *The Hopf algebra structure of renormalizable quantum field theory*, in Encyclopedia of Mathematical Physics, eds. J.-P. Francoise, G. L. Naber, S. T. Tsou. Oxford: Elsevier 2006 (ISBN 978-0-1251-2666-3).

[5] D. J. Broadhurst and D. Kreimer, *Exact solutions of Dyson-Schwinger equations for iterated one-loop integrals and propagator-coupling duality,* Nucl. Phys. B **600** (2001) 403, hep-th/0012146.

[6] D. Kreimer, K. Yeats, *An etude in non-linear Dyson-Schwinger equations,* arXiv:hep-th/0605096, Nucl. Phys. **B** Proc. Suppl., in press.

[7] C. Bergbauer, D. Kreimer, *Hopf Algebras in Renormalization Theory: Locality and Dyson-Schwinger Equations from Hochschild Cohomology*, in *IRMA lectures in Mathematics and Theoretical Physics*, Vol. **10**, Physics and Number Theory, European Mathematical Society, Eds. V. Turaev, L. Nyssen; hep-th/0506190.

[8] D. Kreimer, *Factorization in quantum field theory: An exercise in Hopf algebras and local singularities,* arXiv:hep-th/0306020, contributed to Les Houches School of Physics: Frontiers in Number Theory, Physics and Geometry, Les Houches, France, 9-21 Mar 2003, in press.

[9] D. Kreimer, *An étude in linear Dyson–Schwinger equations,* preprint, IHES/P-06-23. MPI-Math, Bonn, Proceedings, to appear.

[10] S. Bloch, H. Esnault, D. Kreimer, *On Motives Associated to Graph Polynomials,* Comm. Math. Phys. **267** (2006) 181-225; arXiv:math.ag/0510011.

[11] A. Kizilersu, M. Reenders and M. R. Pennington, *One loop QED vertex in any covariant gauge: Its complete analytic form,* Phys. Rev. D **52** (1995) 1242 [arXiv:hep-ph/9503238].

[12] D. J. Gross, *Application of the renormalization group to high energy physics,* in *Les Houches 1975: Methods in field theory,* edited by R. Balian and J. Zinn-Justin (North-Holland/World Scientific, Amsterdam/Singapore, 1976/1981).

[13] D. Kreimer, C. Bergbauer, *The Hopf algebra of rooted trees in Epstein-Glaser renormalization,* Annales Henri Poincaré **6** (2005) 343 [arXiv:hep-th/0403207].

[14] D. Kreimer, *On the Hopf algebra structure of perturbative quantum field theories,* Adv. Theor. Math. Phys. **2** (1998) 303 [arXiv:q-alg/9707029].

[15] A. Connes, D. Kreimer, *Hopf algebras, renormalization and noncommutative geometry,* Commun. Math. Phys. **199** (1998) 203 [arXiv:hep-th/9808042].

[16] A. Connes, D. Kreimer *Renormalization in quantum field theory and the Riemann-Hilbert problem. I: The Hopf algebra structure of graphs and the main theorem,* Commun. Math. Phys. **210** (2000) 249 [arXiv:hep-th/9912092].

[17] A. Connes, D. Kreimer, *Renormalization in quantum field theory and the Riemann-Hilbert problem. II: The beta-function, diffeomorphisms and the renormalization group,* Commun. Math. Phys. **216** (2001) 215, hep-th/0003188.

[18] I. Todorov, *Constructing conformal field theory models*, this volume.

[19] D. Kreimer, *What is the trouble with Dyson-Schwinger equations?*, Nucl. Phys. Proc. Suppl. **135** (2004) 238, hep-th/0407016.

[20] W. van Suijlekom, *The Hopf algebra of Feynman graphs in QED,* Lett. Math. Phys. **77** (2006) 265; arXiv:hep-th/0602126.

[21] D. Kreimer, *On QED*, in preparation.

[22] D. Kreimer, *Structures in Feynman graphs: Hopf algebras and symmetries,* Proc. Symp. Pure Math. **73** (2005) 43 [arXiv:hep-th/0202110].

[23] M. Markl, S. Shnider, J. Stasheff, *Operads in Algebra, Topology and Physics*, Oxford Univ. Press 2002.

[24] A. R. Its, B.-Q. Jin, V. E. Korepin, *Entropy of XY spin chain and block Toeplitz determinants*, this volume.

[25] D. G. C. McKeon, *Using the renormalization group*, this volume.

[26] D. J. Broadhurst, R. Delbourgo and D. Kreimer, *Unknotting the polarized vacuum of quenched QED,* Phys. Lett. B **366** (1996) 421 [arXiv:hep-ph/9509296].

[27] M. Baker, K. Johnson, *Quantum electrodynamics at small distances,* Phys. Rev. **183** (1969) 1292.

[28] J. Bell, S. Burris, and K. Yeats, *Counting Rooted Trees: The Universal Law* $t(n)C\rho^{-n}n^{-3/2}$, Elect. J. Comb. **13** (2006) # R63, math.CO/0512432.

[29] S. Weinberg, *High-energy behavior in quantum field theory,* Phys. Rev. **118** (1960) 838.

[30] D. Broadhurst, D. Kreimer, *Knots and numbers in* Φ^4 *theory to 7 loops and beyond,* Int. J. Mod. Phys. C **6** (1995) 519 [arXiv:hep-ph/9504352].

[31] D. Broadhurst, D. Kreimer, *Association of multiple zeta values with positive knots via Feynman diagrams up to 9 loops,* Phys. Lett. B **393** (1997) 403 [arXiv:hep-th/9609128].

[32] J. Gracey, *Practicalities of renormalizing quantum field theories*, this volume; hep-th/0605-037.

[33] D. Broadhurst, J. Gracey, D. Kreimer, *Beyond the triangle and uniqueness relations: Non-zeta counterterms at large N from positive knots,* Z. Phys. C **75** (1997) 559 [arXiv:hep-th/9607174].

[34] S. Weinzierl, *The art of computing loop integrals*, this volume; hep-th/0604068.

Fields Institute Communications
Volume **50**, 2007

Two-sided $SLE_{8/3}$ and the Infinite Self-Avoiding Polygon

Gregory F. Lawler
Department of Mathematics
University of Chicago
Chicago, IL 60637 USA
lawler@math.uchicago.edu

Joan R. Lind
Department of Mathmatics and Computer Science
Belmont University
Nashville, TN 37212-3757 USA
lindj@mail.belmont.edu

Abstract. In this paper we construct two-sided $SLE_{8/3}$ and describe why it is a model of the infinite self-avoiding polygon.

1 Introduction

The Schramm-Loewner evolution, SLE_κ, as introudced in [9], is a candidate for scaling limits of random paths at criticality in two dimensions. Different values of κ correpond to different systems. One value of particular importance is $\kappa = 8/3$, and the corresponding system is conjectured to be the limit of the self-avoiding walk. Trying to understand this led to the definition of the restriction property [6], and then the (nonrigorous) identification of the limit for self-avoiding walks.

The scaling limit can be considered a probability measure on curves $\gamma : [0, \infty) \to \mathbb{C}$ with $\gamma(0) = 0$. The point 0 is special on the curve. If we look locally at any other point on the curve, then locally we see two curves at that point (the "past" and the "future"). To understand this, one might consider the limit as $r \to \infty$ of the curves $\gamma^{(r)}(s) = \gamma(s + r) - \gamma(r)$. Assuming this limit exists, we should have a limiting measure on curves $\gamma : (-\infty, \infty) \to \mathbb{C}$ with $\gamma(0) = 0$. Equivalently, we can consider this as a measure on pairs of nonintersecting (one-sided) curves.

In this paper, we complete the picture in [7] by describing the measure on two-sided curves. We can think of a two-sided curve as a simple loop that goes through both the origin and infinity. For this reason, we conjecture that this measure is the scaling limit for self-avoiding polygons.

2000 *Mathematics Subject Classification.* Primary 60G50; Secondary 60J65, 82B41.

Research of the first author is supported by the National Science Foundation grant DMS-0405021.

Let us outline this paper. We start by discussing the discrete model, the infinite self-avoiding polygon (ISAP). While we do not know how to prove the scaling limit of ISAP exists, we do use the heuristics from this model to derive the defintion of the two-sided $SLE_{8/3}$. In particular, our approach is to make precise the idea that two-sided radial $SLE_{8/3}$ is obtained by taking two independent radial $SLE_{8/3}$ and conditioning them not to intersect. This is conditioning on an event of probability zero, so one must take a limit. Much of this paper deals with the justifying this limit.

In Section 4, we consider the probability that two radial $SLE_{8/3}$ paths do not intersect, one running to time ∞ and the other to time t. This tends to zero as $t \to \infty$, and we give the asymptotic behavior. The calculation uses the restriction property which reduces the problem to a derivative estimate for radial $SLE_{8/3}$. We do this calculation in detail although the argument is similar to arguments that have appeared in previous papers. After doing this, we weight a path by a corresponding martingale to give a process that we call one-side of radial two-sided $SLE_{8/3}$. It is a radial analogue of the $SLE_{8/3}(\kappa.\rho)$ processes as introduced in [6].

This definition is not obviously symmetric in the two paths. In the next section, we describe an alternate defintion that is obviously symmetric. This shows that the two-sided radial $SLE_{8/3}$ can be grown in any order, that is, we grow one side for a while and then the other. The fact that the order does not make a difference is an example of commutation. See [1] for a much more detailed discussion of commutation properties of SLE_κ paths. In Section 6, we prove the restriction property for two-sided radial $SLE_{8/3}$.

We then consider the chordal analogues of two-sided $SLE_{8/3}$. This is an example of a chordal $SLE(\kappa, \rho)$ process and is also an example of the kind of processes discussed in [1].

Our final section gives a proof that chordal SLE_κ is the limit of radial SLE_κ in the following sense. Suppose that $\eta : (0, \infty) \to \mathbb{D}$ is a simple curve with $\eta(0+) \in \partial\mathbb{D} \setminus \{1\}$ and $\eta(t) \to 0$ as $t \to \infty$. For every $t < \infty$, we can consider radial SLE_κ from 1 to 0 in $\mathbb{D} \setminus \eta(0, t]$. As $t \to \infty$, this measure approaches that of chordal SLE_κ from 1 to 0 in $\mathbb{D} \setminus \eta(0, \infty]$. This fact, which we prove for $\kappa \leq 4$, is used in one of our constructions of two-sided radial $SLE_{8/3}$. We give a more preicse formulation of the limit in Section 8.

Except for the final section, we restrict our consideration to $\kappa = 8/3$ for simplicity. The ideas can be generalized to other $\kappa \leq 4$, but in these cases the measure is not obtained by "two SLE_κ paths conditioned not to intersect".

We assume that the reader knows the basic facts about complex variables and conformal transformations (Schwarz lemma, Schwarz reflection, Koebe-(1/4) lemma, Bieberbach estimate, Beurling estimate). See [2] or [4] for references.

2 The infinite self-avoiding polygon

In this section we will discuss the discrete model whose limit we are trying to describe. A *two-sided self-avoiding walk (2-SAW)* in $\mathbb{Z}^2 = \mathbb{Z} + i\mathbb{Z}$ of lengths j, k (centered at the origin) is a nearest neighbor path

$$\omega = [\omega_{-j}, \omega_{-j+1}, \ldots, \omega_{k-1}, \omega_k], \quad \omega_l \in \mathbb{Z}^2,$$

with $\omega_0 = 0$ and $\omega_l \neq \omega_m, -j \leq l < m \leq k$. Let $\mathcal{L}_{j,k}$ denote the set of such walks and write just $\mathcal{L}_k$ for $\mathcal{L}_{0,k}$. Walks in $\mathcal{L}_k$ are called *(one-sided) self-avoiding walks (SAW)* of length k (rooted at the origin). Note that there is a natural one-to-one

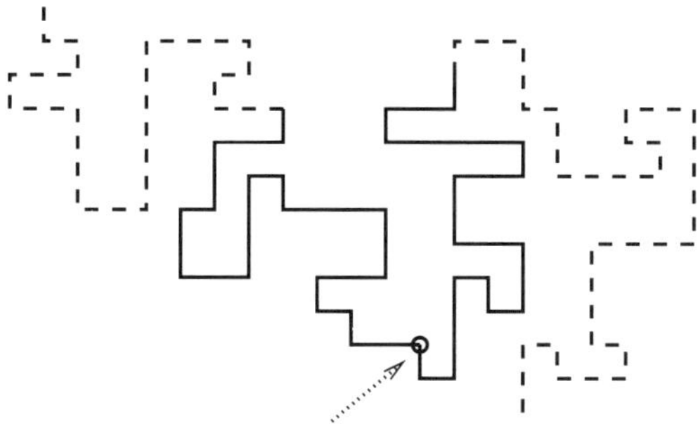

correspondence between $\mathcal{L}_{j,k}$ and $\mathcal{L}_{j+k}$. For any $n \geq j, m \geq k$, there is a probability measure $\mu_{j,k,n,m}$ on $\mathcal{L}_{j,k}$ given by

$$\mu_{j,k,n,m}(\omega) = \frac{\#\{\tilde{\omega} \in \mathcal{L}_{n,m} : [\tilde{\omega}_{-j}, \ldots, \tilde{\omega}_k] = \omega\}}{\#(\mathcal{L}_{n,m})}.$$

It is conjectured but has not been proven that the limits

$$\mu_k^*(\omega) = \lim_{m\to\infty} \mu_{0,k,0,m}(\omega),$$

$$\mu_{j,k}(\omega) = \lim_{n,m\to\infty} \mu_{j,k,n,m}(\omega),$$

exist and the second limit is independent of the way that n, m go to infinity. Assuming this conjecture, the measures $\{\mu_k^* : 0 \leq k < \infty\}$ and $\{\mu_{j,k} : 0 \leq j, k < \infty\}$ must be consistent. Hence we get a probability measure μ^* on infinite (one-sided) SAWs

$$\omega = [\omega_0 = 0, \omega_1, \omega_2, \ldots]$$

and a probability measure μ on infinite two-sided self-avoiding walks

$$\omega = [\ldots, \omega_{-2}, \omega_{-1}, \omega_0 = 0, \omega_1, \omega_2, ...],$$

with projection measures $\mu_k^*, \mu_{j,k}$. We call these measures the *(whole plane) infinite self-avoiding walk (ISAW)* and *(whole plane) infinite self-avoiding polygon (ISAP)*, respectively. Assuming the conjecture, the measure μ must be stationary, i.e., if

$$\theta_n \omega = [\ldots, \omega_{n-2} - \omega_n, \omega_{n-1} - \omega_n, 0, \omega_{n+1} - \omega_n, \omega_{n+2} - \omega_n, \cdots],$$

then for each integer n, $\theta_n \mu = \mu$. We can also consider μ as a measure on one-sided infinite self-avoiding walks, by looking at the projection

$$\omega \mapsto [\omega_0 = 0, \omega_1, \omega_2, \ldots].$$

We call this *one side of ISAP*. Note that this is *not* the same measure as the ISAW μ^*.

There are two important *critical exponents* for SAWs. While these are usually defined in terms of uniform measures on $\mathcal{L}_n$, they can also be defined in terms of the measure μ. The *mean-square displacement exponent* ν is defined by saying that under the measure μ, the expected value of $|\omega_n|^2$ grows like $n^{2\nu}$ as $n \to \infty$. The intersection exponent ζ (this is the same as the exponent $\gamma - 1$ as in [8]) is defined by saying that if two one-sided walks

$$\omega^1 = [\omega_0^1, \omega_1^1, \ldots], \qquad \omega^2 = [\omega_0^2, \omega_1^2, \ldots],$$

are chosen independently using μ^*, then the probability that

$$\{\omega_1^1, \ldots, \omega_n^1\} \cap \{\omega_2^1, \ldots, \omega_n^2\} = \emptyset$$

decays like $n^{-\zeta}$. The existence of these exponents has not been proved but there is very strong evidence for the values $\nu = 3/4, \zeta = 11/32$ (see [8]). Combining these two conjectures, we can say that the probability that two independent SAWs reach distance R without intersecting decays like $R^{-11/24}$.

We will consider continuum limits of these measures. Assuming that the exponent ν exists, we could expect that we can scale the walks by $n^{-\nu}$ to get a measure on continuous curves. There are actually four measures on continuous curves:

- m^*: the scaling limit of μ^* which gives a measure on simple curves $\gamma : [0,\infty) \to \mathbb{C}$ with $\gamma(0) = 0$.
- m: the scaling limit of μ which gives a measure on simple curves $\gamma : (-\infty,\infty) \to \mathbb{C}$ with $\gamma(0) = 0$. Equivalently, it can be considered as a measure on ordered pairs of curves (γ^1, γ^2) where $\gamma^j : [0,\infty) \to \mathbb{C}$ with $\gamma^j(0) = 0$ and
$$\gamma^1(0,\infty) \cap \gamma^2(0,\infty) = \emptyset.$$
- The marginal measure on γ^1 in the measure m.
- The conditional measure on γ^2 given γ^1 in m.

The first and fourth of these measures were considered in [7] where it was shown that there is only one possibility for the scaling limit assuming that the limit exists and is conformally invariant. The fourth measure was considered first. Under the assumption of conformal invariance, given γ^1 we can map $\mathbb{C} \setminus \gamma^1[0,\infty)$ to the upper half-plane $\mathbb{H}$ mapping 0 to 0 and ∞ to ∞. Then the stationarity property of the ISAP implies that the measure on γ^2, appropriately parametrized, satisfies the conformal Markov property. From this it was derived that the distribution must be that of chordal SLE_κ as introduced by Schramm [9]. The nature of the scaling limit also implied that the limit would satisfy a certain property which was denoted the *restriction property.* In [6] it was shown that this implies that κ must be 8/3. A similar argument established that the only possibility for m^* is that of whole-plane $SLE_{8/3}$, which is really a version of radial $SLE_{8/3}$.

At the moment there is no proof of the existence of the scaling limit or of its conformal invariance. However, there is strong numerical evidence [3] that the limit of SAWs is given by $SLE_{8/3}$. Moreover, the analogues of the exponents $\nu = 4/3, \zeta = 11/32$ can be computed for $SLE_{8/3}$ which gives very strong evidence for their correctness.

In this paper we will complete the picture by considering the other two measures. This requires considering two $SLE_{8/3}$ at one time. In summary the conjectured scaling limits of the four measures above are

- whole-plane $SLE_{8/3}$, in other words, the distribution of $\gamma[t,\infty)$ given $\gamma[0,t]$ is radial $SLE_{8/3}$.
- two-sided whole-plane $SLE_{8/3}$, in other words, the distribution of $\gamma(-\infty,\infty)$ given $\gamma[-s,t]$ is two-sided radial $SLE_{8/3}$.
- one side of two-sided whole-plane $SLE_{8/3}$,
- chordal $SLE_{8/3}$.

3 Radial $SLE_{8/3}$ and restriction

In this section, we remind the reader of the definition of radial SLE_κ and the restriction property, which $SLE_{8/3}$ satisfies. The restriction property will be crucial in our development of two-sided $SLE_{8/3}$. See [4] for more details.

Let B_t be a standard Brownian motion. Then radial SLE_κ is the solution to the Loewner equation with driving function $\sqrt{\kappa}B_t$. That is to say, it is the family of conformal maps $\tilde{g}_t$ solving the initial value problem

$$\partial_t \tilde{g}_t(z) = \tilde{g}_t(z)\frac{e^{i\sqrt{\kappa}B_t} + \tilde{g}_t(z)}{e^{i\sqrt{\kappa}B_t} - \tilde{g}_t(z)}, \quad \tilde{g}_0(z) = z,$$

for $z \in \mathbb{D}$. These maps satisfy the normalization that $\tilde{g}'(0) = e^t$. It will be convenient for us to change this parametrization by a factor of $1/\kappa$. This is equivalent to considering solutions to the initial value problem

$$\partial_t g_t(z) = \frac{a}{2}\, g_t(z)\, \frac{e^{iB_t} + g_t(z)}{e^{iB_t} - g_t(z)}, \quad g_0(z) = z, \tag{3.1}$$

for $z \in \mathbb{D}$, where $a = 2/\kappa$. Here the conformal maps g_t are normalized so that $g'(0) = e^{at/2}$. Although this change may make some of the exponents in our computations a little less friendly, we prefer to work with a standard Brownian motion rather one multiplied by $\sqrt{\kappa}$.

If $\kappa \leq 4$, then radial SLE_κ gives a measure on simple curves. The radial SLE_κ path is the function $\gamma : [0,\infty) \to \overline{\mathbb{D}}$ with the following properties: $\gamma(0) = e^{iB_0}$, $\gamma(0,\infty) \subset \mathbb{D} \setminus \{0\}$, $\lim_{t\to\infty} \gamma(t) = 0$, and g_t is the unique conformal transformation of $\mathbb{D} \setminus \gamma[0,t]$ onto $\mathbb{D}$ with $g_t(0) = 0$ and $g_t'(0) > 0$. We will often refer to this path as radial $SLE_{8/3}$ (starting at e^{iB_0}) rather than the family of maps g_t.

Simply stated, the restriction property says that $SLE_{8/3}$ in a subdomain of $\mathbb{D}$ is $SLE_{8/3}$ in $\mathbb{D}$ conditioned to stay in the subdomain. To describe this more fully, let $\mathcal{A}$ denote the set of $A \in \mathbb{D}$ such that $\mathbb{D} \setminus A$ is a simply connected domain containing the origin. If $x \in \mathbb{R}$, let $\mathcal{A}(x) = \mathcal{A}_x = \{A \in \mathcal{A} : \operatorname{dist}(e^{ix}, A) > 0\}$. For $A \in \mathcal{A}$, let $D_A = \mathbb{D} \setminus A$ and let $\Psi_A : D_A \to \mathbb{D}$ be the unique conformal transformation with $\Psi_A(0) = 0$ and $\Psi_A'(0) > 0$. If $A \in \mathcal{A}_x$, then Ψ_A has an analytic extension in a neighborhood of e^{ix}, and hence $\Psi_A'(e^{ix})$ is well defined. Suppose $A \in \mathcal{A}_x$ and γ is radial $SLE_{8/3}$ started at e^{ix}. On the event $\{\gamma(0,\infty) \cap A = \emptyset\}$ let $\eta(t) = \Psi_A \circ \gamma(t)$. Then the restriction property states that the conditional distribution of η given $\{\gamma(0,\infty) \cap A = \emptyset\}$ is the same (modulo time reparametrization) as radial SLE from $\Psi_A(e^{ix})$ to 0 in $\mathbb{D}$.

The following computation is at the heart of the restriction property:

$$\mathbb{P}\{\gamma[0,\infty) \cap A = \emptyset\} = |\Psi_A'(e^{iB_0})|^{5/8}\, \Psi_A'(0)^{5/48}, \tag{3.2}$$

and we will often refer to this simply as the restriction property. To establish (3.2), one must show that M_t, as defined below, is a bounded martingale with $\lim_{t\to\infty} M_t = 1\{\gamma(0,\infty) \cap A = \emptyset\}$. This will allow us to conclude that $M_t = \mathbb{P}[\gamma(0,\infty) \cap A = \emptyset \mid \mathcal{F}_t]$, since the latter is also a bounded martingale with the same limit at infinity. We define

$$M_t = 1\{\gamma(0,t] \cap A = \emptyset\}|\Psi_{A_t}'(e^{iB_t})|^{5/8}\, \Psi_{A_t}'(0)^{5/48},$$

where $A_t = g_t(A) \cap \mathbb{D}$. Notice that $M_0 = |\Psi_A'(e^{iB_0})|^{5/8}\, \Psi_A'(0)^{5/48}$. For the details, see Section 6.5 of [4]. In Section 5, we will prove the restriction property for two-sided radial $SLE_{8/3}$, and the proof of this will follow the same general argument.

Whole-plane SLE_κ. If D is a simply connected domain, $z \in \partial D, w \in D$, then radial SLE_κ from z to w in D is the conformal image of radial SLE_κ in $\mathbb{D}$ from 1 to 0 by the conformal transformation of $\mathbb{D}$ onto D mapping 1 to z and 0 to w. This is considered a measure on paths modulo time reparametrization.

Whole plane SLE_κ, $0 < \kappa \le 4$, is the measure on simple curves $\gamma : [0,\infty) \to \mathbb{C}$ with $\gamma(0) = 0$ that has the property that given $\gamma[0,t]$ the conditional distribution of $\gamma(t,\infty)$ is that of radial SLE_κ from $\gamma(t)$ to ∞ in $\mathbb{C} \setminus \gamma[0,t]$. Standard results about conformal maps can be used to see that this is well defined; see, e.g., [4, Section 6.6].

4 One side of two-sided radial $SLE_{8/3}$

The measure on two-sided SAWs of lengths n, n is exactly the same as the measure of two independent (one-sided) SAWs of length n conditioned not to intersect. Hence, we can think of the infinite ISAP as the measure on two independent ISAWs given by condtioning that they do not intersect. This description does not make precise sense because this is conditioning on an event of probability zero. However, we could hope to make rigorous sense by a limiting argument.

Using this as an analogy, we will try to build up two-sided radial $SLE_{8/3}$ by taking two (one-sided) radial $SLE_{8/3}$ paths and conditioning them not to intersect. Again, this is conditioning on an event of probability zero so we must take a limiting argument. We begin our study of two-sided radial $SLE_{8/3}$ by using the restriction property to understand the probability that a $SLE_{8/3}$ path will avoid the beginning of another, independent $SLE_{8/3}$ path. We will obtain a particular martingale M_s, and then then we will define one side of two-sided $SLE_{8/3}$ by weighting a $SLE_{8/3}$ path by M_s/M_0. The process so obtained is also referred to as "$SLE_{8/3}$ conditioned to avoid another $SLE_{8/3}$." We conclude the section by discussing an alternate definition of the process, derived using Girsanov's Theorem.

Let B_t and $\hat{B}_t$ be independent standard Brownian motions with $z := e^{iB_0} \neq e^{i\hat{B}_0} := \hat{z}$, and let γ and $\hat{\gamma}$ denote the $SLE_{8/3}$ paths generated by these Brownian motions (with corresponding functions g_t and $\hat{g}_t$) by solving (3.1) with $a = 3/8$. The restriction property tells us that

$$\mathbb{P}\{\gamma[0,t] \cap \hat{\gamma}[0,\infty) = \emptyset \mid \gamma[0,t]\} = e^{5t/128}\, |g_t'(\hat{z})|^{5/8}.$$

(Recall that $g_t'(0) = e^{3t/8}$ and hence $e^{5t/128} = g_t'(0)^{5/48}$.) Therefore,

$$\mathbb{P}\{\gamma[0,t] \cap \hat{\gamma}[0,\infty) = \emptyset\} = e^{5t/128}\, \mathbb{E}[|g_t'(\hat{z})|^{5/8}], \tag{4.1}$$

where the expectation is over the first Brownian motion B_t. As we wish to understand what happens with (4.1) as t approaches infinity, we must examine $\mathbb{E}[|g_t'(\hat{z})|^{5/8}]$.

Our first step is to introduce $h_t(z) = -i \log g_t(e^{iz})$. Here the branch of the logarithm is chosen so that $-i \log e^{iB_0} = B_0$. For fixed $t < \infty$, this is well defined in a neighborhood of $\gamma[0,t]$. Note that $|g_t'(\hat{z})| = h_t'(\hat{B}_0)$, allowing us to study $\phi(t,x) = \mathbb{E}[h_t'(x)^{5/8}]$ instead. Equation (3.1) implies that

$$\dot{h}_t(z) = \frac{3}{8} \cot\left(\frac{h_t(z) - B_t}{2}\right), \quad h_0(z) = z. \tag{4.2}$$

Differentiating this gives

$$h_t'(z) = \exp\left\{-\frac{3}{16}\int_0^t \csc^2\left(\frac{h_s(z) - B_s}{2}\right) ds\right\}.$$

If we let $x = \hat{B}_0$ and $V_t = h_t(x) - B_t$, then V_t satisfies

$$dV_t = \frac{3}{8} \cot\left(\frac{V_t}{2}\right) dt - dB_t,$$

and

$$h_t'(x) = \exp\left\{-\frac{3}{16}\int_0^t \csc^2\left(\frac{V_s}{2}\right)\,ds\right\}.$$

We will assume for ease that $B_0 = 0$. Let $\mathcal{F}_t$ denote the filtration generated by $\{B_s : 0 \le s \le t\}$. Then if $s < t$,

$$\mathbb{E}[h_t'(x)^{5/8} \mid \mathcal{F}_s] = h_s'(x)^{5/8}\,\phi(t-s, V_s).$$

Since this is a martingale, Itô's formula implies that

$$-\dot{\phi}(t,x) + \frac{1}{2}\,\phi''(t,x) + \frac{3}{8}\,\cot\left(\frac{x}{2}\right)\,\phi'(t,x) - \frac{15}{128}\,\csc^2\left(\frac{x}{2}\right)\,\phi(t,x) = 0. \tag{4.3}$$

We could also have obtained the differential equation for $\phi(t,x)$ by appealing to the Feynman-Kac formula.

Let L be the differential operator described by (4.3), that is

$$L(f) := -\dot{f} + \frac{1}{2}\,f'' + \frac{3}{8}\,\cot\left(\frac{x}{2}\right)\,f' - \frac{15}{128}\,\csc^2\left(\frac{x}{2}\right)\,f.$$

To solve a differential equation like $L(f) = 0$, one often attempts to find a solution of the form $e^{-\beta t}F(x)$, by solving an ordinary differential equation for F. In our case, we would need to solve

$$\frac{1}{2}\,F''(x) + \frac{3}{8}\,\cot\left(\frac{x}{2}\right)\,F'(x) + \left[\beta - \frac{15}{128}\,\csc^2\left(\frac{x}{2}\right)\right]\,F(x) = 0. \tag{4.4}$$

We notice that

$$F(x) = c\,\sin^{3/4}(x/2),$$

is a solution when $\beta = 27/128$.

Although we now know $c_*\,e^{-27t/128}\sin^{3/4}(x/2)$ is a solution to $L(f) = 0$, this cannot be equal $\phi(t,x)$, since they do not have the same initial conditions. In particular, $\phi(0,x) = 1$. We wish to compare $\phi(t,x)$ to this solution, however, and we will be especially interested in the behavior of these two functions as t goes to infinity. Although the choice seems arbitrary at this point, we will take

$$c_* = \frac{\int_0^{2\pi}\sin^{9/4}(x/2)\,dx}{\int_0^{2\pi}\sin^3(x/2)\,dx}.$$

In the subsection below, we explain how we obtained the constant c_*. However, the exact value of this constant will not matter for our development of two-sided $SLE_{8/3}$.

To show that $\phi(t,x) \sim c_*\,e^{-27t/128}\sin^{3/4}(x/2)$ as $t \to \infty$, we construct functions $F_\epsilon(t,x)$ and $G_\epsilon(t,x)$ as follows. To begin, for $\epsilon > 0$ set

$$F_\epsilon(t,x) = c_*\,e^{-27t/128}\sin^{3/4}(x/2) + e^{-at/128}\left(1 - c_*\,\sin^{3/4}(x/2)\right) + \epsilon - \phi(t,x),$$

where $a > 27$. Then

$$L(F_\epsilon(x,t)) = \frac{e^{-at}}{128}\left(a - 15\csc^2(x/2) - (a-27)c_*\sin^{3/4}(x/2)\right) - \frac{15\epsilon}{128}\csc^2(x/2),$$

which is negative for an appropriate choice of a, such as $a = 54$. Looking at the boundary conditions, note that $F_\epsilon(0,x) = \epsilon > 0$, and $F_\epsilon(t,0) = F_\epsilon(t,2\pi) > 0$. Suppose that $F_\epsilon(t,x) < 0$ for some $(t,x) \in [0,\infty)\times[0,2\pi]$. Then there is some point $(t_0,x_0) \in [0,\infty)\times[0,2\pi]$ with $F_\epsilon(t_0,x_0) = 0$ and with $F_\epsilon(t,x) - \phi(t,x) > 0$ for all $t < t_0$. It follows that we must have $\dot{F}_\epsilon(t_0,x_0) \le 0$, $F_\epsilon'(t_0,x_0) = 0$, and $F_\epsilon''(t_0,x_0) \ge 0$.

This, however, contradicts the fact that $L(F_\epsilon(t_0, x_0)) < 0$. Therefore, $F_\epsilon \geq 0$, and by letting ϵ go to zero, we have that $c_* \, e^{-27t/128} \sin^{3/4}(x/2) + c\, e^{-at/128} \geq \phi(t,x)$. In particular, notice that we have

$$\lim_{t\to\infty} e^{27t/128} \phi(t,x) \leq c_* \, \sin^{3/4}(x/2).$$

We also wish to obtain the opposite inequality. We can accomplish this with a similar argument in which we utilize

$$G_\epsilon(t,x) = c_* \, e^{-27t/128} \sin^{3/4}(x/2) \left(1 + e^{-(a_1-27)t/128}\right) - 2c_* \, e^{-a_2 t/128} - \epsilon - \phi(t,x),$$

with $a_2 > a_1 > 27$ appropriately chosen so that $L(G_\epsilon) > 0$. In this way, we obtain

$$\lim_{t\to\infty} e^{27t/128} \phi(t,x) = c_* \, \sin^{3/4}(x/2) = 2^{-3/4} c_* \, |e^{ix} - 1|^{3/4}. \tag{4.5}$$

We have now established the following proposition.

Proposition 4.1 Suppose $\gamma, \hat{\gamma}$ are independent radial $SLE_{8/3}$ curves started at $z, \hat{z}$ respectively. Then,

$$\lim_{t\to\infty} e^{11t/64} \, \mathbb{P}\{\gamma[0,t] \cap \hat{\gamma}[0,\infty) = \emptyset\} = C_* \, |z - \hat{z}|^{3/4},$$

where

$$C_* = 2^{-3/4} c_* = 2^{-3/4} \frac{\int_0^{2\pi} \sin^{9/4}(x/2) \, dx}{\int_0^{2\pi} \sin^3(x/2) \, dx}.$$

Proof Assume without loss of generality that $z = 1, \hat{z} = e^{ix}$, and as before, let $\phi(t,x) = \mathbb{E}[h'(x)^{5/8}]$. Then,

$$\mathbb{P}\{\gamma[0,t] \cap \hat{\gamma}[0,\infty) = \emptyset\} = e^{5t/128} \, \phi(t,x).$$

The proposition then follows from (4.5).

□

We continue to use the notation $x = \hat{B}_0$ and $V_s = h_s(x) - B_s$, and now we wish to consider

$$M_s = C_* \, e^{27s/128} \, |g_s'(e^{ix})|^{5/8} |e^{iB_s} - g_s(e^{ix})|^{3/4} = e^{27s/128} \, h_s'(x)^{5/8} F(V_s),$$

where $F(x) = c_* \sin^{3/4}(x/2)$ satisfies (4.4) with $\beta = 27/128$. Using this, we can compute that

$$dM_s = -\frac{3}{8} \cot\left(\frac{V_s}{2}\right) M_s \, dX_s,$$

where $X_s = B_s$. Therefore M_s is a martingale with $|M_s| \leq c' \, e^{27s/128}$.

We claim that

$$M_s = \lim_{t\to\infty} e^{11t/64} \, \mathbb{P}\{\gamma[0,t] \cap \hat{\gamma}[0,\infty) = \emptyset \mid \mathcal{F}_s\}.$$

By using the restriction property, we have

$$\begin{aligned}
&\mathbb{P}\{\gamma[0,t] \cap \hat{\gamma}[0,\infty) = \emptyset \mid \mathcal{F}_s\} \\
&\quad = \mathbb{P}\{\gamma[0,s] \cap \hat{\gamma}[0,\infty) = \emptyset \text{ and } g_s \circ \gamma[s,t] \cap g_s \circ \hat{\gamma}[0,\infty) = \emptyset \mid \mathcal{F}_s\} \\
&\quad = e^{5s/128} \, h_s'(x)^{5/8} \, \mathbb{P}\{g_s \circ \gamma[s,t] \cap g_s \circ \hat{\gamma}[0,\infty) = \emptyset \mid \mathcal{F}_s\}
\end{aligned}$$

Proposition 4.1 implies that

$$\lim_{t\to\infty} e^{11(t-s)/64} \, \mathbb{P}\{g_s \circ \gamma[s,t] \cap g_s \circ \hat{\gamma}[0,\infty) = \emptyset \mid \mathcal{F}_s\} = C_* |e^{iB_s} - g_s(e^{ix})|,$$

which proves the claim.

This second view of M_s leads us to define one side of two-sided $SLE_{8/3}$, otherwise called $SLE_{8/3}$ conditioned to avoid another $SLE_{8/3}$, by weighting a $SLE_{8/3}$ path by M_s/M_0. In particular, let $\mathbf{Q}$ denote the probability measure on paths induced by this positive martingale, and let $\mathbf{Q}_s$ denote this measure restricted to $X_t, 0 \leq t \leq s$. Then $d\mathbf{Q}_s/d\mathbf{P} = M_s/M_0$.

By making use of Girsanov's Theorem, we obtain an alternate viewpoint of this object: we can consider one side of two-sided $SLE_{8/3}$ to be the solution to the Loewner equation where the driving term has an appropriate drift. Girsanov's Theorem states that with respect to the measure $\mathbf{Q}$,

$$W_s := X_s + \frac{3}{8}\int_0^s \cot\left(\frac{V_t}{2}\right)\, dt,$$

is a standard Brownian motion, or in other words,

$$dX_s = -\frac{3}{8}\,\cot\left(\frac{V_s}{2}\right)\, ds + dW_s.$$

Thus one side of two-sided $SLE_{8/3}$ is the process generated by the Loewner equation with driving term X_s, where W_s is a Brownian motion,

$$dX_s = -\frac{3}{8}\,\cot\left(\frac{\hat{X}_s - X_s}{2}\right)\, ds + dW_s,$$

and

$$d\hat{X}_s = \frac{3}{8}\cot\left(\frac{\hat{X}_s - X_s}{2}\right)\, ds.$$

Note that we have replaced $h_s(x)$ by $\hat{X}_s$ in anticipation of the notation we will use in the next section, which will reflect the fact that the two sides of two-sided $SLE_{8/3}$ are symmetric.

The constant c_*. In this subsection, we briefly describe how we obtained the constant c_* found in our previous calculations. It arises naturally when finding a certain invariant density. The basic ideas used here are discussed in Section 1.11 of [4].

Recall that

$$dV_t = \frac{3}{8}\cot\left(\frac{V_t}{2}\right)\, dt - dB_t,$$

and

$$h_t'(x)^{5/8} = \exp\left\{-\frac{15}{128}\int_0^t \csc^2(V_s/2)\, ds\right\}.$$

Let $p(t,x,y)$ denote the transition probability density defined by

$$\mathbb{E}^x[f(V_t)\, h_t'(x)^{5/8}] = \int_0^{2\pi} p(t,x,y)\, f(y)\, dy,$$

and notice that for $\phi(t,x) = \mathbb{E}^x[h_t'(x)^{5/8}]$, we have that

$$\phi(t,x) = \int_0^{2\pi} p(t,x,y)\, dy.$$

There are two differential equations that $p(t,x,y)$ must satisfy, one for when y is fixed and the other for x fixed. If we had started in a simpler situation and were interested in the transition probability density for $\mathbb{E}^x[f(V_t)]$, we could easily obtain the two differential equations from Kolmogorov's backward and forward equations. Although our situation is slightly more complicated, it is not difficult to find the desired equations. We have already seen that $\phi(t,x)$ satisfies the differential equation (4.3), and one can show that p must satisfy this as well. That is,

$$\dot{p} = \frac{1}{2}p_{xx} + \frac{3}{8}\cot(x/2)\, p_x - \frac{15}{128}\csc^2(\frac{x}{2})\, p.$$

By considering the adjoint, we can find our second differential equation for p:

$$\dot{p} = \frac{1}{2}p_{yy} - \partial_y\,[\frac{3}{8}\cot(y/2)\, p] - \frac{15}{128}\csc^2(y/2)\, p.$$

See section 1.11 of [4] for further details.

As we did for ϕ, we can solve these differential equations to understand the behavior of p as t approaches infinity. In particular, we will find positive functions ψ_1 and ψ_2 satisfying the ordinary differential equations

$$\frac{1}{2}\psi_1''(x) + \frac{3}{8}\cot(x/2)\,\psi_1'(x) + [\beta - \frac{15}{128}\csc^2(x/2)]\,\psi_1(x) = 0,$$

$$\frac{1}{2}\psi_2''(y) - \partial_y\,[\frac{3}{8}\cot(y/2)\,\psi_2(y)] + [\beta - \frac{15}{128}\csc^2(y/2)]\,\psi_2(y) = 0,$$

so that $p(t,x,y) \sim c\,e^{-\beta t}\,\psi_1(x)\,\psi_2(y)$ as $t \to \infty$. The desired solutions are

$$\psi_1(x) = \sin^{3/4}(x/2), \qquad \psi_2(y) = \sin^{9/4}(y/2),$$

with $\beta = 27/128$. We therefore get

$$p(t,x,y) \sim c_1\, e^{-27t/128}\,\sin^{3/4}(x/2)\,\sin^{9/4}(y/2), \qquad t \to \infty. \tag{4.6}$$

We will now compute the constant c_1 as well as our previous constant c_*. The function $\sin^{9/4}(y/2)$ can be considered an invariant density in the sense that

$$\int_0^{2\pi} \sin^{9/4}(x/2)\, p(t,x,y)\, dx = e^{-27t/128}\,\sin^{9/4}(y/2).$$

Plugging in (4.6) gives

$$c_1 = \left[\int_0^{2\pi} \sin^3(x/2)\, dx\right]^{-1}.$$

Using (4.6) with

$$\phi(t,x) = \mathbb{E}^x[h_t'(x)^{5/8}] = \int_0^{2\pi} p(t,x,y)\, dy,$$

gives again that

$$\phi(t,x) \sim c_*\, e^{-27t/128}\,\sin^{3/4}(x/2),$$

and here it is clear that

$$c_* = \frac{\int_0^{2\pi} \sin^{9/4}(x/2)\, dx}{\int_0^{2\pi} \sin^3(x/2)\, dx}.$$

5 Two-sided radial $SLE_{8/3}$

In this section we will define two-sided radial $SLE_{8/3}$ by weighting two independent $SLE_{8/3}$ paths by a two-parameter martingale $N_{s,r}/N_{0,0}$. We will show that $N_{s,r}$ is symmetric in s and r, and this will give commutation, meaning that we can "grow" the two curves in either order.

We begin by establishing the notation we will use. Let γ and $\hat{\gamma}$ be independent radial $SLE_{8/3}$ paths starting at z and $\hat{z}$ with corresponding conformal maps g_t and $\hat{g}_t$. Let $\mathcal{F}_{s,r}$ denote the σ-algebra generated by $\{B_t : 0 \leq t \leq s\} \cup \{\hat{B}_t : 0 \leq t \leq r\}$, let $E_{t,r}$ denote the event

$$E_{t,r} = \{\gamma[0,t] \cap \hat{\gamma}[0,r] = \emptyset\},$$

and let $E_t = E_{t,\infty}$. In Proposition 4.1 we showed that

$$\lim_{t\to\infty} e^{11t/64}\, \mathbb{P}(E_t) = C_*\, |z - \hat{z}|^{3/4}.$$

On the event $E_{s,r}$, we let $v_{s,r}$ denote the unique conformal transformation of $\mathbb{D} \setminus (\gamma(0,s] \cup \hat{\gamma}(0,r])$ onto $\mathbb{D}$ with $v_{s,r}(0) = 0$ and $v'_{s,r}(0) > 0$. Note that $g_s = v_{s,0}$ and $\hat{g}_r = v_{0,r}$. For $U_{s,r} := v_{s,r}(\gamma(s))$ and $\hat{U}_{s,r} := v_{s,r}(\hat{\gamma}(r))$, we observe that $U_{s,0} = e^{iB_s}$, $\hat{U}_{s,0} = g_s(\hat{z})$, $U_{0,r} = \hat{g}_r(z)$, and $\hat{U}_{0,r} = e^{i\hat{B}_r}$. Finally, define $g_{s,r}$ and $\hat{g}_{s,r}$ by the relations $v_{s,r} = g_{s,r} \circ \hat{g}_r = \hat{g}_{s,r} \circ g_s$. See Figure 1.

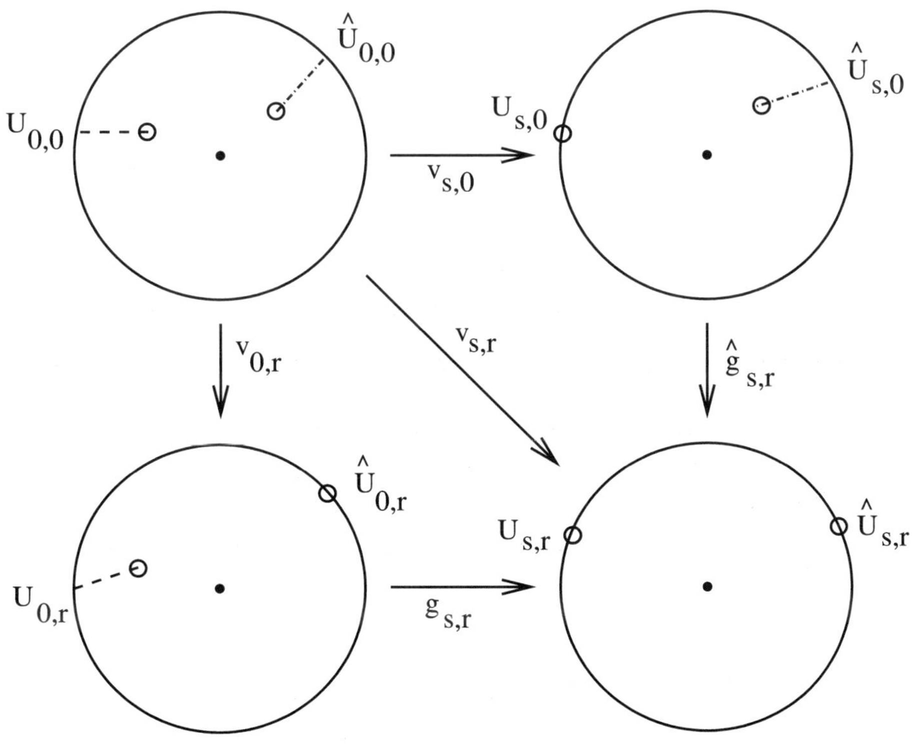

Figure 1 The maps $v_{s,r}, g_{s,r}, \hat{g}_{s,r}$. Note that $g_s = g_{s,0} = v_{s,0}$ and similarly $\hat{g}_r = \hat{g}_{0,r} = v_{0,r}$.

Suppose $0 \leq s \leq t, 0 \leq r < \infty$, and let

$$N_{s,r} := \lim_{t \to \infty} e^{11t/64}\, \mathbb{P}(E_t \mid \mathcal{F}_{s,r}).$$

Although it is not immediately obvious that the limit exists and that the definition of $N_{s,r}$ is symmetric, that is, it does not depend the way we order the two $SLE_{8/3}$'s, the next lemma establishes these facts.

Lemma 5.1

$$\begin{aligned} N_{s,r} \quad &= \quad C_* \, 1_{E_{s,r}} \, g_{s,r}'(0)^{5/48} \, |g_{s,r}'(\hat{U}_{0,r})|^{5/8} \, \hat{g}_{s,r}'(0)^{5/48} \, |\hat{g}_{s,r}'(U_{s,0})|^{5/8} \\ &\qquad \times |U_{s,r} - \hat{U}_{s,r}|^{3/4} \, v_{s,r}'(0)^{11/24}. \end{aligned}$$

In particular, $N_{s,r}$ is a two-parameter martingale in the sense that if $s \leq s'$ and $r \leq r'$,

$$\mathbb{E}[N_{s',r'} \mid \mathcal{F}_{s,r}] = N_{s,r}. \tag{5.1}$$

Moreover, there is a c such that for all s and r, $N_{s,r} \leq c\, e^{11(r+s)/64}$.

Proof We first write

$$N_{s,r} = \lim_{t \to \infty} e^{11t/64}\, \mathbb{E}[\mathbb{P}(E_t \mid \mathcal{F}_{t,r}) \mid \mathcal{F}_{s,r}].$$

The restriction property implies that

$$\mathbb{P}[E_t \mid \mathcal{F}_{t,r}] = 1_{E_{t,r}} \, g_{t,r}'(0)^{5/48} \, |g_{t,r}'(\hat{U}_{0,r})|^{5/8},$$

and hence,

$$\begin{aligned} N_{s,r} &= \lim_{t \to \infty} e^{11t/64}\, \mathbb{E}[1_{E_{t,r}} \, g_{t,r}'(0)^{5/48} \, |g_{t,r}'(\hat{U}_{0,r})|^{5/8} \mid \mathcal{F}_{s,r}] \\ &= \lim_{t \to \infty} \mathbb{P}[E_{t,r} \mid \mathcal{F}_{s,r}] \, e^{11t/64}\, \mathbb{E}[g_{t,r}'(0)^{5/48} \, |g_{t,r}'(\hat{U}_{0,r})|^{5/8} \mid \mathcal{F}_{s,r}, \, E_{t,r}]. \end{aligned}$$

Another application of the restriction property gives

$$\lim_{t \to \infty} \mathbb{P}[E_{t,r} \mid \mathcal{F}_{s,r}] = \mathbb{P}[E_{\infty,r} \mid \mathcal{F}_{s,r}] = 1_{E_{s,r}} \, \hat{g}_{s,r}'(0)^{5/48} \, |\hat{g}_{s,r}'(U_{s,0})|^{5/8}.$$

On the event $E_{t,r}$, let $u_{t,s,r} = v_{t,r} \circ v_{s,r}^{-1}$, so that $g_{t,r} = u_{t,s,r} \circ g_{s,r}$. Therefore,

$$\begin{aligned} N_{s,r} =& 1_{E_{s,r}} \, \hat{g}_{s,r}'(0)^{5/48} \, |\hat{g}_{s,r}'(U_{s,0})|^{5/8} \, g_{s,r}'(0)^{5/48} \, |g_{s,r}'(\hat{U}_{0,r})|^{5/8} \\ &\cdot \lim_{t \to \infty} e^{11t/64}\, \mathbb{E}[u_{t,s,r}'(0)^{5/48} \, |u_{t,s,r}'(\hat{U}_{s,r})|^{5/8} \mid \mathcal{F}_{s,r}, \, E_{t,r}]. \end{aligned}$$

Using the restriction property a third time, we can rewrite the above expectation as

$$\mathbb{P}\{\text{a } SLE_{8/3} \text{ path from } \hat{U}_{s,r} \text{ to } 0 \text{ avoids } v_{s,r} \circ \gamma[s,t]\}.$$

If we reparametrize the curve $v_{s,r} \circ \gamma[s,\tau]$ so that $u_{\tau,s,r}'(0) = e^{3\tau/8}$, we will be able to apply Proposition 4.1 to obtain

$$\begin{aligned} &\lim_{\tau \to 0} e^{11\tau/64}\, \mathbb{P}\{\text{a } SLE_{8/3} \text{ path from } \hat{U}_{s,r} \text{ to } 0 \text{ avoids } v_{s,r} \circ \gamma[s,\tau]\} \\ &\qquad = C_* \, |U_{s,r} - \hat{U}_{s,r}|^{3/4}. \end{aligned}$$

Thus,

$$\begin{aligned} &\lim_{t \to \infty} e^{11t/64}\, \mathbb{E}[u_{t,s,r}'(0)^{5/48} \, |u_{t,s,r}'(\hat{U}_{s,r})|^{5/8} \mid \mathcal{F}_{s,r}, \, E_{t,r}] \\ &\qquad = C_* \, |U_{s,r} - \hat{U}_{s,r}|^{3/4} \lim_{t \to \infty} e^{11t/64} \, \left(u_{t,r,s}'(0)\right)^{-11/24} \\ &\qquad = C_* \, |U_{s,r} - \hat{U}_{s,r}|^{3/4} \, v_{s,r}'(0)^{11/24}, \end{aligned}$$

since $u'_{t,r,s}(0) = v'_{t,r}(0)/v'_{s,r}(0)$ and $v'_{t,r}(0) \sim g'_t(0) = e^{3t/8}$ as t approaches infinity with r fixed. The exponent $11/24$ results from $(11/64)/(3/8)$. We have now shown that $N_{s,r}$ has the form we claimed.

From the restriction property, we can see that

$$[g'_{s,r}(0)^{5/48}\,|g'_{s,r}(\hat{U}_{0,r})|^{5/8}]\,[\hat{g}'_{s,r}(0)^{5/48}\,|\hat{g}'_{s,r}(U_{s,0})|^{5/8}] \leq 1,$$

since the terms in brackets each represent probabilities of events. Also $v'_{s,r}(0) \leq c\,e^{3(r+s)/8}$, implying that $|N_{s,r}| \leq c\,e^{11(r+s)/64}$. The relation (5.1) is immediate. □

Notice that if $r = 0$, $v_{s,0} = g_{s,0} = g_s$ and $\hat{g}_{s,0}$ is the identity. Hence,

$$N_{s,0} = C_*\,e^{27s/128}\,|g'_s(\hat{z})|^{5/8}\,|e^{iB_s} - g_s(\hat{z})|^{3/4},$$

which is equal to the martingale M_s that we considered in the previous section.

With Lemma 5.1 behind us, we will now define two-sided radial $SLE_{8/3}$:

Definition. If $z, \hat{z} \in \partial\mathbb{D}$ are distinct points, then two-sided radial $SLE_{8/3}$ in $\mathbb{D}$ starting at $(z, \hat{z})$ is the measure on ordered pairs of paths $(\gamma, \hat{\gamma})$ such that for each $s, r < \infty$, the distribution of

$$\gamma(s'), 0 \leq s \leq s', \qquad \hat{\gamma}(r'), 0 \leq r \leq r'$$

is given by saying that the Radon-Nikodym derivative of this distribution with respect to that of *independent* radial SLE's starting at $z, \hat{z}$ is $N_{s,r}/N_{0,0}$, which is equal to

$$1_{E_{s,r}}\, g'_{s,r}(0)^{5/48}\,|g'_{s,r}(\hat{U}_{0,r})|^{5/8}\,\hat{g}'_{s,r}(0)^{5/48}\,|\hat{g}'_{s,r}(U_{s,0})|^{5/8}\,\frac{|U_{s,r} - \hat{U}_{s,r}|^{3/4}}{|z - \hat{z}|^{3/4}}\,v'_{s,r}(0)^{11/24}.$$

Note that we can write

$$\frac{N_{s,r}}{N_{0,0}} = \frac{N_{s,0}}{N_{0,0}}\,\frac{N_{s,r}}{N_{s,0}} = \frac{N_{0,r}}{N_{0,0}}\,\frac{N_{s,r}}{N_{0,r}}. \tag{5.2}$$

In other words we can grow the first path and then the second or the second path then the first, and either way we get the same distribution. Since

$$N_{s,0} = M_s = C_*\,e^{27s/128}\,|g'_s(\hat{z})|^{5/8}\,|e^{iB_s} - g_s(\hat{z})|^{3/4},$$

growing one of the paths corresponds to growing one side of two-sided $SLE_{8/3}$, which was the topic of the previous section. Again, we can describe this in terms of the Loewner equation. Let $U_{s,0} = e^{iX_s}, \hat{U}_{s,0} = e^{i\hat{X}_s}$. If X_s and $\hat{X}_s$ satisfy

$$dX_s = -\frac{3}{8}\cot\left(\frac{\hat{X}_s - X_s}{2}\right)ds + dB_s, \qquad d\hat{X}_s = \frac{3}{8}\cot\left(\frac{\hat{X}_s - X_s}{2}\right)ds, \tag{5.3}$$

then the Loewner chain driven by X_s gives the two-sided $SLE_{8/3}$ up to time $(s, 0)$. To get the second path we map down by g_s and then proceed similarly, interchanging the roles of the two paths. In this case we need to go until a time that depends on the path γ.

We will use the notation $(\gamma, \hat{\gamma})$ to denote two-sided $SLE_{8/3}$ starting at $(z, \hat{z})$. By a slight abuse of notation, we will write just γ for $\gamma[0, \infty]$ and $(\gamma, \hat{\gamma})$ for $\gamma[0, \infty] \cup \hat{\gamma}[0, \infty]$. Note that these curves are defined modulo reparametrization.

Conditional distribution of $\hat{\gamma}$ given γ. Suppose that t is large and we have generated $\gamma[0,t]$ according to the distribution of one side of two-sided radial $SLE_{8/3}$. What is the conditional distribution of $\hat{\gamma}[0,1]$ given this? By (5.2), we can see that the Radon-Nikodym derivative of this conditional measure with respect to that of radial $SLE_{8/3}$ run until time 1 is $N_{t,1}/N_{t,0}$, i.e.,

$$1_{E_{t,1}} \frac{g_{t,1}'(0)^{5/48}\,|g_{t,1}'(\hat{U}_{0,1})|^{5/8}\,\hat{g}_{t,1}'(0)^{5/48}\,|\hat{g}_{t,1}'(U_{t,0})|^{5/8}\,|U_{t,1}-\hat{U}_{t,1}|^{3/4}\,v_{t,1}'(0)^{11/24}}{e^{27t/128}\,|g_t'(\hat{z})|^{5/8}\,|e^{iB_t}-g_t(\hat{z})|^{3/4},}. \tag{5.4}$$

Given γ, we can consider three distributions of $\hat{\gamma}[0,1]$:

- radial $SLE_{8/3}$ in $\mathbb{D}$ weighted by $N_{t,1}/N_{t,0}$;
- radial $SLE_{8/3}$ in $\mathbb{D}\setminus\gamma[0,t]$ from $\hat{z}$ to 0;
- chordal $SLE_{8/3}$ in $\mathbb{D}\setminus(\gamma[0,\infty)\cup\{0\})$ from $\hat{z}$ to $\{0\}$.

We would like to show that these three distributions on $\hat{\gamma}[0,1]$ are asymptotically the same as $t\to\infty$.

Let us first consider the second distribution. The restriction property states that the distribution of radial $SLE_{8/3}$ in $\mathbb{D}\setminus\gamma[0,t]$ from $\hat{z}$ to 0 is exactly the same as that of radial $SLE_{8/3}$ in $\mathbb{D}$ from $\hat{z}$ to 0 conditoned on the event $E=\{\gamma[0,t]\cap\hat{\gamma}[0,\infty)=\emptyset\}$. From this we see that the Radon-Nikodym derivative of the second distribution with respect to that of radial $SLE_{8/3}$ in $\mathbb{D}$ from $\hat{z}$ to 0 is given by

$$\frac{1\{\hat{\gamma}[0,1]\cap\gamma[0,t]=\emptyset\}\;\mathbb{P}\{\hat{\gamma}[1,\infty)\cap\gamma[0,t]=\emptyset\mid\hat{\gamma}[0,1]\}}{\mathbb{P}\{\gamma[0,t]\cap\hat{\gamma}[0,\infty)=\emptyset\}},$$

which equals

$$1_{E_{t,1}}\frac{g_{t,1}'(0)^{5/48}\,|g_{t,1}'(\hat{U}_{0,1})|^{5/8}}{g_{t,0}'(0)^{5/48}\,|g_{t,0}'(\hat{z})|^{5/8}} = 1_{E_{t,1}}\frac{g_{t,1}'(0)^{5/48}\,|g_{t,1}'(\hat{U}_{0,1})|^{5/8}}{e^{5t/128}\,|g_{t,0}'(\hat{z})|^{5/8}}. \tag{5.5}$$

If we recall that $v_{t,1}'(0)=\hat{g}_{t,1}'(0)\,g_{t,0}'(0)=e^{3t/8}\,\hat{g}_{t,1}'(0)$, we can see that the quantity in (5.4) equals (5.5) times

$$\hat{g}_{t,1}'(0)^{27/48}\,|\hat{g}_{t,1}'(U_{t,0})|^{5/8}\,\frac{|U_{t,1}-\hat{U}_{t,1}|^{3/4}}{|U_{t,0}-\hat{U}_{t,0}|^{3/4}}.$$

The following lemma implies that this expression approaches one as as $t\to\infty$.

Lemma 5.2 For every $\epsilon>0$, there exists an $r>0$ such that the following holds. Suppose $\gamma:(0,1]\to\mathbb{D},\hat{\gamma}:(0,1]\to\mathbb{D}$ are two simple curves with $\gamma(0+),\hat{\gamma}(0+)\in\partial\mathbb{D}$ and $\gamma(0,1]\cap\hat{\gamma}(0,1]=\emptyset$. Suppose $|\hat{\gamma}(t)|\geq 1/4$ for all t; $0\notin\gamma(0,1]$ and $|\gamma(1)|\leq r$. Let g denote the unique conformal transformation of $\mathbb{D}\setminus\gamma(0,1]$ onto $\mathbb{D}$ with $g(0)=0, g'(0)>0$. Let $A=g(\hat{\gamma}(0,1])$. Let h denote the unique conformal transformation of $\mathbb{D}\setminus A$ onto $\mathbb{D}$ with $h(0)=0,h'(0)>0$. Let $z=g(\gamma(1)),w=g(\hat{\gamma}(0)),z^*=h(z),w^*=h(g(\hat{\gamma}(1)))$. Then,

$$1\leq h'(0)\leq 1+\epsilon,\qquad \big|\,|h'(z)|-1\,\big|\leq\epsilon,\qquad\text{and}\qquad \left|\frac{|z-w|}{|z^*-w^*|}-1\right|\leq\epsilon.$$

We will not give the details of this proof, but we note that the key estimate is the Beurling estimate (see, e.g., [4, Section 3.8]). This can be used to show that there is a c such that

$$\mathrm{diam}(A)\leq c\,r^{1/2}\,\mathrm{dist}(z,A).$$

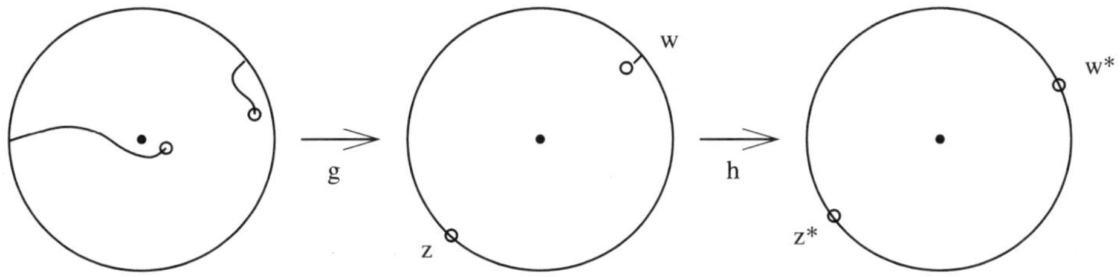

Figure 2 The maps g and h in Lemma 5.2.

A similar argument with more details is given in Section 8, where we also discuss the asymptotic equivalence of the second and third distributions.

6 Restriction property for two-sided radial $SLE_{8/3}$

As one might expect, two-sided radial $SLE_{8/3}$ satisfies the restriction property, which means that two-sided radial $SLE_{8/3}$ conditioned to stay in a subdomain of $\mathbb{D}$ is two-sided radial $SLE_{8/3}$ in that subdomain. More precisely, the restriction property states that the conditional distribution of $\Psi_A(\gamma, \hat{\gamma})$ given $\{(\gamma, \hat{\gamma}) \cap A = \emptyset\}$ is the same as two-sided $SLE_{8/3}$ from $(\Psi_A(e^{ix}), \Psi_A(e^{i\hat{x}}))$ to 0 in $\mathbb{D}$, where $(\gamma, \hat{\gamma})$ is a two-sided $SLE_{8/3}$ started at $(e^{ix}, e^{i\hat{x}})$ and $A \in \mathcal{A}(x) \cap \mathcal{A}(\hat{x})$. Here we have continued to use our notation from the second section, that is, $\mathcal{A}(x)$ denotes the set of $A \in \mathbb{D}$ such that $\mathbb{D} \setminus A$ is a simply connected domain containing the origin with $\text{dist}(e^{ix}, A) > 0$, D_A denotes $\mathbb{D} \setminus A$, and $\Psi_A : D_A \to \mathbb{D}$ is the unique conformal transformation with $\Psi_A(0) = 0$, and $\Psi_A'(0) > 0$.

In this section we prove the restriction property for two-sided radial $SLE_{8/3}$. The proof follows the same general outline as in the case of one-sided radial or chordal $SLE_{8/3}$. The main difference is that the martingale we will consider is more complicated, and hence our calculations using Itô calculus will be more involved. The following theorem contains the essential result for the restriction property.

Theorem 1 Suppose $x, \hat{x} \in \mathbb{R}$ and $A \in \mathcal{A}(x) \cap \mathcal{A}(\hat{x})$. If $(\gamma, \hat{\gamma})$ denotes two-sided radial $SLE_{8/3}$ from $(e^{ix}, e^{i\hat{x}})$ to 0, then

$$\mathbb{P}\{(\gamma, \hat{\gamma}) \cap A = \emptyset\} = \Psi_A'(0)^{2/3} \, |\Psi_A'(e^{ix}) \, \Psi_A'(e^{i\hat{x}})|^{5/8} \left| \frac{\Psi_A(e^{i\hat{x}}) - \Psi_A(e^{ix})}{e^{i\hat{x}} - e^{ix}} \right|^{3/4}.$$

If $x = \hat{x}$, this is to be interpreted as

$$\Psi_A'(0)^{2/3} \, |\Psi_A'(e^{ix})|^2.$$

We first discuss how this implies the restriction property. To specify the distribution of a pair of simple, non-intersecting paths (η_1, η_2) from (e^{x_1}, e^{x_2}) to 0, it suffices to give $\mathbb{P}\{(\eta_1, \eta_2) \cap K = \emptyset\}$ for each $K \in \mathcal{A}(x_1) \cap \mathcal{A}(x_2)$. Thus, to prove the restriction property from the previous theorem, we need to show that for $A \in \mathcal{A}(x_1) \cap \mathcal{A}(x_2)$, then $P_0 := \mathbb{P}\{\Psi_A(\gamma, \hat{\gamma}) \cap K = \emptyset \mid (\gamma, \hat{\gamma}) \cap A = \emptyset\}$ is

$$\Psi_K'(0)^{2/3} \, |\Psi_K'(z) \, \Psi_K'(\hat{z})|^{5/8} \left| \frac{\Psi_K(\hat{z}) - \Psi_K(z)}{\hat{z} - z} \right|^{3/4}, \tag{6.1}$$

where $z = \Psi_A(e^{ix})$ and $\hat{z} = \Psi_A(e^{i\hat{x}})$. We first note that P_0 is equal to

$$\frac{\mathbb{P}\{(\gamma,\hat{\gamma}) \cap (A \cup \Psi_A^{-1}(K)) = \emptyset\}}{\mathbb{P}\{(\gamma,\hat{\gamma}) \cap A = \emptyset\}}.$$

Since $\Psi_{A\cup\Psi_A^{-1}(K)} = \Psi_K \circ \Psi_A$, we obtain (6.1) from another application of Theorem 1.

In order to prove Theorem 1, we start with a simple lemma.

Lemma 6.1 There is a $c < \infty$ such that if $A, x, \hat{x}$, and Ψ_A are as in the theorem, then

$$\Psi_A'(0)^{2/3}\,|\Psi_A'(e^{ix})|^{5/8}\,|\Psi_A'(e^{i\hat{x}})|^{5/8}\left|\frac{\Psi_A(e^{ix}) - \Psi_A(e^{i\hat{x}})}{e^{ix} - e^{i\hat{x}}}\right|^{3/4} \leq c\,r^{1/3},$$

where $r = \text{inrad}(\mathbb{D} \setminus A) = \text{dist}(0, A)$.

Proof If $f : \mathbb{D} \to f(\mathbb{D})$ is a conformal transformation with $f(0) = 0$, the Koebe (1/4)-theorem and the Schwarz lemma imply that $1 \leq f'(0)/\text{dist}[0, \partial f(\mathbb{D})] \leq 4$. Applying this to $f = \Psi_A^{-1}$ gives

$$\frac{1}{4r} \leq \Psi_A'(0) \leq \frac{1}{r}.$$

Suppose $I \subset \partial\mathbb{D}$ and let $h(A; I)$ denote the harmonic measure of I in $\mathbb{D} \setminus A$ from 0; in other words, $h(A; I)$ is the probability that a Brownian motion starting at the origin leaves $\mathbb{D}$ at I. The Beurling estimate implies that there is a c such that the probability that a Brownian motion starting at 0 reaches $\{|z| = 1/2\}$ without leaving $\mathbb{D}$ is at most $c\,r^{1/2}$. The probability that a Brownian motion starting at $\{|z| = 1/2\}$ leaves $\mathbb{D}$ at I is bounded by $c\,l(I)$ where l denotes length. Therefore

$$h(A; I) \leq c\,r^{1/2}\,l(I)$$

which implies

$$\left|\frac{\Psi_A(z) - \Psi_A(w)}{z - w}\right| \leq c\,r^{1/2}.$$

and $|\Psi_A'(z)| \leq c\,r^{1/2}$. □

The proof continues as in the one-sided case. The basic idea is to show the equality of two random variables: the first is a martingale $\tilde{M}_t$ that is equal to $\mathbb{P}\{(\gamma,\hat{\gamma}) \cap A = \emptyset\}$ when $t = 0$, and the second is our "martingale candidate" M_t, which has initial value

$$\Psi_A'(0)^{2/3}\,|\Psi_A'(e^{ix})\,\Psi_A'(e^{i\hat{x}})|^{5/8}\left|\frac{\Psi_A(e^{i\hat{x}}) - \Psi_A(e^{ix})}{e^{i\hat{x}} - e^{ix}}\right|^{3/4}. \tag{6.2}$$

We will think of generating $(\gamma, \hat{\gamma})$ in two steps. First we obtain γ by solving the Loewner equation with the driving term X_t, where X_t is described by (5.3). Then we take $\hat{\gamma}$ to be chordal $SLE_{8/3}$ in $\mathbb{D} \setminus \gamma$. As discussed in the previous subsection, this is one of the ways of obtaining our two-sided process. Let $\mathcal{F}_t$ denote the filtration generated by X_t, and set

$$\tilde{M}_t := \mathbb{E}\left[1_{\{\gamma\cap A=\emptyset\}}\,\mathbb{P}\{\hat{\gamma} \cap A = \emptyset \mid \gamma\} \mid \mathcal{F}_t\right].$$

Note that $\tilde{M}_0 = \mathbb{P}\{(\gamma,\hat{\gamma}) \cap A = \emptyset\}$ and $\tilde{M}_t$ is a continuous, bounded martingale. Additionally,

$$\lim_{t\to\infty} \tilde{M}_t = 1_{\{\gamma\cap A=\emptyset\}}\mathbb{P}\{\hat{\gamma} \cap A = \emptyset \mid \gamma\}.$$

In what follows, we will define our "martingale candidate" M_t, which will satisfy M_0 equal to (6.2), and we will show that M_t is also a continuous, bounded martingale with the same limit at infinity as $\tilde{M}_t$. This will imply that $M_t = \tilde{M}_t$, completing the proof. The most tedious part of the work, which is contained in the subsection below, is the calculation to show that M_t actually is a martingale.

The martingale calculation. As usual, we begin by establishing the notation we will use. Set $X_0 = e^{ix}$ and $\hat{X}_0 = e^{i\hat{x}}$, and let X_t and $\hat{X}_t$ satisfy

$$dX_t = -\frac{3}{8}\,K_t\,dt + dB_t, \qquad d\hat{X}_t = \frac{3}{8}\,K_t\,dt,$$

where we write

$$K_t = \cot\left[\frac{\hat{X}_t - X_t}{2}\right].$$

Recall that the Loewner equation with driving term X_t generates one side of two-sided $SLE_{8/3}$, and we will refer to this curve as γ. Let g_t be the conformal maps associated with this Loewner chain. That is,

$$\dot{g}_t(z) = \frac{3}{8}\,g_t(z)\,\frac{e^{iX_t} + g_t(z)}{e^{iX_t} - g_t(z)}.$$

As before, we let $h_t(z) = -i\log g_t(e^{iz})$. Then, as is discussed in Section 4.6 of [4], on the event $\{\gamma[0,t]\cap A = \emptyset\}$ we take Ψ_t and Φ_t to be conformal perturbations of the Loewner chains with $\Psi_0 = \Psi_A$. More specifically, if g_t^* is the the conformal map from $\mathbb{D}\setminus\Psi_A\circ\gamma[0,t]$ onto $\mathbb{D}$ with $g_t^*(0) = 0$ and $(g_t^*)'(0) > 0$, then $\Psi_t = g_t^*\circ\Psi_A\circ g_t^{-1}$ and $\Phi_t(z) = -i\log\Psi_t(e^{iz})$. Finally, let $\hat{X}_t^* = \Phi_t(\hat{X}_t)$ and $X_t^* = \Phi_t(X_t)$, and take

$$K_t^* = \cot\left[\frac{\hat{X}_t^* - X_t^*}{2}\right], \qquad G_t = \csc^2\left[\frac{\hat{X}_t - X_t}{2}\right], \qquad G_t^* = \csc^2\left[\frac{\hat{X}_t^* - X_t^*}{2}\right].$$

We will make use of the following five equations from Section 4.6 of [4]:

$$\dot{\Phi}_t(X_t) = -\frac{9}{8}\Phi_t''(X_t),$$

$$\dot{\Phi}_t(\hat{X}_t) = \frac{3}{8}\left[\Phi_t'(X_t)^2\,K_t^* - \Phi_t'(\hat{X}_t)\,K_t\right],$$

$$\dot{\Phi}_t'(X_t) = \frac{3}{8}\,\Phi_t'(X_t)\left[\frac{\Phi_t''(X_t)^2}{2\,\Phi_t'(X_t)^2} - \frac{4\,\Phi_t'''(X_t)}{3\,\Phi_t'(X_t)} + \frac{1-\Phi_t'(X_t)^2}{6}\right],$$

$$\dot{\Phi}_t'(\hat{X}_t) = \frac{3}{8}\left[-\frac{1}{2}\,\Phi_t'(X_t)^2\,\Phi_t'(\hat{X}_t)\,G_t^* - \Phi_t''(\hat{X}_t)\,K_t + \frac{1}{2}\,\Phi_t'(\hat{X}_t)\,G_t\right],$$

$$\dot{\Psi}_t'(0) = \frac{3}{8}(\Phi_t'(X_t)^2 - 1)\,\Psi_t'(0).$$

Note that these differ from the results in [4] by a factor of 3/8 because of our choice of parametrization for γ.

We can now state our martingale candidate:

$$M_t := 1_{\{\gamma[0,t]\cap A=\emptyset\}}\,\Psi_t'(0)^{2/3}\,\Phi_t'(X_t)^{5/8}\,\Phi_t'(\hat{X}_t)^{5/8}\,\frac{F(\hat{X}_t^* - X_t^*)}{F(\hat{X}_t - X_t)},$$

where $F(x) = \sin^{3/4}(x/2)$. Equiped with the tools of Itô calculus, we wish to show that M_t is a continuous martingale.

We begin by considering the three derivative terms in M_t. We first compute that

$$d[\Psi_t'(0)^{2/3}] = \frac{1}{4}\,\Psi_t'(0)^{2/3}\,(\Phi_t'(X_t)^2 - 1)\,dt.$$

Using the standard chain rule for functions of two variables, we obtain next that

$$\begin{aligned} d[\Phi_t'(\hat{X}_t)^b] &= b\,\Phi_t'(\hat{X}_t)^b\,\frac{\dot{\Phi}_t'(\hat{X}_t)\,dt + \Phi_t''(\hat{X}_t)\,d\hat{X}_t}{\Phi_t'(\hat{X}_t)} \\ &= \Phi_t'(\hat{X}_t)^b\left[-\frac{3b}{16}\,\Phi_t'(X_t)^2\,G_t^* + \frac{3b}{16}\,G_t\right]dt, \end{aligned}$$

and so,

$$d[\Phi_t'(\hat{X}_t)^{5/8}] = \Phi_t'(\hat{X}_t)^{5/8}\left[-\frac{15}{128}\,\Phi_t'(X_t)^2\,G_t^* + \frac{15}{128}\,G_t\right]dt.$$

For the X_t term we need to use Itô's formula, which tells us that

$$d[\Phi_t'(X_t)] = \dot{\Phi}_t'(X_t)\,dt + \Phi_t''(X_t)\,dX_t + \frac{1}{2}\Phi_t'''(X_t)\,dt.$$

Hence,

$$\begin{aligned} &d[\Phi_t'(X_t)] \\ &= \left[\frac{3}{16}\frac{\Phi_t''(X_t)^2}{\Phi_t'(X_t)} + \frac{1}{16}\,\Phi_t'(X_t)\,[1 - \Phi_t'(X_t)^2] - \frac{3}{8}K_t\,\Phi_t''(X_t)\right]dt + \Phi_t''(X_t)\,dB_t. \end{aligned}$$

From another use of Itô's formula, we obtain

$$\begin{aligned} &d[\Phi_t'(X_t)^{5/8}] \\ &= \Phi_t'(X_t)^{5/8}\left\{\left[\frac{5}{128}\,[1 - \Phi_t'(X_t)^2] - \frac{15}{64}\,K_t\,\frac{\Phi_t''(X_t)}{\Phi_t'(X_t)}\right]dt + \frac{5}{8}\,\frac{\Phi_t''(X_t)}{\Phi_t'(X_t)}\,dB_t\right\}. \end{aligned}$$

Combining all of this gives that

$$d[\Psi_t'(0)^{2/3}\,\Phi_t'(\hat{X}_t)^{5/8}\,\Phi_t'(X_t)^{5/8}]$$

is $\Psi_t'(0)^{2/3}\,\Phi_t'(\hat{X}_t)^{5/8}\,\Phi_t'(X_t)^{5/8}$ times

$$\begin{aligned} &\left[-\frac{27}{128}\,(1 - \Phi_t'(X_t)^2) - \frac{15}{64}\,K_t\,\frac{\Phi_t''(X_t)}{\Phi_t'(X_t)} - \frac{15}{128}\,\Phi_t'(X_t)^2\,G_t^* + \frac{15}{128}\,G_t\right]dt \\ &\qquad + \frac{5}{8}\,\frac{\Phi_t''(X_t)}{\Phi_t'(X_t)}\,dB_t. \end{aligned}$$

We now turn our attention to the terms involving $F(x) = \sin^{3/4}(x/2)$. Note that

$$F'(x) = \frac{3}{8}\,\cot(x/2)\,F(x), \qquad F''(x) = \left[-\frac{3}{64}\,\csc^2(x/2) - \frac{9}{64}\right]F(x).$$

If $f(x) = 1/F(x) = \sin^{-3/4}(x/2)$, then

$$f'(x) = -\frac{3}{8}\,\cot(x/2)\,f(x), \qquad f''(x) = \left[\frac{21}{64}\csc^2(x/2) - \frac{9}{64}\right]f(x).$$

Let $Z_t = \hat{X}_t - X_t$, and $Z_t^* = \hat{X}_t^* - X_t^*$. Then,

$$dZ_t = \frac{3}{4}K_t\,dt - dB_t,$$

and we have

$$d[f(Z_t)] = f'(Z_t)\, dZ_t + \frac{1}{2} f''(Z_t)\, dt$$
$$= f(Z_t) \left\{ \left[-\frac{15}{128}\, G_t + \frac{27}{128} \right] dt + \frac{3}{8} K_t dB_t \right\},$$

where we have made use of the trig identity $1 + K_t^2 = G_t$.

The last term we need to compute is $d[F(Z_t^*)]$, and to do this, we must first compute $d\hat{X}^*$ and dX^*:

$$d\hat{X}_t^* = \dot{\Phi}_t(\hat{X}_t)\, dt + \Phi_t'(\hat{X}_t)\, d\hat{X}_t = \frac{3}{8}\, \Phi_t'(U_t)^2\, K_t^*\, dt.$$

$$dX_t^* = \dot{\Phi}_t(X_t)\, dt + \Phi_t'(X_t)\, dX_t + \frac{1}{2} \Phi_t''(X_t)\, dt$$
$$= \left[-\frac{5}{8}\, \Phi_t''(X_t) - \frac{3}{8}\, K_t\, \Phi_t'(X_t) \right] dt + \Phi_t'(X_t)\, dB_t.$$

Therefore,

$$dZ_t^* = \left[\frac{3}{8} \Phi_t'(X_t)^2\, K_t^* + \frac{5}{8}\, \Phi_t''(X_t) + \frac{3}{8}\, K_t\, \Phi_t'(X_t) \right] dt - \Phi_t'(X_t)\, dB_t,$$

and so,

$$d[F(Z_t^*)] = F'(Z_t^*)\, dZ_t^* + \frac{1}{2} F''(Z_t^*)\, \Phi_t'(X_t)^2\, dt.$$

Using that $1 + (K_t^*)^2 = G_t^*$, this simplifies to give

$$\begin{aligned} d[F(Z_t^*)] \;=\; F(Z_t^*) \Bigg\{ \bigg[&\frac{15}{128}\, G_t^*\, \Phi_t'(X_t)^2 - \frac{27}{128}\, \Phi_t'(X_t)^2 + \frac{15}{64}\, K_t^*\, \Phi_t''(X_t) \\ &+ \frac{9}{64}\, K_t\, K_t^*\, \Phi_t'(X_t) \bigg] dt - \frac{3}{8}\, K_t^*\, \Phi_t'(X_t)\, dB_t \Bigg\}. \end{aligned}$$

Now that we have computed $d[\Psi_t'(0)^{2/3}\, \Phi_t'(\hat{X}_t)^{5/8}\, \Phi_t'(X_t)^{5/8}]$, $d[f(Z_t)]$, and $d[F(Z_t^*)]$, we are ready to compute the drift of M_t on the event that $\{\gamma[0,t] \cap A = \emptyset\}$. We find that this is equal to zero, as desired.

To show that M_t is a continuous martingale, one must also check that the limit of M_t as t approaches t_A from below is zero, where t_A is the first time that $\gamma[0,t] \cap A \neq \emptyset$. We leave this to the reader. Note that Lemma 6.1 implies that M_t is bounded.

The behavior of the martingale at infinity. To finish our proof of the restriction property for two-sided radial $SLE_{8/3}$, we must show that M_t approaches $1_{\{\gamma \cap A = \emptyset\}} \mathbb{P}\{\hat{\gamma} \cap A = \emptyset \mid \gamma\}$ as $t \to \infty$. We first note that when t is large, A has small harmonic measure as viewed from zero in the domain $\mathbb{D} \setminus (\gamma[0,t] \cup A)$, and this implies that the harmonic measure of $g_t(A)$ in $\mathbb{D} \setminus g_t(A)$ is also small. Therefore, away from $g_t(A)$, the maps Ψ_t and Φ_t will be close to the identity, and so $\Psi_t'(0)^{2/3}$ and $\Phi_t'(X_t)^{5/8}$ will be close to 1. Since $e^{i\hat{X}_t}$ will be near to $g_t(A)$, we do not have immediate control over the derivative of Φ_t at $\hat{X}_t$. However, the small harmonic measure of $g_t(A)$ does imply that Φ_t cannot move $\hat{X}_t$ much, which gives that $F(Z_t^*)\, f(Z_t) = \frac{F(\hat{X}_t^* - X_t^*)}{F(\hat{X}_t - X_t)}$ is also close to 1.

The last step is to show that the remaining term, $\Phi_t'(\hat{X}_t)^{5/8}$, approaches $\mathbb{P}\{\hat{\gamma} \cap A = \emptyset \mid \gamma\}$ as $t \to \infty$. The underlying idea here is that if we take t to be large and just look at the part of the boundary of the disk near $g_t(A)$ and $e^{i\hat{X}_t}$, then our picture will look roughly like the upper halfplane, and in this setting $\Phi_t'(\hat{X}_t)^{5/8}$ gives the probability that a chordal $SLE_{8/3}$ started at $\hat{X}_t$ avoids $g_t(A)$. To make things more precise, we will use our result from Section 8 that tells us that the limit as $t \to \infty$ of radial $SLE_{8/3}$ in $\mathbb{D} \setminus \gamma[0,t]$ from $e^{i\hat{X}_0}$ to 0 is a chordal $SLE_{8/3}$ from $e^{i\hat{X}_0}$ to 0 in $\mathbb{D} \setminus (\gamma[0,\infty) \cup \{0\})$. Note that

$$\begin{aligned}\mathbb{P}\{&\text{a radial } SLE_{8/3} \text{ in } \mathbb{D} \setminus \gamma[0,t] \text{ from } e^{i\hat{X}_0} \text{ to 0 avoids } A \mid \gamma[0,t]\} \\ &= \mathbb{P}\{\text{a radial } SLE_{8/3} \text{ in } \mathbb{D} \text{ from } e^{i\hat{X}_t} \text{ to 0 avoids } g_t(A)\} \\ &= \Psi_t'(0)^{5/48}\Phi_t(\hat{X}_t)^{5/8}.\end{aligned}$$

Since $\Psi_t'(0)$ is approaching 1 and $\hat{\gamma}$ is precisely a chordal $SLE_{8/3}$ in $\mathbb{D}\setminus(\gamma[0,\infty)\cup\{0\})$ from $e^{i\hat{X}_0}$ to 0, we have that

$$\lim_{t\to\infty} \Phi_t(\hat{X}_t)^{5/8} = \mathbb{P}\{\hat{\gamma} \cap A = \emptyset \mid \gamma\},$$

as desired.

A connection to Brownian motion. We end our discussion by mentioning an alternate way to view two-sided radial $SLE_{8/3}$ started at $(1,1)$. We can obtain this process by taking the outer boundary of two independent Brownian motions from 0 to 1 in $\mathbb{D}$ that are conditioned not to disconnect 0 from $\partial\mathbb{D}$.

7 Two-sided chordal $SLE_{8/3}$

Although we have previously concerned ourselves only with the radial case, one can also define two-sided chordal $SLE_{8/3}$ and two-sided whole plane $SLE_{8/3}$. The chordal version of this process is actually one of the $SLE(8/3,\rho)$ processes, which were introduced in [6]. We will describe this connection after constructing the process and discussing the restriction property that it satisfies. Here we follow the same general outline as our discussion of the radial case: Given γ and $\hat{\gamma}$, two independent chordal $SLE_{8/3}$ processes, we first wish to understand $\mathbb{P}\{\gamma(0,t] \cap \gamma(0,\infty) = \emptyset\}$ as $t \to \infty$. From this we obtain a martingale, M_s, and weighting an $SLE_{8/3}$ by M_s/M_0 gives one side of the two-sided chordal process. Girsanov's Theorem allows us to describe this process via Loewner's equation. We finish with the definition of the two-sided process.

Let B_t and $\hat{B}_t$ be two independent standard Brownian motions with $B_0 < \hat{B}_0$. Let $\gamma, \hat{\gamma}, g_t$, and $\hat{g}_t$ be the corresponding $SLE_{8/3}$. In other words, g_t is the conformal transformation of $\mathbb{H} \setminus \gamma(0,t]$ onto $\mathbb{H}$ such that $g_t(z) - z \to 0$ as $z \to \infty$. It satisfies the Loewner equation

$$\dot{g}_t(z) = \frac{3/4}{g_t(z) - B_t}, \qquad g_0(z) = z. \tag{7.1}$$

Here γ has been parametrized so that $\text{hcap}(\gamma(0,t]) = 3t/4$, instead of $2t$. All the same holds of $\hat{\gamma}$ and $\hat{g}_t$.

By the restriction property for chordal $SLE_{8/3}$, we know that

$$\mathbb{P}\{\gamma(0,t] \cap \hat{\gamma}(0,\infty) = \emptyset \mid \gamma[0,t]\} = g_t'(\hat{B}_0)^{5/8},$$

and therefore,

$$\mathbb{P}\{\gamma(0,t] \cap \hat{\gamma}(0,\infty) = \emptyset\} = \mathbb{E}[g_t'(\hat{x})^{5/8}].$$

By differentiating (7.1), we see that

$$g_t'(\hat{B}_0)^{5/8} = \exp\left\{-\frac{15}{32}\int_0^t \frac{ds}{Y_s^2}\right\},$$

where $Y_t = g_t(\hat{B}_0) - B_t$ satisfies

$$dY_t = \frac{3/4}{Y_t}\,dt - dB_t.$$

We assume for ease that $B_0 = 0$, and we let $\phi(t,x) = \mathbb{E}[g_t'(x)^{5/8}]$ for $x > 0$. Since

$$\phi(T-t, Y_t)\,\exp\left\{-\frac{15}{32}\int_0^t \frac{ds}{Y_s^2}\right\}, \quad 0 \le t < T$$

is a martingale, Itô's formula shows that ϕ must satisfy

$$\dot{\phi}(t,x) = \frac{1}{2}\phi''(t,x) + \frac{3}{4x}\,\phi'(t,x) - \frac{15}{32\,x^2}\,\phi(t,x).$$

One could also obtain this differential equation from the Feynman-Kac formula. If $\psi(x) = \phi(1,x)$, then scaling implies that $\phi(t,x) = \psi(x/\sqrt{t})$. Letting $y = x/\sqrt{t}$, we see that

$$\psi''(y) + \left(y + \frac{3}{2y}\right)\,\psi'(y) - \frac{15}{16\,y^2}\,\psi(y) = 0.$$

We must have boundary conditions $\psi(\infty) = 1$ and $\psi(0) = 0$. The solution to this initial value problem, discussed in Appendix B.2 of [4], is $\psi(x) = x^{3/4} f(x)$, where

$$f(x) = e^{-x^2/2}\,\frac{\Gamma(13/8)}{2^{3/8}\,\Gamma(2)}\,\Phi(13/8, 2; x^2/2)$$

and Φ denotes the confluent hypergeometric function (of the first kind). The actual expression for f is unneeded, as all we will use is that $c_0 = f(0)$ is well-defined and non-zero. We have now established that

$$\mathbb{E}[g_t'(\hat{B}_0)^{5/8}] = \hat{B}_0^{3/4}\,t^{-3/8}\,f(\hat{B}_0/\sqrt{t}),$$

and therefore

$$\lim_{t\to\infty} t^{3/8}\,\mathbb{P}\{\gamma(0,t] \cap \hat{\gamma}(0,\infty) = \emptyset\} = c_0\,(\hat{B}_0 - B_0)^{3/4}. \tag{7.2}$$

Now we define

$$M_s := c_0\,g_s'(\hat{B}_0)^{5/8}\,Y_s^{3/4},$$

and a simple calculation show us that

$$dM_s = -\frac{3/4}{Y_s}\,M_s\,dB_s.$$

Equation (7.2) allows us to conclude that

$$M_s = \lim_{t\to\infty} t^{3/8}\,\mathbb{P}\{\gamma(0,t] \cap \hat{\gamma}(0,\infty) = \emptyset \mid \mathcal{F}_s\},$$

since

$$\begin{aligned}
&\lim_{t\to\infty} t^{3/8}\,\mathbb{P}\{\gamma(0,t] \cap \hat{\gamma}(0,\infty) = \emptyset \mid \mathcal{F}_s\} \\
&\quad = g_s'(\hat{B}_0)^{5/8}\,\lim_{t\to\infty}(t-s)^{3/8}\,\mathbb{P}\{g_s \circ \gamma(s,t] \cap g_s \circ \hat{\gamma}(0,\infty) = \emptyset\} \\
&\quad = g_s'(\hat{B}_0)^{5/8}\,c_0\,|g_s(\hat{B}_0) - B_s|^{3/4}.
\end{aligned}$$

It is this latter view of M_s that leads us to define one side of two-sided chordal $SLE_{8/3}$ as $SLE_{8/3}$ weighted by M_s/M_0. By making use of Girsanov's Theorem, this is the same as saying that one side of two-sided chordal $SLE_{8/3}$ is the process obtained from the chordal Loewner equation with driving term X_s, where W_s is a standard Brownian motion (with respect to the probability measure induced by M_s/M_0) and

$$dX_s = -\frac{3/4}{\hat{X}_s - X_s}\,ds + dW_s, \qquad d\hat{X}_s = \frac{3/4}{\hat{X}_s - X_s}\,ds.$$

Next we would like to define the general two-sided chordal process. As in the radial case, we will do so by weighting two independent chordal $SLE_{8/3}$ processes by a two-parameter martingale $N_{s,r}/N_{0,0}$. We define

$$N_{s,r} = \lim_{t\to\infty} t^{3/8}\,\mathbb{P}\{\gamma(0,t] \cap \hat{\gamma}(0,\infty) = \emptyset \mid \mathcal{F}_{s,r}\}.$$

In the lemma below, we will show that this limit exits and that $N_{s,r}$ is symmetric in s and r.

First we introduce some notation. On the event $E_{s,r} := \{\gamma(0,s] \cap \hat{\gamma}(0,r] = \emptyset\}$, let $v_{s,r}$ be the unique conformal transformation of $\mathbb{H} \setminus (\gamma(0,s] \cup \hat{\gamma}(0,r])$ such that $v_{s,r}(z) - z \to 0$ as $z \to \infty$. Let $U_{s,r} = v_{s,r}(\gamma(s))$ and $\hat{U}_{s,r} = v_{s,r}(\hat{\gamma}(r))$, and define $g_{s,r}$ and $\hat{g}_{s,r}$ by the relations $v_{s,r} = g_{s,r} \circ \hat{g}_r = \hat{g}_{s,r} \circ g_s$.

Lemma 7.1

$$N_{s,r} = c_0\, 1_{E_{s,r}}\, g'_{s,r}(\hat{U}_{0,r})^{5/8} \hat{g}'_{s,r}(U_{s,0})^{5/8}\, |\hat{U}_{s,r} - U_{s,r}|^{3/4}.$$

Proof We first write $N_{s,r}$ as

$$\lim_{t\to\infty} t^{3/8}\,\mathbb{E}[\mathbb{P}\{\gamma(0,t] \cap \hat{\gamma}(0,\infty) = \emptyset \mid \mathcal{F}_{t,r}\} \mid \mathcal{F}_{s,r}].$$

Then the restriction property implies that

$$\mathbb{P}\{\gamma(0,t] \cap \hat{\gamma}(0,\infty) = \emptyset \mid \mathcal{F}_{t,r}\} = 1_{E_{t,r}}\, g'_{t,r}(\hat{U}_{0,r})^{5/8},$$

and

$$\lim_{t\to\infty} \mathbb{E}[E_{t,r} \mid \mathcal{F}_{s,r}] = 1_{E_{s,r}}\, \hat{g}'_{s,r}(U_{s,o})^{5/8}.$$

Therefore,

$$N_{s,r} = 1_{E_{s,r}}\, \hat{g}'_{s,r}(U_{s,o})^{5/8} \lim_{t\to\infty} t^{3/8}\,\mathbb{E}[g'_{t,r}(\hat{U}_{0,r})^{5/8} \mid \mathcal{F}_{s,r}, E_{t,r}].$$

On the event $E_{t,r}$, we define $u_{t,s,r}$ by $g_{t,r} = u_{t,s,r} \circ g_{s,r}$, or equivalently by $v_{t,r} = u_{t,s,r} \circ v_{s,r}$, and we set $\eta(\tau) = v_{s,r} \circ \gamma(\tau + s)$ for $\tau \geq 0$. Notice that

$$\mathbb{E}[u'_{t,s,r}(U_{s,r})^{5/8} \mid \mathcal{F}_{s,r}, E_{t,r}] = \mathbb{P}\{\text{ a } SLE_{8/3} \text{ started at } \hat{U}_{s,r} \text{ avoids } \eta(0,t-s]\ \}$$

by a third use of the restriction property. In order to use (7.2) to conclude that

$$\lim_{t\to\infty} t^{3/8}\,\mathbb{E}[u'_{t,s,r}(U_{s,r})^{5/8} \mid \mathcal{F}_{s,r}, E_{t,r}] = c_0\, |\hat{U}_{s,r} - U_{s,r}|^{3/4},$$

we must have that $\lim_{t\to\infty} \frac{\text{hcap}(\eta(0,t-s])}{3t/4} = 1$. This, however, follows from the fact that $\text{hcap}(\eta(0,t-s]) = \text{hcap}(\gamma(0,t] \cup \hat{\gamma}(0,r]) - \text{hcap}(\gamma(0,s] \cup \hat{\gamma}(0,r])$ and $\lim_{t\to\infty} \frac{\text{hcap}(\gamma(0,t] \cup \hat{\gamma}(0,r])}{3t/4} = 1$. Thus we have established that

$$N_{s,r} = c_0\, 1_{E_{s,r}}\, g'_{s,r}(\hat{U}_{0,r})^{5/8} \hat{g}'_{s,r}(U_{s,0})^{5/8}\, |\hat{U}_{s,r} - U_{s,r}|^{3/4}.$$

□

Definition. If $x < \hat{x}$, then two-sided chordal $SLE_{8/3}$ in $\mathbb{H}$ starting at $(x, \hat{x})$ is the measure on ordered pairs of paths $(\gamma, \hat{\gamma})$ such that for each $s, r < \infty$, the distribution of

$$\gamma(s'), 0 \le s \le s', \qquad \hat{\gamma}(r'), 0 \le r \le r'$$

is given by saying that the Radon-Nikodym derivative of this distribution with respect to that of *independent* chordal SLE's starting at $x, \hat{x}$ is

$$\frac{N_{s,r}}{N_{0,0}} = 1_{E_{s,r}}\, g'_{s,r}(\hat{U}_{0,r})^{5/8} \hat{g}'_{s,r}(U_{s,0})^{5/8} \left(\frac{\hat{U}_{s,r} - U_{s,r}}{\hat{x} - x}\right)^{3/4}.$$

Notice that as in the radial case,

$$\frac{N_{s,r}}{N_{0,0}} = \frac{N_{s,0}}{N_{0,0}} \frac{N_{s,r}}{N_{s,0}} = \frac{N_{0,r}}{N_{0,0}} \frac{N_{s,r}}{N_{0.r}},$$

which implies that we can grow some of the first curve and then some of the second, or vice versa. Again we can make sense of what this means using Girsanov's Theorem. Since $M_s/M_0 = N_{s,0}/N_{0,0}$, we can obtain part of the first curve by running the Loewner equation with driving term X_t until time s, where $X_0 = x, \hat{X}_0 = \hat{x}$, and

$$dX_t = -\frac{3/4}{\hat{X}_t - X_t}\, dt + dW_t, \qquad d\hat{X}_t = \frac{3/4}{\hat{X}_t - X_t}\, dt. \tag{7.3}$$

To obtain a piece of the second curve, we map the first curve down by g_s and then proceed as before, switching the roles of X_t and $\hat{X}_t$

Alternately, we could create the two-sided chordal $SLE_{8/3}$ process in two steps. First grow one complete curve γ as above by using the Loewner equation with driving term X_t described by (7.3). Then the second curve $\hat{\gamma}$ is chordal $SLE_{8/3}$ from $\hat{x}$ to infinity in the smaller domain D_γ, where D_γ is the simply connected component of $\mathbb{H} \setminus \gamma(0, \infty)$ that has $\hat{x}$ on the boundary. This is a consequence of the fact that

$$\lim_{s \to \infty} \frac{N_{s,r}}{N_{0.r}} = 1_{E_{\infty,r}}\, \tilde{\Phi}'_r(\hat{X}_r)^{5/8},$$

where $\tilde{\Phi}_r$ is the conformal perturbation of a $SLE_{8/3}$ Loewner chain with $\tilde{\Phi}_0$ a conformal map from D_γ onto $\mathbb{H}$. Weighting a chordal $SLE_{8/3}$ by $1_{E_{\infty,r}}\, \tilde{\Phi}'_r(\hat{X}_r)^{5/8}$ gives $SLE_{8/3}$ in the domain D_γ.

Restriction property for two-sided chordal $SLE_{8/3}$. Two-sided chordal $SLE_{8/3}$ satisfies the restriction property: if $(\gamma, \hat{\gamma})$ is two-sided chordal $SLE_{8/3}$ starting at $(x, \hat{x})$, then the conditional distribution of $\Phi_A(\gamma, \hat{\gamma})$ given $\{(\gamma, \hat{\gamma}) \cap A = \emptyset\}$ is the same as two-sided chordal $SLE_{8/3}$ starting at $(\Psi_A(x), \Psi_A(\hat{x}))$. Here A is a compact set in $\mathbb{H}$ such that $\mathbb{H} \setminus A$ is simply connected and $\mathrm{dist}(\{x, \hat{x}\}, A) > 0$, and Φ_A denotes a conformal map from $\mathbb{H} \setminus A$ onto $\mathbb{H}$ with $\Phi_A(z) \sim z$ for z near infinity. The restriction property follows from the following theorem.

Theorem 2 If $(\gamma, \hat{\gamma})$ denotes two-sided chordal $SLE_{8/3}$ starting at $(x, \hat{x})$, then

$$\mathbb{P}\{(\gamma, \hat{\gamma}) \cap A = \emptyset\} = \Phi'_A(x)^{5/8}\, \Phi'_A(\hat{x})^{5/8} \left(\frac{\Phi_A(\hat{x}) - \Phi_A(x)}{\hat{x} - x}\right)^{3/4}.$$

If $x = \hat{x}$, this is to be interpreted as $\Phi'_A(x)^2$.

This theorem is proved in the same manner as Theorem 1. We use our third method of obtaining a two-sided chordal $SLE_{8/3}$: γ is generated by the Loewner equation with driving term X_t, where X_t satisfies (7.3), and $\hat{\gamma}$ is chordal $SLE_{8/3}$ in D_γ. Let $\mathcal{F}_t$ denote the filtration generated by X_t, and let Φ_t be the conformal perturbation of this Loewner chain with $\Phi_0 = \Phi_A$. Then, one must show that the martingale

$$\tilde{M}_t := \mathbb{E}[1_{\{\gamma\cap A=\emptyset\}}\mathbb{P}\{\hat{\gamma}\cap A=\emptyset \mid \gamma\} \mid \mathcal{F}_t]$$

is equal to the "martingale candidate"

$$M_t := 1_{\{\gamma(0,t]\cap A=\emptyset\}}\,\Phi_t'(X_t)^{5/8}\,\Phi_t'(\hat{X}_t)^{5/8}\left(\frac{\Phi_t(\hat{X}_t)-\Phi_t(X_t)}{\hat{X}_t-X_t}\right)^{3/4}.$$

This is done by showing that M_t is a bounded martingale and that $\tilde{M}_t$ and M_t have the same limit at infinity. We omit the details.

We end by noting a connection between two-sided chordal $SLE_{8/3}$ and the $SLE(\kappa,\rho)$ processes. The latter processes can be defined as solutions to the Loewner equation driven by a random function having the appropriate drift. In particular, if (O_t, U_t) are a pair of processes satisfying

$$dO_t = \frac{a}{O_t - U_t}\,dt, \qquad dU_t = \frac{-v}{O_t-U_t} + dB_t, \qquad O_0 = U_0 = 0,$$

then the solution to the Loewner equation with driving term U_t is the $SLE(2/a, 2v/a)$ process. (See Section 9.3 of [4] for a brief introduction to the $SLE(\kappa,\rho)$ processes.) Therefore, one side of two-sided chordal SLE started from $(0,0)$ is the same as $SLE(8/3,2)$. This also follows from the restriction exponent, since both processes satisfy the restriction property with exponent 2.

8 Chordal SLE as the limit of radial SLE

In our previous discussion of two-sided $SLE_{8/3}$, we used the fact that chordal SLE can be obtained as a limit of radial SLE. We will be more precise about this here. Since it is no more difficult, we will discuss $\kappa \leq 4$, and as before we let $a = 2/\kappa$.

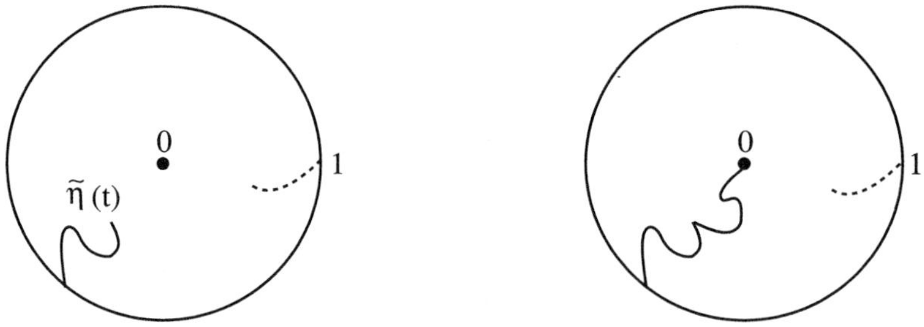

Figure 3 A comparison of radial and chordal SLE.

Suppose $\tilde{\eta} : (0,\infty) \to \mathbb{D}\setminus\{0\}$ is a simple curve with $\tilde{\eta}(0+) \in \partial\mathbb{D}\setminus\{1\}$ and $\tilde{\eta}(t) \to 0$ as $t \to \infty$. Define the following measures on paths (modulo reparametrization) $\tilde{\gamma} : [0,\infty) \to \mathbb{D}$:

- ν_t: Radial SLE_κ in $\mathbb{D} \setminus \tilde{\eta}(0,t]$ from 1 to 0;

- ν_∞: Chordal SLE_κ in $\mathbb{D} \setminus \tilde{\eta}(0,\infty)$ from 1 to 0.

In this section, we will give a precise version of the result that as $t \to \infty$, ν_t approaches ν_∞. By considering $\gamma(t) = -i \log \tilde{\gamma}(t)$, we can can consider ν_t, ν_∞ as measures on paths (modulo reparametrization) $\gamma : (0,\infty) \to \mathbb{H}$ with $\gamma(0+) = 0$. We choose the parametrization to be the half-plane parametrization. To be more precise, if g_t denotes the conformal transformation of $\mathbb{H} \setminus \gamma(0,t]$ onto $\mathbb{H}$ with $g_t(z) - z = o(1)$ as $z \to \infty$, then g_t has expansion

$$g_t(z) = z + \frac{at}{z} + O(|z|^{-2}), \qquad z \to \infty.$$

In this case, g_t satisfies the chordal Loewner equation

$$\dot{g}_s(z) = \frac{a}{g_s(z) - U_s}, \qquad g_0(z) = z, \tag{8.1}$$

with "driving function" $U_s = g_s(\gamma(s))$. For fixed $r < \infty$, let $\nu_{t,r}, \nu_{\infty,r}$ denote these measures on paths stopped at time r. We write $\eta(t) = -i \log \tilde{\eta}(t)$ where the branch of the logarithm is chosen so that $-2\pi < \eta(0+) < 0$.

Proposition 8.1 Suppose $\kappa \leq 4$, η is a curve as in the previous paragraph, and $0 < r < \infty$. Let $\nu_{t,r}, \nu_{\infty,r}$ be ν_t, ν_∞ restricted to curves up to time r, $\gamma(s), 0 \leq s \leq r$. There exists a $T = T(\eta, r)$ such that for $t \geq T$, $\nu_{t,r}$ and $\nu_{\infty,r}$ are mutually absolutely continuous with respect to each other. Moreover, with probability one with respect to $\nu_{\infty,r}$, the Radon-Nikodym derivative has a limit of 1 as t approaches infinity, i.e.

$$\lim_{t\to\infty} \frac{d\nu_{t,r}}{d\nu_{\infty,r}} = 1.$$

We start by giving the basic idea for the proof. Without loss of generality we will assume that $r = 1$; other values of r can be handled by scaling. Let g_s denote the conformal transformation of $\mathbb{H} \setminus \gamma(0,s]$ onto $\mathbb{H}$ satisfying $g_s(z) - z = o(1)$ as $z \to \infty$. To give a measure on the maps g_s (or, equivalently, on the curve γ) we give a measure on the driving function U_s. As we will see, the measure $\nu_{\infty,1}$ can be obtained by solving (8.1) where the driving function U_s satisfies a stochastic differential equation

$$dU_s = R_s\, ds + dB_s. \tag{8.2}$$

The drift term R_s depends on η and is adapted to the Brownian motion. Similarly, the measure $\nu_{t,1}$ can be obtained from the Loewner equation using the driving function $U_{s,t}$ satisfying

$$dU_{s,t} = R_{s,t}\, ds + dB_s = [R_{s,t} - R_s]\, ds + dU_s. \tag{8.3}$$

Let $\mathcal{W}$ denote the standard Wiener measure, i.e., the measure on paths $B_s, 0 \leq s \leq 1$ that gives the standard Brownian motion. Then the Girsanov transformation tells us that the measure whose Radon-Nikodym deriviative with respect to $\mathcal{W}$ is

$$\exp\left\{\int_0^1 R_s\, dB_s - \frac{1}{2}\int_0^1 R_s^2\, ds\right\}$$

is the same as the measure satisfying the differential equation (8.2). Similarly, if we choose Radon-Nikodym derivative

$$\exp\left\{\int_0^1 R_{s,t}\, dB_s - \frac{1}{2}\int_0^1 R_{s,t}^2\, ds\right\},$$

the measure satisfies (8.3). In other words, we can define the measures $\nu_{t,1}$ and $\nu_{\infty,1}$ on the same probability space so that

$$\frac{d\nu_{t,r}}{d\nu_{\infty,r}} = \exp\left\{\int_0^1 (R_{s,t} - R_s)\,dB_s - \frac{1}{2}\int_0^1 [R_{s,t}^2 - R_s^2]\,ds\right\}.$$

We will let $T = T(\eta, 1) = \sigma_{4\sqrt{a}}$ where $\sigma_r = \sup\{t : \mathrm{Im}[\eta(t)] \leq r\}$. By properties of half-plane capacity, $\gamma(0,1]$ is contained in $\{z : \mathrm{Im}(z) \leq 2\sqrt{a}\}$. For $T(\eta,1) \leq t \leq \infty$, $\nu_{t,1}$ is supported on those paths γ with $\gamma(0,1] \cap \eta(0,\infty) = \emptyset$ and $(\gamma(0,1] \cap \eta(0,\infty) + 2\pi) = \emptyset$; this shows the mutual absolute continuity. Therefore, to prove the proposition it suffices to show that with $\nu_{\infty,1}$ probability one,

$$\lim_{t\to\infty} \sup_{0\leq s\leq 1} |R_{s,t} - R_s| = 0. \tag{8.4}$$

Chordal and radial SLE in subdomains of $\mathbb{H}$. Chordal SLE_κ from 0 to ∞ in $\mathbb{H}$ is defined by solving the Loewner equation (8.1) where the driving function is Brownian motion. Similarly radial SLE_κ in $\mathbb{D}$ is defined by solving the radial equation. The standard way to define chordal and radial SLE_κ in simply connected subdomains is by conformal transformation. In this section we describe a different way of obtaining radial SLE_κ in $\mathbb{H}$ and chordal and radial SLE_κ in subdomains by solving the Loewner equation (8.1) with a driving function with appropriate drift.

For this section we let $\kappa \leq 4$ and set $a = 2/\kappa$, $b = (3a-1)/2$. Suppose $D \subset \mathbb{H}$ is a domain contaning $\{z \in \mathbb{H} : |z| < \epsilon\}$ for some ϵ and suppose $w \in \partial D \setminus \{0\}$. Then chordal SLE_κ from 0 to w in D is defined (modulo time reparametrization) to be the image of SLE_κ in $\mathbb{H}$ from 0 to infinity under a conformal map taking 0 to 0 and ∞ to w. We begin by constructing this measure in a different way.

If $\gamma : (0,t] \to \mathbb{H}$ is a simple curve with $\gamma(0+) = 0$, let g_t denote the conformal transformation of $\mathbb{H}\setminus\gamma(0,t]$ onto $\mathbb{H}$ satisfying $g_t(z) - z = o(1)$ as $z \to \infty$. If $\gamma(0,t] \subset D$, let $D_t = g_t(D \setminus \gamma(0,t])$, $U_t = g_t(\gamma(t))$, $w_t = g_t(w)$. Let F_t denote a conformal transformation of $\mathbb{H}$ onto D_t with $F_t(\infty) = w_t$, $F_t(0) = U_t$. Let $\Phi_t = F_t^{-1}$, which is a conformal transformation of D_t onto $\mathbb{H}$. Then (see, e.g., [5]) chordal SLE_κ in D can be given by solving the chordal Loewner equation (8.1) with driving function U_t satisfying the SDE

$$dU_t = b\,\frac{\Phi_t''(U_t)}{\Phi_t'(U_t)}\,dt + dB_t,$$

where B_t is a standard Brownian motion.

A similar construction can be given for radial SLE. Suppose $\tilde{\gamma}(t)$ denotes radial SLE_κ in $\mathbb{D}$ and $\tilde{g}_t$ denotes the unique conformal transformation of $\mathbb{D} \setminus \tilde{\gamma}(0,t]$ onto $\mathbb{D}$ with $\tilde{g}_t(0) = 0, \tilde{g}_t'(0) > 0$. Then (under a suitable parametrization), the maps $h_t := -i \log \tilde{g}_t$ satisfy

$$\dot{h}_t(z) = \frac{a}{2}\cot\left(\frac{h_t(z) - B_t}{2}\right),$$

where B_t is a standard Brownian motion. Let $\gamma = -i\log\tilde{\gamma}(t)$ where a branch of the logarithm is chosen with $\log 1 = 0$. Note that for t very small, γ grows almost like chordal SLE_κ (at time 0 it is growing exactly like this).

To see the difference between radial and chordal, suppose that the path has produced $\gamma(0,t]$. For radial SLE, the path has also produced all the 2π translates of $\gamma(0,t]$. Therefore, locally the path is now growing like chordal SLE from $\gamma(t)$ to

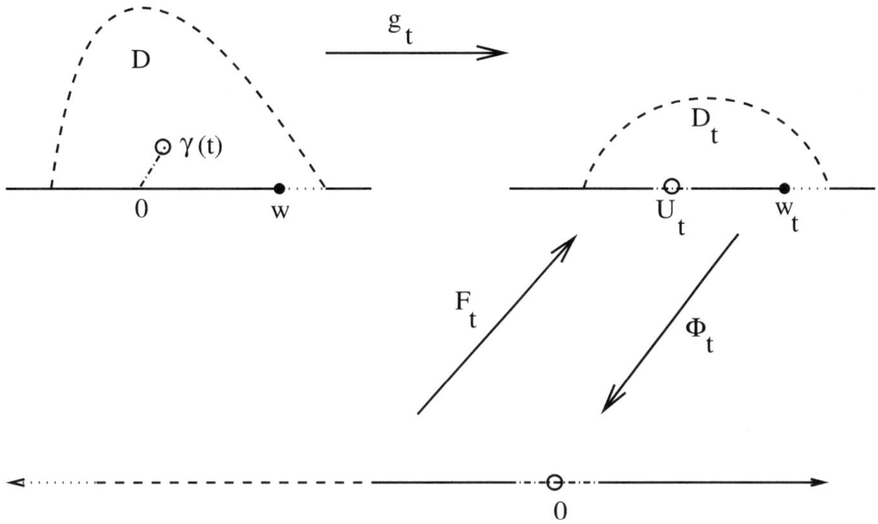

Figure 4 The domain D_t and the maps F_t, Φ_t.

∞ in the domain

$$\hat{D}_t := \mathbb{H} \setminus \left[\bigcup_{k=-\infty}^{\infty} (2\pi k + \gamma(0,t]) \right].$$

Let $D_t = g_t(\hat{D}_t), U_t = g_t(\gamma(t))$. Although $\hat{D}_t$ is periodic, the domain D_t is not periodic. By conformal invariance, radial SLE_κ is the process that acts locally like chordal SLE_κ from U_t to ∞ in the domain D_t. Let Ψ_t denote a conformal transformation of D_t onto $\mathbb{H}$ with $\Psi_t(\infty) = \infty$. This transformation is not unique, but if $\tilde{\Psi}_t$ is another such transformation, then $\tilde{\Psi}_t = c\,\Psi_t + x$ for some $c > 0, x \in \mathbb{R}$. If we parametrize the curve γ so that $g_t(z) = z + (at/z) + O(|z|^{-2})$ as $z \to \infty$, then the maps g_t satisfy (8.1) where U_t satisfies the SDE

$$dU_t = b\,\frac{\Psi_t''(U_t)}{\Psi_t'(U_t)}\,dt + dB_t,$$

and B_t is a standard Brownian motion. Note that Ψ''/Ψ' is independent of the choice of Ψ.

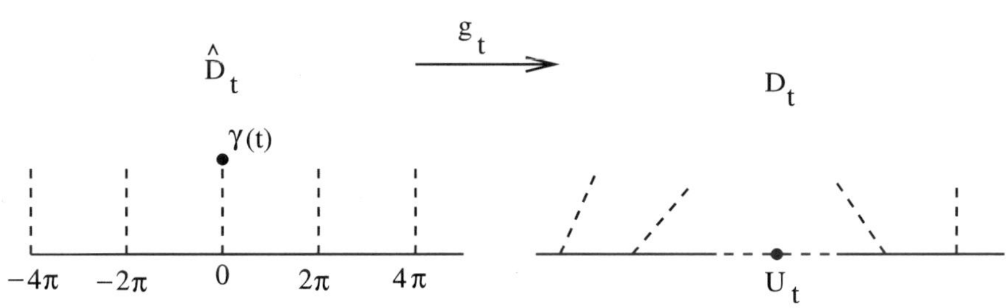

Figure 5 The domains $\hat{D}_t$ and D_t for radial SLE.

Now suppose that $A \subset \overline{\mathbb{D}}$ is a closed set not containing 0 or 1 such that $\mathbb{D} \setminus A$ is simply connected. Let $D_A = -i \log(\mathbb{D} \setminus A)$ which is the upper half plane with a peridoic set removed. Suppose $\Theta = \Theta_A$ is a conformal transformation of $\mathbb{D}$ onto $\mathbb{D} \setminus A$ with $\Theta(0) = 0, \Theta(1) = 1$ and let $\theta(z) = -i \log \Theta(e^{iz})$ which is a conformal transformation of $\mathbb{H}$ onto D_A. The image of radial SLE_κ from 1 to 0 under Θ is radial SLE_κ in $\mathbb{D} \setminus A$ from 1 to 0. Therefore radial SLE_κ in D_A can be obtained as the image under θ of the measure described in the previous paragraph. By combining, we see that (the image under the logarithm map of) radial SLE_κ in $\mathbb{D} \setminus A$ looks locally like chordal SLE in D_A. Let

$$\hat{D}_{t,A} = D_A \setminus \left[\bigcup_{k=-\infty}^{\infty} (2\pi k + \gamma(0,t]) \right],$$

and $D_{t,A} = g_t(\hat{D}_{t,A})$. Then locally (the image under the logarithm map of) radial SLE_κ in D_A looks like chordal SLE_κ from 0 to infinity in the domain $D_{t,A}$. In particular, it satisfies (8.1) with a driving process U_t satisfying

$$dU_t = b \frac{\hat{\Psi}_t''(U_t)}{\hat{\Psi}_t'(U_t)} dt + dB_t,$$

where $\hat{\Psi}_t$ is a conformal transformation of $D_{t,A}$ onto $\mathbb{H}$ fixing infinity.

Finally, suppose that $\eta : (0, \infty) \to \mathbb{H}$ is a simple curve with $-2\pi < \eta(0-) < 0$ and $\mathrm{Im}[\eta(s)] \to \infty$. Assume also that

$$\eta(0,\infty) \cap [2\pi + \eta(0,\infty)] = \emptyset.$$

For each $r > 0$, let σ_r denote the largest s with $\mathrm{Im}[\eta(s)] \leq r$. Let D denote the domain bounded by $[\eta(0+), \eta(0+) + 2\pi], \eta(0, \infty)$, and $[2\pi + \eta(0, \infty)]$. For each $s < \infty$, let $D^{(s)}$ denote the domain

$$D^{(s)} = \mathbb{H} \setminus \left[\bigcup_{k=-\infty}^{\infty} (2\pi k + \eta(0,s]) \right].$$

We need to compare chordal SLE_κ in D from 0 to ∞ to radial SLE_κ in $D^{(s)}$ from 0 to ∞. Both processes can be considered as measures on paths $\gamma : (0, \infty) \to \mathbb{H}$ with $\gamma(0+) = 0$. In both cases, the measures can be obtained by solving (8.1) with a driving function U_t; the difference comes in the SDE that U_t satisfies. Let $D_t = g_t(D)$ and $D_t^{(s)} = g_t(D^{(s)})$. Let Φ_t be a conformal transformation of D_t onto $\mathbb{H}$ fixing infinity and let $\Phi_{t,s}$ be a conformal transformation of $D_t^{(s)}$ onto $\mathbb{H}$ fixing infinity. Then the driving processes, $U_t, U_{t,s}$ satisfy

$$dU_t = b \frac{\Phi_t''(U_t)}{\Phi_t'(U_t)} dt + dB_t, \qquad dU_{t,s} = b \frac{\Phi_{t,s}''(U_{t,s})}{\Phi_{t,s}'(U_{t,s})} dt + dB_t,$$

Let I_r be the open interval $(b_1 + ri, b_2 + 2\pi + ri)$ where $b_1 = \max\{x : x + ri \in \eta(0, \sigma_r]\}$, $b_2 = \min\{x : x + ri \in \eta(0, t_r]\}$. Let D_r denote the Jordan domain bounded by I_r, $(\eta(0+), \eta(0+) + 2\pi)$, $\eta(0, t_1]$, $2\pi + \eta(0, t_2]$ where $\eta(t_1) = b_1 + ri$, $\eta(t_2) = b_2 + ri$. Note that if $s \geq \sigma_r$, then any curve from 0 to $D^{(s)} \setminus D$ in $D^{(s)}$ must go through I_r.

We will restrict to $0 \leq t \leq 1$. From (8.1) we can see that if $|z| \geq 2\sqrt{a}$ and $t \leq 1$, then $|g_t(z) - z| \leq 2a/|z|$. In particular, if $r \geq 2\sqrt{a}$, then $g_t(I_r) \subset \{\mathrm{Im}(z) \geq r - \sqrt{a}\}$. Note that any curve from U_t to $g_t(D^{(s)})$ in $g(D^{(s)})$ must go through $g_t(I_r)$. We list some other properties here.

- There exists an $\epsilon > 0$ such that for all $0 \leq t \leq 1$,

$$\{z \in \mathbb{H} : |z - U_t| < \epsilon\} \subset g_t(D_r).$$

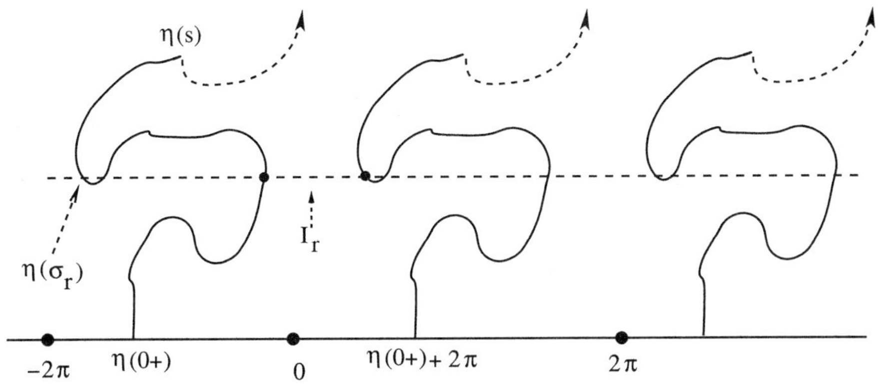

Figure 6 The domain $D^{(s)}$ is $\mathbb{H}$ with the solid curves removed. The domain D is the connected component of $\mathbb{H} \setminus (\eta(0,\infty) \cup [2\pi + \eta(0,\infty)])$ with $0 \in \partial D$.

- There exists a $c < \infty$ such that the probability that a Brownian motion starting at U_t reflected off the real axis into $\mathbb{H}$ reaches $g_t(I_r)$ before leaving $g_t(D_r)$ is bounded above by $c\, e^{-r/2}$. This follows from the Beurling estimate (see, e.g., [4, Section 3.8]) and the fact that $g_t(I_r) \subset \{\mathrm{Im}(z) \geq r - \sqrt{a}\}$.

We point out that the constant c in the second statement depends only on a while the ϵ in the first statement depends on γ and η.

To prove (8.4) it suffices to establish an estimate on conformal maps. We do this in the next subsection. Once we obtain (8.5), we can conclude that if

$$s \geq \sup\{s' : \mathrm{Im}(\eta(s')) \leq r\},$$

then

$$|R_{s,t} - R_t| \leq c\,\epsilon^{-1}\, e^{-r/2}.$$

This implies (8.4).

Lemmas about conformal maps. Here we will discuss some of the necessary estimates about conformal maps. We start with some setup.

Suppose $\gamma^1, \gamma^2 : (0,1] \to \mathbb{H}$ are simple curves satisfying

- $x_1 = \gamma^1(0+) < 0 < \gamma^2(0+) = x_2$
- $\gamma^1(0,1] \cap \gamma^2(0,1] = \emptyset$.
- If $I = (\gamma^1(1), \gamma^2(1))$ denote the open line segment connecting the endpoints, then

$$I \cap (\gamma^1(0,1] \cup \gamma^2(0,1]) = \emptyset.$$

Let $\hat{D}$ denote the Jordan domain bounded by the curves $\gamma^1(0,1], \gamma^2(0,1], I$, and $[x_1, x_2]$. Let

$$\hat{D}^* = D \cup (x_1, x_2) \cup \{z : \overline{z} \in D\},$$

be the extension of $\hat{D}$ by Schwarz reflection. Finally, let $q = q(\gamma^1, \gamma^2)$ denote the harmonic measure of $I \cup I^*$ in $\hat{D}^*$ from 0. Equivalently, q is the probability that a Brownian motion starting at 0 leaves $\hat{D}^*$ at I or I^*. By symmetry, the probability of leaving at I is $q/2$.

Lemma 8.2 There is a $c < \infty$ such that the following holds. Assume γ^1, γ^2 are given as above. Suppose D is a simply connected domain with

$$\hat{D} \cup I \subset D \subset \mathbb{H} \setminus (\gamma^1(0,1] \cup \gamma^2(0,1]).$$

Let $F : D \to \hat{D}$ be the unique conformal transformation with $F(x_1) = x_1, F(0) = 0, F(x_2) = x_2$. Suppose $\hat{D}^*$ contains the open ball of radius ϵ about 0. Then,

$$|F'(0) - 1| \leq cq, \qquad |F''(0)| \leq c\,\epsilon^{-1}\, q.$$

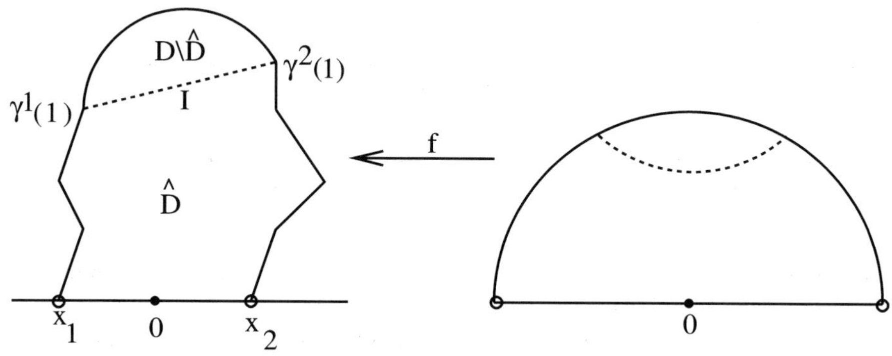

Proof By scaling we may assume that $\epsilon = 1$. The Koebe-1/4 and the Bieberbach estimate give $1/4 \leq F'(0) \leq 4, |F''(0)| \leq 2\,F'(0)$, and so it suffices to prove the result for q sufficiently small.

The Riemann mapping theorem states that there is a unique conformal transformation $f : \mathbb{D} \cap \mathbb{H} \to D$ with $f(-1) = x_1, f(0) = 0, f(1) = x_2$. By Schwarz reflection, this can be extended to a conformal transformation $f : \mathbb{D} \to D^*$ where $D^* = D \cup (-x_1, x_2) \cup \{z : \overline{z} \in D\}$.

Let $U = f^{-1}(\hat{D}^*)$. Then U is a simply connected subdomain of $\mathbb{D}$ with the property that the probability that a Brownian motion starting at the origin leaves U before leaving $\mathbb{D}$ equals q. Since U is simply connected (and, hence, ∂U is connected), we can see that there is a c such that

$$(1 - c\,q)\,\mathbb{D} \subset U.$$

We will assume that q is sufficient small so that $cq < 1/2$ and write $\delta = cq$.

Let $h : \mathbb{D} \to U$ be the unique conformal transformation with $h(0) = 0, h'(0) > 0$. The Schwarz lemma tells us that $(1-\delta) \leq h'(0) \leq 1$. We will show that $|h''(0)| \leq c\,\delta$. Let $g(z) = \log(h(z)/z)$ which is a well-defined analytic function since $h'(0) > 0$ and $h(z) \neq 0$ for $z \neq 0$. The maximum principle implies that

$$|\mathrm{Re}\, g(z)| \leq \sup\{|\mathrm{Re}\, g(z)| : |z| = 1\} \leq |\log(1-\delta)| \leq c\,\delta.$$

Since $\mathrm{Re}\, g$ is a harmonic function, this implies that the partial derivatives of $\mathrm{Re}\, g(z)$ are $O(\delta)$ for $|z| \leq 1/4$. Hence, by the Cauchy-Riemann equations, $|g'(z)| = O(\delta)$ for $|z| \leq 1/4$. Since $g(0) = \log h'(0) = O(\delta)$, we conclude that $|g(z)| \leq c\,\delta$ for $|z| \leq 1/4$, and hence

$$|h(z) - z| \leq c\delta.$$

From this we conclude that $|h''(0)| = |(h-z)''(0)| \leq c\,\delta$.

Since h is unique, we can see that $h = f^{-1} \circ F \circ f$. The chain rule gives

$$F'(0) = h'(0),$$

$$F''(0) = \frac{1}{f'(0)} \left[h''(0) - h'(0)\,[h'(0) - 1]\, \frac{f''(0)}{f'(0)} \right].$$

The Koebe-1/4 and the Bieberbach estimate give $|f'(0)| \geq 1/4, |f''(0)| \leq 2\, f'(0)$. Therefore,

$$F'(0) - 1 = h'(0) - 1, \qquad |F''(0)| \leq 8\,[|h''(0)| + |h'(0)|\,|h'(0) - 1|].$$

However we have seen that

$$|h'(0) - 1|, |h''(0)| \leq c\, q,$$

at least if q is sufficiently small.

□

Lemma 8.3 Suppose $\hat{D}$ is as above and $D, \tilde{D}$ are two domains satisfying the conditions (on D) of the previous lemma and let q, ϵ be as in that lemma. Let $\Phi : D \longrightarrow \mathbb{H}$ denote the unique conformal transformation with $\Phi(0) = 0, \Phi(\infty) = \infty, \Phi'(0) = 1$; and let $\tilde{\Phi}$ denote the corresponding transformation for $\tilde{D}$. Then

$$|\Phi''(0) - \tilde{\Phi}''(0)| \leq c\, q\, \epsilon^{-1}. \tag{8.5}$$

Proof Let F be the unique conformal transformation of D onto $\tilde{D}$ with $F(x_1) = x_1, F(0) = 0, F(x_2) = x_2$. By applying the previous lemma we cam see that

$$|F'(0) - 1| \leq c\, q, \qquad |F''(0)| \leq c\, \epsilon^{-1}\, q.$$

Let

$$\Psi(z) = \frac{\tilde{\Phi} \circ F(z)}{F'(0)}.$$

Then Ψ is a conformal transformation of D onto $\mathbb{H}$ with $\Psi(0) = 0, \Psi'(0) = 1$. Also,

$$\Psi''(0) = \tilde{\Phi}''(0)\, F'(0) + \frac{F''(0)}{F'(0)}.$$

The transformation Ψ might not equal Φ since $\Psi(\infty)$ might not equal ∞. However, it is easy to check that

$$\Phi(z) = \frac{\Psi(\infty)\, \Psi(z)}{\Psi(\infty) - \Psi(z)},$$

where $\Psi(\infty)/[\Psi(\infty) - \Psi(z)]$ is interpreted to equal 1 if $\Psi(\infty) = \infty$. Note that $\Phi'(0) = 1$ and

$$\Phi''(0) = \Psi''(0) + \frac{2}{\Psi(\infty)} = \tilde{\Phi}''(0)\, F'(0) + \frac{F''(0)}{F'(0)} + \frac{2}{\Psi(\infty)}.$$

Therefore,

$$|\Phi''(0) - \tilde{\Phi}''(0)] \leq |\tilde{\Phi}''(0)|\, |F'(0) - 1| + \left| \frac{F''(0)}{F'(0)} \right| + \frac{2}{|\Psi(\infty)|}.$$

Applying the Bieberbach estimate to (the Schwarz reflection extension of) $z \mapsto \tilde{\Phi}(\epsilon z)/\epsilon$ gives $|\tilde{\Phi}''(0)| \leq 2/\epsilon$. We have already bounded $|F'(0) - 1|$ and $|F''(0)|$. We now need to estimate $|\Psi(\infty)|$. Note that $\Psi(I)$ is a curve in $\mathbb{H}$ connecting the negative real axis to the positive real axis. Let d be the distance of this curve from the origin. Using the gambler's ruin estimate, it is not difficult to show that the probability that a Brownian motion starting at δi hits this image before leaving $\mathbb{H}$ is bounded below by $c\,\delta/d$. [In fact, if $\delta < d/2$, and $z \in \mathbb{H}$ with $|z| \leq 2d$; B_t is a Brownian motion starting at $i\delta$; and T denotes the first t with $B_t \in \mathbb{R}$, then with probability at least $O(\delta/d)$ the point z will be in a bounded component of

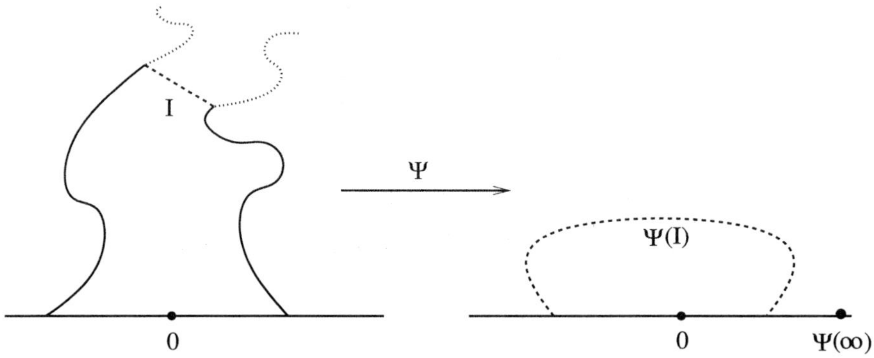

$\mathbb{H} \setminus (B[0,T] \cup [0, \delta i])$. In this case we must have $B(0,T)$ intersecting the image curve.] Note that $\Psi(\infty)$ lies outside this curve so $|\Psi(\infty)| \geq d$. The probability starting at δi that a Brownian motion leaves D at I is bounded above by $c\,(\delta/\epsilon)\,q$. [Here, $c\,\delta/\epsilon$ bounds the probability to reach the sphere of radius $\epsilon/2$ and the Harnack inequality implies that the probability of reaching I given this is bounded by $c\,q$.] Hence, since $\Psi'(0) = 1$, we get that

$$|\Psi(\infty)| \geq c\,\epsilon\,q^{-1}.$$

This establishes (8.5). □

References

[1] J. Dubédat, Some remarks on commuation relations for *SLE*, preprint.

[2] P. Duren (1983). *Univalent Functions*, Springer-Verlag.

[3] T. Kennedy (2003), Monte Carlo tests of *SLE* predictions for 2*D* self-avoiding walks, Phys. Rev. Lett. **88**, 130601.

[4] G. Lawler (2005). *Conformally Invariant Processes in the Plane*, Amer. Math. Soc.

[5] G. Lawler, The Laplacian-*b* walk and the Schramm-Loewner evolution, to appear in Illinois J. Math.

[6] G. Lawler, O. Schramm, W. Werner (2003), Conformal restriction: the chordal case, J. Amer. Math. Soc. **16**, 917–955.

[7] G. Lawler, O. Schramm, W. Werner (2004), On the scaling limit of planar self-avoiding walk, in *Fractal Geometry and Applications: A Jubilee of Benoit Mandelbrot*, Vol II., M. Lapidus, M. van Frankenhuijsen, ed., Amer. Math. Soc., 339–364.

[8] N. Madras and G. Slade (1993). *The Self-Avoiding Walk*, Birkhäuser.

[9] O. Schramm (2000), Scaling limits of loop-erased random walks and uniform spanning trees, Israel J. Math. **118**, 221–288.

Fields Institute Communications
Volume **50**, 2007

Using the Renormalization Group

D. G. C. McKeon
Department of Applied Mathematics
University of Western Ontario
London N6A 5B7 Canada
dgmckeo2@uwo.ca

In computing quantum effects, it is necessary to perform a sum over all intermediate states consistent with prescribed initial and final states. Divergences arising in the course of evaluating this sum forces one to "renormalize" parameters characterizing the system. An ambiguity inherent in this rescaling is parameterized by a dimensionful parameter μ^2 which serves to set a scale for the process. Requiring that the explicit and implicit dependence of a physical quantity on μ^2 conspire to cancel leads to the so-called "renormalization group" equation [1-10]. It has proved possible to extract a lot of useful information from this equation; we will enumerate a number of these in this report.

The process of renormalization is not inherently mysterious; indeed it can occur even in simple mechanical systems. For example, the Hamiltonian for a simple harmonic oscillator $H = \frac{p^2}{2m} + \frac{k}{2}(x - x_0)^2$ can be supplemented by a term $-mgx$ arising from being placed in a constant external gravitational field. These two terms together lead to the Hamiltonian being given by $H' = \frac{p^2}{2m} + \frac{k}{2}(x - x_0')^2 + C$ where $x_0' = x_0 + \frac{mg}{k}$ and C is an irrelevant constant $-\left(mgx_0 + \frac{(mg)^2}{2k}\right)$. The form of H and H' is the same; the interaction with the external gravitational field has only served to shift, or "renormalize", the equilibrium position of the oscillator from x_0 to x_0'. A physical process has been taken into account by simply altering the parameter x_0 that characterizes the theory. Much the same sort of thing happens in quantum field theory. Once again, a physical process serves to simply shift the value of the physical parameters that define a theory – these parameters being the masses, couplings and external field strengths that appear in the initial Lagrangian. The non-linear nature of these models means that the interactions are self interactions, rather than the result of an external force, as is the case with the simple harmonic oscillator in a constant gravitational field noted above.

There is a large degree of arbitrariness in the way in which the renormalization procedure can be effected; so long as renormalization serves to remove the divergences noted above it is possible to vary the amount absorbed into the "renormalized" parameters by an arbitrary amount. Quite often one chooses to adopt

---

2000 *Mathematics Subject Classification.* Primary 81T17; Secondary 81T15.

some physically motivated criterion for determining the renormalization procedure, but it must be noted that there is nothing that forces one to adopt any particular renormalization scheme. Indeed, the "mass independent" subtractions of refs. [4-6] are technically the most easy to apply, but in this case the parameters appearing in the finite answers arrived at after renormalization are not easily related to physical parameters such as pole masses. A further finite renormalization is required to relate renormalized parameters, defined through some mass independent renormalization procedure, to the physically measured parameters. (As an aside, we note that it is possible to entirely circumvent the appearances of divergences at all stages of a calculation, even prior to renormalization, through analytically continuing the sums over all intermediate states that normally lead to these divergences. This is what happens in zeta-function and its extension, operator regularization.)

There are, in fact, two distinct types of ambiguities occuring in the renormalization that occurs in quantum field theory. First of all, there is the appearance of an unphysical scale parameter μ^2 within any particular renormalization scheme such as minimal subtraction [4]. Different values of μ^2 can be chosen; reconciling perturbative results using the ambiguous value of μ^2 can be accomplished by altering the value of the parameters in the theory. This is the usual "renormalization group"; a particularly interesting discussion of the role of μ^2 appears in ref. [10]. A second ambiguity is related to how exactly one chooses to renormalize. This has nothing to do with how μ^2 is selected, rather it amounts to redefining the parameters in a theory; a coupling λ is replaced by $\lambda' = \sum_{n=0}^{\infty} \alpha_n \lambda^n$. This "renormalization scheme ambiguity" is a problem not discussed in this review; we will only make use of information that can be obtained about physical processes through examining the role played by μ^2 in renormalization. **(I)** The first instance we will analyze to show the utility of this approach is the relationship between the bare and renormalized parameters in the context of dimensional regularization and mass independent renormalization [4,11-13].

In this approach, bare quantities appearing in the $n = 4 - 2\epsilon$ dimensional Lagrangian are expanded in terms of poles at $\epsilon = 0$. In particular, the bare coupling g_B and the renormalized coupling g are related by

$$g_B = \mu^\epsilon \sum_{\nu=0}^{\infty} \frac{a_\nu(g)}{\epsilon^\nu} \tag{1}$$

where μ is a renormalization induced scalar parameter. In minimal subtraction,

$$a_0(g) = g. \tag{2}$$

Explicit calculation leads to

$$a_\nu(g) = \sum_{k=\nu}^{\infty} a_{k,\nu} g^{2k+1} \tag{3}$$

with

$$a_{0,0} = 1. \tag{4}$$

The sum in eq. (1) can now be reorganized to result in

$$g_B = \mu^\epsilon \sum_{k=0}^{\infty} g^{2k+1} S_k \left(\frac{g^2}{\epsilon}\right) \tag{5}$$

where

$$S_k(\xi) = \sum_{\ell=0}^{\infty} a_{k+\ell,\ell}\, \xi^\ell \tag{6}$$

and

$$S_k(0) = a_{k,0}. \tag{7}$$

Since g_B is independent of μ,

$$\mu \frac{dg_B}{d\mu} = \left(\mu \frac{\partial}{\partial \mu} + \mu \frac{dg}{d\mu}\frac{\partial}{\partial g}\right) g_B = 0. \tag{8}$$

Equations (8) and (1) are consistent at order ϵ^0 provided

$$a_1 - \frac{a_0}{a_0'} a_1' + \beta a_0' = 0 \tag{9}$$

where

$$\mu \frac{dg}{d\mu} = \beta(g) - \epsilon \frac{a_0}{a_0'}, \tag{10}$$

so that

$$\beta(g) = \frac{a_0^2}{a_0'^2} \frac{d}{dg}\left(\frac{a_1}{a_0}\right). \tag{11}$$

Terms of order ϵ^{-k} result in the equation

$$-\frac{a_0^2}{a_0'} \frac{d}{dg}\left(\frac{a_k}{a_0}\right) + \beta a_{k-1}' = 0 \tag{12}$$

so that a_k is determined by a_{k-1} and $\beta (k \geq 2)$, as well as $a_0(g)$. We also find that together eqs. (1) and (6) result in

$$\sum_{n=0}^{\infty} g^{2n}\left[\frac{g^3}{\xi} S_n(\xi) + \left(-\frac{a_0}{a_0'}\frac{g^2}{\xi} + \beta\right)\left((2n+1)S_n(\xi) + 2\xi S_n'(\xi)\right)\right] = 0 \tag{13}$$

where $\xi = \frac{g^2}{\epsilon}$. Upon making the expansions

$$\frac{a_0(g)}{a_0'(g)} = \frac{g + a_{1,0}g^3 + a_{2,0}g^5 + \ldots}{1 + 3a_{1,0}g^2 + 5a_{2,0}g^4 + \ldots} = g + \alpha_3 g^3 + \alpha_5 g^5 + \ldots \tag{14}$$

and

$$\beta(g) = \sum_{k=1}^{\infty} B_{2k+1} g^{2k+1} \tag{15}$$

then by looking at contributions to eq. (13) in ascending powers of g,

$$(1 - \xi B_3)\, S_0' - \frac{1}{2} B_3 S_0 = 0 \tag{16}$$

$$\frac{1}{\xi} S_1 + \left(-\frac{1}{\xi} + B_3\right)(3S_1 + 2\xi S_1') + \left(-\frac{\alpha_3}{\xi} + B_5\right)(S_0 + 2\xi S_0') = 0 \tag{17}$$

etc., whose solutions subject to eq. (7) are

$$S_0(\xi) = w^{-1/2} \tag{18}$$

$$S_1(\xi) = \frac{1}{w^{1/2}(1-w)}\left[\left(\frac{\alpha_3}{2} - \frac{B_5}{2B_3}\right)(-1 + w^{-1}) - \frac{B_5}{2B_3} \ln w\right], \tag{19}$$

where $w = 1 - B_3\xi$. It is apparent that in the limit $\epsilon \to 0 (w \to \infty)$,

$$S_0 \to 0 \qquad S_1 \to 0 \tag{20}$$

which is consistent with $S_k \to 0 (k \geq 2)$. This indicates that the bare coupling g_B vanishes as ϵ vanishes, which is not what might be expected from eq. (1). One can also derive this result from eq. (12), as upon multiply eq. (12) by ϵ^{-k+1} and summing over k we arrive at

$$\epsilon \left(g_B - \frac{a_0(g)}{a_0'(g)} \frac{\partial}{\partial g} g_B \right) + \beta(g) \frac{\partial g_B}{\partial g} = 0 \tag{21}$$

so that

$$g_B = \mu^\epsilon \exp \left(- \int_K^g dx \frac{\epsilon}{\left(\beta(x) - \epsilon \frac{a_0(x)}{a_0'(x)} \right)} \right) \tag{22}$$

where K is a cut off. Consistency with eqs. (18) and (19) is achieved by having

$$g_B = \mu^\epsilon \exp \left(- \int_0^g dx \left[\frac{\epsilon}{\beta(x) - \epsilon \frac{a_0(x)}{a_0'(x)}} + \frac{1}{x} \right] \right). \tag{23}$$

In the limit $\epsilon \to 0$,

$$g_B \to g \exp \left(- \lim_{\delta \to 0^+} \int_\delta^g \frac{dx}{x} \right) = 0 \tag{24}$$

which is consistent with eq. (20).

The function $a_0(g)$ characterizes the choice of renormalization scheme; minimal subtraction uses a renormalized coupling $\widetilde{g}$ where

$$\widetilde{g} = a_0(g). \tag{25}$$

By eq. (10),

$$\mu \frac{d\widetilde{g}}{d\mu} = \widetilde{\beta}(\widetilde{g}) - \epsilon \widetilde{g} = \left(\frac{d\widetilde{g}}{dg} \right) \left(\mu \frac{dg}{d\mu} \right) = a_0'(g) \left(\beta(g) - \epsilon \frac{a_0(g)}{a_0'(g)} \right) \tag{26}$$

so that

$$\widetilde{\beta}(\widetilde{g}) = a_0'(g) \beta(g). \tag{27}$$

(II) The "method of characteristics" can also be used to analyze the renormalization group equation, in particular eq. (8) [14,9,12,15-17]. We shall examine a simple example of how this technique can be used to extract information from a first order partial differential equation.

Suppose $A_0(x, y)$ is a solution to the equation

$$f(x,y) \frac{\partial A(x,y)}{\partial x} + g(x,y) \frac{\partial A(x,y)}{\partial y} + h(x,y) A(x,y) = 0 \tag{28}$$

where f, g and h are prescribed functions of x and y. If now we have "characteristic functions" $\overline{x}(t)$ and $\overline{y}(t)$ defined by

$$\frac{d\overline{x}(t)}{dt} = f(\overline{x}(t), \overline{y}(t)); \quad \overline{x}(0) = x \tag{29}$$

$$\frac{d\overline{y}(t)}{dt} = g(\overline{x}(t), \overline{y}(t)); \quad \overline{y}(0) = y \tag{30}$$

then

$$A(\overline{x}(t), \overline{y}(t)) = A_0(\overline{x}(t), \overline{y}(t)) \exp \int_0^t h(\overline{x}(t'), \overline{y}(t')) dt' \tag{31}$$

reduces to $A_0(x, y)$ when $t = 0$ and $\frac{d}{dt} A(\overline{x}(t), \overline{y}(t)) = 0$.

For example, if $f = x$, $g = y^2$ and $h = 0$, a solution to eq. (28) is

$$A_0(x,y) = xe^{1/y} \tag{32}$$

while by eqs. (29) and (30),

$$\overline{x}(t) = xe^t \tag{33}$$

$$\overline{y}(t) = \frac{y}{1-yt}. \tag{34}$$

We indeed find that for all t

$$A_0(\overline{x}(t), \overline{y}(t)) = \left(xe^t\right) \exp\left[\frac{1}{y/(1-yt)}\right] = xe^{1/y}. \tag{35}$$

However, if we knew only an approximate solution

$$A_0^{(1)}(x,y) = x\left(1 + \frac{1}{y}\right) \tag{36}$$

then we might make the ansatz

$$A_0(x,y) = x\sum_{n=0}^{\infty} \alpha_n \left(\frac{1}{y}\right)^n, \tag{37}$$

with

$$\alpha_0 = \alpha_1 = 1. \tag{38}$$

The differential equation

$$x\frac{\partial A_0}{\partial x} + y^2\frac{\partial A_0}{\partial y} = 0 \tag{39}$$

shows that

$$\alpha_n = (n+1)\alpha_{n+1} = \frac{1}{n!} \tag{40}$$

and so

$$A_0(x,y) = x\sum_{n=0}^{\infty} \frac{1}{n!}\left(\frac{1}{y}\right)^n = xe^{1/y}. \tag{41}$$

Eq. (37) is analogous to eq. (5); in both cases we have knowledge of some part of a solution to a first order partial differential equation, and the remaining portion of the solution can be determined by a recursion relation. In the equation

$$\left(\mu\frac{\partial}{\partial\mu} + (-\epsilon g + \beta(g))\frac{\partial}{\partial g}\right) g_B = 0 \tag{42}$$

we in general have only a partial knowledge of $\beta(g)$ itself; knowing $\beta(g)$ to k-loop order, as well using the fact that in the MS scheme $a_0(g) = g$, allows one to determine $S_0(\xi)\dots S_k(\xi)$ in eq. (5) using eq. (42).

One could also use the characteristic function of eqs. (33) and (34) in conjunction with the approximate solution of eq. (36) to recover the exact solution of eq. (32). If we were to examine

$$A_0^{(1)}(\overline{x}(t), \overline{y}(t)) = \left(xe^t\right)\left[1 + \frac{1}{(y/(1-yt))}\right] \tag{43}$$

then it is apparent that at a particular value of t (namely $t = \frac{1}{y}$) we have

$$A_0^{(1)}\left(\overline{x}\left(\frac{1}{y}\right), \overline{y}\left(\frac{1}{y}\right)\right) = A_0(x,y). \tag{44}$$

(Furthermore, the ansatz of eq. (37) yields

$$A_0(\overline{x}(t),\overline{y}(t)) = xe^t \sum_{n=0}^{\infty} \alpha_n \left(\frac{1-yt}{y}\right) \tag{45}$$

which is independent of t only if eq. (40) is satisfied, in which case we also recover eq. (32).)

An analogous treatment of eq. (42) using the characteristic functions $\overline{\mu}(t)$ and $\overline{g}(t)$ can also be performed. We first define

$$\frac{d\overline{\mu}(t)}{dt} = \overline{\mu}(t) \qquad (\overline{\mu}(0) = \mu) \tag{46}$$

$$\frac{d\overline{g}(t)}{dt} = -\epsilon\overline{g}(t) + \beta(\overline{g}(t)) \quad (\overline{g}(0) = g). \tag{47}$$

Equations (46) and (47) are solved iteratively using an approach outlined in . A rescaling

$$\epsilon \to \epsilon\lambda \quad t \to t/\lambda \quad \overline{g}(t) \to \overline{g}(t)\sqrt{\lambda} \tag{48}$$

followed by an expansion

$$\overline{g}(t) = \overline{g}_0(t) + \overline{g}_1(t)\lambda + \overline{g}_2(t)\lambda^2 + \ldots \tag{49}$$

with

$$\overline{g}_n(0) = g\delta_{n0} \tag{50}$$

leads to

$$\overline{\mu}(t) = \mu e^{t/\lambda} \tag{51}$$

as well as

$$\overline{g}_0(t) = g^2 \left[\frac{B_3}{\epsilon}\left(1 - 2e^{2\epsilon t}\right) + e^{2\epsilon t}\right]^{-1} \tag{52}$$

$$\overline{g}_1(t) = \frac{B_5 e^{2\epsilon t} g^5}{2\epsilon\left[\frac{B_3 g^2}{\epsilon} + \left(1 - \frac{B_3 g^2}{\epsilon}\right)e^{2\epsilon t}\right]^{3/2}} \left\{\frac{1 - \frac{B_3}{\epsilon}g^2}{\left(\frac{B_3 g^2}{\epsilon}\right)^2} \ln\left(\frac{\frac{B_3 g^2}{\epsilon} + \left(1 - \frac{B_3 g^2}{\epsilon}\right)e^{2\epsilon t}}{e^{2\epsilon t}}\right) + \frac{1}{\left(\frac{B_3}{\epsilon}g^2\right)}\left(1 - \frac{1}{e^{2\epsilon t}}\right)\right\} \tag{53}$$

etc. Together, eqs. (1), (3), (48) and (49) lead to

$$\begin{aligned} g_B(\overline{\mu}(t),\overline{g}(t)) = \overline{\mu}^{\epsilon\lambda} \Bigg\{ &\lambda^{1/2}\left[\overline{g}_0 + \frac{a_{11}\overline{g}_0^3}{\epsilon} + \frac{a_{22}\overline{g}_0^5}{\epsilon^2} + \ldots\right] \\ &+ \lambda^{3/2}\left[\left(\overline{g}_1 + \frac{3a_{11}\overline{g}_0^2\,\overline{g}_1}{\epsilon} + \frac{5a_{22}\overline{g}_0^4\,\overline{g}_1}{\epsilon^2} + \ldots\right) \right. \\ &\left. + \left(\frac{a_{21}\overline{g}_0^5}{\epsilon} + \frac{a_{31}\overline{g}_0^7}{\epsilon^2} + \ldots\right)\right] + O(\lambda^{5/2})\Bigg\}. \end{aligned} \tag{54}$$

In the limit $t \to \infty$ we find from eqs. (52) and (53) that

$$\overline{g}_0 \to g\left(1 - \frac{B_3 g^2}{\epsilon}\right)e^{-\epsilon t} \tag{55}$$

$$\overline{g}_1 \to \frac{B_5 g^5}{2\epsilon}\left(1 - \frac{B_3 g^2}{\epsilon}\right)^{-3/2} \left\{\frac{1 - \frac{B_3 g^2}{\epsilon}}{\left(\frac{B_3 g^2}{\epsilon}\right)^2} \ln\left(1 - \frac{B_3 g^2}{\epsilon}\right) + \frac{1}{\left(\frac{B_3 g^2}{\epsilon}\right)}\right\} e^{-et} \tag{56}$$

while

$$\overline{\mu}^{\epsilon\lambda} \to \mu^{\epsilon\lambda} e^{\epsilon t}. \tag{57}$$

From eqs. (54 - 57) we find that as $t \to \infty$

$$g_B = \mu^{\epsilon\lambda} \left\{ \lambda^{1/2} \left[\frac{g}{\sqrt{1 - \frac{B_3 g^2}{\epsilon}}} \right] + \lambda^{3/2} \left[\frac{B_5 g^5}{2\epsilon} \frac{1}{\left(1 - \frac{B_3 g^2}{\epsilon}\right)^{3/2}} \right.\right.$$
$$\left.\left. \times \left(\frac{1 - \frac{B_3 g^2}{\epsilon}}{\left(\frac{B_3 g^2}{\epsilon}\right)^2} \ln \left(1 - \frac{B_3 g^2}{\epsilon}\right) + \frac{1}{\left(\frac{B_3 g^2}{\epsilon}\right)} \right) \right] \right\}, \tag{58}$$

which when $\lambda = 1$ reproduces what was found in eqs. (5), (18) and (19). The method of characteristics should reproduce all of the $S_n(\xi)$ appearing in eq. (5) as $t \to \infty$. This is analogous to eq. (43) reducing to eq. (44) when $t = \frac{1}{y}$.

(III) Having illustrated how the renormalization group equation works by considering the relationship between the bare and renormalized couplings when using dimensional regularization, we now will examine how it can be employed in conjunction with the perturbative calculation of a physical quantity [19-20].

In particular, if we consider $R(s)$, the ratio of the total cross section σ ($e^+e^- \to$ (hadrons)) to the total cross section $\sigma(e^+e^- \to \mu^+\mu^-)$, we find that a perturbative calculation of $R(s)$ yields

$$R(S) = 3 \sum_f Q_f^2 S \left[x(\mu^2), \ln\left(\frac{\mu^2}{s}\right) \right] \tag{59}$$

where s is the centre of mass energy, μ^2 is the renormalization scale and $x(\mu^2)$ is the coupling $\alpha_s(\mu^2)/\pi$. As μ^2 is unphysical, we have the renormalization group equation

$$\mu^2 \frac{dR(s)}{d\mu^2} = 0 = \left(\mu^2 \frac{\partial}{\partial \mu^2} + \beta(x) \frac{\partial}{\partial x} \right) R(s) \tag{60}$$

where

$$\beta(x(\mu^2)) = \mu^2 \frac{dx(\mu^2)}{d\mu^2}. \tag{61}$$

Perturbation theory results in the expansions

$$\beta(x) = -x^2 \sum_{k=0}^{\infty} \beta_k x^k \tag{62}$$

$$S[x, L] = 1 + \sum_{n=1}^{\infty} \sum_{m=0}^{n-1} T_{n,m} x^n L^m \tag{63}$$

where $x = x(\mu^2)$ and $L = \ln\left(\frac{\mu^2}{s}\right)$. Reorganizing the sum in eq. (63) so that in analogy with eq. (5)

$$S[x, L] = 1 + \sum_{n=1}^{\infty} x^n S_n(xL) \tag{64}$$

where

$$S_n(\xi) = \sum_{k=0}^{\infty} T_{n+k,k} \xi^k \tag{65}$$

leads to

$$(1-\beta_0\xi)\frac{dS_k}{d\xi} - k\beta_0 S_k = (1-\delta_{k,1})\sum_{\ell=1}^{k-1}\beta_\ell\left(\xi\frac{d}{d\xi} - \ell + k\right)S_{k-\ell} \quad (S_n(0) = T_{n,0}) \tag{66}$$

upon substitution of eq. (64) into eq. (60). Solving these equations sequentially we obtain [20]

$$S_1(\xi) = w^{-1} \quad (w \equiv 1-\beta_0\xi) \tag{67}$$

$$S_2(\xi) = \left[T_{2,0} - \frac{\beta_1}{\beta_0}\ln w\right] w^{-2} \tag{68}$$

$$S_3(\xi) = \left(\frac{\beta_1^2}{\beta_0^2} - \frac{\beta_2}{\beta_0}\right)w^{-2} + \left(T_{3,0} - \left(\frac{\beta_1^2}{\beta_0^2} - \frac{\beta_2}{\beta_0}\right) - \frac{\beta_1}{\beta_0}\left(2T_{2,0} + \frac{\beta_1}{\beta_0}\right)\ln w + \frac{\beta_1^2}{\beta_0^2}\ln^2 w\right)w^{-3} \tag{69}$$

and

$$\begin{aligned} S_4(\xi) = &-\frac{1}{2}\left[\frac{\beta_1}{\beta_0}\left(\frac{\beta_1^2}{\beta_0^2} - 2\frac{\beta_2}{\beta_0}\right) + \frac{\beta_3}{\beta_0}\right]w^{-2} + \left(2T_{2,0} + \frac{\beta_1}{\beta_0}\right)\left(\frac{\beta_1^2}{\beta_0^2} - \frac{\beta_2}{\beta_0}\right)w^{-3} \\ &+2\left(\frac{\beta_1}{\beta_0}\right)\left(\frac{\beta_2}{\beta_0} - \frac{\beta_1^2}{\beta_0^2}\right)w^{-3}\ln w + \left[T_{4,0} + \frac{\beta_3}{2\beta_0} - \frac{1}{2}\frac{\beta_1^3}{\beta_0^3}\right. \\ &\left.-2T_{2,0}\left(\frac{\beta_1^2}{\beta_0^2} - \frac{\beta_2}{\beta_0}\right)\right]w^{-4} + \frac{\beta_1}{\beta_0}\left[2\frac{\beta_1^2}{\beta_0^2} - 3\frac{\beta_3}{\beta_0} - 2T_{2,0}\frac{\beta_1}{\beta_0} - 3T_{3,0}\right]w^{-4}\ln w \\ &+\frac{\beta_1}{\beta_0}\left[\frac{5\beta_1^2}{2\beta_0^2} + 3T_{2,0}\frac{\beta_1}{\beta_0}\right]w^{-4}\ln^2 w - \frac{\beta_1^3}{\beta_0^3}w^{-4}\ln^3 w. \end{aligned} \tag{70}$$

$S_k(\xi)$ for $k > 4$ could be determined if $T_{k,0}$ and β_{k-1} were known. S_1 is the "leading log" (LL) sum; S_2 is the "next to leading log" (NLL) sum etc.

The approximation to $R(s)$ given by

$$S[x(\mu), L(\mu)] = 1 + \sum_{n=1}^{4} x^n(\mu) S_n(x(\mu)L(\mu)) \tag{71}$$

where $x(\mu)$ is a solution to

$$\mu^2\frac{dx(\mu)}{d\mu^2} = -x^2(\mu)\sum_{k=0}^{3}\beta_k x^k(\mu) \tag{72}$$

is virtually independent of μ; using the purely perturbative result

$$S[x(\mu), L(\mu)] = 1 + \sum_{n=1}^{4}\sum_{m=0}^{3} T_{n,m} x^n(\mu) L^m(\mu) \tag{73}$$

has a pronounced dependence on μ. This is not surprising, as eq. (72) is a solution of the renormalization group equation, with $\beta(x)$ truncated at four-loop order. This serves to demonstrate that the renormalization group improved expression for $R(s)$ given by eq. (71) has enhanced predictive power over the purely perturbative result of eq. (73), as physical results should be independent of the unphysical parameter μ^2.

One can also employ the method of characteristics to recover the result of eq.(71) [17] by following the techniques used in deriving eq. (54) when the relationship between g_B and g was discussed. In association with eq. (60) we define characteristic functions $\overline{\mu}^2(t)$ and $\overline{x}(t)$ satisfying

$$\frac{d}{dt}\overline{\mu}^2(t) = \overline{\mu}^2(t) \qquad (\overline{\mu}^2(0) = \mu^2) \tag{74}$$

$$\frac{d}{dt}\overline{x}(t) = \beta(\overline{x}(t)) \qquad (\overline{x}(0) = x(\mu^2))\,. \tag{75}$$

It is apparent from eqs. (61) and (75) that $x(\mu^2)$ and $\overline{x}(t)$, the running and characteristic functions, satisfy the same equations. They are however distinct functions with the running coupling serving as a boundary condition to the characteristic function.

We begin by a rescaling

$$t \to t/\lambda \tag{76}$$

$$\overline{x} \to \overline{x}\lambda \tag{77}$$

where λ is a book keeping parameter, so that

$$\lambda\frac{d\overline{\mu}(t)}{dt} = \overline{\mu}^2(t) \tag{78}$$

$$\lambda^2\frac{d\overline{x}(t)}{dt} = -\lambda^2\overline{x}^2(t)\sum_{n=0}^{\infty}\overline{x}^n(t)\lambda^n\beta_n \tag{79}$$

and

$$S[\overline{x},\overline{L}] = 1 + \sum_{n=1}^{\infty}\sum_{m=0}^{n-1}T_{n,m}\lambda^n\overline{x}^n\overline{L}^m \tag{80}$$

where $\overline{L} \equiv \ln\left(\overline{\mu}^2/S\right)$. Perturbatively expanding $\overline{x}(t)$

$$\overline{x}(t) = \sum_{n=0}^{\infty}\lambda^n\overline{x}_n(t) \qquad (\overline{x}_n(0) = x(\mu^2)\delta_{n0}) \tag{81}$$

means that eq. (79) can be satisfied order-by-order in λ provided that

$$\overline{x}_0(t) = x\overline{w}^{-1} \quad (x \equiv x(\mu^2), \quad \overline{w} \equiv 1 + \beta_0 x(\mu^2)t) \tag{82}$$

$$\overline{x}_1(t) = -\frac{\beta_1}{\beta_0}\left(\frac{x}{\overline{w}}\right)^2\ln\overline{w} \tag{83}$$

$$\overline{x}_2(t) = \left(\frac{x}{\overline{w}}\right)^3\left[\left(\frac{\beta_1^2}{\beta_0^2} - \frac{\beta_2}{\beta_0}\right)\beta_0 xt - \frac{\beta_1^2}{\beta_0^2}\ln\overline{w} + \frac{\beta_1^2}{\beta_0^2}\ln^2\overline{w}\right] \tag{84}$$

$$\begin{aligned}\overline{x}_3(t) = \left(\frac{x}{\overline{w}}\right)^4\Bigg[&\left(\frac{-\beta_1^3}{2\beta_0^3} + \frac{\beta_1\beta_2}{\beta_0^2} - \frac{\beta_3}{2\beta_0}\right)\overline{w}^2 + \left(\frac{\beta_1^3}{\beta_0^3} - \frac{\beta_1\beta_2}{\beta_0^2}\right)\overline{w}(1 - 2\ln\overline{w}) \\ &+\left(-\frac{\beta_1^3}{2\beta_0^3} + \frac{\beta_3}{2\beta_0}\right) + \left(\frac{2\beta_1^3}{\beta_0^3} - \frac{3\beta_1\beta_2}{\beta_0^2}\right)\ln\overline{w} \\ &+\frac{5\beta_1^3}{2\beta_0^3}\ln^2\overline{w} - \frac{\beta_1^3}{\beta_0^3}\ln^3\overline{w}\Bigg]\,.\end{aligned} \tag{85}$$

Together, eqs. (80) and (81) lead to

$$\begin{aligned}S[\overline{x}(t),\overline{L}(t)] = &1 + \lambda\left[T_{1,0}\overline{x}_0\right] + \lambda^2\left[T_{1,0}\overline{x}_1 + \left(T_{2,0} + T_{2,1}\overline{L}\right)\overline{x}_0^2\right] \\ &+\lambda^3\left[T_{1,0}\overline{x}_2 + \left(T_{2,0} + T_{2,1}\overline{L}\right)(2\overline{x}_0\overline{x}_1)\right.\end{aligned}$$

$$
\begin{aligned}
&+\left(T_{3,0}+T_{3,1}\overline{L}+T_{3,2}\overline{L}^2\right)\overline{x}_0^3\Big] \\
&+\lambda^4\left[T_{1,0}\overline{x}_3+\left(T_{2,0}+T_{2,1}\overline{L}\right)\left(\overline{x}_1^2+2\overline{x}_0\overline{x}_2\right)\right. \\
&+\left(T_{3,0}+T_{3,1}\overline{L}+T_{3,2}\overline{L}^2\right)\left(3\overline{x}_0\overline{x}_1\right) \\
&\left.+\left(T_{4,0}+T_{4,1}\overline{L}+T_{4,2}\overline{L}^2+T_{4,3}\overline{L}^3\right)\overline{x}_0^4\right]+\dots
\end{aligned}
\tag{86}
$$

As was shown in the discussion leading up to eq. (31), eq. (86) is independent of t, provided one were to sum all contributions to $S[\overline{x}(t),\overline{L}(t)]$. This independence of t can be exploited by writing t as $t=\lambda\ln k$ (so that by eq. (78) $\overline{\mu}^2(t)=k\mu^2$) and then expanding eq. (86) in powers of k; the resulting expression reduces to eq. (63) provided [17]

$$
\begin{aligned}
T_{2,1}&=\beta_0T_{1,0} \\
T_{3,1}&=2T_{2,0}\beta_0+\beta_1T_{1,0} \\
T_{3,2}&=\beta_0^2T_{1,0} \\
T_{4,1}&=3\beta_0T_{3,0}+2\beta_1T_{1,0}+\beta_2T_{1,0} \\
T_{4,2}&=3\beta_0^2T_{2,0}+\frac{5}{2}\beta_0\beta_1T_{1,0} \\
T_{4,3}&=\beta_0^3T_{1,0}
\end{aligned}
\tag{87}
$$

etc. These relationships are precisely what one obtains if eqs. (62) and (63) were substituted directly into eq. (60). Furthermore, eqs. (87) and (65) can be shown to lead to eqs. (67-70). A more direct way of recovering eq. (71) is to choose $t=\lambda\ln\left(\frac{s}{\mu^2}\right)$ so that $\overline{\mu}^2=s$, $\overline{L}=0$ and $\overline{w}=w$; taking $\lambda=1$ results in eq. (86) when truncated at $O\left(\lambda^4\right)$ reducing to eq. (71). This is much like what happened in eq. (44); in both cases a perturbative approximation to the solution of a linear partial differential equation becomes an exact solution when evaluated at the characteristic functions for a certain value of the characteristic parameter.

A number of other physical processes [19], among them the semi-leptonic decay rate of b mesons, QCD contributions to vector and scalar correlation functions, the Higgs decay rate into two gluons and the perturbative static potential, can also be analyzed using the renormalization group. In each case, physical quantities when computed perturbatively, exhibit a dependence on the renormalization μ^2; this dependence is virtually eliminated when a kth order perturbative calculation is improved by making use of the renormalization group equation to sum N^kLL contributions.

(IV) Renormalization group improvement can also be employed in thermal field theory calculations [21]. In QCD, if there are n_f quark flavours, the thermodynamic free energy for $T>>0$ is [22,23,24]

$$
\begin{aligned}
F=\frac{-8\pi^2}{45}T^4&\left\{\left(1+\frac{21}{32}n_f\right)+\frac{-15}{4}\left(1+\frac{5}{12}n_f\right)\frac{\alpha_s}{\pi}+30\left[\left(1+\frac{n_f}{6}\right)\left(\frac{\alpha_s}{\pi}\right)\right]^{3/2}\right. \\
&+\left\{237.2+15.97n_f-.413n_f^2+\frac{135}{2}\left(1+\frac{n_f}{6}\right)\ln\left[\frac{\alpha_s}{\pi}\left(1+\frac{n_f}{6}\right)\right]\right. \\
&\left.-\frac{165}{8}\left(1+\frac{5}{12}n_f\right)\left(1-\frac{2}{33}n_f\right)\ln\frac{\overline{\mu}}{2\pi T}\right\}\left(\frac{\alpha_s}{\pi}\right)^2+\left(1+\frac{n_f}{6}\right)^{1/2} \\
&\left[-799.2-21.96n_f-1.926n_f^2+\frac{495}{2}\left(1+\frac{n_f}{6}\right)\left(1-\frac{2}{33}n_f\right)\right.
\end{aligned}
\tag{88}
$$

$$\ln \frac{\overline{\mu}}{2\pi T}\bigg] \left(\frac{\alpha_s}{\pi}\right)^{5/2} + O\left(\alpha_s^3 \ln \alpha_s\right)\bigg\}$$

where $\overline{\mu}^2$ is the $\overline{MS}$ renormalization scale, and $x(\overline{\mu}^2) = \frac{\alpha_s(\overline{\mu}^2)}{\pi}$ satisfies the two-loop renormalization group equation

$$\mu^2 \frac{dx(\mu^2)}{d\mu^2} = b_2 x^2(\mu^2) + b_3 x^3(\mu^2). \tag{89}$$

Changes in $\overline{\mu}^2$ in eq. (88) (with corresponding changes in α_s dictated by eq. (89)) lead to large variations in F. To overcome this deficiency in the perturbative calculation of F, we look for a solution to the renormalization group equation

$$\left(\overline{\mu}^2 \frac{\partial}{\partial \overline{\mu}^2} + \beta(x)\frac{\partial}{\partial x}\right)\left(\frac{F}{F_0}\right) = 0 \tag{90}$$

that is of the form

$$F/F_0 = 1 + \sum_{n=0}^{\infty} \left(R_n(\xi)x^{n+1} + S_n(\xi)x^{n+3/2} + T_n(\xi)x^{n+2}\ln x\right) \tag{91}$$

where F_0 is the ideal-gas value of F, $\xi = xL = x\ln\left(\frac{\overline{\mu}^2}{(2\pi T)^2}\right)$.

The general form of the functions $R_n(\xi)$, $S_n(\xi)$ and $T_n(\xi)$ is

$$R_n(\xi) = \sum_{m=0}^{\infty} A_{n+m,m}\xi^n, \quad S_n(\xi) = \sum_{m=0}^{\infty} B_{n+m,m}\xi^m, \quad T_n(\xi) = \sum_{m=0}^{\infty} C_{n+m,m}\xi^m \tag{92}$$

with the renormalization group equation (90) allowing us to determine $A_{n+m,m}$, $B_{n+m,m}$ and $C_{n+m,m}$ $(m > 0)$ in terms of $A_{n,0}$, $B_{n,0}$, $C_{n,0}$ and the β-function coefficients. From eqs. (89-91) it follows that

$$\begin{aligned} 0 = \sum_{n=0}^{\infty} \{ & \left[R_n' x^{n+2} + \left(b_2 x^2 + \ldots\right)\left(\xi R_n' + (n+1)R_n + xT_n\right)x^n\right] \\ & + \left[S_n' x^{n+5/2} + \left(b_2 x^2 + \ldots\right)\left(\xi S_n' + \left(n+\frac{3}{2}\right)S_n\right)x^{n+1/2}\right] \\ & + \left[T_n' x^{n+3} + \left(b_2 x^2 + \ldots\right)\left(\xi T_n' + (n+2)\,T_n\right)x^{n+1}\ln x\right]\}. \end{aligned} \tag{93}$$

It is possible to read off $A_{0,0}$, $A_{1,0}$, $B_{0,0}$, $B_{1,0}$ and $T_{0,0}$ from eq. (88); these values combined with the requirement that eq. (93) be satisfied at each order in x results in

$$R_0(\xi) = A_{0,0}\, w^{-1}, \quad S_0(\xi) = B_{0,0}\, w^{-3/2}, \quad T_0(\xi) = C_{0,0}\, w^{-2} \tag{94}$$

and

$$\begin{aligned} R_1(\xi) &= w^{-2}\left[A_{1,0} - \left(\frac{b_3}{b_2} A_{0,0} + C_{0,0}\right)\ln w\right], \\ S_1(\xi) &= w^{-5/2}\left[B_{1,0} - \frac{3}{2}\frac{b_3}{b_2} B_{0,0}\ln w\right] \end{aligned} \tag{95}$$

where $w = 1 + b_2\xi$. With these functions, the approximation to F/F_0 is given by

$$F/F_0 = 1 + \sum_{n=0}^{1}\left(R_n(\xi)x^{n+1} + S_n(\xi)x^{n+3/2}\right) + T_0(\xi)x^2\ln x \tag{96}$$

is virtually independent of $\overline{\mu}^2$ and has a value which coincides with the results of lattice calculations of this quantity. The effect of absorbing changes if $\overline{\mu}^2$ by altering

$A_{0,0}$, $B_{0,0}$ and $C_{0,0}$ rather than having it simply change α_s according to eq. (89) is considered in [21], although this is contrary to the renormalization group approach.

The method of characteristics can also be used to arrive at this result [12]. As has been done in previous examples, characteristic functions $\overline{\mu}^2(t)$ and $\overline{x}(t)$ are introduced

$$\frac{d\overline{\mu}^2(t)}{dt} = \overline{\mu}^2(t) \quad \left(\overline{\mu}^2(0) = \mu^2\right) \tag{97}$$

$$\frac{d\overline{x}(t)}{dt} = b_2\overline{x}^2(t) + b_3\overline{x}^3(t) + \ldots (\overline{x}(0) = x(\mu^2)) \tag{98}$$

and then a rescaling

$$t \to t/\lambda, \quad \overline{x}(t) \to \lambda\overline{x}(t) \tag{99}$$

is performed, followed by the perturbative expansion

$$\overline{x}(t) = \sum_{n=0}^{\infty} \lambda^n \overline{x}_n(t) \quad \left(\overline{x}_n(0) = x(\mu^2)\delta_{n,0}\right). \tag{100}$$

Solving for $\overline{x}_n(t)$ iteratively and then substitution of the results into the perturbative result

$$F/F_0 = 1 + \left(A_{0,0}x + A_{1,0}x^2 + A_{1,1}x^2 L \right.$$
$$\left. + B_{0,0}x^{3/2} + B_{1,0}x^{5/2} + B_{1,1}x^{5/2}L + C_{0,0}x^2 \ln x\right) \tag{101}$$

(which is eq. (88)) reduces to what was obtained in eq. (96) when $t = \lambda \ln\left(\frac{(2\pi T)^2}{\overline{\mu}^2}\right)$.

(V) When the renormalization group function β is known exactly, information can be extracted about the effective action. This is the converse of the analysis of ref. [25] where computation of the two-loop effective action in both scalar and spinor quantum electrodynmaics in the presence of a self-dual background field strength was used to determine the two-loop β-function.

It has been shown that the effective action L and the β-function are related by the equation

$$L = \frac{-1}{4}\frac{g^2}{\overline{g}^2(t,g)}\Phi \tag{102}$$

where Φ is related to the constant background field strength $F^a_{\mu\nu}$ by

$$\Phi = F^a_{\mu\nu}F^{a\mu\nu} \tag{103}$$

and

$$t = \frac{1}{4}\ln\left(\frac{g^2\Phi}{\mu^4}\right) \tag{104}$$

$$= \int_g^{\overline{g}(t,g)} \frac{dx}{\beta(x)}. \tag{105}$$

Eq. (102) follows from the trace anomaly condition

$$\langle \theta^\mu{}_\mu \rangle = \frac{\beta(\overline{g})}{2\overline{g}}\left(\frac{g}{\overline{g}}\right)^2 \Phi \tag{106}$$

and the definition of the expectation value of the energy-momentum tensor

$$\langle \theta^{\mu\nu} \rangle = -g^{\mu\nu}L + 2\frac{\delta L}{\delta g_{\mu\nu}}. \tag{107}$$

L in eq. (102) satisfies the renormalization group equation

$$\left(\mu\frac{\partial}{\partial\mu}+\beta(g)\frac{\partial}{\partial g}+\gamma(g)F^a_{\alpha\beta}\frac{\partial}{\partial F^a_{\alpha\beta}}\right)L=0 \tag{108}$$

provided

$$\beta(g)=-g\gamma(g). \tag{109}$$

Eq. (109) follows from the fact that [26,27] gA^a_μ is not renormalized in order to preserve gauge invariance in the background field.

We note that from eqs. (104) and (105),

$$\frac{\partial\overline{g}(t,g)}{\partial t}=\beta(\overline{g}(t,g))=\frac{\partial\overline{g}(t,g)}{\partial t}\beta(g) \tag{110}$$

and

$$\frac{\partial t}{\partial\mu}=-1. \tag{111}$$

In the $SU(3) N=1$ super Yang-Mills theory, the β-function in a supersymmetric renormalization scheme is given by [28,29]

$$\beta(g)=\frac{-9g^3/(4\pi)^2}{1-6g^2/(4\pi)^2}. \tag{112}$$

(It is not immediately apparent how this renormalization scheme is related to minimal subtraction [30].) If $\overline{y}=-(4\pi)^2/6\overline{g}^2(t,g)$ and $y=-(4\pi)^2/6g^2$, then by eq. (105)

$$\overline{y}e^{\overline{y}}=e^{-3t}ye^y, \tag{113}$$

so that if $W(\eta)e^{W(\eta)}=\eta$ defines the Lambert W function [31], then eq. (102) becomes

$$L=\frac{-1}{4y}W\left[\left(\frac{g^2\Phi}{\mu^4}\right)^{-3/4}ye^y\right]\Phi. \tag{114}$$

If we consider an $SU(3)$ $N=2$ super Yang-Mills theory, then the β-function is an entirely one-loop quantity,

$$\beta(g)=-\frac{6}{(4\pi)^2}g^3; \tag{115}$$

with this it follows from eqs. (102)and (105) that

$$L=\frac{-1}{4}\left[1+\frac{3g^2}{(4\pi)^2}\ln\left(\frac{g^2\Phi}{\mu^4}\right)\right]\Phi. \tag{116}$$

(VI) Instanton contributions to the effective action will now be examined; renormalization group considerations again show how a purely perturbative calculation can be improved. The one instanton contribution to the effective action in an $SU(2)$ gauge theory with n_f flavours is of the form [32]

$$L_{eff}=K\int d\rho\,\rho^{-5+3n_f}\exp\left\{\frac{-8\pi^2}{g^2(\mu)}S\right\} \tag{117}$$

where S is a power series

$$S=1+\sum_{n=1}^{\infty}\sum_{m=0}^{n}T_{n,m}g^{2n}(\mu)\ln^m(\mu\rho). \tag{118}$$

The integral over ρ, the "size" of the instanton, converges in the ultraviolet limit ($\rho \to 0$) but diverges in the infrared limit ($\rho \to \infty$) when the one-loop contributions $T_{1,0}$ and $T_{1,1}$ computed in [32] are included.

Resumming the two series in eq. (118) [18] so that

$$S = \sum_{k=0}^{\infty} g^{2k}(\mu) \left(\sum_{\ell=k}^{\infty} T_{\ell,\ell-k} \left(g^2(\mu)\ln(\mu\rho)\right)^{\ell-k} \right)$$
$$\equiv \sum_{k=0}^{\infty} g^{2k}(\mu) S_k \left(g^2(\mu)\ln(\mu\rho)\right) \tag{119}$$

and applying the renormalization group equation

$$\left(\mu\frac{\partial}{\partial\mu} + \beta(g)\frac{\partial}{\partial g}\right) S = 0 \tag{120}$$

with the β-function being given by

$$\beta(g) = \sum_{\ell=1}^{\infty} b_{2\ell+1} g^{2\ell+1} \tag{121}$$

we find that

$$S_0 = 1 + 2b_3 g^2(\mu)\ln(\mu\rho) \equiv w \tag{122}$$

and for $k > 1$

$$\frac{dS_k}{dw} + \frac{k-1}{w} S_k = \frac{-1}{2b_3 w}\left[\sum_{\ell=0}^{k-1} b_{3+2(k-\ell)}\left(2(\ell-1)S_\ell + 2(w-1)\frac{dS_\ell}{dw}\right)\right] \tag{123}$$

with $S_k(w=1) = T_{k,0}$. From eq. (123) it follows that

$$S_1 = T_{1,0} + \frac{b_5}{b_3}\ln|w| \tag{124}$$

$$S_2 = \left(\frac{b_7}{b_3} - \frac{b_5^2}{b_3^2}\right) + \frac{T_{2,0} - \left(\frac{b_7}{b_3} - \frac{b_5^2}{b_3^2}\right) + \frac{b_5^2}{b_3^2}\ln|w|}{w}. \tag{125}$$

In general, from eq. (123) it follows that for $k \geq 2$,

$$S_k(w) = C_k + O\left(\frac{1}{w}\right) \tag{126}$$

where C_k is dependent only on the β-function coefficients. Consequently, by eq. (119),

$$\rho^{-5+3n_f}\exp\left(\frac{-8\pi^2}{g^2}S\right) \underset{(\ln\rho\to\infty)}{\longrightarrow} \rho^{\frac{7}{3}(1+n_f)}|\ln\rho|^{-8\pi^2 b_5/b_3}. \tag{127}$$

We consequently see that even upon including the contribution of all terms in eq. (119) to S, the integral over ρ still suffers from an infrared divergence as $\rho \to \infty$.

(VII) It is also possible to use the renormalization group to improve a perturbative calculation of the effective potential [33-37]. This procedure has been applied in both scalar electrodynamics and the standard model at leading-log order [38-39] and beyond [40]. We will illustrate this procedure by considering leading-log corrections to the one-loop calculation of the effective potential in massless scalar electrodynamics.

In this model, the classical Lagrangian is given by

$$L = \frac{1}{2}\left(\partial_\mu \phi_1 - eA_\mu \phi_2\right)^2 + \frac{1}{2}\left(\partial_\mu \phi_2 + eA_\mu \phi_1\right)^2 - \frac{\lambda}{4!}\left(\phi_1^2 + \phi_2^2\right)^2. \tag{128}$$

One-loop corrections to $V = \frac{\lambda}{4!}\left(\phi_1^2 + \phi_2^2\right)^2 \equiv \frac{\lambda\phi^4}{4!}$ in the Landau gauge result in [33]

$$V = \phi^4\left[\frac{\lambda}{4!} + \left(\frac{5\lambda^2}{1152\pi^2} + \frac{3e^4}{64\pi^2}\right)\left(\ln\frac{\phi^2}{\mu^2} + k\right) + O(\lambda^3, e^6)\right] \tag{129}$$

where k is a constant whose value is fixed by a renormalization condition. If we chose this condition to be

$$\frac{d^4V}{d\phi^4} = \lambda \tag{130}$$

when $\phi^2 = \mu^2$, then

$$k = -\frac{25}{6}. \tag{131}$$

If now $<\phi>$ is the vacuum expectation value of ϕ, then upon choosing $\mu^2 =<\phi>^2$, it follows from eq. (129) that if

$$\left.\frac{dV}{d\phi}\right|_{<\phi>} = 0 \tag{132}$$

then at the scale $\mu^2 =<\phi>^2$

$$\lambda = \frac{33e^4}{8\pi^2} \tag{133}$$

so that eq. (129) becomes

$$V = \frac{3e^4}{64\pi^2}\phi^4\left[\ln\left(\frac{\phi^2}{<\phi>^2}\right) - \frac{1}{2}\right] + O(e^6). \tag{134}$$

Upon identifying thescalar and vector masses with

$$m_\phi^2 = V''(\phi =<\phi>) \qquad m_A^2 = e^2 <\phi>^2 \tag{135}$$

respectively, we find from eq. (134) that [33]

$$\frac{m_\phi^2}{m_A^2} = \frac{3e^2}{8\pi^2} \tag{136}$$

where in eq. (136) (as in eq. (134)) e^2 is evaluated at $\mu^2 =<\phi>^2$.

The general form of the effective potential is

$$V = \frac{\pi^2\phi^4}{6}S(\lambda, e^2, L) \tag{137}$$

where $L = \ln\left(\frac{\phi^2}{\mu^2}\right)$. The renormalization group equation $\mu\,\frac{dV}{d\mu} = 0$ then gives rise to the equation

$$\left[(-2+2\gamma)\frac{\partial}{\partial L} + \beta_e\frac{\partial}{\partial e^2} + \beta_\lambda\frac{\partial}{\partial\lambda} + 4\gamma\right]S = 0 \tag{138}$$

with at one-loop order [33]

$$\gamma = \frac{3e^2}{16\pi^2}, \tag{139}$$

$$\beta_e = \frac{e^4}{24\pi^2}, \tag{140}$$

$$\beta_\lambda = \frac{5\lambda^2}{24\pi^2} - \frac{3\lambda e^2}{4\pi^2} + \frac{9e^4}{4\pi^2}. \tag{141}$$

S in eq. (137) can now be expressed as a power series in λ, e^2 and L with the leading-log contribution (which is fixed entirely by the one-loop calculation) being given by those terms in which the power L is raised to being one less than the sum of the powers to which e^2 and λ are raised. If $x = 4\pi^2 e^2$, $y = 4\pi^2\lambda$ the Leading-Log contribution to S is of the form

$$S_{LL} = \sum_{n=0}^{\infty} \left(R_{n,n-1} y^n L^{n-1} + \sum_{k=0}^{\infty} T_{n,k} x^n y^k L^{n+k-1} \right) \tag{142}$$

Eq. (138) now becomes

$$\left[-2\frac{\partial}{\partial L} + \left(\frac{5}{6}y^2 - 3xy + 9x^2 \right) \frac{\partial}{\partial y} + \frac{x^2}{6}\frac{\partial}{\partial x} + 3x \right] S_{LL}(x, y, L) = 0. \tag{143}$$

Together, eqs. (142) and (143) are satisfied at order,

$$y^p L^{p-1} \text{ if } \quad -2(p-1)R_{p,p-1} + \frac{5}{6}(p-1)R_{p-1,p-2} = 0; \tag{144}$$

$$xy^p L \text{ if } \quad -2pT_{1,p} + \frac{5}{6}(p-1)T_{1,p-1} - 3(p-1)R_{p,p-1} = 0; \, (p \geq 1) \tag{145}$$

$$\begin{aligned} x^2 y^p L^p \text{ if } \quad & -2(p+1)T_{2,p} + \frac{5}{6}(p-1)T_{2,p-1} - 3pT_{1,p} \\ & +9(p+1)R_{p+1,p} + \frac{19}{6}T_{1,p} = 0 \;\; (p \geq 1) \end{aligned} \tag{146}$$

$$\begin{aligned} x^n y^p L^{n+p-2} \text{ if } \quad & -2(p+n-1) + \frac{5}{6}(p-1)T_{n,p-1} - 3pT_{n-1,p} \\ & +9(p+1)T_{n-2,p+1} + \frac{n+17}{6}T_{n-1,p} = 0 \;\; (n \geq 3, p \geq 1) \end{aligned} \tag{147}$$

and

$$-2(n-1)T_{n,0} + 9T_{n-2,1} + \frac{n+17}{6}T_{n-1,0} = 0 \;\; (n \geq 3). \tag{148}$$

From eq. (129), we have the values of $T_{0,1}$, $R_{1,0}$, $T_{2,0}$, $T_{0,2}$ as well as $T_{1,1} = 0$. These values are consistent with eqs. (144-148), and furthermore, if

$$S_0(yL) = \sum_{n=1}^{\infty} R_{n,n-1}(yL)^{n-1}, \tag{149}$$

$$S_j(yL) = \sum_{n=0}^{\infty} T_{j,n}(yL)^n \tag{150}$$

we find that these recursion relations serve to fix the functions $S_n(yL)$. It can be shown that in particular

$$S_0(yL) = \frac{1}{1 - \frac{5}{12}yL} \equiv \frac{1}{w} \tag{151}$$

$$S_1(w) = -\frac{9}{5}\left(\frac{w-1}{w} \right)^2 \tag{152}$$

$$S_2(w) = \frac{1}{20w^3}\left(20w^3 + 77w^2 - 34w + 27 \right) \tag{153}$$

and

$$S_3(w) = \frac{1}{240w^4}\left(580w^4 + 760w^3 - 323w^2 + 126w - 243 \right). \tag{154}$$

With

$$V^{LL} = \frac{\pi^2\phi^4}{6}\left[yS_0(yL) + \sum_{n=1}^{\infty} x^n L^{n-1} S_n(yL)\right] + K\phi^4 \tag{155}$$

the condition of eq. (130) leads to

$$\begin{aligned}\frac{6K}{\pi^2} = -\left(\frac{125}{72}y^2 + \frac{75}{4}x^2\right) - \left(\frac{2625y^3 + 1875y^4 + 625y^5}{1296}\right) \\ + \frac{x\left(175y^2 + 250y^3 + 125y^4\right)}{48} \\ - \frac{x^2\left(9450y + 10275y^2 + 5275y^3\right)}{432},\end{aligned} \tag{156}$$

upon including S_1 and S_2 in the sum of eq. (155). (At lowest order, this is consistent with eq. (131).) If at $\mu^2 =< \phi >^2$, where $L = 0$, the condition of eq. (132) leads to a generalization of eq. (133)

$$\begin{aligned}0 = \Bigg[\left(\frac{1296y - 1980y^2 - 2625y^3 - 1875y^4 - 625y^5}{1944}\right) \\ + x\left(\frac{175y^2 + 250y^3 + 125y^4}{72}\right) \\ - x^2\left(\frac{7128 + 9450y + 10275y^2 + 5275y^3}{648}\right)\Bigg].\end{aligned} \tag{157}$$

This defines a relationship between x and y when these running couplings are evaluated at $\mu^2 =< \phi >^2$. As eq. (157) is quadratic in x, there are two values of y for each value of x, one large and the other small. If $y \leq .01$, then eq. (157) reduces to eq. (133).

We also find that if we use the potential of eq. (155) (with only S_1 and S_2 included) and K given by eq. (156), then the analogue of eq. (136) is

$$\begin{aligned}\frac{m_\phi^2}{m_A^2} = \left(\frac{y}{2x} - \frac{27x}{4}\right) - \frac{1}{2592x}\big[1620y^2 + 2475y^3 + 1875y^4 \\ + 625y^5 - x\left(4455y^2 + 6750y^3 + 3375y^4\right) \\ + x^2\left(26730y + 30825y^2 + 15825y^3\right)\big].\end{aligned} \tag{158}$$

In the region in which eq. (157) reduces to eq. (133), eq. (158) reduces to eq. (136).

The analysis that has just been used to examine the effective potential in massless scalar electrodynamics can also be applied in the standard model [38-39]. The dominant couplings that must be included are the top quark Yukawa coupling, the quartic scalar coupling and the strong gauge coupling; the effects of the $SU(2) \times U(1)$ gauge couplings are treated as being secondary. If the tree level mass of the scalar is taken to vanish, then upon including leading log contributions to the effective potential and using experimental values for the top quark Yukawa coupling and the strong gauge coupling, it is possible to show that the Higgs scalar would have a mass of about 216 GeV. Furthermore, the value of the quartic scalar coupling is considerably enhanced, leading to an increase in the scattering cross section $\sigma(W_L^+ W_L^- \to Z_L^0 Z_L^0)$ by a factor of about 30 over conventional expectations. These results are quite stable when the accessible higher order effects are included [40].

(VIII) We now turn to an exact solution of the renormalization group equation, focusing on a massless $\lambda\phi_4^4$ theory first of all, then considering the effect of including a mass for this scalar field.

The renormalization group equation for the effective potential V in this model is [33,15,16,41-43]

$$\mu\frac{dV(\mu,\lambda,\phi)}{d\mu} = \left(\mu\frac{\partial}{\partial\mu} + \beta(\lambda)\frac{\partial}{\partial\lambda} + \gamma(\lambda)\phi\frac{\partial}{\partial\phi}\right)V(\mu,\lambda,\phi) = 0, \tag{159}$$

where $\lambda = \lambda(\mu)$ and $\phi = \phi(\mu)$ satisfy

$$\mu\frac{d\lambda}{d\mu} = \beta(\lambda) \tag{160}$$

and

$$\mu\frac{d\phi}{d\mu} = \gamma(\lambda)\phi. \tag{161}$$

Upon taking

$$V(\mu,\lambda,\phi) = Y(\lambda,L)\phi^4 \tag{162}$$

where $L \equiv \ln\left(\frac{\lambda\phi}{\mu}\right)$, eq. (159) becomes

$$\left[\frac{\partial}{\partial L} - \tilde{\beta}(\lambda)\frac{\partial}{\partial\lambda} - 4\tilde{\gamma}(\lambda)\right]Y(\lambda,L) - 0 \tag{163}$$

where

$$\tilde{\beta} = \beta/(1-\gamma-\beta/2\lambda), \quad \tilde{\gamma} = \gamma/(1-\gamma-\beta/2\lambda). \tag{164}$$

(The factor of λ in the argument of L arises if one uses dimensional regularization [43].) An auxiliary function $\overline{\lambda}(L,\lambda)$ is defined by

$$L = \int_\lambda^{\overline{\lambda}(L,\lambda)} \frac{dx}{\tilde{\beta}(x)} \tag{165}$$

satisfies

$$\overline{\lambda}(L=0,\lambda) = \lambda \tag{166}$$

$$\frac{\partial\overline{\lambda}(L,\lambda)}{\partial L} = \tilde{\beta}(\overline{\lambda}(L,\lambda)) \tag{167}$$

$$\frac{\partial\overline{\lambda}(L,\lambda)}{\partial\lambda} = \frac{-\tilde{\beta}(\overline{\lambda}(L,\lambda))}{\tilde{\beta}(\lambda)}. \tag{168}$$

Often, eq. (163) is taken to imply [33,5,10,44]

$$Y(\lambda,L) = f(\overline{\lambda}(L,\lambda))\exp\left[4\int_0^L \tilde{\gamma}(\overline{\lambda}(x,\lambda))dx\right]; \tag{169}$$

however, substitution of eq. (169) into the left side of eq. (163) results in

$$\begin{aligned} I = \Bigg\{ & f'(\overline{\lambda}(L,\lambda))\left[\frac{\partial\overline{\lambda}(L,\lambda)}{\partial L} - \tilde{\beta}(\lambda)\frac{\partial\overline{\lambda}(L,\lambda)}{\partial\lambda}\right] \\ & + f(\overline{\lambda}(L,\lambda))\Big[4\tilde{\gamma}(\overline{\lambda}(L,\lambda)) - 4\tilde{\gamma}(\lambda) \\ & + 4\int_0^L \frac{\partial\tilde{\gamma}(\overline{\lambda}(x,\lambda))}{\partial\lambda}dx\Big]\Bigg\}\exp\left[4\int_0^L \tilde{\gamma}(\overline{\lambda}(x,L)))dx\right] \end{aligned} \tag{170}$$

which only vanishes if $L = 0$ (where eq. (163) is satisfied by eq. (169) because of eqs. (166-168)).

There is, however, a way of obtaining a formal solution to eq. (163) [12]. If

$$Y(\lambda, L) = \sum_{m=1}^{\infty} \sum_{n=0}^{m-1} a_{mn} \lambda^m L^n \tag{171}$$

$$\equiv \sum_{m=0}^{\infty} A_m(\lambda) L^m$$

then from eq. (163) we find that

$$A_{n+1}(\lambda) = \frac{1}{2(n+1)} \left[\tilde{\beta}(\lambda) \frac{\partial}{\partial \lambda} + 4\tilde{\gamma}(\lambda) \right] A_n(\lambda). \tag{172}$$

If now

$$A_n(\lambda) = \exp \left[-4 \int_{\lambda_0}^{\lambda} dx \frac{\tilde{\gamma}(x)}{\tilde{\beta}(x)} \right] \tilde{A}_n(\lambda) \tag{173}$$

and

$$\eta(\lambda) = 2 \int_{\lambda_0}^{\lambda} \frac{dx}{\tilde{\beta}(x)} \tag{174}$$

then eq. (172) becomes

$$\tilde{A}_{n+1}(\eta) = \frac{1}{n+1} \frac{\partial}{\partial \eta} \tilde{A}_n(\eta). \tag{175}$$

As a result of eqs. (171-175) we find that

$$Y(\lambda, L) = \exp \left[-4 \int_{\lambda_0}^{\lambda} dx \frac{\tilde{\gamma}(x)}{\tilde{\beta}(x)} \right] \sum_{n=0}^{\infty} \tilde{A}_n(\lambda) L^n$$

$$= \exp \left[-4 \int_{\lambda_0}^{\lambda} dx \frac{\tilde{\gamma}(x)}{\tilde{\beta}(x)} \right] \sum_{n=0}^{\infty} \frac{L^n}{n!} \left(\frac{d}{d\eta} \right)^n \tilde{A}_0(\lambda(\eta))$$

$$= \exp \left[-4 \int_{\lambda_0}^{\lambda} dx \frac{\tilde{\gamma}(x)}{\tilde{\beta}(x)} \right] \tilde{A}_0(\lambda(\eta + L))$$

$$= \exp \left[-4 \int_{\lambda(\eta+L)}^{\lambda} dx \frac{\tilde{\gamma}(x)}{\tilde{\beta}(x)} \right] A_0(\lambda(\eta + L)). \tag{176}$$

Eqs. (169) and (176) are distinct.

We have succeeded in expressing the effective potential in terms of the function A_0; we can now find A_0 itself by imposing a second condition on the effective potential. This condition is provided by eq. (132). Together, eqs. (132) and (171) lead to

$$\sum_{n=0}^{\infty} (4A_n(\lambda(\mu)) + 2(n+1)A_{n+1}(\lambda(\mu))) \left(\ln \frac{\lambda < \phi >^2}{\mu^2} \right)^n < \phi >^3 = 0. \tag{177}$$

When dealing with massless scalar electrodynamics, the analogue of this result at one-loop order leads to eq. (133) [33] which serves to relate the value of the quartic scalar coupling to the gauge coupling; this is extended to the leading log effective potential in eq. (157) above [38-39]. In the purely scalar $\lambda\phi_4^4$ model, the one-loop effective potential gives an inconsistent result [33]. A different interpretation of eq. (132) [45] is to take it to define a relationship between the function $A_0 \left(\lambda \left(\mu^2 = \lambda(\mu^2) < \phi >^2 \right) \right)$ and $A_1 \left(\lambda \left(\mu^2 = \lambda(\mu^2) < \phi >^2 \right) \right)$, as if $\mu^2 = \lambda(\mu^2) < \phi >^2$ is the value chosen for μ^2, then all terms in the sum appearing in eq. (177) vanish except for the $n = 0$ term. We take the resulting equation

$$A_1\left(\lambda\left(\mu^2\right)\right) = -2A_0\left(\lambda\left(\mu^2\right)\right) \tag{178}$$

to hold for all μ^2 as the actual value of $\lambda(\mu^2)$, for any value of μ^2, is contingent upon the boundary condition imposed on the equation for running coupling and is hence arbitrary.

Together, eqs. (172) with $n = 0$ and (178) imply that

$$A_0'(\lambda) + \left(\frac{4}{\beta(\lambda)} - \frac{2}{\lambda}\right) A_0(\lambda) = 0 \tag{179}$$

and so

$$A_0(\lambda) = K \exp\left[-\int_{\lambda_0}^{\lambda}\left(\frac{4}{\beta(x)} - \frac{2}{x}\right) dx\right] \tag{180}$$

$$= K'\lambda^{(2+4b_3/b_2^2)} \exp\left(\frac{4}{b_2\lambda}\right) \exp\left\{-\int_0^{\lambda}\left[\frac{4}{\beta(x)} - 4\left(\frac{1}{b_2x^2} - \frac{b_3}{b_2^2x}\right)\right] dx\right\}. \tag{181}$$

(Here $K = A_0(\lambda_0)$ is a boundary value; the divergence appearing in eq. (180) as $\lambda_0 \to 0$ is absorbed into K' in eq. (181) with $\beta(\lambda) = b_2\lambda^2 + b_3\lambda^3 + \dots$.)

Substitution of eq. (180) into eq. (176) leads to

$$V = A_0(\lambda) \exp\left(\int_{\lambda}^{\lambda(\eta+L)}\left(\frac{4\gamma(x) - 4 + 2\beta(x)/x}{\beta(x)}\right) dx\right)\phi^4. \tag{182}$$

By eqs. (164) and (174), eq. (182) collapses to

$$\begin{aligned} V &= A_0(\lambda) \exp -2\left[\eta\left(\lambda(\eta + L)\right) - \eta\right]\phi^4 \\ &= A_0(\lambda)\exp(-2L)\phi^4 = \frac{A_0(\lambda)\mu^4}{\lambda^2}. \end{aligned} \tag{183}$$

This result is independent of ϕ (and is hence "trivial" [46]) and has non-analytic dependence on λ, but still satisfies eq. (159). Eq. (183) can also be shown to follow if one were to use counterterm renormalization so that L appearing in eq. (161) is given by $L = \ln\left(\frac{\phi^2}{\mu^2}\right)$ without any dependence on λ.

It appears that the only way to escape the conclusion that V is independent of ϕ is to have eq. (177) satisfied not by eq. (178), but rather $< \phi >= 0$, in which case it is not possible to determine the function A_0, but there is no radiatively driven breaking to the symmetry $\phi \to -\phi$ in the original Lagrange density.

An alternate way of arriving at eq. (183) is to have eq. (177) satisfied order by order in $\ln\left(\frac{\lambda<\phi>^2}{\mu^2}\right)$, so that

$$A_{n+1}(\lambda(\mu)) = \frac{-2}{(n+1)}A_n(\lambda(\mu)). \tag{184}$$

Together, eqs. (163) and (183) show that

$$A_n(\lambda) = K_n \exp\left[-\int_{\lambda_0}^{\lambda}\left(\frac{4}{\beta(x)} - \frac{2}{x}\right) dx\right] \tag{185}$$

with $K_{n+1} = -2K_n/(n+1)$, with

$$A_n(\lambda) = \frac{(-2)^n}{n!}A_0(\lambda) \tag{186}$$

so that eq. (171) becomes

$$Y(\lambda, L) = \sum_{n=0}^{\infty} \frac{(-2)^n}{n!} L^n A_0(\lambda), \tag{187}$$

reproducing eq. (183).

The Lagrange density of massless scalar electrodynamics, given by eq. (128), is characterized by the two couplings λ and e^2, so the effective potential now takes the form

$$V(\lambda, e^2, \phi, \mu) = \sum_{n=0}^{\infty} A_n(\lambda, e^2) L^n \phi^4 \tag{188}$$

where again $L = \ln\left(\frac{\lambda\phi^2}{\mu^2}\right)$. Substitution of eq. (188) into the renormalization group equation

$$\left[\mu\frac{\partial}{\partial\mu} + \beta_\lambda(\lambda, e^2)\frac{\partial}{\partial\lambda} + \beta_{e^2}(\lambda, e^2)\frac{\partial}{\partial e^2} + \gamma(\lambda, e^2)\phi\frac{\partial}{\partial\phi}\right] V = 0 \tag{189}$$

gives rise to an analogue of eq. (171)

$$A_{n+1} = \frac{1}{2(n+1)}\left(\tilde{\beta}_\lambda\frac{\partial}{\partial\lambda} + \tilde{\beta}_{e^2}\frac{\partial}{\partial e^2} + 4\tilde{\gamma}\right) A_n \tag{190}$$

where

$$\tilde{\beta}_\lambda = \frac{\beta_\lambda}{1-\gamma-\beta_\lambda/2\lambda}, \qquad \tilde{\beta}_{e^2} = \frac{\beta_{e^2}}{1-\gamma-\beta_\lambda/2\lambda}, \qquad \tilde{\gamma} = \frac{\gamma}{1-\gamma-\beta_\lambda/2\lambda}. \tag{191}$$

Furthermore, if as in eq. (132),

$$\frac{d}{d\phi}V(\lambda, e^2, \phi = <\phi>, \mu) = 0 \tag{192}$$

then, much as we obtained eq. (177), we find that

$$\sum_{n=0}^{\infty} \left[4A_n(\lambda, e^2) + 2(n+1)A_{n+1}(\lambda, e^2)\right]\left(\ln\frac{\lambda <\phi>^2}{\mu^2}\right)^n <\phi>^3 = 0. \tag{193}$$

If this is to be satisfied order by order in $\ln\left(\frac{\lambda<\phi>^2}{\mu^2}\right)$ when $<\phi> \neq 0$, then we see that

$$A_{n+1}(\lambda, e^2) = -\frac{2}{n+1}A_n(\lambda, e^2), \tag{194}$$

so that a generalization of eq. (183) follows

$$V = \frac{A_0(\lambda, e^2)\mu^4}{\lambda^2}. \tag{195}$$

An argument presented in ref. [45] also can be used to establish eq. (195) using only the $n = 0$ contribution to eq. (194) with the renormalization group equation of eq. (190).

So also, the massless $\lambda\phi_3^6$ model can be shown by using the techniques which have led to eq. (183) to have an effective potential which is independent of the background field ϕ and to have a non-analytic dependence on λ unless the vacuum expectation value $<\phi>$ vanishes.

Finally we shall apply [47] the renormalization group to examine a massive $\lambda\phi_4^4$ model in which the Lagrange density is

$$L = \frac{1}{2}(\partial_\mu \phi)^2 - \frac{1}{2}m^2\phi^2 - \frac{1}{4!}\lambda\phi^4. \tag{196}$$

The effective potential V for this model has also been considered by [41-42]. In ref. [41] it was noted that the general form of V is

$$V = \left\{ \left(a + \frac{b}{x}\right) + \sum_{\ell=1}^{\infty} \lambda^{\ell+1} \sum_{m=0}^{\ell-1} x^{m-2} \sum_{n=0}^{\ell} y^n a_{\ell mn} \right. \tag{197}$$

$$\left. + \left(\frac{1-x}{x}\right)^2 \sum_{k=1}^{\infty} b_k \lambda^k \right\} \phi^4.$$

where $a = \frac{-5}{24}$, $b = \frac{1}{4}$ and

$$x = \frac{1}{1 + \frac{2m^2}{\lambda\phi^2}}, \qquad y = \ln\left(\frac{\lambda\phi^2}{2\mu^2}\frac{1}{x}\right) = \ln\left(\frac{m^2 + \frac{\lambda\phi^2}{2}}{\mu^2}\right). \tag{198}$$

This motivates making an expansion in line with those of eqs. (171) and (188) [47],

$$V(\lambda, x, y, \mu, \phi) = \sum_{n=0}^{\infty} A_n(\lambda, x) y^n \phi^4. \tag{199}$$

Substitution of eq. (199) into the renormalization group equation

$$\left(\mu^2\frac{\partial}{\partial\mu^2} + \beta(\lambda)\frac{\partial}{\partial\lambda} + \gamma_m(\lambda)m^2\frac{\partial}{\partial m^2} + \gamma_\phi(\lambda)\phi^2\frac{\partial}{\partial\phi^2}\right) V = 0 \tag{200}$$

results in

$$A_{n+1}(\lambda, x) = \frac{1}{n+1}\left[f(\lambda, x)\frac{\partial}{\partial x} + g(\lambda, x)\frac{\partial}{\partial\lambda} + h(\lambda, x)\right] A_n(\lambda, x) \tag{201}$$

where

$$f(\lambda, x) = \left(\frac{\beta}{\lambda} - \gamma_m + \gamma_\phi\right) x(1-x)/D(\lambda, x)$$

$$g(\lambda, x) = \beta/D(\lambda, x)$$

$$h(\lambda, x) = 2\gamma_\phi/D(\lambda, x)$$

$\left(D(\lambda, x) = 1 - \left(\frac{\beta}{\lambda} + \gamma_\phi\right) x - \gamma_m(1-x)\right)$ if the renormalization group equation is satisfied at each order in y.

Furthermore, since

$$\frac{dV}{d\phi^2} = \sum_{n=0}^{\infty}\left[\left(x(1-x)\frac{\partial A_n}{\partial x} + 2A_n\right) y^n + nxA_n y^{n-1}\right]\phi^2 \tag{202}$$

V has an extremum at $\phi = <\phi>$ if at each order in y

$$A_{n+1}(\lambda, x_0) = \frac{1}{n+1}\left[-\frac{2}{x} - (1-x)\frac{\partial}{\partial x}\right] A_n(\lambda, x_0) \tag{203}$$

provided $<\phi>\neq 0$ $\left(x_0 = \left(1+\frac{2m^2}{\lambda<\phi>^2}\right)^{-1}\right)$. Taking eq. (203) to hold for all x, then it together with eq. (199) results in

$$V = \sum_{n=0}^{\infty} \frac{y^n}{n!}\left[-\frac{2}{x} - (1-x)\frac{\partial}{\partial x}\right]^n A_0(\lambda, x)\phi^4. \tag{204}$$

Upon setting

$$z = \ln\left(\frac{m^2}{\mu^2}\right) - y = \ln(1-x) \tag{205}$$

and

$$B_0(\lambda, z) = \exp\left(-2\int_{z_0}^{z}\frac{dt}{1-e^t}\right) A_0(\lambda, x) \tag{206}$$

eq. (204) becomes

$$\begin{aligned} V &= \exp\left(2\int_{z_0}^{z}\frac{dt}{1-e^t}\right)\sum_{n=0}^{\infty}\frac{y^n}{n!}\left(\frac{\partial}{\partial z}\right)^n B_0(\lambda, z)\phi^4 \\ &= \exp\left(-2\int_{z}^{z+y}\frac{dt}{1-e^t}\right) A_0(\lambda, z+y)\phi^4. \end{aligned} \tag{207}$$

By eq. (205), eq. (207) collapses down to an expression for V which is independent of ϕ, much like eqs. (183) and (195),

$$V = \frac{4(m^2-\mu^2)^2}{\lambda^2} A_0\left(\lambda, \ln\frac{m^2}{\mu^2}\right). \tag{208}$$

The function $A_0(\lambda, x)$ satisfies an equation that follows from eqs. (201) and (203) when $n = 0$,

$$\left(-\frac{2}{x} - (1-x)\frac{\partial}{\partial x}\right) A_0(\lambda, x) = \left(f(\lambda, x)\frac{\partial}{\partial x} + g(\lambda, x)\frac{\partial}{\partial \lambda} + h(\lambda, x)\right) A_0(\lambda, x). \tag{209}$$

There is an alternate way of deriving eq. (208) that does not require that eq. (202) holds order by order in y when $\phi = <\phi>$. This involves setting μ^2 equal to a value μ_0^2 so that $y = 0$, $\lambda = \lambda_0$ and $x = x_0$; in this case only the y^0 contribution to eq. (202) survives,

$$A_1(\lambda_0, x_0) = \left(-\frac{2}{x} - (1-x)\frac{\partial}{\partial x}\right) A_0(\lambda_0, x_0). \tag{210}$$

As the actual value of λ_0 is contingent upon the unspecified value of the boundary condition on the equation for this running coupling constant, eq. (210) can be taken to be a functional relation between $A_0(\lambda, x)$ and $A_1(\lambda, x)$.

We now follow an approach modeled on the method of characteristics. First we define

$$a_n(\overline{\lambda}(t), \overline{x}(t), t) = \exp\left(\int_{t_0}^{t} d\tau\, h(\overline{\lambda}(\tau), \overline{x}(\tau))\right) A_n(\overline{\lambda}(t), \overline{x}(t)) \tag{211}$$

where

$$\frac{d\overline{x}(t)}{dt} = f\left(\overline{\lambda}(t), \overline{x}(t)\right) \quad (\overline{x}(t_0) = x) \tag{212}$$

$$\frac{d\overline{\lambda}(t)}{dt} = g\left(\overline{\lambda}(t), \overline{x}(t)\right) \quad \left(\overline{\lambda}(t_0) = \lambda\right). \tag{213}$$

Eqs. (201), (211-213) together lead to

$$\frac{d}{dt}a_n\left(\overline{\lambda}(t),\overline{x}(t),t\right)=(n+1)a_{n+1}\left(\overline{\lambda}(t),\overline{x}(t),t\right) \tag{214}$$

Upon taking

$$\overline{y}(t)=\ln\left(\frac{\overline{\lambda}(t)\overline{\phi}^2(t)}{2\overline{\mu}^2(t)}\frac{1}{\overline{x}(t)}\right) \tag{215}$$

where

$$\frac{d\ln\overline{\phi}^2(t)}{dt}=\gamma_\phi\left(\overline{\lambda}(t)\right)/D\left(\overline{\lambda}(t),\overline{x}(t)\right) \tag{216}$$

$$\frac{d\ln\overline{\mu}^2(t)}{dt}=1/D\left(\overline{\lambda}(t),\overline{x}(t)\right) \tag{217}$$

$$\left(\overline{\phi}^2(t_0)=\phi^2,\quad \overline{\mu}^2(t_0)=\mu^2\right)$$

then by eqs. (212, 213, 216, 217) we see that

$$\frac{d\overline{y}(t)}{dt}=-1\qquad(\overline{y}(t_0)=y) \tag{218}$$

so that

$$V=\sum_{n=0}^{\infty}a_n\left(\overline{\lambda}(t),\overline{x}(t),t\right)\overline{y}^n(t)\phi^4 \tag{219}$$

satisfies

$$\frac{dV(t)}{dt}=0\qquad(V(t_0)=V(\lambda,x,y,\mu,\phi)). \tag{220}$$

Eqs. (214) and (219) now lead to

$$\begin{aligned}V(t)&=\sum_{n=0}^{\infty}\frac{\overline{y}^n(t)}{n!}\left(\frac{d}{dt}\right)^n a_0\left(\overline{\lambda}(t),\overline{x}(t),t\right)\phi^4\\&=a_0\left(\left(\overline{\lambda}(t+\overline{y}(t)\right),\overline{x}\left(t+\overline{y}(t)\right),t+\overline{y}(t)\right)\phi^4.\end{aligned} \tag{221}$$

We now define

$$\begin{aligned}\tilde{A}_0\left(\overline{\lambda}(t),\overline{x}(t),t\right)=&\left[\exp\int_{t_0}^{t}\left(\frac{2}{\overline{x}(\tau)}+h(\overline{\lambda}(\tau),\overline{x}(\tau))\right)d\tau\right]\\&A_0\left(\overline{\lambda}(t),\overline{x}(t)\right)\end{aligned} \tag{222}$$

which by eqs. (211-213) shows that

$$\tilde{A}_0\left(\overline{\lambda}(t_0),\overline{x}(t_0),t_0\right)=A_0(\lambda,x)=a_0\left(\overline{\lambda}(t_0),\overline{x}(t_0),t_0\right) \tag{223}$$

while eqs. (209, 212, 213) ensure that

$$\frac{d}{dt}\tilde{A}_0\left(\overline{\lambda}(t),\overline{x}(t),t\right)=-(1-\overline{x}(t))\frac{\partial}{\partial\overline{x}(t)}\tilde{A}_0\left(\overline{\lambda}(t),\overline{x}(t),t\right). \tag{224}$$

The definitions of eqs. (211) and (222) allow us to rewrite eq. (220) as

$$V(t)=\sum_{n=0}^{\infty}\frac{\overline{y}^n(t)}{n!}\left(\frac{d}{dt}\right)^n\left\{\left[\exp-\int_{t_0}^{t}\frac{2d\tau}{\overline{x}(\tau)}\right]\tilde{A}_0\left(\overline{\lambda}(t),\overline{x}(t),t\right)\right\}\phi^4 \tag{225}$$

which by eq. (224) becomes

$$=\left[\exp-\int_{t_0}^{t}\frac{2d\tau}{\overline{x}(\tau)}\right]\sum_{n=0}^{\infty}\frac{\overline{y}^n(t)}{n!}\left[\frac{-2}{\overline{x}(t)}-(1-\overline{x}(t))\frac{\partial}{\partial\overline{x}(t)}\right]^n\tilde{A}_0\left(\overline{\lambda}(t),\overline{x}(t),t\right)\phi^4 \tag{226}$$

Just as eq. (204) led to eq. (207), eq. (226) becomes

$$V(t) = \left[\exp - \int_{t_0}^{t} \frac{2d\tau}{\overline{x}(\tau)}\right] \exp\left(-2\int_{\overline{z}(t)}^{\overline{z}(t)+\overline{y}(t)} \frac{d\tau}{1-e^{\tau}}\right) \tilde{A}_0\left(\overline{\lambda}(t), \overline{z}(t)+\overline{y}(t), t\right)\phi^4. \tag{227}$$

In eq. (227) we can set $t = t_0$; eq. (220) ensures that eq. (208) is recovered.

(IX) It has been demonstrated in this review that the renormalization group is an effective tool for obtaining information about quantum processes beyond what is learned from purely perturbative calculations carried out to finite order. Further applications, such as an examination of the effective potential in massive scalar electrodynamics, are currently being considered.

The author would like to thank his many collaborators, especially M. Ahmady, F. Brandt, F. Chishtie, V. Elias, A. Fariborz, R. Mann, A. Rebhan, T. Sherry and T. Steele. This work was supported by NSERC. R. Macleod was helpful in the final formulation.

References

[1] E.C.G. Stueckelberg and A. Peterman, "Normalization of Constants in a Quantum Theory", Helv. Phys. Acta 26, 499 (1953).

[2] N.N. Bogoliubov and D.V. Shirkov, Introduction to the Theory of Quantized Fields, Interscience, New York 1959.

[3] M. Gell-Mann and F. Low, "Quantum Electrodynamics at Small Distances", Phys. Rev. 95, 1300 (1954).

[4] G. 't Hooft, "Dimensional Regularization and the Renormalization Group", Nucl. Phys. B61, 455 (1973).

[5] S. Weinberg, "New Approach to the Renormalization Group", Phys. Rev. D8, 3497 (1973).

[6] J.C. Collins and A.J. MacFarlane, "New Methods for the Renormalization Group", Phys. Rev. D10, 1201 (1974).

[7] C. Callan, "Broken Scale Invariance and Asymptotic Behaviour", Phys. Rev. D2, 1541 (1970).

[8] K. Symanzik, "Small-Distance Behaviour Analysis and Wilson Expansions", Comm. Math. Phys. 18, 227 (1970).

[9] A. Peterman, "Renormalization Group and the Deep Structure of the Proton", Phys. Rep. 53C, 157 (1979).

[10] P.M. Stevenson, "Dimensional Analysis in Field Theory", Ann. of Phys. 132, 383 (1981).

[11] V. Elias and D.G.C. McKeon, "Vanishing of the Bare Coupling in Four Dimensions", Int. J. Mod. Phys. A18, 2395 (2003).

[12] V. Elias, D.G.C. McKeon and T.N. Sherry, "Summation of Higher Order Effects Using the Renormalization Group Equation", Int. J. Mod. Phys. A20, 1065 (2005).

[13] V. Elias and D.G.C. McKeon, "The Renormalization Group with Exact Beta Functions", Can. J. Phys. (to be published) hep-th 0509095.

[14] R. Courant and D. Hilbert, Methods of Mathematics Physics Vol. II (Interscience N.Y.) 1966 Ch. II.

[15] S.M. Chung and B.K. Chung, "Renormalization Group Improvement of the Effective Potential in Massive ϕ^4 Theory", Phys. Rev. D60, 105001 (1999).

[16] C. Ford, "Multiscale Renormalization Group Improvement of the Effectve Potential", Phys. Rev. D50, 7531 (1994).

[17] V. Elias, D.G.C. McKeon and T.G. Steele, "Optimal Renormalization Group Improvement of R(s) via the Method of Characteristics", Int. J. Mod. Phys. A18, 3417 (2003).

[18] M.R. Ahmady, V. Elias, D.G.C. McKeon, A. Squires, T.G. Steele, "Renormalization group improvement of Effective Actions Beyond Summation of Leading Logarithms", Nucl. Phys. B655, 221 (2003).

[19] M.R. Ahmady, F.A. Chishtie, V. Elias, A.H. Fariborz, N. Fattahi, D.G.C. McKeon, T.N. Sherry, T.G. Steele, "Closed Form Summation of Renormalization Group Accessible Logarithmic Contributions to Semileptonic B Decays and Other Perturbative Processes", Phys. Rev. D66, 014010 (2002).

[20] M.R. Ahmady, F.A. Chishtie, V. Elias, A.H. Fariborz, D.G.C. McKeon, A. Squires, T.G. Steele, "Optimal Renormalization Group Improvment of the Perturbative Series for the e^+e^- Annihilation Cross Section", Phys. Rev. D67 034017 (2003).
[21] D.G.C. McKeon, A. Rebhan, "Renormalization Group Summation and the Free Energy of Hot QCD", Phys. Rev. D67, 027701 (2003).
[22] P. Arnold, C.X. Zhai, "Three Loop Free Energy for High Temperature QED and QCD with Fermions", Phys. Rev. D51, 1906 (1995).
[23] C.X. Zhai, B. Kastening, "Free Energy of Hot Guage Theories with Fermions Through g^5", Phys. Rev. D52, 7232 (1995).
[24] E. Braaten, A. Nieto, "Free Energy of QCD at High Temperature", Phys. Rev. D53, 3421 (1996).
[25] G.V. Dunne, H. Gies, C. Schubert, "Zero Modes, Beta Functions and IR/UV interplay in Higher Loop QED", JHEP 0211, 032 (2002).
[26] S.G. Matinyan, G.V. Savvidy, "Vacuum Polarization Induced by the Intense Gauge Field", Nucl. Phys. B134, 539 (1978).
[27] L. Abbott, "The Background Field Method Beyond One Loop", Nucl. Phys. B185, 189 (1981).
[28] V. Novikov, V. Shifman, A. Vainshtein and V. Zakharov, "Exact Gell-Mann-Low Function of Supersymmetric Yang-Mills Theory from Instanton Calculus", Nucl. Phys. B229, 381 (1983).
[29] D.R.T. Jones, "More on the Axial Anomaly in Supersymmetric Yang-Mills Theory", Phys.Lett. 123B, 45 (1983).
[30] V. Elias, "Infrared Consistency of NSVZ and DRED Supersymmetric Gluodynamics", J. Phys. G27, 217 (2001).
[31] R.M. Corless, G.H. Gonnet, D.E.G. Hare, D.J. Jeffrey, D.E. Knuth, "On the Lambert W Function", Adv. Comp. Math. 5, 329 (1996).
[32] G. 't Hooft, "Computation of a Quantum Effect Due to a Four Dimensional Pseudo Particle", Phys. Rev. D14, 3432 (1976).
[33] S. Coleman and E. Weinberg, "Radiative Corrections as the Origin of Spontaneous Symmetry Breaking", Phys. Rev. D7, 1888 (1973).
[34] R. Jackiw, "Functional Evaluation of the Effective Potential", Phys. Rev. D9, 1686 (1974).
[35] S. Weinberg, "Perturbative Calculations of Symmetry Breaking", Phys. Rev. D7, 2887 (1973).
[36] A. Salam and J. Strathdee, "Comment on the Computation of Effective Potentials", Phys. Rev. D9, 1129 (1974).
[37] M. Sher, "Electroweak Higgs Potentials and Vacuum Stability", Phys. Rep. 179, 273 (1989).
[38] V. Elias, R.B. Mann, D.G.C. McKeon, T.G. Steele, "Radiative Electroweak Symmetry Breaking Revisited", Phys. Rev. Lett. 91, 251601 (2003).
[39] V. Elias, R.B. Mann, D.G.C. McKeon, T.G. Steele, "Optimal Renormalization Group Improvement of Two Radiatively Broken Theories", Nucl. Phys. B678, 147 (2004) (E B703, 413 (2004)).
[40] V. Elias, R.B. Mann, D.G.C. McKeon, T.G. Steele, "Higher Order Stability of Radiatively Induced 220 Gev Higgs", Phys. Rev. D72, 037902 (2005).
[41] B.M. Kastening, "Renormalization Group Improvement of the Effective Potential in Massive ϕ^4 Theory", Phys. Lett. B283, 287 (1992).
[42] M. Bando, T. Kugo, N. Maekawa, H. Nakano, "Improving the Effective Potential", Phys. Lett. B301, 83 (1993).
[43] C. Ford, D.R.T. Jones, "The Effective Potential and the Differential Equations Method for Feynman Integrals", Phys. Lett. B274, 409 (1992).
[44] H. Yamagishi, "Coupling Constant Flows and Dynamical Symmetry Breaking", Phys. Rev. D23, 1880 (1981).
[45] F. Brandt, F. Chishtie, D.G.C. McKeon, "The Renormalization Group Improved Effective Potential in Massless Models", Mod. Phys. Lett. A20, 2215 (2005).
[46] D.J.E. Callaway, "Triviality Pursuit: Can Elementary Scalar Particles Exist?", Phys. Rep. 167, 241 (1988).
[47] F. Brandt, F. Chishtie, D.G.C. McKeon, "The Effective Potential in the Massive ϕ_4^4 model", UWO report (2005).

Fields Institute Communications
Volume **50**, 2007

Short Distance Behavior of Scaling Functions for the Planar Ising Model

John Palmer
Department of Mthematics
The University of Arizona
Tuscon, AZ 95721-0089 USA
palmer@math.arizona.edu

The planar Ising model in absence of a magnetic field is a solvable model, and the phase transition that occurs in this model can be looked at under a microscope, with a mathematical resolution that is not possible for any other local model in statistical physics. Here we describe the way in which one can understand the short distance behavior of the scaling functions for this model using the machinery that arises from the Sato, Miwa, and Jimbo analysis of scaling. More details for this approach can be found in [4] and [5] and in a forthcoming book on planar Ising correlations. For general backround concerning the importance of this model in Physics we refer the reader to [1].

The Gibbs measure for the two dimensional Ising model assigns a probability to spin configurations on finite subsets Λ of $\mathbf{Z}^2$. Each such configuration is naturally a map,

$$\sigma : \Lambda \to \{-1, 1\},$$

and the probability measure is expressed in terms of the nearest neighbor sum for the energy,

$$E_\Lambda(\sigma) = \sum_{\langle ij \rangle} J_{ij}\sigma_i\sigma_j$$

where $J_{ij} > 0$ (ferromagnetic) and is translation invariant. The probability of the configuration σ is,

$$P_\Lambda(\sigma) = \frac{e^{-E_\Lambda(\sigma)/kT}}{Z_\Lambda}, \text{ where,}$$

$$Z_\Lambda = \sum_{\sigma \in \mathcal{C}(\Lambda)} e^{-E_\Lambda(\sigma)/kT},$$

is a sum over all spin configurations $\mathcal{C}(\Lambda)$. To make this assignment well defined one must specify boundary conditions – make some assignment of spins to sites which are nearest neighbors of sites in Λ, but not in Λ (assigning $+1$ to all such sites gives "plus" boundary conditions), or perhaps drop those terms in the energy function associated with bonds that connect sites in Λ to sites outside of Λ (open

---

2000 *Mathematics Subject Classification.* Primary 82B20; Secondary 82C20.

boundary conditions). Correlation functions for a boundary condition B (e.g., +, open) are given by sums,

$$\langle \sigma_{a_1} \cdots \sigma_{a_n} \rangle_{\Lambda,T}^{B} = \sum_{\sigma \in \mathcal{C}_B(\Lambda)} \sigma_{a_1} \cdots \sigma_{a_n} P_\Lambda(\sigma),$$

where $\mathcal{C}_B(\Lambda)$ is a subset of the configurations appropriate to the boundary condition B.

The infinite volume or thermodynamic limit of the correlations,

$$\langle \sigma_{a_1} \cdots \sigma_{a_n} \rangle_{T}^{B} = \lim_{\Lambda \uparrow \mathbf{Z}^2} \langle \sigma_{a_1} \cdots \sigma_{a_n} \rangle_{\Lambda,T}^{B},$$

is a setting in which it is possible to study phase transitions, and spontaneous symmetry breaking. For the two dimensional Ising model there is a critical temperature T_c below which the correlations depend on the boundary condition. For example,

$$\langle \sigma \rangle_T^+ > 0 \text{ for } T < T_c$$

$$\langle \sigma \rangle_T^+ = 0 \text{ for } T > T_c,$$

but the one point function $\langle \sigma \rangle_T^{\text{open}} = 0$ for all temperatures. The spin flip symmetry in the energy function is broken by plus boundary conditions below T_c in a way that survives the infinite volume limit. To discuss the exact solution of the Ising model we first recall some facts about:

1 Spin representations of $\mathrm{SO}_{res}(W)$

Suppose W a complex Hilbert space with Hermitian inner product, $\langle \cdot, \cdot \rangle$. and a distiguished bilinear form

$$(x, y) = \langle \bar{x}, y \rangle,$$

where $x \longrightarrow \bar{x}$ is a conjugation (a complex anti linear involution). The Clifford algebra of W is an associative algebra with unit e generated by the elements of W subject to the relations,

$$xy + yx = (x, y)e.$$

A polarization of W is a splitting,

$$W = W_+ \oplus W_-,$$

where $W_\pm$ are isotropic subspaces. A subspace $W_\pm$ is isotropic if the distinguished bilinear form on W vanishes identically on it. Polarizations of W are associated with representations of the Clifford algebra in which the elements of W_+ are creation operators and the elements of W_- are annihilation operators. Such a representation of the Clifford relations on the alternating tensor algebra, $\mathrm{Alt}(W_+)$, is given by,

$$\mathrm{Alt}(W_+) = \mathbf{C} \oplus W_+ \oplus W_+ \wedge W_+ \oplus \cdots$$

$$x = a^*(x_+) + a(\bar{x}_-),$$

where a^* and a are the usual creation and annihilation operators.

This representation space of the Clifford relations supports a projective representation of the restricted special orthogonal group, $\mathrm{SO}_{res}(W)$. The restricted special orthogonal group consists of those elements in the connected component of the identity in the orthogonal group whose matrix elements with respect to the splitting $W = W_+ \oplus W_-$ are restricted so that,

$$G \in \mathrm{SO}_{res}(W) \text{ iff } G = \begin{bmatrix} A & B \\ C & D \end{bmatrix} \text{ with } B, C \in \text{Schmidt class.}$$

There is a covering group Spin(W),

$$\mathbf{C}^* \to \mathrm{Spin}(W) \xrightarrow{T} \mathrm{SO}_{res}(W)$$

with the covering homomorphism, T, defined by an action on the Clifford relations,

$$gxg^{-1} = T(g)x \text{ for } g \in \mathrm{Spin}(W),\ x \in W.$$

In general $g \in \mathrm{Spin}(W)$ acts on a dense invariant domain (it is not a bounded operator). This representation is important for the planar Ising model since in one manifestation of the models "exact solvability" both the transfer matrix and the spin operators are elements of Spin(W) for a suitable choice of W and a suitable polarization. The infinite volume correlations for the planar Ising model have representations as vacuum expections,

$$\langle \sigma_{a_1} \cdots \sigma_{a_n} \rangle = \langle 0 | g_1 \cdots g_n | 0 \rangle,$$

where $g_j \in \mathrm{Spin}(W)$, and $|0\rangle$ is the vacuum vector $1 \oplus 0 \oplus 0 \cdots$ in $\mathrm{Alt}(W_+)$.

There is a generalized Wick theorem,

$$\langle 0 | g_1 \cdots g_n | 0 \rangle = \textstyle\prod_j \langle 0 | g_j | 0 \rangle \mathrm{Pf} \begin{bmatrix} I & \mathcal{LB} \\ \mathcal{UC} & I \end{bmatrix}$$

$\mathcal{L}, \mathcal{B}, \mathcal{U}$ and $\mathcal{C}$ act on the n fold direct sum of $W_\pm$ with itself, which allows one to study the correlations directly in terms of infinite Pfaffians (Pf). These infinite Pfaffians are mathematically well defined and behave much the same as infinite determinants [2]. To be more explicit we look at:

2 Ising correlations in the thermodynamic limit

The infinite volume planar Ising correlations have a group theoretic representation

$$\langle \sigma_{a_1} \cdots \sigma_{a_n} \rangle_T^{+} = \langle 0 | \mathcal{T} \sigma_{a_1} \cdots \sigma_{a_n} | 0 \rangle \ T \neq T_c$$
$$\mathrm{Spin}(W) \ni \sigma_a = V^{-y} \sigma_x V^y, \text{ for } a = (x, y).$$
$$W = \ell^2(\mathbf{Z} + \frac{1}{2}, \mathbf{C}^2).$$

The linear transformation V is called the transfer matrix and the induced rotations and polarization appropriate to this representation are:

$$T(V) = T_+ z + T_0 + T_- z^{-1} \ z = \text{translation by 1 on } \mathbf{Z} + \frac{1}{2}$$

$$T(\sigma_x)\delta_{i,k} = -\mathrm{sgn}(x-k)\delta_{i,k}, \text{ for } k \in \mathbf{Z} + \frac{1}{2}, i = 1, 2,$$

where $T_\pm$, and T_0 are 2×2 matrices with entries that depend on the interaction strengths of the model and,

$$\delta_{i,k}(\ell) = \delta_{k,\ell} \begin{bmatrix} \delta_{1,i} \\ \delta_{2,i} \end{bmatrix}.$$

The polarization $W = W_+ \oplus W_-$ is given by the spectral subspaces for the induced rotation $T(V)$,

$$\mathrm{spec}(T(V)|W_+) \subset [0, 1), \text{ and } \mathrm{spec}(T(V)|W_-) \subset (1, \infty).$$

Thus the restriction of $T_+(V)$ to W_+ is a contraction. Also,

$$V = 1 \oplus T_+(V) \oplus T_+(V) \otimes T_+(V) \oplus \cdots,$$

so $|0\rangle$ is the eigevector for V associated with the largest eigenvalue (in the infinite volume limit $|0\rangle$ is the only eigenvector for V period). Incidentally, this group theoretic representation of the correlations fails at the critical point where 1 is in the spectrum of $T(V)$ ([6]).

The generalized Wick's theorem allows one to write down Pfaffian formulas for the correlations,

$$\langle\sigma_{a_1}\cdots\sigma_{a_n}\rangle_T^+ = \langle\sigma\rangle_T^{+n}\mathrm{Pf}\begin{bmatrix} I & \mathcal{LB} \\ \mathcal{UC} & I \end{bmatrix} \text{ for } 0 < T < T_c$$

$$\langle\sigma_{a_1}\cdots\sigma_{a_n}\rangle_T = \langle\sigma\rangle_{T^*}^{+n}\langle\sigma_{a_1}\cdots\sigma_{a_n}\rangle_{T^*}^+\mathrm{Pf}(\mathcal{T}) \text{ for } T > T_c$$

where $T^* < T_c$ is the Kramers-Wannier dual temperature. We won't bother to write down the explicit formulas for $\mathcal{L}$, $\mathcal{B}$, $\mathcal{U}$, $\mathcal{C}$, and $\mathcal{T}$ here. The first four are $n \times n$ operator valued matrices whose entries can all be computed in terms of the induced rotation for the spin operators σ_{a_j} and $\mathcal{T}$ is an $n \times n$ matrix whose entries are matrix elements of the $2n \times 2n$ operator valued matrix,

$$\begin{bmatrix} I & \mathcal{LB} \\ \mathcal{UC} & I \end{bmatrix}^{-1}.$$

Technically, these formulas allow us to control the wave function renormalization needed for taking a continuum limit by factoring out the one point functions, $\langle\sigma\rangle_{T(*)}^n$. The remaining Pfaffians are continuous in the strong operator topology for $\mathcal{L}$ and $\mathcal{U}$ and the Schmidt norm topology for $\mathcal{B}$ and $\mathcal{C}$. This continuity also allows one to explicitly control an important continuum limit:

3 Scaling limits for Ising correlations

The correlation length at temperature T is the inverse $m(T)^{-1}$ of the "mass" at temperature T. For symmetric interaction strengths on the lattice the correlation length describes the exponential clustering of the two point function in the limit of large separation,

$$\langle\sigma_a\sigma_b\rangle_T^+ - \langle\sigma\rangle_T^{+2} \simeq C|a-b|^{-\frac{1}{2}}e^{-m(T)|a-b|}$$

$$m(T)^{-1} = \text{correlation length}$$

$$\lim_{T\to T_c} m(T) = 0$$

The convential wisdom in the statistical physics of phase transitions identifies three scaling regions in the neighborhood of the critical temperature T_c:

(1) Gaussian regime (site separation much greater than the correlation length)
(2) Conformal regime (site separation much larger than the lattice spacing but much smaller than the correlation length)
(3) Scaling regime (site separation on the order of the correlation length)

It is this last regime that we are interested in here. It is possible to summarize the behavior of the correlations in this scaling regime by introducing continuum scaling functions defined by (here $m = m(T)$ for brevity),

$$\tau^{[-]}(a) = \lim_{T\uparrow T_c} \frac{\langle 0|\mathcal{T}\sigma_{m^{-1}a_1}\cdots\sigma_{m^{-1}a_n}|0\rangle_T^+}{\langle\sigma\rangle_T^{+n}},$$

$$\tau^{[+]}(a) = \lim_{T\downarrow T_c} \frac{\langle 0|\mathcal{T}\sigma_{m^{-1}a_1}\cdots\sigma_{m^{-1}a_n}|0\rangle_T}{\langle\sigma\rangle_{T^*}^{+n}}.$$

The scaling functions $\tau^{[\pm]}(a)$ are different for scaling from below $[-]$ or scaling from above $[+]$ the critical temperature T_c. The Pfaffian formulas on the lattice scale in a simple way to new Pfaffian formulas in the continuum,

$$\tau^{[-]}(a) = \mathrm{Pf}\begin{bmatrix} I & \mathcal{LB} \\ \mathcal{UC} & I \end{bmatrix},$$

$$\tau^{[+]}(a) = \mathrm{Pf}\begin{bmatrix} I & \mathcal{LB} \\ \mathcal{UC} & I \end{bmatrix} \mathrm{Pf}(\mathcal{T}).$$

The large scale behavior of the correlations on the lattice can equivalently be thought of in terms of correlations at fixed separation in a lattice with a lattice spacing between sites that tends to 0 – a picture that helps understand the continuum nature of the limit. These scaling functions are the objects of a famous analysis by McCoy, Tracy, Wu, and Barouch (for the two point functions) that was later generalized by Sato, Miwa and Jimbo. The Wu, McCoy, Tracy, and Barouch formulas ([13]) are,

$$\tau^{[-]}(0,a) = \cosh\left(\frac{\psi}{2}\right) \exp \frac{1}{2}\int_r^\infty r dr \left(m^2 \sinh^2 \psi - (\psi')^2\right),$$

and

$$\tau^{[+]}(0,a) = \sinh\left(\frac{\psi}{2}\right) \exp \frac{1}{2}\int_r^\infty r dr \left(m^2 \sinh^2 \psi - (\psi')^2\right),$$

where ψ is an appropriate solution to the "sinh Gordon" equation,

$$\frac{1}{r}\left(r\psi'(r)\right)' = \frac{m^2}{2}\sinh^2 2\psi.$$

The function $\eta = e^{-\psi}$ is a solution to a Painlevé equation of the third kind. The Sato, Miwa, and Jimbo extension of this analysis can be found in ([7, 8, 9, 10, 11]).

4 Sato Miwa and Jimbo analysis of the scaling functions

It is straightforward to see that the Clifford algebra generators for the Ising model scale to a Free Fermi field, $\psi(x)$ which satisfies,

$$(1-\mathcal{D})\psi(x) = 0$$

where,

$$\mathcal{D} = \begin{bmatrix} 0 & 2\partial_z \\ 2\bar{\partial}_z & 0 \end{bmatrix} = \text{Euclidean Dirac Operator.}$$

One way to understand the actors in the SMJ analysis is to introduce the Green functions (mixed correlations),

$$G^{[\pm]}(x,y;a) = \frac{\langle 0|\mathcal{T}\psi(x)\psi(y)\sigma_{a_1}\cdots\sigma_{a_n}|0\rangle_{[\pm]}}{\tau^{[\pm]}(a)},$$

where $\mathcal{T}$ is a "time ordering prescription" that orders the operators in the product in increasing order of the second coordinates from *right to left*. The time ordering prescription has consequences for the spin operators and Fermi fields that follow from the action of the spin fields on the Clifford algebra,

$$\lim_{x_2\downarrow a_2} \sigma_a\psi(x) = -\mathrm{sgn}(x_1-a_1)\lim_{x_2\uparrow a_2}\psi(x)\sigma_a.$$

As a consequence of this $G(x,y;a)$ is the Green function for a Dirac operator acting on the sections of a vector bundle over the punctured plane,

$$\mathbf{C}\backslash\{a_1,\ldots,a_n\},$$

with smooth sections that undergo a sign change in a circuit of each of the sites a_j.

Local operator product expansions for $\mathcal{T}\psi(x)\sigma_a$ show that,

$$G^{[\pm]}(x,y;a)=\sum_{n=0}^{\infty}c_n^i w_n(x-a_i)+c_n^{*i}w_n^*(x-a_i),\ \text{with}\ c_0^i=\mp c_0^{*i}.$$

with

$$w_n(x)=\left[\begin{array}{c} I_{n_-}(r)e^{in_-\theta} \\ I_{n_+}(r)e^{in_+\theta}\end{array}\right],\ \text{and}\ w_n^*(x)=\left[\begin{array}{c} I_{n_+}(r)e^{-in_+\theta} \\ I_{n_-}(r)e^{-in_-\theta}\end{array}\right]$$

for $n_\pm = n\pm\frac{1}{2}$. The $w_n(x)$ and $w_n^*(x)$ are solutions to the Dirac equation that are also eigenfunctions for the infinitesimal rotation that commutes with the Dirac operator and that have appropriate monodromy in a circuit of 0. The functions $I_k(r)$ are modified Bessel functions of the second kind. We now introduce a geometric way to think about the SMJ analysis that is an analogue of the Segal-Wilson geometrization of the KdV analysis ([12]):

5 Localization, Grassmannians and τ functions

We first localize $I-\mathcal{D}$ away from the singularities at a_j. Choose a base point $a^0\in\mathbf{C}^n$ and disjoint disks D_j centered at a_j^0.

$$C_j=\partial D_j$$

$$f\in \text{Domain}(I-\mathcal{D})_a^{[\pm]}\Leftrightarrow f|_{C_j}\in W_{int}^{[\pm]}(a)\ \text{for}\ a\in D_1\times\cdots\times D_n$$

$$W_{int}^{[\pm]}(a)=\text{boundary values of solutions to the Dirac equation in}\ D_j$$
$$\text{that have appropriate local expansions in a neighborhood of}\ a_j.$$

Note that the subspaces $W_{int}^{[\pm]}(a)$ are spaces of smooth sections of a bundle over $\cup_j C_j$ that has -1 "monodromy" around each C_j. There is a Segal-Wilson type Grassmannian, Gr_0, and an associated bundle $\det^*$ over Gr_0 so that,

$$W_{int}^{[\pm]}(a)\in \text{Gr}_0=\text{subspaces of}\ H^{\frac{1}{2}}(C_j)\ \text{close to}\ W_{int}^{[\pm]}(a^0)$$

$$\begin{array}{ccc} \det^* & & \det^* \\ \downarrow & & \downarrow \\ W_{int}^{[\pm]}(a) & \rightarrow & \text{Gr}_0 \end{array}.$$

Trivializations of the restriction of the $\det^*$ bundle to the family of subspaces $W_{int}^{[\pm]}(a)$ and τ functions are determined by the projections,

$$\text{P}_{int}^{[\pm]}(a)f(x)=\lim_{D_j\ni x'\to x}\int_{C_j}G_1(x',y;a)f_1(y)dy+G_2(x',y;a)f_2(y)d\bar{y}$$

$$\text{Pr}(a)=\bigoplus_j \text{P}_{int}^{[-]}(a_j).$$

The projection $\text{P}_{int}^{[\pm]}(a)$ determines what is called the canonical section, σ, of $\det^*$ over the family of subspaces $W_{int}^{[\pm]}(a)$ ([3]). The direct sum of the "one point projections" $\text{Pr}(a)$ determines a second trivialization, δ. The ratio of these two sections is well defined (since det* is a complex line bundle) and following Segal and Wilson we can define τ functions by,

$$\tau\left(W_{int}^{[\pm]}(a)\right) = \frac{\sigma\left(W_{int}^{[\pm]}(a)\right)}{\delta\left(W_{int}^{[\pm]}(a)\right)}.$$

In particular for the [-] subspace boundary condition it turns out that,

$$\tau\left(W_{int}^{[-]}(a)\right) = \det \mathrm{P}_{int}^{[-]-1}\mathrm{Pr}(a).$$

For [+] boundary conditions there is a similar but slightly more complicated formula. These tau functions and the scaling functions introduced earlier are related:

6 Scaling functions as τ functions

The relation between τ functions just described and the scaling functions introduced earlier is,

$$\tau\left(W_{int}^{[-]}(a)\right) = \left(\frac{\tau^{[-]}(a)}{\tau^{[-]}(a^0)}\right)^2,$$

$$\tau\left(W_{int}^{[+]}(a)\right) = \left(\frac{\tau^{[+]}(a)}{\tau^{[-]}(a^0)}\right)^2.$$

The reason for the square in the relation is that the det* bundle used to define the tau functions is not quite the right bundle. There is a Pfaffian bundle that eliminates this square in the relation ([2]) but it is a little more involved to introduce and because our interest is mainly in the relation between the log derivative of the scaling function and the log derivative of the tau function,

$$d_a \log \tau^{[-]}(a) = \tfrac{1}{2} d_a \log \tau\left(W_{int}^{[-]}(a)\right),$$

$$d_a \log \tau\left(W_{int}^{[-]}(a)\right) = -\mathrm{Tr}\left(\mathrm{Pr}(a) d_a \mathrm{P}_{int}^{[-]}(a)\right),$$

not much is gained. We've just recorded the result for $[-]$ boundary conditions but there is a similar result for $[+]$ boundary conditions. From this point of view the fundamental result that leads to the deformation analysis of $\tau^{[\pm]}(a)$ is the fact that the derivative of the Green function $G^{[-]}(x,y;a)$ with respect to a is *finite rank*,

$$\frac{\partial}{\partial a_j} G^{[-]}(x,y;a) = -\frac{im}{2}\left(\mathcal{W}_j(x)\frac{\partial \mathcal{W}_j^{\tau}(y)}{\partial a_j} - \frac{\partial \mathcal{W}_j(x)}{\partial a_j}\mathcal{W}_j^{\tau}(y)\right),$$

$$\frac{\partial}{\partial \bar{a}_j} G^{[-]}(x,y;a) = \frac{im}{2}\left(\mathcal{W}_j(x)\frac{\partial \mathcal{W}_j^{\tau}(y)}{\partial \bar{a}_j} - \frac{\partial \mathcal{W}_j(x)}{\partial \bar{a}_j}\mathcal{W}_j^{\tau}(y)\right),$$

where $\mathcal{W}_j(x)$ is a solution to the Dirac equation which is uniquely specified by the behavior of the coefficients $c_0^i(\mathcal{W}_j)$ in the local expansion about a_i. In fact $\mathcal{W}_j(x)$ is a solution to the Dirac equation away from the points a_j, is in L^2 near $x = \infty$ and,

$$c_0^i(\mathcal{W}_j) - c_0^{i*}(\mathcal{W}_j) = \delta_{ij}.$$

This derivative formula and a residue type calculation now leads to the SMJ formula for the log derivative of the τ function

$$d_a \log \tau\left(W_{int}^{[-]}(a)\right) = -\frac{i}{2}\sum_{j=1}^{n} c_1^j(\mathcal{W}_j) da_j - \overline{c_1^j(\mathcal{W}_j)} d\bar{a}_j.$$

The $c_1^j(\mathcal{W}_j)$ are the local expansion coefficients for the wave function $\mathcal{W}_j$ at level 1. The wave function $[\mathcal{W}_1, \ldots, \mathcal{W}_n]$ turn out to be flat section for a Dirac compatible integrable connection (which is a consequence of $d_a^2 G = 0$). In the two point case this leads to an expression for $d \log \tau^{[\pm]}(a)$ in terms of Painlevé functions of the third kind–the celebrated result of McCoy, Tracy, Wu and Barouch.

Introducing mass, m, into the Dirac operator allows one to give a formula for the scaling functions with an additional scaling parameter m:

$$(m - \mathcal{D})\mathcal{W}_j = 0$$

$$d_a \log \tau^{[-]}(ma) = -\frac{im}{2} \sum_{j=1}^{n} c_1^j(\mathcal{W}_j) da_j - \overline{c_1^j(\mathcal{W}_j)} d\bar{a}_j$$

We can then study the short distance behavior of the scaling functions $\tau^{[\pm]}(ma)$ as $m \to 0$ by studying the:

7 Mass 0 limit for the Dirac operator

The short distance scaling can be studied by looking at the mass 0 limit for the solution to,

$$(m - \mathcal{D})\mathcal{W}_j = 0$$

$$d_a \log \tau^{[-]}(ma) = -\frac{im}{2} \sum_{j=1}^{n} c_1^j(\mathcal{W}_j)\, da_j - \overline{c_1^j(\mathcal{W}_j)}\, d\bar{a}_j.$$

$$c_0^i(\mathcal{W}_j) - c_0^{i*}(\mathcal{W}_j) = \delta_{ij},$$

and $\mathcal{W}_j \in L^2(\mathbf{C})$. The wave function $\mathcal{W}_j$ may be characterized as the solution to a boundary value problem,

$$\delta\mathcal{W}_j = \mathcal{W}_j - \varphi\hat{w}_j \text{ (local subtraction at } a_j)$$

$$(m - \mathcal{D})\delta\mathcal{W}_j = -(m - \mathcal{D})\varphi\hat{w}_j = f_j$$

$$c_0^{*j}(\delta\mathcal{W}_j) - c_0^j(\delta\mathcal{W}_j) = 0.$$

$$\delta\mathcal{W}_j|_{C_\epsilon(a_j)} \in W_{int}^{[-]}(a_j), \text{ and } \delta\mathcal{W}_j|_{C_R} \in W_{int}^{[-]}(\infty).$$

The "Fourier coefficient" $c_1^j(\mathcal{W}_j)$ is given by,

$$mc_1^j(\mathcal{W}_j) = \frac{\sqrt{m}}{2\pi I_{\frac{1}{2}}(m\epsilon)} \int_0^{2\pi} (\delta\mathcal{W}_j)_1 \left(\epsilon e^{i\theta_j}\right) e^{-i\frac{\theta_j}{2}} d\theta_j.$$

The mass zero limit for the subspace boundary conditions is simple,

$$W_{int}^{[-]}(a) \stackrel{m\to 0}{\longrightarrow} W_0(a) = \text{span}_{k\geq\frac{1}{2}} \left\{ \begin{bmatrix} e^{ik\theta} \\ 0 \end{bmatrix}, \begin{bmatrix} 0 \\ e^{-ik\theta} \end{bmatrix} \right\} \cup \mathbf{C} \begin{bmatrix} e^{-i\frac{\theta}{2}} \\ e^{i\frac{\theta}{2}} \end{bmatrix}$$

$$W_{int}^{[-]} \stackrel{m\to 0}{\longrightarrow} \text{span}_{n\geq 1} \begin{bmatrix} e^{-in\theta} \\ 0 \end{bmatrix} + \text{span}_{n\geq 1} \begin{bmatrix} 0 \\ e^{in\theta} \end{bmatrix}.$$

The simple form for the limiting subspace boundary conditions make it possible to guess the form for the mass 0 limit of the Green function $G(x, y; a)$.

8 The $m = 0$ Green Function

Recall that,

$$\mathcal{D} = \begin{bmatrix} 0 & 2\partial_z \\ 2\bar{\partial}_z & 0 \end{bmatrix},$$

The mass 0 Green function is,

$$G_0(x,y) = -\frac{1}{4\pi}\begin{bmatrix} \frac{\sum_j u_j(x)\overline{v_j(y)}}{g(x,y)} & \frac{g(x,y)}{\sum_j u_j(x)v_j(y)} \end{bmatrix},$$

where,

$$u_j(x) = (x-a_j)^{-\frac{1}{2}} \prod_{k\neq j} \frac{(x-a_k)^{\frac{1}{2}}}{(a_j-a_k)^{\frac{1}{2}}}$$

$$g(x,y) = \sum_{|\epsilon|=0} c(\epsilon) \frac{\prod_j (x-a_j)^{\epsilon_j}(y-a_j)^{-\epsilon_j}}{x-y},$$

with $\epsilon = (\epsilon_1, \ldots, \epsilon_n)$ and $\epsilon_j = \pm\frac{1}{2}$. Also,

$$|\epsilon| = \sum_j \epsilon_j$$

$$c(\epsilon) = \frac{\prod_{j<k} |a_j - a_k|^{2\epsilon_j\epsilon_k}}{\sum_{|\epsilon|=0}\prod_{j<k} |a_j - a_k|^{2\epsilon_j\epsilon_k}}$$

and

$$v_j(x) = (x-a_j)^{-\frac{1}{2}} \sum_{|\epsilon|=0, \epsilon_j=\frac{1}{2}} c(\epsilon) \prod_{k\neq j} \frac{(x-a_k)^{\epsilon_k}}{(a_j-a_k)^{\epsilon_k}}.$$

Chiral symmetry breaking terms appear because of boundary conditions. To verify that this formula is correct one needs the homogeneous function identity,

$$\sum_j u_j(x)\overline{v_j(y)} = \sum_j v_j(x)\overline{u_j(y)},$$

to check the boundary conditions at ∞.

9 Perturbation scheme and the Luther-Peschel formula

To calculate the 0 mass limit we introduce a perturbation scheme,

$$\delta\mathcal{W}_j = \delta_1\mathcal{W}_j + \delta_2\mathcal{W}_j + \delta_3\mathcal{W}_j$$

$$\delta_1\mathcal{W}_j = G_0(I + mG_0)^{-1} f_j$$

$\delta_1\mathcal{W}_j$ satisfies,

$$(m - \mathcal{D})\delta_1\mathcal{W}_j = 0,$$

but with $\delta_1\mathcal{W}_j \in W_0$ (the $m = 0$ boundary space). $\delta_2\mathcal{W}_j$ straightens out the boundary conditions at the expense of an additional inhomogeneous term which is $O(m)$. $\delta_3\mathcal{W}_j$ fixes the extra inhomogeneous term. $\delta_2\mathcal{W}_j$ and $\delta_3\mathcal{W}_j$ tend to 0 in an appropriate sense as $m \to 0$ and $\delta_1\mathcal{W}_j \to G_0 f_j$.

The upshot of these calculations are Luther-Peschel asymptotics,

$$\lim_{m\to 0} d_a \log \tau^{[-]}(ma) = \frac{1}{2} d_a \log \sum_{|\epsilon|=0} \prod_{i<j} |a_i - a_j|^{2\epsilon_i\epsilon_j} \text{ for } n \text{ even}$$

and

$$\lim_{m\to 0} d_a \log \tau^{[-]}(ma) = \frac{1}{2} d_a \log \sum_{|\epsilon|=\pm\frac{1}{2}} \prod_{i<j} |a_i - a_j|^{2\epsilon_i\epsilon_j} \text{ for } n \text{ odd.}$$

The Luther-Peschel result applies to the supercritical short distance behavior for $\tau^{[+]}(ma)$ as well. Only the even scaling functions are non zero in the supercritical case so it is the first result that is relevant. The details for this calculation can be found in ([4]) and ([5]).

References

[1] B. M. McCoy and T. T. Wu. *The Two-Dimensional Ising Model.* Harvard Univ. Press, Cambridge, Mass., 1973.

[2] J. Palmer. Pfaffian bundles and the ising model. *Comm. Math. Phys.*, 120:547–574, 1989.

[3] J. Palmer. Tau functions for the Dirac operator in the Euclidean plane. *Pacific J. Math.*, 160(2):259–342, 1993.

[4] J. Palmer. Ising model scaling functions at short distance. In *Isomonodromic deformations and applications in physics (Montréal, QC, 2000)*, volume 31 of *CRM Proc. Lecture Notes*, pages 195–205. Amer. Math. Soc., Providence, RI, 2002.

[5] J. Palmer. Short distance asymptotics of Ising correlations. *J. Math. Phys.*, 43(2):918–953, 2002.

[6] J. Palmer and C. A. Tracy. Two-dimensional Ising correlations: convergence of the scaling limit. *Adv. in Appl. Math.*, 2(3):329–388, 1981.

[7] M. Sato, T. Miwa, and M Jimbo. Holonomic quantum fields I. *Publ. RIMS Kyoto Univ.*, 14:223–267, 1978.

[8] M. Sato, T. Miwa, and M. Jimbo. Holonomic quantum fields II. *Publ. RIMS Kyoto Univ.*, 15:201–278, 1979.

[9] M. Sato, T. Miwa, and M. Jimbo. Holonomic quantum fields III. *Publ. RIMS Kyoto Univ.*, 15:577–629, 1979.

[10] M. Sato, T. Miwa, and M. Jimbo. Holonomic quantum fields IV. *Publ. RIMS Kyoto Univ.*, 15:871–972, 1979.

[11] M. Sato, T. Miwa, and M. Jimbo. Holonomic quantum fields V. *Publ. RIMS Kyoto Univ.*, 16:531–584, 1980.

[12] G. Segal and G. Wilson. Loop groups and equations of KdV type. *Inst. Hautes Études Sci. Publ. Math.*, (61):5–65, 1985.

[13] T. T. Wu, B. M. McCoy, C. A. Tracy, and E. Barouch. Spin-spin correlation functions for the two dimensional Ising model: Exact theory in the scaling region. *Physical Review B*, 13:316–374, 1976.

Fields Institute Communications
Volume **50**, 2007

Constructing Conformal Field Theory Models

Ivan Todorov
Institut für Theoretische Physik der Universität Göttingen
Friedrich-Hund-Platz 1, D-37077 Göttingen, Germany
and
Institut des Hautes Études Scientifiques, 35 route de Chartres
F-91440 Bures-sur-Yvette, France
todorov@inrne.bas.bg

Abstract. Defining a mathematically consistent (non-free) quantum field theory in four space-time dimensions is a continuing challenge faced by mathematical physicists for nearly eighty years. It justifies attempts at constructing not entirely realistic conformal invariant models. We review two complementary projects of this type based on two inequivalent notions, infinitesimal and global, of conformal symmetry. The first, based on joint work with Gerhard Mack in the 1970's, culminated in rearranging the perturbation series in terms of divergence free skeleton graphs with dressed conformal invariant vertex functions and propagators [49], and in solving the Schwinger-Dyson equations by a conformal partial wave expansion [22]. The second started (in joint work with N. Nikolov [57]) by introducing the notion of *global conformal invariance* (GCI) and proving rationality of GCI Wightman functions. Recent and ongoing work with B. Bakalov, N. Nikolov, K.-H. Rehren and Ya. Stanev on the local field positive energy representations of the $\mathrm{sp}(\infty, \mathbb{R})$ Lie algebra is (p)reviewed and discussed.

Contents

2000 *Mathematics Subject Classification.* Primary 81T05, 81R10; Secondary 81T10, 20G05.
Permanent address: Institute for Nuclear Research and Nuclear Energy, Tsarigradsko Chaussee 72, BG-1784 Sofia, Bulgaria.

1 Introduction

1.1 Local causal automorphisms versus global conformal transformations. Conformal invariance can be viewed as a local version of dilation symmetry[1]: a *rational* (real) coordinate *transformation* $g : x \to y(x)$ (with singularities) of Minkowski space $M = \mathbb{R}^{3,1}$ is said to be *conformal* if it multiplies the Lorentzian interval by a positive factor (which may depend on x):

$$dy^2 = \frac{dx^2}{\omega^2(x,g)}\,, \quad dx^2 = d\boldsymbol{x}^2 - (dx^0)^2\,, \quad d\boldsymbol{x}^2 = \sum_{i=1}^{3} (dx^i)^2\,. \tag{1.1}$$

The general *conformal factor* $\omega(x,g)$ satisfying the *cocycle condition* $\omega(x, g_1\, g_2) = \omega(g_2\, x, g_1)\, \omega(x, g_2)$ is found to be a quadratic function of x. In fact, according to (an extension to four dimensions of) *Liouville's theorem* the conformal group of M is compounded by Poincaré transformations, uniform dilations ($y = \rho x$) for which ω is constant ($\omega = \rho^{-1}$) and *special conformal transformations* $x \to y(x,c)$ which can be defined as translations sandwiched between two conformal inversions ($x \to x/x^2$):

$$y\,(= y(x,c)) = \left(\frac{x}{x^2} + c\right) \left[\left(\frac{x}{x^2} + c\right)^2\right]^{-1} = \frac{x + c\, x^2}{\omega(x,c)}\,, \; \omega(x,c) = 1 + 2\, c\, x + c^2\, x^2 \tag{1.2}$$

($c\, x = \boldsymbol{c}\, \boldsymbol{x} - c^0\, x^0$). Clearly, for any $x \neq 0$, $\omega(x,c)$ changes sign with varying $c \in M$ – and vanishes on a 3-cone (which degenerates into a hyperplane for $x^2 = 0$). It follows that, unlike the infinitesimal interval (1.1), the square of a finite interval may change sign under special conformal transformation:

$$y_{12}^2 := (y_1 - y_2)^2 = \frac{x_{12}^2}{\omega(x_1,c)\, \omega(x_2,c)} \text{ for } x_{12} = x_1 - x_2\,,\; y_i = y(x_i,c)\,,\; i = 1,2\,. \tag{1.3}$$

The 15-parameter group of rational coordinate transformations thus described is isomorphic to the factor group $SO_0(4,2)/\mathbb{Z}_2$ of the connected group of pseudo-rotations $SO_0(4,2)$ with respect to its centre $\mathbb{Z}_2$ (involving the reflection of all six axes). We shall use instead the spinorial *conformal group*

$$\mathcal{C} = SU(2,2) \tag{1.4}$$

that is a double cover of $SO_0(4,2)$ (and thus a four-fold cover of $SO_0(4,2)/\mathbb{Z}_2$). $\mathcal{C}$ is not simply connected and has an infinite sheeted universal covering group $\tilde{\mathcal{C}}$ that is not a matrix group.

We thus have two inequivalent notions of conformal symmetry: (1) *local causal automorphisms* of M [29] which preserve the causal order of events (and in particular, the sign of the square interval) and are compounded by Poincaré transformations, dilations and infinitesimal special conformal transformations; (2) *finite (global) conformal transformations* which only act without singularities on the conformal compactification $\bar{M}$ of Minkowski space. It was Dirac[2] [20] who realized $\bar{M}$

[1]This point of view was expounded by Hermann Weyl (1885-1955) in his 1918 paper *Gravitation und Elektrizität* which first introduced the notion of gauge symmetry – in the wrong place. The attempt to unify gravity and electromagnetism on this basis was rightly dismissed by physicists starting with Einstein. In the words of Michael Atiyah [4] the precious "infant" (gauge theory) "was nearly thrown out with the bath water".

[2]P.A.M. Dirac (1902-1984) was fascinated by projective geometry ever since his student years in Bristol. He was probably the only one who praised the dismissed Weyl theory 55 years later, [21], saying that it "remains as the outstanding one, unrivaled by its simplicity and beauty".

as a projective quadric in six dimensions

$$\bar{M} \simeq Q/\mathbb{R}^* \simeq \mathbb{S}^3 \times \mathbb{S}^1/\mathbb{Z}^2\,, \quad Q = \left\{ \vec{\xi} \in \mathbb{R}^{4,2}\,;\; \vec{\xi}^2 := \sum_{\alpha=1}^{4} \xi_\alpha^2 - \xi_0^2 - \xi_{-1}^2 = 0 \right\}, \tag{1.5}$$

providing a manifestly conformal invariant description of massless wave equations. In fact, the conformal invariance of the classical Maxwell equations was recognized earlier by Cunningham and Bateman[3] [17] [9] soon after the notion of a symmetry group entered physics with the work of Poincaré and Einstein on relativity.

1.2 Conformal invariance and critical behaviour in QFT. Discrete masses of atoms and elementary particles violate "the great principle of similitude"[4] (*i.e.* scale and, *a fortiori*, conformal invariance). The situation in *quantum field theory* (QFT) is still more involved – and more interesting: dimensional parameters arise in the process of renormalization even if they are absent in the classical theory. Dilation and conformal invariance can only be preserved for a renormalization group fixed point – *i.e.*, for a critical theory, the QFT counterpart of a point of phase transition. Modern mathematical study of Feynman graphs (and renormalization) often neglect masses as an inessential complication (see *e.g.* [11]). Perhaps the study of an idealized critical theory with no dimensional parameters will prove to be an essential step in understanding QFT – just as Galilei's law of inertia, that neglects friction, has been crucial in formulating and understanding classical mechanics?

What does matter is the fact that conformal invariance opens new avenues for attacking the central theoretical problem: constructing a consistent 4-dimensional QFT model. First, it allows one to compute (dressed) propagators and vertex functions [63] [68] and to write down a skeleton graph expansion without divergences [49] which obeys a conformally invariant (bootstrap) form of the Schwinger-Dyson equations [50] [60] – see Sect. 2. Secondly, soon after Wilson [86] introduced the notion of (small distance) *operator product expansion* (OPE) two Italian groups, [12] and [28], proposed a way to write exact (global) conformal invariant OPE. This was further developed leading to the notion of *conformal partial wave expansion* and applied to the solution of the dynamical (Schwinger-Dyson) equations, [45] [22] [23] as well as (later) to constructing conformal composite operators in $(\varphi^3)_6$ and in QCD, [16]. (The work of the 1970's is reviewed in [83] and in [30], for (alternative) subsequent developments – see [31].)

All the above developments use local causal automorphisms (see Sect. 1.1) (what is called weak conformal invariance in [38]); it implies global Euclidean conformal invariance (and allows to construct a conformal QFT on the universal cover $\tilde{M} = \mathbb{S}^3 \times \mathbb{R}$, of compactified Minkowski space, [44]). The notion of *global conformal invariance* (GCI) in Minkowski space [57], on the other hand, is much stronger. It yields a higher dimensional counterpart of a chiral *vertex algebra* (which became, in the 1-dimensional context, a rich mathematical theory starting with [13] and the subject of textbooks and monographs – see *e.g.* [40] [32] as well as the recent

---

[3]Ebenezer Cunningham (1881-1977) outlived his friend Harry Bateman (1882-1946) by over 30 years. A keenly religious man and a music lover, he did not return to major research projects after World War I.

[4]See Lord Rayleigh, The principle of similitude, *Nature* **95**:2368 (March 1915) 66-68 and 644. John William Strutt – Lord Rayleigh (1842-1919) was awarded the 1904 Nobel Prize for his discovery of the inert gas argon.

expository paper [18]). The work of the last five years, which explores GCI [57] [55] [56] [58] [52] [59] [53], and the possibility to construct a conformal QFT model on this basis, is reviewed in Sect. 3.

Let us stress from the outset that even the weaker assumption of local conformal invariance excludes the postulate of asymptotic completeness (except for the uninteresting case of free massless fields). Thus the usual criterion for a non-free model, the existence of a non-trivial scattering matrix, is not applicable to a conformal QFT. It seems likely that a GCI model will also share the local property of normal products of free fields discussed, say, in [14] and [15].

In order to make this survey relatively self-contained we collect some background material on the conformal Lie algebra and on the all important complex variable realization of compactified Minkowski space ([82] [59] [58]) in Appendices A and B, respectively.

To summarize: the paper reviews old and new efforts to construct non-trivial conformal QFT models. The organization of material can be read off the table of contents.

2 Skeleton graph expansion for a critical Yukawa theory

2.1 Anomalous dimensions. Conformal 2- and 3-point functions. Positivity constraint. Feynman integrals produce (poly)logarithms, an observation that led to the popular catch phrase "no logs – no physics". Conformal QFT models induce us rethink this "perturbative wisdom". Indeed the leading logs in a 2-point function (or in the "wave function renormalization") sum up to a power precisely when the Callan-Symanzik β-function [79] vanishes (*i.e.* when the renormalized charge $\bar{g}(g,\mu)$ does not change under variation of the mass scale μ). This was discovered in the context of the 2D Thirring model [86] [43] and led to the introduction of anomalous dimensions[5] in QFT. It signals a change of representation, compared to the classical (or free quantum) field of canonical dimension.

Conformal invariance allows to approach the construction of a critical theory without appealing to standard perturbative expansions. The knowledge of the field dimensions allows to construct conformally invariant propagators and vertex functions ([63], [68]). These are particularly simple in a scalar field theory. The time-ordered propagator of a (hermitean) scalar conformal field ϕ of dimension d is given (for an appropriately chosen normalization) by

$$\langle 0 \mid T\phi(x_1)\,\phi(x_2) \mid 0\rangle = \frac{\Gamma(d)}{(4\pi)^2}\left(\frac{x_{12}^2}{4}+i0\right)^{-d} = -i\int \frac{\Gamma(2-d)}{(p^2-i0)^{2-d}}\,e^{ipx_{12}}\,\frac{d^4p}{(2\pi)^4}\,,$$

$$x_{12}=x_1-x_2\,,\quad x^2=\boldsymbol{x}^2-(x^0)^2\quad (\boldsymbol{x}^2=x_1^2+x_2^2+x_3^2)\,. \tag{2.1}$$

The general conformal 3-point Green's function of a triple of scalar fields $\phi_i(x)$ of dimensions d_i, $i=1,2,3$ is written as a product of three "infraparticle" propagators,

$$\langle 0 \mid T\phi_1(x_1)\,\phi_2(x_2)\,\phi_3(x_3) \mid 0\rangle = C_{123}\,(x_{12}^2+i0)^{-\delta_3}\,(x_{13}^2+i0)^{-\delta_2}\,(x_{23}^2+i0)^{-\delta_1}\,, \tag{2.2}$$

[5] According to W. Nahm [51], "In 1964 Wilson conjectured that perturbations just introduce some logarithmic corrections" to the canonical scaling, but then K. "Johnson familiarized him with the Thirring model" which led to [86]. The notion of anomalous dimension (if not the term) was encountered earlier in the critical Ising model, solved in 1944 by L. Onsager (1903-1976). It was given a field theoretic interpretation by L. Kadanoff [41].

where the exponents δ_i obey the conservation of dimension law

$$\delta_1+\delta_2=d_3\,,\quad \delta_1+\delta_3=d_2\,,\quad \delta_2+\delta_3=d_1\quad (\delta_3=\tfrac{1}{2}(d_1+d_2-d_3)\text{ etc.})\,. \tag{2.3}$$

The conformal invariant 2-point function of arbitrary irreducible spin-tensor fields is uniquely determined (and explicitely constructed – see [46] [82]). For reducible representations the uniqueness can be restored by demanding invariance under some discrete transformations. Thus the propagator of a conformal Dirac field ψ of dimension d' is determined (up to normalization) if we also demand invariance under space reflection: $\psi(x^0,\boldsymbol{x})\to\gamma^0\psi(x^0,-\boldsymbol{x})$, $\tilde{\psi}(x^0,\boldsymbol{x})\to\tilde{\psi}(x^0,-\boldsymbol{x})\,\gamma^{0*}$, $\tilde{\psi}$ standing for the Dirac conjugate field:

$$\langle 0\mid T\psi(x_1)\,\tilde{\psi}(x_2)\mid 0\rangle = i\,\not{\partial}_1\,\frac{\Gamma\left(d'-\frac{1}{2}\right)}{(4\pi)^2\left(\frac{x_{12}^2}{4}+i0\right)^{d'-\frac{1}{2}}}\qquad \left(\not{\partial}_1=\gamma^\mu\,\frac{\partial}{\partial x_1^\mu}\right)\,. \tag{2.4}$$

For a pseudoscalar field ϕ the Yukawa type 3-point function in a space-reflection invariant conformal theory (for $\phi(x^0,\boldsymbol{x})$ going to $-\phi(x^0,-\boldsymbol{x})$ under space reflections) is given by (see Appendix B to [49])

$$\begin{aligned}&\langle 0\mid T(\phi(x_1)\,\psi(x_2)\,\tilde{\psi}(x_3))\mid 0\rangle\\ &=\; g\,\not{x}_{12}\,\gamma_5\,\not{x}_{13}\,(x_{23}^2+i0)^{\frac{d}{2}-d'}\left[(x_{12}^2+i0)(x_{13}^2+i0)\right]^{-\frac{d+1}{2}}\,.\end{aligned} \tag{2.5}$$

Remark 2.1 In general, the 3-point function of a tensor field is not unique (even after imposing natural discrete symmetries). For instance, the 3-point function of the stress energy tensor involves three independent structures corresponding to three free models: the massless scalar field, the massless Dirac (or the 2-component Weyl) field, the free Maxwell field – see [73].

Hilbert space positivity constraints the Wightman 2-point functions and hence the dimensions of all fields in the theory. We shall illustrate what is going on with the example of a scalar field and will then formulate the general result.

The x-space Wightman function, $w(x_{12})$, is a boundary value of an analytic function $w(x+iy)$ holomorphic in the backward *tube* $\mathcal{T}_-$ where

$$\mathcal{T}_\pm=\left\{x+iy\in\mathbb{C}^4\,;\ \pm y^0>|\boldsymbol{y}|=\sqrt{y_1^2+y_2^2+y_3^2}\right\}\,. \tag{2.6}$$

In particular, $w(x_{12})=\langle 0\mid\phi(x_1)\,\phi(x_2)\mid 0\rangle$ differs from (2.1) just by the "$i\varepsilon$-prescription" (its Fourier transform having support in the future cone):

$$w(x)=\frac{\Gamma(d)}{(4\pi)^2}\left(\frac{x^2}{4}+i0x^0\right)^{-d}=\frac{2\pi}{\Gamma(d-1)}\int\theta(p^0)(-p^2)_+^{d-2}\,e^{ipx}\,\frac{d^4p}{(2\pi)^4} \tag{2.7}$$

where

$$(a)_+=\begin{cases}a & \text{for } a>0\\ 0 & \text{for } a<0\end{cases}\,,\quad \lim_{d\to 1}\frac{(-p^2)_+^{d-2}}{\Gamma(d-1)}=\delta(p^2)\,. \tag{2.8}$$

Wightman positivity means that the Fourier transform $\tilde{w}(p)$ of w should be a positive measure. This is satisfied if the integral (2.7) converges at the lower limit (for $p_\mu\to 0$), that is, for d not smaller than its canonical value, $d\geq 1$.

In general, for (irreducible) conformal fields, 2-point positivity amounts to unitarity of the corresponding representations of the universal cover $\tilde{C}$ of the conformal group. The positive energy irreducible representations of $\tilde{C}$ have been classified by Mack [46]. They are induced by irreducible finite-dimensional representations of the (covering of the) maximal compact subgroup $S(U(2)\times U(2))$ of $G=\mathrm{SU}(2,2)$, labeled by triples $(d; j_1, j_2)$ where $d \geq 0$ stands for the minimal eigenvalue of the conformal Hamiltonian H (A.9) (see Appendix A) $j_{1,2}$ ($\in \frac{1}{2}\mathbb{N}$) are the SU(2) spins, the dimension of the inducing representation being $(2j_1+1)(2j_2+1)$. For Minkowski space fields d coincides with the scale dimension, while (j_1, j_2) label the corresponding finite-dimensional representation of SL$(2,\mathbb{C})$ (the quantum mechanical Lorentz group). For a non-trivial representation ($(d; j_1, j_2) \neq (0; 0, 0)$) the unitarity (Wightman positivity) condition reads

$$d \geq 1 + j_1 + j_2 + \theta(j_1 j_2)\,, \quad \theta(a) = 1 \text{ for } a > 0\,, \; \theta(0) = 0\,. \tag{2.9}$$

The resulting representations are single-valued on $\mathcal{C}$ for integer *twist* $d - j_1 - j_2$. The fields corresponding to the boundary of the positivity domain (where the equality sign in (2.9) takes place) satisfy free field equations for $j_1 j_2 = 0$ or conservation laws – for $j_1 j_2 > 0$.

2.2 Vertex functions, skeleton graphs, bootstrap equations. In order to write down a *skeleton graph* expansion – with no self-energy and vertex function corrections – we need the "amputated" vertex function, obtained from the 3-point Green's function ((2.3) or (2.5)) by taking the convolution with the inverse external lines' propagators. The result can be reproduced (up to a proportionality constant) by substituting the conformal dimensions d_i by the corresponding "shadow[6] dimensions" $4 - d_i$, or equivalently, replacing the exponents δ_i in (2.3) by $2 - \delta_i$ (see Sect. IIB and Appendix B of [49]).

The inequalities (2.9) for the conformal dimensions of interacting fields suggest that the resulting dressed propagators have worse ultraviolet behaviour than the free ones. It is all the more remarkable that the contribution corresponding to an arbitrary skeleton graph in the above Yukawa model of meson-nucleon interaction is given by a convergent integral for non-exceptional momenta provided

$$1 < d < 3 \;(d \neq 2) \quad \frac{3}{2} < d' < \frac{5}{2}\,. \tag{2.10}$$

This is the central result (proven in Sect. III) of [49]. It shows, in particular, that the skeleton graphs with four pseudoscalar external meson lines are convergent, in spite of the fact that the corresponding canonical Feynman graphs are logarithmically divergent (the dressed conformal vertex functions providing effectively an ultraviolet cutoff, which compensates for the slow decrease of propagators).

The Schwinger-Dyson equations are implemented by Migdal's bootstrap equation [50] for the vertex function and by the generalized unitarity relation of Parisi and Peliti [64] for the propagator. In our derivation ([49] Sect. IV) we start with Symanzik's[7] form [77] of the integro-differential equations of the theory. After differentiation with respect to an external momentum (which annihilates the constant $g\gamma_5$-term) the resulting renormalized Schwinger-Dyson equation for the vertex function is the same for all values of coupling constant (and masses) and is therefore valid in the renormalization group fixed point (also called Gell-Man-Low limit

[6] A terminology, introduced in [28] in the context of conformal OPE.

[7] Kurt Symanzik (1923-1983) has contributed more to this field (as well as to many others) than usually recognized: see (besides the above lectures) [78] [79] (where his β-function is introduced), [80], [48] (where generalized unitarity and Ward identities involving the stress energy tensor are considered).

theory). In the presence of anomalous dimensions the requirement of dilation invariance fixes the integration constant to zero and one recovers Migdal bootstrap equation.

We thus end up with three equations – one for the vertex function and one for each of the two (meson and nucleon) propagators – for the three dimensionless constants, g, d and d' in the theory. One can, therefore, hope that the resulting conformal theory will involve no free parameters (or that there will be a discrete set of solutions) – a hope that cannot be verified as long as we are unable to solve the system of (highly non-linear) integral equations.

2.3 Conformal partial wave expansion. Discussion. Ordinary partial wave expansion is nothing but the tensor product expansion of two irreducible (Wigner) particle representations of the Poincaré group. *Conformal partial wave expansion* is, by definition, the tensor product expansion of two irreducible positive energy representations of $\tilde{C}$. For the Wightman function of a pair of conjugate scalar fields ϕ and ϕ^* of dimension d and $n-2$ arbitrary local fields ϕ_i this expansion assumes the form

$$\begin{aligned}&\langle 0 \mid \phi^*(x_1)\,\phi(x_2)\,\phi_3(x_3)\ldots\phi_n(x_n)\mid 0\rangle\\ &\quad= \langle 0 \mid \phi^*(x_1)\,\phi(x_2)\mid 0\rangle\langle 0\mid \phi_3(x_3)\ldots\phi_n(x_n)\mid 0\rangle\\ &\qquad+\sum_{\kappa\geq 1}\sum_{\ell=0}^{\infty}\langle 0\mid\phi^*(x_1)\,\phi(x_2)\prod_{\kappa\ell}\phi_3(x_3)\ldots\phi_n(x_n)\mid 0\rangle \end{aligned} \tag{2.11}$$

where $\prod_{\kappa\ell}$ is the orthogonal projection operator on the subspace of symmetric traceless tensors of rank ℓ and twist 2κ (*i.e.* of dimension $2\kappa+\ell$). For a free massless field φ the twists are even integers so that κ takes positive integer values. We assume that the "compact Hamiltonian" H (A.9) in a conformal QFT has a discrete spectrum so that (2.11) involves no integral in κ. (This is derived as a Corollary in [47] of an appropriate assumption about the existence of OPE.)

The applicability of the expansion (2.11) is enhanced by the fact that one can write rather explicitly the projected 2-particle state:

$$\langle 0\mid \phi^*(x_1)\,\phi(x_2)\prod_{\kappa\ell} = (x_{12}^2)^{\kappa-d}\,K_{\kappa\ell}(x_{12}\,\partial, x_{12}^2\,\Box)\,O_{2\kappa,\ell}(x;x_{12})\mid_{x=x_2} \tag{2.12}$$

where $O_{2\kappa,\ell}(x;\zeta)$ is a rank ℓ symmetric traceless tensor field contracted with ζ:

$$O_{2\kappa,\ell}(x;\zeta) = O_{2\kappa}^{\mu_1\ldots\mu_\ell}(x)\,\zeta_{\mu_1}\ldots\zeta_{\mu_\ell}\,,\quad \Box_\zeta\, O_{2\kappa,\ell}(x;\zeta)=0\,, \tag{2.13}$$

and

$$K_{\kappa\ell}(t_1,t_2)=\sum_{n=0}^{\infty}\int_0^1\frac{[\alpha(1-\alpha)]^{\ell+\kappa+n-1}\,e^{\alpha t_1}\,(-t_2)^n\,d\alpha}{4^n\,B(\ell+\kappa,\ell+\kappa)\,n!(2\ell+2\kappa-1)_n} \tag{2.14}$$

where

$$(a)_n=\frac{\Gamma(a+n)}{\Gamma(a)}\,,\qquad B(a,b)=\frac{\Gamma(a)\,\Gamma(b)}{\Gamma(a+b)}\,,$$

the normalization being chosen in such a way that $K_{\kappa\ell}(0,0)=1$.

It was Gerhard Mack [45] who realized that the infinite set of dynamical integro-differential equations for the conformal Euclidean Green's functions (incorporating generalized-off-shell-unitarity) can be solved by the Euclidean (Spin (5,1)-) partial wave expansion (*cf.* also [64]). For a scalar QFT model this program is implemented in [22] [23]. The Yukawa model is dealt with (at least formally) in [30].

The above discussion does not apply directly to gauge theories. One cannot expect correlation functions of gauge dependent fields to be conformal invariant. For instance, the general conformal invariant 2-point function of a gauge potential (a vector field A_μ of dimension 1) is purely longitudinal, proportional to $\frac{\partial}{\partial x_1^\mu} \frac{x_{12\nu}}{x_{12}^2 + i0\, x_{12}^0}$, thus giving rise to a zero field strength. Furthermore, the general conformal and space-reflection invariant 2-point function of the Maxwell field,

$$\begin{aligned}
&\langle 0 \mid F_{\lambda_1\mu_1}(x_1)\, F_{\lambda_2\mu_2}(x_2) \mid 0 \rangle \\
&= \{\partial_{\lambda_1}(\partial_{\lambda_2}\, \eta_{\mu_1\mu_2} - \partial_{\mu_2}\, \eta_{\mu_1\lambda_2}) \\
&\quad - \partial_{\mu_1}(\partial_{\lambda_2}\, \eta_{\lambda_1\mu_2} - \partial_{\mu_2}\, \eta_{\lambda_1\lambda_2})\}\, \frac{1}{4\pi^2 (x_{12}^2 + i0\, x_{12}^0)}
\end{aligned} \tag{2.15}$$

satisfies the free field equations implying that the electromagnetic current vanishes (*cf.* [1] [7]; for – so far inconclusive – attempts to use indecomposable representations of $\mathcal{C}$ in conformal quantum electrodynamics – see [62] [74]). One can hope, however, to find conformal correlation functions of gauge invariant composite operators, a hope that appears to be fulfilled in the $N = 4$ super Yang-Mills theory (see, *e.g.* [19] for a review, and [3] [34] [37] [66] for a sample of more recent publications).

Composite conserved currents and the stress-energy tensor appear in the OPE of mutually conjugate basic fields (like in (2.12) for $\kappa = 1$). Their properties are studied in [48] [73] [61] [2] among others. Such (observable) integer dimension fields are not appropriate substitutes for basic fields appearing in a skeleton expansion since the convergence result (cited in Sect. 2.2) does not hold for integer d. By contrast, we shall argue in the next section that the GCI postulate applies precisely to such fields and thus provides an alternative tool for model building.

3 Four-dimensional conformal field theories with an infinite number of conserved tensor fields

3.1 From GCI in Minkowski space to higher dimensional vertex algebras. The notion of global conformal invariance (GCI) in Minkowski space (not to be confused with the weaker property of GCI on the infinite sheeted cover $\tilde{M}$ of $\bar{M}$ [44]) has been around for many years (and even appeared sporadically in the literature [35] [38]) but was not fully explored until recently. Realizing that a special conformal transformation (1.2) of a pair of points (x_1, x_2) with non-lightlike x_{12} may change the sign of the interval x_{12}^2 (according to (1.3)) makes one skeptical on the applicability of the stronger notion of GCI to a causal field theory. The story of our first paper [57] on the subject has been a story of how my young collaborator Nikolay Nikolov, who took GCI seriously, was overcoming my skepticism.

The first important consequence of GCI exploits precisely the counterintuitive fact that space like and time like pairs of points can be exchanged by special conformal transformations.

Proposition 3.1. (a) *The conformal group acts transitively on compactified Minkowski space* $\bar{M}$. *The stability subgroup of a point is conjugate to the Poincaré subgroup* $\mathcal{P}$ *with dilations, which can be identified with the* 11*-parameter automorphism group* $\operatorname{Aut}\mathcal{P}$ *of* $\mathcal{P}$, *so that* $\bar{M}$ *can be viewed as the coset space*

$$\bar{M} = \mathcal{C}/\operatorname{Aut}\mathcal{P}\,. \tag{3.1}$$

(b) *Any pair* ($p_0 = \{\lambda \vec{\xi}_0\}$, $p_1 = \{\lambda \vec{\xi}_1\}$) *of non-isotropic points of* $\bar{M}$ ($\vec{\xi}_0^{\,2} = 0 = \vec{\xi}_1^{\,2}$ – *see* (1.5); $\vec{\xi}_0 \cdot \vec{\xi}_1 \neq 0$) *can be mapped onto any other such pair by a conformal transformation.*

A complete *proof* of this Proposition is given in the PhD thesis of Nikolov. For a proof of (b) assuming (a) – see [57] Proposition 1.1.

Remark 3.1. The tip $p_\infty = \{\lambda \vec{\xi}_\infty\}$ of the cone at infinity ($\vec{\xi}_\infty = (-1,0,0,0,0,1)$ for the embedding (B.9) of M in $\bar{M}$ – see Appendix B) is left invariant precisely by the Poincaré transformations and dilations of Aut $\mathcal{P}$. The stability subgroup of the origin $x = 0$ in Minkowski space M (corresponding to $p_0 = \lambda(1,0,0,0,0,1)$ in $\bar{M}$) is the 11-parameter subgroup spanned by Lorentz transformations, dilations and the special conformal transformations (1.2) which are conjugate to translations by the Weyl inversion w of Appendix A.

Local commutativity combined with the fact that time like and space like pairs of points can be transformed into each other (Proposition 3.1(b)) implies the *Huygens principle:* local GCI fields commute for non-light-like separations. For a field ψ of conformal weight $(d; j_1, j_2)$ (see Sect. 2.1) the Huygens principle within the Wightman framework [75] assumes the form of a *strong locality condition:*

$$(z_{12}^2)^N \{\psi(z_1)\,\psi^*(z_2) - (-1)^{2j_1+2j_2}\,\psi^*(z_2)\,\psi(z_1)\} = 0\,, \text{ for } N \geq d + j_1 + j_2\,. \tag{3.2}$$

Here $z_{1,2}$ refer to the global complex variable parametrization of compactified Minkowski space $\bar{M}$ reviewed in Appendix B. In fact, as the z's are related to the real Minkowski space coordinates x by the complex conformal transformation, (B.1), so that

$$z_{12}^2 := \sum_{\alpha=1}^{4} (z_1^\alpha - z_2^\alpha)^2 = \frac{x_{12}^2}{\omega(x_1)\,\omega(x_2)}\,, \quad x_{12}^2 = (\boldsymbol{x}_1 - \boldsymbol{x}_2)^2 - (x_1^0 - x_2^0)^2 \tag{3.3}$$

where the conformal factor $\omega(x) = \frac{1}{2}\,(x^2+1) - ix^0$ does not vanish for real x, an equation identical to (3.2) would be also valid for the fields $\psi(x)$ in M. What does make the z-picture special is the fact that each field can be represented by a formal power series of the type

$$\psi(z) = \sum_{n \in \mathbb{Z}} \sum_{m \geq 0} (z^2)^n\, \psi_{\{n,m\}}(z)\,, \tag{3.4}$$

where $\psi_{\{n,m\}}(z)$ are (in general, multicomponent) operator valued homogeneous (of degree m) harmonic polynomials. (The uniqueness of such a decomposition and its relation to the representation theory of the rotation group are displayed in [8].) Strong locality for such series can be interpreted as an algebraic relation.

The expansion (3.4) diagonalizes the conformal Hamiltonian H (A.9):

$$[H, \psi(z)] = \left(z \cdot \frac{\partial}{\partial z} + d \right) \psi(z) \Rightarrow [H, \psi_{\{n,m\}}(z)] = (2n + m + d)\, \psi_{\{n,m\}}(z)\,. \tag{3.5}$$

The relativistic spectral conditions (including energy positivity), which ensure stability of the vacuum, imply the analyticity of the vector valued function $\psi(z) \mid 0\rangle$ in the compact picture *tube domain* T_+ (B.5). It follows that

$$\psi_{\{n,m\}}(z) \mid 0\rangle = 0 \text{ for } n < 0\,. \tag{3.6}$$

The invariance of the vacuum vector under (complex) z-translations T_α,

$$[T_\alpha, \psi(z)] = \frac{\partial}{\partial z^\alpha} \psi(z)\,,\ T_\alpha \mid 0\rangle = 0\,, \tag{3.7}$$

together with (3.6) allows to establish the *state field correspondence:* there is a one-to-one correspondence between finite energy states v and translation covariant strongly local fields ("vertex operators") $Y(v,z)$ such that

$$Y(v,z) \mid 0\rangle = e^{T\cdot z}\, v \quad (v = Y(v,0) \mid 0\rangle)\,. \tag{3.8}$$

The properties (3.2) (3.4) (3.6) (3.7) allow a straightforward derivation of the central result of [57] (Theorem 3.1).

Proposition 3.2. *For any set $\phi_1, \ldots, \phi_n$ of strongly local GCI fields the formal power series*

$$F_{1\ldots n}(z_1,\ldots,z_n) = \prod_{1\le i<j\le n} (z_{ij}^2)^{n_{ij}} \langle 0 \mid \phi_1(z_1)\ldots\phi_n(z_n) \mid 0\rangle\,,\ n_{ij} \geq \Delta_{ij} \tag{3.9}$$

is a (translation invariant, homogeneous) polynomial in $z_1,\ldots,z_n$, Δ_{ij} being natural numbers only depending on the conformal weights of the fields $\phi_i(z)$ and $\phi_j(z)$.

The fact that the Huygens principle together with the Wightman axioms yields rationality of correlation functions has been noted earlier by Baumann [10]. A more precise statement and a proof of Proposition 3.2 in the vertex algebra formalism is given in Theorem 3.3 of [59] (see also [52]).

The mathematical properties of the above defined higher dimensional vertex algebras are investigated in [52] and [5]. Thermal elliptic correlation functions of GCI fields and associated modular forms are introduced and studied in [59].

3.2 Model building: correlation functions and the twist-two bilocal field. As reviewed in Sect. 2.1, 2- and 3-point functions are essentially uniquely determined by conformal invariance. By contrast, the 4-point function may depend, in general, arbitrarily on two conformally invariant cross-ratios

$$s = \frac{z_{12}^2\, z_{34}^2}{z_{13}^2\, z_{24}^2}\,, \quad t = \frac{z_{14}^2\, z_{23}^2}{z_{13}^2\, z_{24}^2}\,. \tag{3.10}$$

(In a D-dimensional space-time there are $\frac{n(n-3)}{2}$ independent n-point cross-ratios for $4 \le n \le D+2$; for all $n \geq D+1$ the number of the algebraically independent ones is $n\,D - \binom{D+2}{2}$.) Due to the rationality of correlation functions (Proposition 3.2), however, including the restrictions on the degree of singularity coming from Wightman positivity (the precise values of Δ_{ij} – see Proposition 4.3 of [57]) all correlation functions are determined within a finite number of free parameters.

We shall illustrate this fact by writing down the 4-point functions of a scalar GCI field of integral dimension d, which involves $\left[\frac{d^2}{3}\right]$ $(= 2d-3$ for $d = 2,3,4)$ independent parameters. In proving this (for $D = 4$ – see [55]) one uses the fact that, as a consequence of Wightman positivity, the truncated n-point functions $(n \geq 3)$ of a scalar field have strictly weaker singularities than its 2-point function. One finds, in particular, that the truncated 4-point function of a $d = 1$ scalar field vanishes identically (without having to use the free field equation and the Reeh-Schlieder theorem [65]).

We are looking for an *irreducible model* of a scalar field, *i.e.* a model whose field algebra cannot be represented as the tensor product of two (commuting) local algebras. In particular, we assume that there is a unique stress-energy tensor – *i.e.*, a unique twist-two, rank-two (necessarily conserved) symmetric traceless tensor $T(z;v) = T_{\alpha\beta}(z)\, v^\alpha\, v^\beta$ (*cf.* (2.13)) that generates its own translations. We then define the *central charge* c by the normalization of its 2-point function ([53]; *cf.* also [2])

$$\langle 0 \mid T(z_1;v_1)\, T(z_2;v_2) \mid 0\rangle = \frac{2\,c}{3(z_{12}^2)^4}\, [4\,(v_1\, r(z_{12})\, v_2)^2 - v_1^2\, v_2^2] \tag{3.11}$$

where

$$v_1\, r(z)\, v_2 = v_1 \cdot v_2 - 2\, \frac{(z\cdot v_1)(z\cdot v_2)}{z^2} \qquad (r^2(z) = \mathbb{1}) \tag{3.12}$$

and the normalization is chosen in such a way that $c=1$ for the stress tensor of a free hermitean massless scalar field φ. With c so defined we fix the normalization of $\phi(z)$ setting

$$\langle 12\rangle \equiv \langle 0 \mid \phi(z_1)\, \phi(z_2) \mid 0\rangle = \frac{c}{d}\, (z_{12}^2)^{-d}\,. \tag{3.13}$$

Eqs. (3.11) (3.13) and the Ward-Takahashi identities for T then imply

$$\langle 0 \mid \phi(z_1)\, \phi(z_2)\, T(z_3,v) \mid 0\rangle = \frac{2\,c\,(z_{12}^2)^{1-d}}{3\; z_{13}^2\, z_{23}^2} \left[4\,(X_{12}^3\cdot v)^2 - \frac{z_{12}^2}{z_{13}^2\, z_{23}^2}\right], \tag{3.14}$$

where

$$X_{12}^3 := \frac{z_{13}}{z_{13}^2} - \frac{z_{23}}{z_{23}^2} \qquad \left(\text{so that } (X_{12}^3)^2 = \frac{z_{12}^2}{z_{13}^2\, z_{23}^2}\right). \tag{3.15}$$

We shall review the study of the simplest model of a scalar field ϕ of dimension $d=2$ ([55] [6]) and will only mention in passing the most promising model with $d=4$. The uniqueness of T implies that the scalar field of dimension 2 necessarily appearing in the OPE of two ϕ's is again ϕ, and moreover[8]

$$\langle 123\rangle \equiv \langle 0 \mid \phi(z_1)\, \phi(z_2)\, \phi(z_3) \mid 0\rangle = \frac{c}{z_{12}^2\, z_{23}^2\, z_{13}^2}\,, \quad i.e.\ c = \frac{8\,\langle 12\rangle\, \langle 13\rangle\, \langle 23\rangle}{|\langle 123\rangle|^2}\,. \tag{3.16}$$

We shall reproduce the truncated 4-point function of a hermitean scalar field of dimension d ($w_4^t(\cdot;d) = \langle 1234\rangle - \langle 12\rangle\,\langle 34\rangle - \langle 14\rangle\,\langle 23\rangle - \langle 13\rangle\,\langle 24\rangle$) for $d = 2,3,4$. For $d=2$, w_4^t is again expressed in terms of the central charge c (as a consequence of the argument referred to above for a unique T):

$$\begin{aligned} w_4^t(\cdot;2) &= \frac{c}{z_{12}^2\, z_{23}^2\, z_{34}^2\, z_{14}^2}\,(1+s+t) \\ &= c\left(\frac{1}{z_{12}^2\, z_{34}^2\, z_{14}^2\, z_{23}^2} + \frac{1}{z_{13}^2\, z_{24}^2\, z_{14}^2\, z_{23}^2} + \frac{1}{z_{12}^2\, z_{34}^2\, z_{13}^2\, z_{24}^2}\right). \end{aligned} \tag{3.17}$$

For $d=3$ the (rational) 3-point function vanishes while

$$w_4^t(\cdot;3) = \left(\prod_{1\le i<j\le 4} z_{ij}^2\right)^{-1} c\,(b_0\,\tilde{J}_0(s,t) + b_1\,\tilde{J}_1(s,t) + b_2) \tag{3.18}$$

[8] See [55], Sect. 2; a more general statement - without the uniqueness assumption - is also valid [54].

where $\tilde{J}_\lambda(s,t)$ are the crossing symmetric rational functions

$$\tilde{J}_0(s,t) = s + t + \frac{t+1}{s} + \frac{s+1}{t},$$
$$\tilde{J}_1(s,t) = \frac{(1-t)(1-t^2)}{st} - t^{-1} - t - s\,(1+t^{-1}) + \frac{s^2}{t}. \qquad (3.19)$$

Finally, for $d = 4$, we can write

$$w_4^t(\cdot\,;4) = \frac{c}{(z_{12}^2\, z_{23}^2\, z_{34}^2\, z_{14}^2)^2} \left\{ \sum_{\nu=0}^{2} a_\nu \, \frac{J_\nu(s,t)}{st} + bD(s,t) + b'Q(s,t) \right\} \qquad (3.20)$$

where $J_\nu(s,t)$ are polynomials in s and t of overall degree 5,

$$\begin{aligned}
\frac{1}{st}\, J_0(s,t) &= \frac{t+t^2}{s} + s(t^{-1}+t^2) + s^2(t^{-1}+t) \\
\frac{1}{st}\, J_1(s,t) &= \frac{(1-t)(1-t^2)}{s} + t^{-1} - 2 - 2\,t^2 + t^3 - s\,(t^{-1}+t^2) \\
&\quad - \frac{s^2}{t}\,(1+t)^2 + s^3(t^{-1}+1) \\
\frac{1}{st}\, J_2(s,t) &= \frac{(1-t)(1-t^2)}{s}\,(t^{-1}-1+t) - 2(t^{-1}+t^3) + 1 + t^2 + s(t^{-1}+t^2) \\
&\quad + s^2(t^{-1}+1+t) - 2\,s^3(t^{-1}+1) + \frac{s^4}{t}\,;
\end{aligned} \qquad (3.21)$$

D and Q are second degree polynomials:

$$D(s,t) = (1-t)^2 - 2s(1+t) + s^2\,, \quad Q(s,t) = s + t + st\,. \qquad (3.22)$$

According to Sect. 4 of [53] the uniqueness of the stress-energy tensor implies the following relations among the coefficients b_ν and a_ν in (3.18) and in (3.20):

$$b_0 + 6b_1 = 1 = a_0 + 6a_1 + 6a_2.$$

These expressions are characterized by three conditions: 1) they are rational GCI functions; 2) their poles in z_{ij}^2 $(i<j)$ have strictly lower degrees than the poles of the corresponding 2-point function $\langle ij \rangle$; 3) they are crossing symmetric – *i.e.*, symmetric under any permutation of $(z_1, \dots, z_4)$.

We shall make explicit the meaning of the third property. The permutation group $\mathcal{S}_4$ has a normal subgroup $\mathbb{Z}_2 \times \mathbb{Z}_2$, generated by the double transpositions $s_{12}\, s_{34}$ and $s_{14}\, s_{23}$ $(s_{12}\, s_{34}\, s_{14}\, s_{23} = s_{13}\, s_{24})$, leaving invariant the cross-ratios s and t. The factor group $\mathcal{S}_4/\mathbb{Z}_2 \times \mathbb{Z}_2$, isomorphic to $\mathcal{S}_3$, acts faithfully on the "amplidues" $\tilde{J}_\lambda$, J_ν, D, Q:

$$s_{12}\, \tilde{J}_\lambda(s,t) := \tilde{J}_\lambda\left(\frac{s}{t}, \frac{1}{t}\right) = \tilde{J}_\lambda(s,t)\,,$$
$$s_{23}\, \tilde{J}_\lambda(s,t) := \tilde{J}_\lambda\left(\frac{1}{s}, \frac{t}{s}\right) = \tilde{J}_\lambda(s,t)\,, \ \lambda = 0,1\,; \qquad (3.23)$$

if $f(s,t)$ is any of the five functions $\frac{1}{st}\, J_\nu(s,t)$, $D(s,t)$, $Q(s,t)$ then

$$(s_{12}\, f)(s,t) := t^2 f\left(\frac{s}{t}, \frac{1}{t}\right) = f(s,t)\,, \ (s_{23}\, f)(s,t) := s^2 f\left(\frac{1}{s}, \frac{t}{s}\right). \qquad (3.24)$$

We shall analyse our models by writing down the OPE of two ϕ's in the form

$$(z_{12}^2)^{d-1} \left[\phi(z_1)\, \phi(z_2) - \langle 12 \rangle\right] = V(z_1, z_2) + z_{12}^2\, \tilde{V}(z_1, z_2) \qquad (3.25)$$

where $V(z_1, z_2) = V(z_2, z_1)$ is a *bilocal conformal scalar field* of dimension $(1,1)$ whose OPE only involves twist-two conserved symmetric traceless tensor currents $T_{2\ell}(z; v)$,

$$T_{2\ell}(z; v) = T_{2\ell}(z)_{\alpha_1 \dots \alpha_{2\ell}} \, v^{\alpha_1} \dots v^{\alpha_{2\ell}} \, , \quad \frac{\partial}{\partial v} \cdot \frac{\partial}{\partial z} \, T_{2\ell}(z; v) = 0 = \frac{\partial}{\partial v} \cdot \frac{\partial}{\partial v} \, T(z, v) \, , \tag{3.26}$$

and every such current enters the OPE of V in local fields (so that the remainder, $\tilde{V}(z_1, z_2)$, is expanded in twist four and higher fields only). This definition of V has two immediate consequences: the orthogonality relations

$$\langle 0 \mid V(z_1, z_2) \, \tilde{V}(z_3, z_4) \mid 0 \rangle = 0 = \langle 0 \mid V(z_1, z_2) \mid 0 \rangle \tag{3.27}$$

hold (as the vacuum expectation value of the product of two fields transforming under two inequivalent irreducible representations of $\mathcal{C}$ vanishes) and V is harmonic in each argument,

$$\Delta_{z_1} V(z_1, z_2) := \frac{\partial}{\partial z_1} \cdot \frac{\partial}{\partial z_1} \, V(z_1, z_2) = 0 = \Delta_{z_2} V(z_1, z_2) \tag{3.28}$$

(as a consequence of the conservation law for all $T_{2\ell}$ (3.26) – see Proposition 2.1 of [56]).

The bilocal field $V(z_1, z_2)$ has thus the properties of a normal product of free fields and is, in fact, a simpler object than the original field ϕ. It is therefore important that for $d = 2$, V can be expressed in terms of ϕ, a result announced in [55] (Proposition 2.3); two versions of the proof were sketched one in Sect. 2 of [55] and another in Appendix A.1 of [56].

Proposition 3.3. (a) *The truncated n-point function of* ϕ $(n \geq 3)$ *is a sum of* $\frac{1}{2}(n-1)!$ *1-loop graph contributions obtained from* $c\,[z_{12}^2 \, z_{23}^2 \dots z_{n-1n}^2 \, z_{1n}^2]^{-1}$ *by permutations* σ *of* $(2, \dots, n)$ *identifying* $\sigma(2) \dots \sigma(n)$ *with* $\sigma(n) \dots \sigma(2)$.
(b) *Given* (a) *for* $3 \leq n \leq 6$ *we can define* $V(z_1, z_2)$ *by*

$$V(z_1, z_2) = \lim_{\substack{z_{13}^2 \to 0 \\ z_{23}^2 \to 0}} \{ z_{13}^2 \, z_{23}^2 \, [\phi(z_1) \, \phi(z_2) \, \phi(z_3) - \langle 123 \rangle - \langle 23 \rangle \, \phi(z_1) - \langle 13 \rangle \, \phi(z_2)] \} \tag{3.29}$$

proving that the limit is independent of z_3.

Sketch of proof. The statement (a) can be proven by induction, using the OPE (3.25) and the harmonicity of V (3.28). We shall indicate here, on the example of the 5-point function, a streamlined way to using the latter. Expanding the 5-point function $\langle 12345 \rangle$ with respect to z_{45} and keeping only the singular terms we find

$$\langle 12345 \rangle = \langle 123 \rangle \, \langle 45 \rangle + \langle 0 \mid \phi(z_1) \, \phi(z_2) \, \phi(z_3) \, \frac{V(z_4, z_5)}{z_{45}^2} \mid 0 \rangle + 0(1) \, .$$

The space of GCI 5-point functions of weight $(2, 2, 2; 1, 1)$ with the properties of

$$F(z_1, z_2, z_3; z_4, z_5) = \langle 0 \mid \phi z_1) \, \phi(z_2) \, \phi(z_3) \, V(z_4, z_5) \mid 0 \rangle \tag{3.30}$$

is spanned by graded monomials of the type

$$\prod_{1 \leq i < j \leq 5} \rho_{ij}^{\mu_{ij}} \, , \quad \rho_{ij} = z_{ij}^2 \, , \quad \mu_{ij} (\equiv \mu_{ji}) \in \mathbb{Z} \, , \quad \sum_{j=1}^{5} \mu_{ij} = - \, d_i \;\; (\mu_{ii} = 0)$$
$$i = 1, \dots, 5 \, , \quad d_1 = d_2 = d_3 = 2 \, , \quad d_4 = d_5 = 1 \, . \tag{3.31}$$

The harmonicity of V further implies

$$\begin{aligned}
&\frac{\partial}{\partial z_k}\cdot\frac{\partial}{\partial z_k}\,F(z_1,z_2,z_3;z_4,z_5)\\
&= \left\{8\sum_{j\neq k}\frac{\partial}{\partial\rho_{jk}}+2\sum_{i,j(\neq k)}(\rho_{ik}+\rho_{jk}-\rho_{ij})\,\frac{\partial^2}{\partial\rho_{ik}\,\partial\rho_{jk}}\right\}F\\
&= -2\sum_{i,j(\neq k)}\rho_{ij}\,\frac{\partial^2 F}{\partial\rho_{ik}\,\partial\rho_{jk}}=0 \quad\text{for}\quad k=4,5\,.
\end{aligned}\tag{3.32}$$

(In the second equation we have used the homogeneity condition (3.31).) We find as a corollary that a monomial of type (3.31) cannot appear if there is a pair (i,j), $i\neq j$ such that $\mu_{ik}<0$ and $\mu_{jk}<0$ for $k=4$ or $k=5$. Taking further the growth and the symmetry conditions into account we find the following general form of F:

$$F(z_1,\dots,z_5)=c\left(\frac{\langle 12\rangle}{\rho_{34}\,\rho_{35}}+\frac{\langle 13\rangle}{\rho_{24}\,\rho_{25}}+\frac{\langle 23\rangle}{\rho_{14}\,\rho_{15}}\right)+\langle 123\rangle\sum_{i,j=1}^{3}\frac{\rho_{ij}}{\rho_{i4}\,\rho_{j5}}\,. \tag{3.33}$$

The statement (b) is established by a direct computation noting that to evaluate $\langle 0\mid V(z_1,z_2)\,V(z_3,z_4)\mid 0\rangle$ from (3.29) one only needs the n-point function of ϕ for $n=2,3,4,6$.

Note that Proposition 3.3 is not valid for 1-dimensional $z_{1,2}$.

3.3 Four-point functions and partial wave expansions of twist-two bilocal fields. In order to study systematically the properties of harmonic bilocal fields we begin with the obvious generalization of the OPE (3.25) to the case of a pair of hermitean conjugate complex scalar fields $\psi(z)$ and $\psi^*(z)$ of (integer) dimension d:

$$(z_{12}^2)^{d-1}[\psi^*(z_1)\,\psi(z_2)-\langle 12\rangle]=W(z_1,z_2)+z_{12}^2\,\tilde{W}(z_1,z_2) \tag{3.34}$$

where $W(z_1,z_2)$ is again harmonic (of dimension $(1,1)$) but no longer symmetric, satisfying instead

$$W^*(z_1,z_2)=W(z_2,z_1)\,. \tag{3.35}$$

The general 4-point function of W can be written in the form

$$\langle 0\mid W(z_1,z_2)\,W(z_3,z_4)\mid 0\rangle=\frac{1}{z_{13}^2\,z_{24}^2}\sum_{\nu=0}^{d-2}a_\nu\,f_\nu(s,t) \tag{3.36}$$

where the *amplitudes* $f_\nu(s,t)$ are solutions of the conformal Laplace equation

$$\left\{s\,\frac{\partial^2}{\partial s^2}+t\,\frac{\partial^2}{\partial t^2}+(s+t-1)\,\frac{\partial^2}{\partial s\,\partial t}+2\left(\frac{\partial}{\partial s}+\frac{\partial}{\partial t}\right)\right\}f_\nu(s,t)=0 \tag{3.37}$$

such that $f_\nu(0,t)=\frac{(1-t)^\nu}{t^{\nu+1}}$. (The upper limit in the sum (3.36) comes from the restriction on the degree of the pole in $z_{14}^2\,z_{23}^2$, allowed by Wightman positivity.) The solution of (3.37) assumes a simple form in the higher dimensional generalization [24] of the *chiral conformal variables*, $u,\bar u$ related to s and t by:

$$s=u\,\bar u\,,\ t=(1-u)(1-\bar u)\,;\ u+\bar u=1+s-t\,,\ (u-\bar u)^2=D(s,t) \tag{3.38}$$

where the discriminant $D(s,t)$ is given by (3.22). Regarded as a function of u and $\bar{u}$, $f_\nu(s,t)$ is given by

$$(u-\bar{u})\, f_\nu(s,t) = \frac{u^{\nu+1}}{(1-u)^{\nu+1}} - \frac{\bar{u}^{\nu+1}}{(1-\bar{u})^{\nu+1}} \qquad \nu = 0,1,2,\dots . \tag{3.39}$$

(The variables u and $\bar{u}$ are complex conjugate to each other for real Euclidean z_i (for which $s \geq 0$, $t \geq 0$, $D(s,t) \leq 0$).)

Proposition 3.4. *The amplitudes f_ν admit an expansion in twist-two conformal partial waves*

$$\beta_L(u,\bar{u}) = \frac{G_{L+1}(u) - G_{L+1}(\bar{u})}{u-\bar{u}}\,, \quad G_\nu(u) = u^\nu\, F(\nu,\nu;2\nu;u) \tag{3.40}$$

computed from the following identities for hypergeometric functions:

$$\frac{u^{\nu+1}}{(1-u)^{\nu+1}} = \sum_{n=0}^{\infty} \frac{(n+2\nu)!}{n!\,(\nu!)^2} \frac{G_{n+\nu+1}(u)}{\binom{2n+2\nu}{n+\nu}}\,. \tag{3.41}$$

The lowest angular momentum contributing to the expansion of f_ν is $L=\nu$.

The *proof* is given in Appendix B to [53].

Remark 3.2 The solution f_ν can be recovered if we identify $W_\nu(z_1,z_2)$ with a product of hermitean conjugate free massless fields of spin $\nu/2$; in particular,

$$W_0(z_1,z_2) =: \varphi^*(z_1)\,\varphi(z_2) :\,, \quad W_1(z_1,z_2) =: \psi^*(z_1)\, \not{z}_{12}\, \psi(z_2) :\,,$$

$$W_2(z_1,z_2) =: F^*_{\dot{A}_1\dot{B}_1}(z_1)\, \not{z}_{12}^{\dot{A}_1 A_2}\, \not{z}_{12}^{\dot{B}_1 B_2}\, F_{A_2B_2}(z_2) :\,, \tag{3.42}$$

where $\psi(z) = \{\psi_A(z),\ A=1,2\}$ is a 2-component Weyl spinor, $\not{z}$ is the 2×2 (quaternionic) matrix (B.14) – see Appendix B, $F_{AB}(z)$ $(= F_{BA}(z))$ is a complex spin 1 field. Bilinear combinations of higher rank free spin-tensor fields do not contain a stress-energy tensor in their OPE. Therefore, we shall not be interested in W_ν (and V_ν) for $\nu > 2$.

The sums $V_\nu(z_1,z_2) = W_\nu(z_1,z_2) + W_\nu(z_2,z_1)$, $\nu = 0,1,\dots$, represent special cases of symmetric bilocal fields with 4-point functions obtained from (3.36) (3.39) by symmetrization in z_1, z_2:

$$\langle 0 \mid V_\nu(z_1,z_2)\, V_\nu(z_3,z_4) \mid 0\rangle = \frac{j_\nu(s,t)}{z_{13}^2\, z_{24}^2}\,, \quad j_\nu(s,t) = f_\nu(s,t) + \frac{1}{t}\, f_\nu\left(\frac{s}{t}, \frac{1}{t}\right)\,, \tag{3.43}$$

which gives

$$(u-\bar{u})\, j_\nu(s,t) = \left(\frac{u}{1-u}\right)^{\nu+1} - (-u)^{\nu+1} - \{u \to \bar{u}\}\,. \tag{3.44}$$

We have, in particular,

$$j_0(s,t) = 1 + t^{-1}\,, \quad j_1(s,t) = (1-t)(t^{-2}-1) - s(t^{-2}+1)\,,$$

$$j_2(s,t) = (1+t^{-3})\,[(1-t)^2 + s^2] - \frac{s}{t^3}\,(2 - t - t^3 + 2\,t^4)\,. \tag{3.45}$$

The three amplitudes (3.45) appear in the general 4-point function of a hermitean scalar field $\mathcal{L}(z)$ of dimension $d = 4 (= D)$ which has the properties of a (gauge invariant) Lagrangian density and provides the most attractive possibility for a physically interesting model [56]. The functions $\frac{1}{st}\, J_\nu(s,t)$ (3.21) appear as symmetrizations of $\frac{t^2}{s}\, j_\nu(s,t)$:

$$\frac{1}{st}\, J_\nu(s,t) = \lambda_\nu (1 + s_{23} + s_{13}) \left(\frac{t^2}{s}\, j_\nu(s,t) \right) \quad \lambda_0 = \lambda_1 = 1\,, \ \lambda_2 = \frac{1}{2} \tag{3.46}$$

where s_{23} is defined in (3.24) and s_{13} $(= s_{12}\, s_{23}\, s_{12})$ exchanges s and t; the values of λ_ν in (3.46) are chosen in such a way that $\frac{t^2}{s}\, j_\nu$ provides the leading term in the small s expansion of $\frac{1}{st}\, J_\nu$.

Remark 3.3 In general, the contribution of all tensor fields of a given twist to the OPE of two scalar fields is *not* a (strongly) bilocal field: its correlation functions are not rational (they involve hypergeometric functions of type (3.40) which can be expressed in terms of logs - see [56]). It is all the more remarkable that the twist two contribution $W(z_1, z_2)$ or $V(z_1, z_2)$ is a bilocal field - as a consequence of the wave equation (3.28) - with rational correlation functions as displayed above. We shall use this fact in our analysis of the $d = 2$ case below.

3.4 The $d = 2$ case; the sp$(\infty, \mathbb{R})$-algebra and its positive energy representations. The theory of a (unique) hermitean scalar field ϕ of dimension two (normalized to satisfy (3.16) (3.17)) is reduced, in view of Proposition 3.3, to the study of a biharmonic bilocal field $V(z_1, z_2)$ such that

$$\phi(z_1)\, \phi(z_2) - \frac{c}{2(z_{12}^2)^2} = \frac{1}{z_{12}^2}\, V(z_1, z_2) + :\phi(z_1)\, \phi(z_2):\,, \quad V(z,z) = 2\,\phi(z)\,, \tag{3.47}$$

and whose $2n$-point function can be read off the sum of 1-loop graphs giving the truncated Wightman functions of ϕ (Proposition 3.3 (a)).

We shall see (following [55]) that V generates the infinite dimensional Lie algebra sp$(\infty, \mathbb{R})$ and that its general vacuum representation corresponds to positive integral central charge $c = N$ for which V has the properties of a sum of normal products of N free massless scalar fields $\varphi_i(z)$:

$$V(z_1, z_2) = \sum_{i=1}^{N} :\varphi_i(z_1)\, \varphi_i(z_2):\,, \quad [\varphi_i(z_1), \varphi_j(z_2)] = 0 \text{ for } i \neq j\,. \tag{3.48}$$

We shall also preview the results of [6] concerning general positive energy representations of the vacuum "algebra of observables" [36] generated by local fields.

As the commutation relations of V are representation independent (except for the value of the central charge) they can be deduced, in view of (3.48), from those of the free massless field φ. The power series expansion (3.4) for φ coincides with its decomposition into irreducible representations of type $(n; j, j)$, $2j + 1 = |n|$, of the maximal compact subgroup $S(U(2) \times U(2))$ of $\mathcal{C}$:

$$\varphi(z) = \sum_{n \in \mathbb{Z}^\times} \varphi_n(z)\,, \quad \varphi_n(\lambda z) = \lambda^{-n-1} \varphi_n(z) \tag{3.49}$$

where φ_n are homogeneous harmonic functions of z:

$$\varphi_{-n}(z) = \sum_{\sigma=1}^{n^2} \varphi_{-n\sigma}\, h_{n\sigma}(z)\,, n \in \mathbb{Z}^\times\,, h_{-n,\sigma}(z) = \frac{h_{n\sigma(z)}}{(z^2)^n} \ \mathit{for} \ n > 0; \tag{3.50}$$

$\{h_{n\sigma}(z), \sigma = 1, \ldots, n^2\}$ is an orthonormal basis of homogeneous of degree $n-1$ harmonic polynomials in z_α, (with real coefficients) and $\varphi_{\pm n\sigma}$ can be expressed in terms of the *residue functional* of [5] and are hermitean conjugate to each other:

$$\varphi_{n\sigma} = Res_z(\frac{\varphi(z)}{z^2} h_{n\sigma}(z)), \ \varphi_{-n\sigma} = \varphi^*_{n\sigma}, \ n \in \mathbb{Z}^\times \tag{3.51}$$

($\mathbb{Z}^\times$ is the set of non-zero integers). Here $Res_z f(z)$ is the unique translation invariant functional normalized by $Res_z(z^2)^{-2} = 1$ so that

$$2\pi^3 i Res_z f(z) = \int_{\bar{M}} f(z) d^4 z \tag{3.52}$$

Let $u_a = (u_{a\alpha})$ $a = 1, 2$ be real Euclidean unit 4-vectors:

$$u_a \in \mathbb{S}^3, \quad i.e. \ u_a^2 = \boldsymbol{u}_a^2 + u_{a4}^2 = 1, \ a = 1, 2, \ u_1 \cdot u_2 = \cos\alpha . \tag{3.53}$$

Then the canonical (Heisenberg) commutation relations for $\varphi_n(u)$ and $\varphi_{n\sigma}$ assume the form:

$$[\varphi_n(u_1), \varphi_{-m}(u_2)] = \delta_{nm} \frac{\sin n\alpha}{\sin\alpha}, \ n[\varphi_{n\sigma}, \varphi^*_{m\tau}] = \delta_{nm}\, \delta_{\sigma\tau} . \tag{3.54}$$

The sum in σ of $h_{n\sigma}(z_1) h_{n\sigma}(z_2)$ is independent of the choice of basis:

$$\sum_{\sigma=1}^{n^2} h_{n\sigma}(z_1) h_{n\sigma}(z_2) = n H_n(z_1, z_2) \tag{3.55}$$

where H_n is the unique $O(4)$-invariant biharmonic bihomgeneopus polynomial of degree $(n-1, n-1)$ normalized by $H_n(u, u) = n$ for $u \in \mathbb{S}^3$. According to (3.54) its restriction to the product of unit 3-spheres is expressed in terms of the hyperspherical (Gegenbauer) polynomial:

$$H_n(u_1, u_2) = C^1_{n-1}(cos\alpha) = \frac{\sin n\alpha}{\sin\alpha} \ (u_1.u_2 = cos\alpha) \tag{3.56}$$

($H_1 = 1$, $H_2 = 2z_1.z_2$, $H_3 = 4(z_1.z_2)^2 - z_1^2 z_2^2$, ...). The free 2-point function of φ provides the generating function for H_n:

$$(z^2)^{-1} \langle 0 \mid \varphi(\frac{z}{z^2}) \varphi(w) \mid 0 \rangle = (1 - 2z.w + z^2 w^2)^{-1} = \sum_{n=1}^{\infty} H_n(z, w) \tag{3.57}$$

Proceeding to the bilocal field V, we shall write its mode expansion in the form

$$\begin{aligned} V(z_1, z_2) = & \sum_{n_1, n_2 > 0} \{ (z_1^2)^{-n_1} (z_2^2)^{-n_2} X_{n_1\sigma_1, n_2\sigma_2} + X^*_{n_1\sigma_1, n_2\sigma_2} + \\ & (z_2^2)^{-n_2} N_{n_1\sigma_1, n_2\sigma_2} + (z_1^2)^{-n_1} N_{n_2\sigma_2, n_1\sigma_1} \} \frac{h_{n_1\sigma_1}(z_1)\, h_{n_2\sigma_2}(z_2)}{\sqrt{n_1 n_2}} . \end{aligned} \tag{3.58}$$

The operator valued coefficients $X^{(*)}_{\chi_1\chi_2}$ and $N_{\chi_1\chi_2}$ where χ_i stands for the pair (n_i, σ_i) span the Lie algebra $\mathrm{sp}(\infty, \mathbb{R})$ characterized by the commutation relations

$$[X_{\chi_1\chi_2}, X_{\chi_3\chi_4}] = 0 = [X^*_{\chi_1\chi_2}, X^*_{\chi_3\chi_4}] , \tag{3.59}$$

$$\begin{aligned} [X_{\chi_1\chi_2}, X^*_{\chi_3\chi_4}] &= \delta_{\chi_2\chi_3}\, E_{\chi_4\chi_1} + \delta_{\chi_2\chi_4}\, E_{\chi_3\chi_1} + \delta_{\chi_1\chi_3}\, E_{\chi_4\chi_2} \\ & + \delta_{\chi_1\chi_4}\, E_{\chi_3\chi_2}, \ E_{\chi_1\chi_2} = N_{\chi_1\chi_2} + \frac{c}{2} \delta_{\chi_1\chi_2} , \end{aligned} \tag{3.60}$$

where $\delta_{\chi_1\chi_2} \equiv \delta_{n_1 n_2}\, \delta_{\sigma_1\sigma_2}$ (for $\chi_i = (n_i, \sigma_i)$);

$$[E_{\chi_1\chi_2}, E_{\chi_3\chi_4}] = \delta_{\chi_2\chi_3}\, E_{\chi_1\chi_4} - \delta_{\chi_1\chi_4}\, E_{\chi_3\chi_2} = [N_{\chi_1\chi_2}, N_{\chi_3\chi_4}] ; \tag{3.61}$$

$$[E_{\chi_1\chi_2}, X^*_{\chi_3\chi_4}] = \delta_{\chi_2\chi_3} X^*_{\chi_1\chi_4} + \delta_{\chi_2\chi_4} X^*_{\chi_1\chi_3},$$
$$[E_{\chi_1\chi_2}, X_{\chi_3\chi_4}] = \delta_{\chi_2\chi_3} X_{\chi_1\chi_4} + \delta_{\chi_1\chi_3} X_{\chi_2\chi_4}. \tag{3.62}$$

Note that the central extension of $sp(\infty, \mathbb{R})$ (hidden in the $\frac{c}{2}$ term in (3.60)) is trivial and is incorporated in the redefinition $N_{\chi_1\chi_2} \to E_{\chi_1\chi_2}$. The operators $X_{\chi_1\chi_2}$ and $X^*_{\chi_1\chi_2}$ are conjugate to each other and symmetric ($X_{\chi_1\chi_2} = X_{\chi_2\chi_1}$), while the E's satisfy

$$(E_{\chi_1\chi_2})^* = E_{\chi_2\chi_1} \tag{3.63}$$

and span an $u(\infty)$ subalgebra. In view of (3.61) and (3.62) the *Cartan elements*

$$h_\chi = N_{\chi\chi} = E_{\chi\chi} - \frac{c}{2}, \ \chi = (\ell, \sigma), \ h_\ell = \sum_{\sigma=1}^{\ell^2} h_{\ell\sigma} \tag{3.64}$$

satisfy

$$[h_\chi, E_{\chi_1\chi_2}] = (\delta_{\chi\chi_1} - \delta_{\chi\chi_2}) E_{\chi_1\chi_2}, \ [h_\chi, X^*_{\chi_1\chi_2}] = (\delta_{\chi\chi_1} + \delta_{\chi\chi_2}) X^*_{\chi_1\chi_2}. \tag{3.65}$$

Their sum, $\sum_{\ell=1}^{\infty} h_\ell$ generates the centre of the $u(\infty)$ subalgebra while the conformal Hamiltonian H is given by

$$H = \sum_{\ell=1}^{\infty} \ell\, h_\ell = \sum_{\chi=(n,\sigma)} n(E_{\chi\chi} - \frac{c}{2}). \tag{3.66}$$

We are interested in unitary irreducible $sp(\infty, \mathbb{R})$ modules $\mathcal{H}_{c\underline{m}}$, $\underline{m} = (m_{11}, m_{21}, \ldots, m_{24}, \ldots, m_{\ell 1}, \ldots, m_{\ell\ell^2}, \ldots)$ that possess a minimal energy normalized ground state $|c; \underline{m}\rangle$ defined as follows. We fix an order in the set of pairs χ setting

$$\chi_1 = (n_1, \sigma_1) < \chi_2 = (n_2, \sigma_2) \text{ if either } n_1 < n_2 \text{ or } n_1 = n_2, \ \sigma_1 < \sigma_2. \tag{3.67}$$

We then require

$$X_{\chi_1\chi_2} \mid c; \underline{m}\rangle = 0 \text{ for all } \chi_1, \chi_2; \ E_{\chi_1\chi_2} \mid c; \underline{m}\rangle = 0 \text{ for } \chi_1 < \chi_2;$$
$$(h_{\ell\sigma} - m_{\ell\sigma}) \mid c; \underline{m}\rangle = 0, \tag{3.68}$$

and convergence of the series $E_{\underline{m}}$ which gives the energy of $|c; \underline{m}\rangle$,

$$E_{\underline{m}} = \sum_{\ell=1}^{\infty} \ell \sum_{\sigma=1}^{\ell^2} m_{\ell\sigma}. \tag{3.69}$$

The following result of [6] is a straightforward consequence of the definition.

Proposition 3.5. *Hilbert space positivity and convergence of the series* (3.69) *imply that* $\underline{m}$ *is a finite decreasing sequence of positive integers:*

$$m_{11} \geq m_{21} \geq m_{22} \geq m_{23} \geq m_{24} \geq m_{31} \geq \cdots \geq 0, \ m_{\ell\sigma} = 0 \text{ for } \ell > r. \tag{3.70}$$

for some finite non-negative r*).*

Proof. For any $\chi = (\ell, \sigma)$ we have

$$0 \leq \| X^*_{\chi\chi} \mid c; \underline{m}\rangle \|^2 = \langle c; \underline{m} \mid X_{\chi\chi} X^*_{\chi\chi} \mid c; \underline{m}\rangle$$
$$= 4 \left\langle c; \underline{m} \middle| h_\chi + \frac{c}{2} \middle| c; \underline{m} \right\rangle = 4\, m_{\ell\sigma} + 2c.$$

On the other hand for $\chi_1 = (n_1, \sigma_1) < \chi_2 = (n_2, \sigma_2)$ we have

$$\|(E_{\chi_2\chi_1})^\ell \mid c; \underline{m}\rangle\|^2 = \ell! \prod_{k=0}^{\ell-1} (m_{n_1\sigma_1} - m_{n_2\sigma_2} - k)\,. \tag{3.71}$$

Unitarity implies that the difference $m_{n_1\sigma_1} - m_{n_2\sigma_2}$ should be a non-negative integer. Convergence of (3.69) says that all but a finite number of $m_{n\sigma}$ should vanish as asserted.

We see as a corollary that the minimal energy $E_{\underline{m}}$ of an unitary irreducible module is non-negative. It is only zero for the *vacuum module* for which all $m_{\ell\sigma}$ vanish. Furthermore, the vacuum spans a 1-dimensional representation of $u(\infty)$:

$$E_{\chi_1\chi_2} \mid c; \underline{0}\rangle = \frac{c}{2}\,\delta_{\chi_1\chi_2} \mid c; \underline{0}\rangle\,. \tag{3.72}$$

The following result of [55] amounts to computing the *Kac determinant* [39] for the vacuum Verma module of $\mathrm{sp}(\infty, \mathbb{R})$ and restricting as a result the central charge c to positive integer values. The null vectors for general highest weight modules of both $\mathrm{sp}(\infty, \mathbb{R})$ and $\mathrm{u}(\infty, \infty)$ are displayed in [6].

Theorem 3.6. *The vacuum irreducible representation* (IR) *(with a minimal energy state* $|c; \underline{0}\rangle$ *satisfying* (3.72)) *is unitary iff* c *is a positive integer.*

Sketch of proof. Let $\chi_1 < \chi_2 < \cdots < \chi_n$; the norm square of the vector

$$\langle c; D(\chi_1, \ldots, \chi_n)| := \langle c; \underline{0}| \begin{vmatrix} X_{\chi_1\chi_1} & \ldots & X_{\chi_1\chi_n} \\ \ldots & \ldots & \ldots \\ X_{\chi_n\chi_1} & \ldots & X_{\chi_n\chi_n} \end{vmatrix} \tag{3.73}$$

has the form

$$\langle c; D(\chi_1, \ldots, \chi_n) \mid c; D(\chi_1, \ldots, \chi_n)\rangle = B(\chi_1, \ldots, \chi_n)\, c(c-1)\ldots(c-n+1) \tag{3.74}$$

where B is a positive constant (independent of c). Indeed, the commutation relations (3.60) imply that the scalar square (3.74) is a polynomial in c of degree not exceeding n. On the other hand, the determinant in (3.73) vanishes identically for $V(z_1, z_2)$ given by (3.48) with $N < n$.

The right hand side of (3.74) is non-negative for all n iff c is a non-negative integer. In that case the "bra" (3.73) is a null vector for $n = c + 1$.

The unitarity of the resulting IR for $c \in \mathbb{N}$ follows from the expression (3.48) of V and from the unitarity of the Fock space representation of free fields.

Thus the vacuum representation of a $d = 2$ field $\phi(z)$ – and hence of a bilocal field $V(z_1, z_2)$ related to ϕ by (3.29) (3.47) – satisfies Wightman positivity for positive integer c only. For $c = N$ the QFT generated by ϕ (or V) is an $O(N)$ invariant subtheory of the N-fold tensor product of the Fock space representations of a free massless scalar field. According to a fundamental result of Doplicher and Roberts [25] one expects that the QFT representations of $\mathrm{sp}(\infty, \mathbb{R})$ for $c = N$ are in one-to-one correspondence with the representations of the *gauge group* $O(N)$. In order to establish a result of this type one needs a polylocal version of the determinant in (3.73). That is achieved [6] by considering an antisymmetrized normal product of $N+1$ V-factors and applying the Reeh-Schlieder theorem [65] that allows to extend the null vector condition to any unitary positive energy representation generated by fields relatively local to the observables (in accord with the philosophy of Haag [36]).

Alternatively, we can recover this result by reintroducing c in the definition of the conformal Hamiltonian H by the second equation (3.66) and looking for positive energy representations of $sp(\infty, \mathbb{R})$ (i.e. demanding positivity of H) and using just Lie algebraic methods [26] [6]. In the mathematical work on lowest weight representations of infinite dimensional groups on Fock spaces (see, e.g. [27] [67] and references therein) the integrality of c is derived from the assumption that the highest weight representations of $sp(\infty, \mathbb{R})$ can be integrated to representations of the metaplectic group.

As a result all QFT representations of $sp(\infty, \mathbb{R})$ provide local extensions of the observable algebra generated by $V(z_1, z_2)$. For instance, for $c = 2$, they are of two sorts: (1) an infinite series of charge field extensions; (2) a d=3 real("neutral") extension. The representations of the first type are generated by a pair of conjugate charged fields $\psi^{(*)}$ of charge $\pm d$ and dimension d:

$$\psi(z) = \frac{1}{d}\sqrt{\frac{2}{(d-1)!}} : \varphi^d(z) :, \ \varphi(z) = \frac{1}{\sqrt{2}} (\varphi_1(z) + i\, \varphi_2(z)) . \tag{3.75}$$

Both ψ and ψ^* generate a unitary IR of $sp(\infty, \mathbb{R})$ with lowest weight vector $|c; \underline{m}\rangle = |2; d, 0, \dots\rangle$ $(= \psi^{(*)}(0) \mid 2; \underline{0}\rangle)$. The single nontrivial neutral extension corresponds to a lowest weight vector $|2; 1, 1, 0, ...\rangle = J_1(0)|2; \underline{0}\rangle$ where $J_\alpha(z)$ is the U(1) current

$$J_\alpha(z) = \varphi_1(z)\partial_\alpha\varphi_2(z) - \varphi_2(z)\partial_\alpha\varphi_1(z). \tag{3.76}$$

An SO(2) rotation $(R_\theta\, \varphi)_1 = \varphi_1 \cos\theta + \varphi_2 \sin\theta$, $(R_\theta\, \varphi)_2 = -\varphi_1 \sin\theta + \varphi_2 \cos\theta$ yields a $U(1)$ gauge transformation $R_\theta\, \psi = e^{-i\theta d}\, \psi$, $R_\theta\, \psi^* = e^{i\theta d}\, \psi^*$. The reflection $(\varphi_1, \varphi_2) \rightarrow (\varphi_1, -\varphi_2)$ exchanges ψ and ψ^* and changes the sign of J. Clearly, the $O(2)$ action commutes with $sp(\infty, \mathbb{R})$ as it leaves $V(z_1, z_2)$ invariant. Note that the field algebra, generated by ψ and ψ^*, involves a wider extension of the charge-zero algebra, which now involves the field $W(z_1, z_2)$ defined by (3.34) and its conjugate on top of the $U(1)$ current J (3.76). The $sp(\infty, \mathbb{R})$ Lie algebra is thus extended to $u(\infty, \infty)$.

4 Outlook and concluding remarks

Conformal QFT is not expected to be a realistic interacting theory with a non-trivial scattering matrix. Yet, it is hoped to display some essential characteristics of such a theory, in particular the small distance behaviour, which may help resolve the omnipresent ultraviolet problem. Not least important, it appears to offer a better chance for constructing a viable non-perturbative QFT model than a direct assault, using constructive QFT methods, at a more realistic asymptotically free theory like quantum chromodynamics. This has served as a motivation for the work on both (complementary) attempts, surveyed here, to construct a conformal QFT.

The developments of the 1970's (triggered at the time by the interest in the Bjorken scaling in deep inelastic scattering[9]), briefly reviewed in Sect. 2, are in our view of more than a historical interest. If for the method of OPE and conformal partial wave expansion this is rather obvious[10], it is perhaps not superfluous to recall the role of anomalous dimensions in making sense of skeleton graph expansion and in solving the conformal (bootstrap) Schwinger-Dyson equations. We refer in

---

[9] For a survey of this work and references to the original papers – see [83], as well as [49].

[10] For applications to (extended) superconformal gauge theories see, e.g. [24] as well as the review [19]; concerning their role in the study of Wightman positivity in GCI models – see [56] and [53].

this connection to D. Kreimer's suggestion (in Sect. 4.2 of [42]) to use the simplified Schwinger-Dyson equation in a critical gauge QFT, at the point where the β-function vanishes. It may be also of interest to apply these methods to the conformal super-Yang-Mills theory, regarding the Konishi field ([19]) of scale dimension in the interval between 2 and 3 as fundamental.

The major part of the paper (Sect. 3 and Appendices A and B) is devoted to a survey of recent developments in GCI QFT and to the ensuing concept of higher dimensional vertex algebras. The chief objection which skeptics voice towards this work is the suspicion that it may turn out to be just an exercise in free field theory (if one insists from the outset on Wightman positivity). As we saw in Sect. 3.4, proving that the local field extensions of the vacuum representation of $sp(\infty, \mathbb{R})$ are generated by normal products of free fields in the tensor product of Fock spaces is a rather non-trivial task related, as it turns out, to a long term mathematical problem[11]. The study of thermal equilibrium states with associated elliptic correlation functions and modular invariant energy mean values [59] (a subject not touched upon in the present survey) provides another topic in which free fields display nontrivial properties that may be of interest to both physicists and mathematicians.

On the other hand, the study of local field extensions of the theory of a dimension-two scalar field solves a problem, set up in [53]: to factor out such a scalar field contribution from the most promising candidate for a nontrivial GCI QFT – the theory of a $d = 4$ Lagrangian field ([56] [53]).

Acknowledgments. I am indebted to all my coauthors for my understanding of the topics reviewed in this lecture. Recent discussions with Nikolay Nikolov and Karl-Henning Rehren have been particularly valuable.

The paper was written during the author's stay at the Institut für Theoretische Physik, Universität Göttingen (sponsored by an Alexander von Humboldt Research Award) and at l'Institut des Hautes Études Scientifiques, Bures-sur-Yvette. It represents an extended version of a lecture given at the Renormalization and Universality in Mathematical Physics Workshop at the Fields Institute in Toronto (at the invitation of Dirk Kreimer). It was also presented at the 17th Workshop "Foundations and Constructive Aspects of QFT" at the Universität Göttingen. The author thanks all these institutions for hospitality and support. This work is supported in part by the Research Training Network of the European Commission under contract MRTN-CT-2004-00514 and by the Bulgarian National Council for Scientific Research under contract Ph-1406.

[11] For the latest development on the Kashiwara-Vergne (1978) decomposition of tensor products of the Segal-Shale-Weil (1962) representations and for references to earlier work – see [27] [67].

Appendix A The conformal Lie algebra. Bases of vector fields

A natural basis of $so(4,2)$ is provided by the pseudorotations X_{ab} in $\mathbb{R}^{4,2}$ obeying the commutation relations (CR)

$$[X_{ab}, X_{cd}] = \eta_{ac}\, X_{bd} - \eta_{bc}\, X_{ad} + \eta_{bd}\, X_{ac} - \eta_{ad}\, X_{bc} \tag{A.1}$$

for $a, b, c, d = -1, 0, 1, 2, 3, 4$ and flat space metric tensor η of signature $(4,2)$:

$$\eta_{11} = \eta_{22} = \eta_{33} = \eta_{44} = 1 = -\eta_{00} = -\eta_{-1-1}\,, \ \eta_{ab} = 0 \text{ for } a \neq b\,. \tag{A.2}$$

Note that the correspondence $X \to O(X)$ between the commutator action of X_{ab} on (operator, spin-tensor valued) fields $\phi(\xi)$ and the differential operators

$$O(X_{ab}) = \xi_a \frac{\partial}{\partial\, \xi^b} - \xi_b \frac{\partial}{\partial\, \xi^a} + S_{ab} \tag{A.3}$$

(where S_{ab} are constant matrices spanning a finite dimensional representation of $su(2,2)$) given by

$$[X, \phi(\xi)] = O(X)\, \phi(\xi) \tag{A.4}$$

provides a Lie algebra *antihomomorphism*

$$[O(X), O(Y)] = -\,O([X,Y]) \tag{A.5}$$

(see for details Appendix B of [58]).

The Minkowski space infinitesimal translations $\frac{\partial}{\partial\, x^\mu}$ and special conformal transformations $x^2 \frac{\partial}{\partial\, x^\mu} - 2\, x_\mu\, x \cdot \frac{\partial}{\partial\, x}$ are images of the $so(4,2)$ generators $i\, P_\mu$, $i\, K_\mu$, respectively,

$$i\, P_\mu = -X_{-1\mu} - X_{\mu 4}\,, \ i\, K_\mu = -X_{-1\mu} + X_{\mu 4} \tag{A.6}$$

under this correspondence (P_μ being the *energy-momentum operator;* both P_μ and K_μ are self-adjoint for a unitary representation of $\mathcal{C}$). The CR (A.1) imply

$$[P_\mu, K_\nu] = 2\, X_{\mu\nu} - 2\, \eta_{\mu\nu}\, X_{-14}\,, \tag{A.7}$$

where X_{-14} is the *dilation generator,*

$$[X_{-14}, P_\mu] = P_\mu\,, \ [X_{-14}, K_\mu] = -K_\mu\,. \tag{A.8}$$

The *conformal Hamiltonian*

$$H := i\, X_{-10} = \frac{1}{2}\, (P_0 + K_0) \tag{A.9}$$

is the generator of the centre of the Lie algebra $so(2) \times so(4) \approx s(u(2) \times u(2))$ of the maximal compact subgroup $S(U(2) \times U(2))$ of $\mathcal{C}$. There is a proper conformal transformation, the *Weyl inversion* $w = e^{\pi X_{-10}}$, that is a rotation on π in the $(-1,0)$-plane, such that $w X_{04}\, w^{-1} = -X_{04}$ and hence $w\, P_0\, w^{-1} = K_0$. It follows that the spectrum of K_0 coincides with the spectrum of P_0. This yields Segal's observation [71] that the conformal Hamiltonian (A.9) is bounded below whenever the Minkowski space energy operator P_0 is. (It can be demonstrated that the converse is also true.) On the other hand, H unlike P_0 has discrete (in fact, integer spaced) spectrum in any unitary irreducible representation of $\tilde{\mathcal{C}}$.

We also use the complex Lie algebra generators T_α and C_α, corresponding to z-translations and special conformal transformations where the z's are the complex variables in the realization of the future tube and the compactified Minkowski space described in Appendix B:

$$T_\alpha = i\, X_{0\alpha} - X_{-1\alpha}\,, \ C_\alpha = -i\, X_{0\alpha} - X_{-1\alpha}\,, \ \alpha = 1,2,3,4\,. \tag{A.10}$$

It is readily verified that T_α, C_β, $X_{\alpha\beta}$ and H satisfy the same commutation relations (A.7) (A.8) as $i\,P_\mu$, $i\,K_\nu$ (A.6) $X_{\mu\nu}$ and X_{-14}:

$$[T_\alpha, C_\beta] = 2\,(\delta_{\alpha\beta}\,H - X_{\alpha\beta})\,, \tag{A.11}$$

$$[H, T_\alpha] = T_\alpha\,,\ \ [H, C_\alpha] = -\,C_\alpha\,. \tag{A.12}$$

In fact, the generators T_α, C_β, H and $X_{\alpha\beta}$ span the real Euclidean conformal Lie algebra spin$\,(5,1)$. They are conjugate in the complex conformal Lie algebra to iP_μ, iK_ν, X_{-14} and $X_{\mu\nu}$ by the same complex conformal transformation g_c (that corresponds to a real Euclidean rotation) which defines the compactification map $x \to z$ of Appendix B.

Appendix B Global complex-variable parametrization of compactified Minkowski space

There is a remarkable complex conformal transformation [84] [82], a rotation on $\frac{\pi}{2}$ in the $(0,4)$ plane of $\mathbb{C}^6$, which plays the role of a (Cayley) compactification map for real Minkowski space:

$$g_c : x = (x^0, \boldsymbol{x}) \to z = (\boldsymbol{z}, z_4)\,,\ \ \boldsymbol{z} = \frac{\boldsymbol{x}}{\omega(x)}\,,\ \ z_4 = \frac{1 - x^2}{2\,\omega(x)}\,, \tag{B.1}$$

$$z^2 := \boldsymbol{z}^2 + z_4^2 = \frac{1 + x^2 + 2\,i\,x^0}{1 + x^2 - 2\,i\,x^0}\,,\ \ 2\,\omega(x) = 1 + x^2 - 2\,i\,x^0\,,\ \ x^2 = \boldsymbol{x}^2 - x_0^2\,. \tag{B.2}$$

The complex conjugation in complexified Minkowski space $M_{\mathbb{C}}$ (which leaves M fixed) corresponds to the involution

$$z \to z^* = \frac{\bar{z}}{\bar{z}^2} \tag{B.3}$$

which keeps $\bar{M}$ fixed. A real parametrization of $\bar{M}$ is provided by the polar decomposition of z:

$$\begin{aligned} \bar{M} &= \{z_\alpha = e^{it}\,u_\alpha\,;\ t, u_\alpha \in \mathbb{R}\,,\ \alpha = 1,2,3,4\,;\ u^2 = \boldsymbol{u}^2 + u_4^2 = 1\} \\ &\cong \mathbb{S}^3 \times \mathbb{S}^1/\mathbb{Z}_2 \end{aligned} \tag{B.4}$$

($\mathbb{Z}_2 = \mathbb{Z}/2\,\mathbb{Z}$). The properties of the transformation (B.1), viewed as a map from $M_{\mathbb{C}}$ to $E_{\mathbb{C}}$ ($= \mathbb{C}^4$ equipped with a complex Euclidean metric) are summarized as follows ([58] [52]).

Proposition B.1. *The rational coordinate transformation* $g_c : M_{\mathbb{C}} \to E_{\mathbb{C}}$ (B.1) *is a complex conformal map such that* $z_{12}^2 = \frac{x_{12}^2}{\omega(x_1)\,\omega(x_2)}$, $dz^2 = \frac{dx^2}{\omega^2(x)}$ *and the image of the tube domain* $\mathcal{T}_+$ (2.6) *is the precompact symmetric space*

$$T_+ := \left\{ z \in \mathbb{C}^4\,;\ |z^2| < 1\,,\ |z|^2 = \sum_{\alpha=1}^{4} |z_\alpha|^2 < \frac{1}{2}\,(1 + |z^2|^2)\,(< 1) \right\}. \tag{B.5}$$

The image of M *in* $\bar{M}$ *under the map* (B.1) *is the open dense subset*

$$2\,\omega_z := 1 + z^2 + 2\,z_4 = 2\,e^{it}(\cos t + u_4) \neq 0\,, \tag{B.6}$$

on which x *is expressed conversely as a function of* z *(or* u *and* t*):*

$$\boldsymbol{x} = \frac{\boldsymbol{z}}{\omega_z} = \frac{\boldsymbol{u}}{\cos t + u_4}\,,\ \ -\,i\,x^0 = \frac{1 - z^2}{2\,\omega_z} = \frac{-\,i \sin t}{\cos t + u_4}$$

$$\left(x^2 = \frac{1+z^2-2\,z_4}{1+z^2+2\,z_4} = \frac{\cos t - u_4}{\cos t + u_4} \right) . \tag{B.7}$$

Note that the complex Euclidean interval $dz^2 = d\mathbf{z}^2 + dz_4^2$ has in fact Lorentzian signature when expressed in terms of real coordinates (like t and the spherical angles for $\mathbf{u} = sin\rho\,(sin\theta cos\phi,\ sin\theta sin\phi,\ cos\theta),\ u_4 = cos\rho$):

$$\frac{dz^2}{z^2} = \frac{dx^2}{|\omega(x)|^2} = d\rho^2 + sin^2\rho\,(d\theta^2 + sin^2\theta\, d\phi^2) - dt^2 . \tag{B.8}$$

Remark B.1. The tube domain T_+ (B.5) is biholomorphically equivalent to the classical Cartan domain of the fourth type - see e.g. [72] (pp. 182-192) and references therein.

Remark B.2. If we define the embedding $M \hookrightarrow \bar{M}$ in the manifestly covariant picture (1.5) of $\bar{M}$ by

$$x^\mu = \frac{\xi^\mu}{\xi^{-1}+\xi^4} = \frac{\xi^\mu}{\xi_4 - \xi_{-1}} \qquad \left(x^2 = \frac{\xi^{-1}-\xi^4}{\xi^{-1}+\xi^4} = \frac{\xi_{-1}+\xi_4}{\xi_{-1}-\xi_4} \right) \tag{B.9}$$

then the complex 4-vector is expressed in terms of ξ as

$$z_\alpha = \frac{\xi_\alpha}{i\,\xi_0 - \xi_{-1}} \Longrightarrow z^2 = e^{4\pi i \zeta} = \frac{\xi_{-1} + i\,\xi_0}{\xi_{-1} - i\,\xi_0} . \tag{B.10}$$

Introducing the Euclidean variable $x_E^0 = -i\,x^0,\ \xi_E^0 = -i\,\xi^0$ and comparing (B.9) and (B.10) we see that (B.1) indeed amounts to a real rotation in $\frac{\pi}{2}$ in the Euclidean plane $(\xi_E^0, \xi_E^4 (= \xi_4))$.

It is a simple corollary of the counterpart of Wightman axioms for vertex algebras (Sect. 3.1) and of Proposition B.1 that the series

$$\psi(z) \mid 0\rangle = \sum_{n,m=0}^{\infty} (z^2)^n\, \psi_{\{n,m\}}(z) \mid 0\rangle , \qquad z \in T_+ \tag{B.11}$$

is norm convergent and analytic.

We shall describe the free (2-component) Weyl spinor thus providing an example of a non-scalar z-picture field. To this end we introduce a 2×2 matrix realization of the quaternion units Q_α $(\alpha = 1,2,3,4)$ satisfying $Q_1\,Q_2 = Q_3 = -\,Q_2\,Q_1$,

$$Q_\alpha^+\,Q_\beta + Q_\beta^+\,Q_\alpha = 2\,\delta_{\alpha\beta} = Q_\alpha\,Q_\beta^+ + Q_\beta\,Q_\alpha^+ \tag{B.12}$$

where $Q_i^+ = -\,Q_i$ for $i = 1,2,3$; $Q_4^+ = Q_4$. In this notation a free Weyl field $\psi(z)$ is characterized by its 2-point function,

$$\langle 0 \mid \psi(z_1)\,\psi^*(z_2) \mid 0\rangle = Q^+ \cdot \frac{\partial}{\partial z_2}\,\frac{1}{z_{12}^2} = 2\,\frac{\not{z}_{12}^+}{(z_{12}^2)^2} \tag{B.13}$$

where $\not{z}^+$ is a short-hand for $Q^+ \cdot z$:

$$\not{z} = Q\cdot z = \begin{pmatrix} z_4 - iz_3 & -iz_1 - z_2 \\ z_2 - iz_1 & z_4 + iz_3 \end{pmatrix} , \ z^+ = Q^+ \cdot z = \begin{pmatrix} z_4 + iz_3 & iz_1 + z_2 \\ iz_1 - z_2 & z_4 - iz_3 \end{pmatrix} . \tag{B.14}$$

As a consequence, $\psi(z)$ satisfies the Weyl equation

$$Q \cdot \frac{\partial}{\partial z}\,\psi(z) = 0 = \frac{\partial}{\partial z_\alpha}\,\psi^*(z)\,Q_\alpha \tag{B.15}$$

which, in turn, implies the conservation of the current

$$J_\alpha(z) = i : \psi^*(z)\,Q_\alpha\,\psi(z) : . \tag{B.16}$$

References

[1] S. Adler, Short distance behaviour of quantum electrodynamics and an eigenvalue condition for α, *Phys. Rev.* **D5** (1972) 3021-3047; Erratum, *ibid.* **D7** (1973) 3821.

[2] D. Anselmi, Central functions and their physical implications, JHEP 9805 (1998) 005; hep-th/9702056.

[3] G. Arutyunov, S. Penati, A. Santabrogio, E. Sokatchev, Four-point correlators of BPS operators in $N = 4$ SYM at order g^4, *Nucl. Phys.* **B670** (2003) 103-147.

[4] M. Atiyah, Hermann Weyl, *Biographical Memoirs* **82** (2002) National Academy of Sciences, Washington D.C.

[5] B. Bakalov, N.M. Nikolov, Jacobi identities for vertex algebras in higher dimensions, math-ph/0601012; *J. Math. Phys.* **47** (2006) 053505 (30 pages).

[6] B. Bakalov, N.M. Nikolov, K.-H. Rehren, I. Todorov, Unitary positive energy representations of scalar bilocal quantum fields, math-ph/0604069; *Commun. Math. Phys.* (to appear).

[7] M. Baker, K. Johnson, Applications of conformal symmetry in quantum electrodynamics, *Physica* **96A** (1979) 120-180.

[8] V. Bargmann, I.T. Todorov, Spaces of analytic functions on a complex cone as carriers for symmetric tensor representations of SO(n), *J. Math. Phys.* **18** (1977) 1141-1148.

[9] H. Bateman, The transformation of electrodynamical equations, *Proc. London Math. Soc.* **8** (1910) 223-264.

[10] K. Baumann, All massless scalar fields with a trivial S-matrix are Wick polynomials, *Commun. Math. Phys.* **86** (1982) 247-256.

[11] S. Bloch, H. Esnault, D. Kreimer, On motives associated to graph polynomials, math.AG/0510011.

[12] L. Bonora, G. Sartori, M. Tonin, Conformal covariant operator product expansions, *Nuovo Cim.* **10A** (1972) 667-681.

[13] R.E. Borcherds, Vertex algebras, Kac-Moody algebras and the Monster, *Proc. Nat. Acad. Sci. USA* **83** (1986) 3068-3071.

[14] R. Brunetti, K. Fredenhagen, Interacting quantum fields on a curved background, *Proceedings of the XII-ICMP*, Briskane 1997, International Press, Singapore; hep-th/9709011.

[15] R. Brunetti, K. Fredenhagen, R. Verch, The generally covariant locality principle – A new paradigm for local quantum physics, *Commun. Math. Phys.* **237** (2003) 31-68.

[16] N.S. Craigie, V.K. Dobrev, I.T. Todorov, Conformal covariant composite operators in quantum chromodynamics, *Ann. Physics* **159** (1985) 411-444.

[17] E. Cunningham, The principle of relativity in electrodynamics and an extension thereof, *Proc. London Math. Soc.* **8** (1910) 77-98.

[18] A. De Sole, V.G. Kac, Finite vs affine W-algebras, math-ph/0511055.

[19] E. D'Hoker, D.Z. Freedman, Supersymmetric gauge theories and the AdS/CFT correspondence, TASI 2001 Lecture Notes, hep-th/0201253.

[20] P.A.M. Dirac, Wave equations in conformal space, *Ann. Math.* **37** (1936) 429-442.

[21] P.A.M. Dirac, Long range forces and broken symmetries, *Proc. Roy. Soc. of London* **A333** (1973) 403-418.

[22] V.K. Dobrev, G. Mack, V.B. Petkova, S.G. Petrova, I.T. Todorov, *Harmonic Analysis on the n-Dimensional Lorentz Group and Its Application to Conformal Quantum Field Theory*, Lecture Notes in Physics **63**, Springer, Berlin 1977.

[23] V.K. Dobrev, V.B. Petkova, S.G. Petrova, I.T. Todorov, Dynamical derivation of vacuum operator product expansion in Euclidean conformal field theory, *Phys. Rev.* **D13** (1976) 887-912.

[24] F.A. Dolan, H. Osborn, Conformal four point functions and operator product expansion, *Nucl. Phys.* **B599** (2001) 459-496; Conformal partial waves and operator product expansion, *Nucl. Phys* **B678** (2004) 491-507.

[25] S. Doplicher, J. Roberts, Why there is a field algebra with a compact gauge group describing the superselection structure in particle physics, *Commun. Math. Phys.* **131** (1990) 51-107.

[26] T.J. Enright, R. Parthasarathy, A proof of a conjecture of Kashiwara and Vergne, in *Non Commutative Harmonic Analysis and Lie Groups*, Lecture Notes in Mathematics **880**, Springer 1981, pp. 74-90.

[27] T. Enright, R. Howe, N. Wallach, A classification of unitary highest weight modules, in *Representation Theory of Reductive Groups*, Progress in Mathematics **40**, Birkhäuser 1983.

[28] S. Ferrara, R. Gatto, A. Grillo, G. Parisi, The shadow operator formalism for conformal algebra vacuum operator values and operator products, *Nuovo Cim. Lett.* **4** (1972) 115-120.

[29] M. Flato, D. Sternheimer, Remarques sur les automorphismes causals de l'espace temps, *Comptes Rendus Acad. Sci., Paris* **263A** (1966) 935-936.

[30] E.S. Fradkin, M.Ya Palchik, Recent developments in conformal invariant QFT, *Phys. Reports* **44C** (1978) 249-349.

[31] E.S. Fradkin, M.Ya. Palchik, New developments in d-dimensional conformal quantum field theory, *Phys. Reports* **300** (1998) 1-112.

[32] E. Frenkel, D. Ben-Zvi, *Vertex Algebras and Algebraic Curves*, AMS, Mathematical Surveys and Monographs **88**, 2001; 2nd ed. 2004.

[33] M. Gell-Mann, F. Low, Quantum electrodynamics at small distances, *Phys. Rev.* **95** (1954) 1300-1312.

[34] L. Genovese, Ya.S. Stanev, Rationality of the anomalous dimensions in $N = 4$ SYM theory, *Nucl. Phys.* **B721** (2005) 212-228.

[35] F. Gürsey, S. Orfanidis, Conformal invariance and field theory in two dimensions, *Phys. Rev.* **D7** (1973) 2414-2437.

[36] R. Haag, *Local Quantum Physics: Fields, Particles, Algebras,* Springer, N.Y. 1992.

[37] J. Henn, C. Jarczak, E. Sokatchev, On twist-two operators in $N = 4$ SYM, *Nucl. Phys.* **B730** (2005) 191-209.

[38] M. Hortascu, R. Seiler, B. Schroer, Conformal symmetry and reverberations, *Phys. Rev.* **D5** (1972) 2519-2534.

[39] V.G. Kac, Contravariant form for infinite dimensional Lie algebras and superalgebras, in: *Lecture Notes in Physics*, Springer, N.Y. 1979, pp. 441-445.

[40] V.G. Kac, *Vertex Algebras for Beginners*, AMS, ULS **10**, Providence, R.I., 2nd ed. 1998.

[41] L.P. Kadanoff, Operator algebra and the determination of critical indices, *Phys. Rev. Lett.* **23** (1969) 1430-1433.

[42] D. Kreimer, Anatomy of gauge theory, hep-th/0509135; *Ann. Physics* (to appear).

[43] J.H. Lowenstein, Normal products in the Thirring model, *Commun. Math. Phys.* **16** (1970) 265-289.

[44] M. Lüscher, G. Mack, Global conformal invariance in quantum field theory, *Commun. Math. Phys.* **41** (1975) 203-234.

[45] G. Mack, Group theoretical approach to conformal invariant quantum field theory, *in: Renormalization and Invariance in Quantum Field Theory*, ed. E.R. Caianiello, Plenum Press, NY 1974, pp. 123-157.

[46] G. Mack, All unitary representations of the conformal group SU (2, 2) with positive energy, *Commun. Math. Phys.* **55** (1977) 1-28.

[47] G. Mack, Convergence of operator product expansions on the vacuum in conformal invariant quantum field theory, *Commun. Math. Phys.* **53** (1977) 155-184.

[48] G. Mack, K. Symanzik, Currents, stress tensor and generalized unitarity in conformal invariant quantum field theory, *Commun. Math. Phys.* **27** (1972) 247-281.

[49] G. Mack, I.T. Todorov, Conformal invariant Green functions without ultraviolet divergences, *Phys. Rev.* **D8** (1973) 1764-1787.

[50] A.A. Migdal, Conformal invariance and bootstrap, *Phys. Lett.* **37B** (1971) 386-388.

[51] W. Nahm, Conformal field theory: a bridge over troubled waters, *in* : A.N. Mitra (ed.) *Quantum Field Theory – A Twentieth Century Profile*, Hindustani Book Agency and Indian Nat. Sci. Acad. 2000, pp. 571-604.

[52] N.M. Nikolov, Vertex algebras in higher dimensions and globally conformal invariant quantum field theory, *Commun. Math. Phys.* **253** (2005) 283-322.

[53] N.M. Nikolov, K.-H. Rehren, I.T. Todorov, Partial wave expansion and Wightman positivity in conformal field theory, *Nucl. Phys.* **B722** (2005) 266-296.

[54] N.M. Nikolov, K.-H. Rehren, I.T. Todorov, Harmonic bilocal fields generated by globally conformal invariant scalar fields (in preparation)

[55] N.M. Nikolov, Ya.S. Stanev, I.T. Todorov, Four dimensional CFT models with rational correlation functions, *J. Phys. A: Math Gen.* **35** (2002) 2985-3007.

[56] N.M. NIKOLOV, YA.S. STANEV, I.T. TODOROV, Globally conformal invariant gauge field theory with rational correlation functions, *Nucl. Phys.* **B670 [FS]** (2003) 373-400.
[57] N.M. NIKOLOV, I.T. TODOROV, Rationality of conformally invariant local correlation functions on compactified Minkowski space, *Commun. Math. Phys.* **218** (2001) 417-436.
[58] N.M. NIKOLOV, I.T. TODOROV, Lectures on elliptic functions and modular forms in conformal field theory, *Proceedings of the 3rd Summer School in Modern Mathematical Physics*, Zlatibor, Serbia and Montenegro (August 20-31, 2004), Eds. B. Dragovich, Z. Rakić, B. Sazdović, Institute of Physics, Belgrade 2005, pp. 1-93; math-ph/0412039.
[59] N.M. NIKOLOV, I.T. TODOROV, Elliptic thermal correlation functions and modular forms in a globally conformal invariant QFT, *Rev. Math. Phys.* **17** (2005) 613-667.
[60] G. PARISI, G. PELITI, Calculation of critical indices, *Nuovo Cim. Lett.* **2** (1971) 627-628.
[61] A.C. PETKOU, Conserved currents, consistency relations and operator product expansions in the conformally invariant $O(N)$ vector model, *Ann. Physics* **249** (1996) 180-221.
[62] V.B. PETKOVA, G.M. SOTKOV, I.T. TODOROV, Conformal gauges and renormalized equations of motion in massless quantum electrodynamics, *Commun. Math. Phys.* **97** (1985) 227-256.
[63] A.M. POLYAKOV, Conformal invariance of critical fluctuations, *ZhETF Pis. Red.* **12** (1970) 538-540 (*JETP Lett.* **12** (1970) 381-383).
[64] A.M. POLYAKOV, Non-Hamiltonian approach in the conformal invariant quantum field theory, *Zh.Eksp. Teor. Fiz* **66** (1974) 23-42 (English transl.: *JETP* **39** (1974) 10-18).
[65] H. REEH, S. SCHLIEDER, Bemerkungen zur Unitäräquivalenz von Lorentz-invarianten Feldern, *Nuovo Cim.* **22** (1961) 1051-1068.
[66] G.C. ROSSI, E. SOKATCHEV, YA.S. STANEV, New results in the deformed $N = 4$ SYM theory, *Nucl. Phys.* **B729** (2005) 581-593; hep-th/0507113.
[67] M.U. SCHMIDT, Lowest weight representations of some infinite dimensional groups on Fock spaces, *Acta Appl. Math.* **18** (1990) 59-84.
[68] S. SCHREIER, Conformal symmetry and three-point functions, *Phys. Rev.* **D3** (1971) 980-988.
[69] B. SCHROER, A necessary and sufficient condition for the softness of the trace of the energy-momentum tensor, *Nuovo. Cim. Lett.* **2** (1971) 867-872.
[70] B. SCHROER, J.A. SWIECA, A.H. VÖLKEL, Global operator expansion in conformally invariant relativistic theory, *Phys. Rev.* **D11** (1975) 1509-1520.
[71] I.E. SEGAL, Causally oriented manifolds and groups, *Bull. Amer. Math. Soc.* **77** (1971) 958-959.
[72] A.G.SERGEEV, V.S. VLADIMIROV, IV. Complex analysis in the future tube, in: *Several Complex Variables II. Function Theory in Classical Domains and Complex Potential Theory*, G.M. Khenkin, A.G. Vitushkin (Eds.)*Encyclopaedia of Mathematical Sciences* Vol. **8**, Springer, Berlin et al. 1994, pp.179-253.
[73] YA.S. STANEV, Stress energy tensor and $U(1)$ current operator product expansion in conformal QFT, *Bulg. J. Phys.* **15** (1988) 93-107.
[74] YA.S. STANEV, I.T. TODOROV, Towards a conformal QED4 with a non-vanishing current 2-point function, *Int. J. Mod. Phys.* **A3** (1988) 1023-1049.
[75] R.F. STREATER, A.S. WIGHTMAN, *PCT, Spin and Statistics, and All That*, W.A. Benjamin, 1964; Princeton Univ. Press, Princeton, N.J. 2000.
[76] J.A. SWIECA, A.H. VÖLKEL, Remarks on conformal invariance, *Commun. Math. Phys.* **29** (1973) 319-342.
[77] K. SYMANZIK, Green's functions method and renormalization of renormalizable quantum field theory, in: *Lectures in High Energy Physics*, ed. by B. Jaksič, Zagreb 1961, Gordon and Breach, N.Y. 1965.
[78] K. SYMANZIK, Small distance behaviour in field theory and power counting, *Commun. Math. Phys.* **18** (1970) 227-246.
[79] K. SYMANZIK, Small distance behaviour and Wilson expansions, *Commun. Math. Phys.* **23** (1971) 49-86.
[80] K. SYMANZIK, On calculations in conformal field theories, *Nuovo Cim. Lett.* **3** (1972) 734-739.
[81] I.T. TODOROV, Local field representations of the conformal group and their applications, in: *Mathematics + Physics, Lectures on Recent Results*, ed. L. Streit, Vol. 1, World Scientific, Singapore 1985, pp. 195-338.

[82] I.T. TODOROV, Infinite dimensional Lie algebras in conformal QFT models. In: A.O. Barut and H.-D. Doebner (eds.) *Conformal Groups and Related Symmetries. Physical Results and Mathematical Background*, *Lecture Notes in Physics* **261**, pp. 387-443, Springer, Berlin et al. 1986.

[83] I.T. TODOROV, M.C. MINTCHEV, V.B. PETKOVA, *Conformal Invariance in Quantum Field Theory*, Scuola Normale Superiore, Pisa 1978.

[84] A. UHLMANN, Remarks on the future tube, *Acta Phys. Pol.* **24** (1963) 293; The closure of Minkowski space, *ibid.* 295-296.

[85] K.G. WILSON, Non-Lagrangian models of current algebra, *Phys. Rev.* **179** (1969) 1499-1512.

[86] K.G. WILSON, Operator product expansions and anomalous dimensions in the Thirring model, *Phys. Rev.* **D2** (1970) 1473-1477; Anomalous dimension and break down of scale invariance in perturbation theory, *ibid*, 1478-1493.

Fields Institute Communications
Volume **50**, 2007

The Art of Computing Loop Integrals

Stefan Weinzierl
Institut für Physik
Universität Mainz
D - 55099 Mainz, Germany
stefanw@thep.physik.uni-mainz.de

Abstract. A perturbative approach to quantum field theory involves the computation of loop integrals, as soon as one goes beyond the leading term in the perturbative expansion. First I review standard techniques for the computation of loop integrals. In a second part I discuss more advanced algorithms. For these algorithms algebraic methods play an important rôle. A special section is devoted to multiple polylogarithms.

I tried to make these notes self-contained and accessible both to physicists and mathematicians.

1 Introduction

In this lecture I will discuss techniques for the computation of loop integrals, which occur in perturbative calculations in quantum field theory. But before embarking onto a journey of integration and special function theory, it is worth recalling the motivation for such an effort.

High-energy physics is successfully described by the Standard Model. The term "Standard Model" has become a synonym for a quantum field theory based on the gauge group $SU(3) \otimes SU(2) \otimes U(1)$. At high energies all coupling constants are small and perturbation theory is a valuable tool to obtain predictions from the theory. For the Standard Model there are three coupling constants, g_1, g_2 and g_3, corresponding to the gauge groups $U(1)$, $SU(2)$ and $SU(3)$, respectively. As all methods which will be discussed below do not depend on the specific nature of these gauge groups and are even applicable to extensions of the Standard Model (like supersymmetry), I will just talk about a single expansion in a single coupling constant. All observable quantities are taken as a power series expansion in the coupling constant, and calculated order by order in perturbation theory.

Over the years particle physics has become a field where precision measurements have become possible. Of course, the increase in experimental precision has to be matched with more accurate calculations from the theoretical side. This is the "raison d'être" for loop calculations: A higher accuracy is reached by including

2000 *Mathematics Subject Classification.* Primary 81Q30, 81T18; Secondary 16W30, 33C20.

more terms in the perturbative expansion. There is even an additional "bonus" we get from loop calculations: Inside the loops we have to take into account all particles which could possibly circle there, even the ones which are too heavy to be produced directly in an experiment. Therefore loop calculations in combination with precision measurements allow us to extend the range of sensitivity of experiments from the region which is directly accessible towards the range of heavier particles which manifest themselves only through quantum corrections. As an example, the mass of top quark has been predicted before the discovery of the top quark from the loop corrections to electro-weak precision experiments. The same experiments predict currently a range for the mass of the yet undiscovered Higgs boson.

It is generally believed that a perturbative series is only an asymptotic series, which will diverge, if more and more terms beyond a certain order are included. However this shall be of no concern to us here. We content ourselves to the first few terms in the perturbative expansion with the implicit assumption, that the point where the power series starts to diverge is far beyond our computational abilities. In fact, our computational abilities are rather limited. The complexity of a calculation increases obviously with the number of loops, but also with the number of external particles or the number of non-zero internal masses associated to propagators. To give an idea of the state of the art, specific quantities which are just pure numbers have been computed up to an impressive fourth or third order. Examples are the calculation of the 4-loop contribution to the QCD β-function [1], the calculation of the anomalous magnetic moment of the electron up to three loops [2], and the calculation of the ratio

$$R = \frac{\sigma(e^+e^- \to \text{hadrons})}{\sigma(e^+e^- \to \mu^+\mu^-)} \tag{1.1}$$

of the total cross section for hadron production to the total cross section for the production of a $\mu^+\mu^-$ pair in electron-positron annihilation to order $O\left(g_3^3\right)$ (also involving a three loop calculation) [3]. Quantities which depend on a single variable are known at the best to the third order. Outstanding examples are the computation of the three-loop Altarelli-Parisi splitting functions [4,5] or the calculation of the two-loop amplitudes for the most interesting $2 \to 2$ processes [6–16]. The complexity of a two-loop computation increases, if the result depends on more than one variable. An example for a two-loop calculation whose result depends on two variables is the computation of the two-loop amplitudes for $e^+e^- \to 3$ jets [17–19]. But in general, if more than one variable is involved, we have to content ourselves with next-to-leading order calculations. An example for the state of the art is here the computation of the electro-weak corrections to the process $e^+e^- \to 4$ fermions [20, 21].

From a mathematical point of view it is an interesting question to ask which type of functions or constants appear in the result of a particular calculation. For one-loop calculations it is known that the result can be expressed in terms of logarithms and dilogarithms. Transcendental constants which do occur are just proportional to π^2, or phrased differently, proportional to $\zeta_2 = \pi^2/6$. Going to higher loops one finds the following: In results which are just numbers higher zeta values do occur. In results which depend on one variable harmonic polylogarithms arise and in results which depend on more than one variable multiple polylogarithms occur. The multiple polylogarithms have nice algebraic properties and contain as special cases (multiple) zeta values and the harmonic polylogarithms. They will be discussed in detail in this lecture.

This paper is organised as follows: In the next section I will discuss basic techniques, which allow us to exchange the integrals over the loop momenta against integrals over Feynman or Schwinger parameters. Sect. 3 shows how tensor integrals, e.g. integrals where the loop momentum occurs also in the numerator, can be reduced to scalar integrals. In sect. 4 I discuss how the Feynman parametrisation for a generic scalar l-loop integral can be read off directly from the underlying Feynman graph. The simplest, but most important loop integrals are the one-loop integrals, whose theory is presented in sect. 5. The main results are that all one-loop integrals can be reduced to one-loop integrals with no more than four external legs. Furthermore, when computed within dimensional regularisation up to the finite part, the only occurring transcendental functions are the logarithm and the dilogarithm. In sect. 6 I continue with the general case of an l-loop integral and present several advanced methods for the calculation of the Feynman parameter integrals. As already mentioned, the multiple polylogarithms are the class of functions, to which the logarithm and the dilogarithm naturally extend. They are discussed in detail in sect 7. Unfortunately we do not yet have a complete theory for loop integrals in quantum theory. There are many unknown integrals and a variety of conjectures related to them. Sect. 8 tries to give an outlook towards open questions and future directions. Finally, sect. 9 contains a summary.

2 Basic techniques

To set the scene, let us start with a brief summary how observables are calculated in perturbative quantum field theory. Let us assume that we are interested in a process with two incoming elementary particles (like an electron and a positron) and n outgoing particles. We study an observable $O\left(p_1, p_2; k_1, ..., k_n\right)$ which depends on the four-momenta p_1 and p_2 of the two incoming particles and the four-momenta k_j of the outgoing particles and we are interested in the value of O integrated over all possible final state momenta k_j. $O\left(p_1, p_2; k_1, ..., k_n\right)$ depends on the experimental set-up and can be an arbitrary complicated function of the four-momenta. In the simplest case this function is just a constant equal to one, corresponding to the situation where we count every event with n particles in the final state. In more realistic situations one takes for example into account that it is not possible to detect particles close to the beam pipe. The function O would then be zero in these regions of phase space.

The expectation value for the observable O is given by

$$\langle O\rangle = \frac{1}{8(p_1+p_2)^2}\sum_n \int d\phi_n\left(p_1, p_2; k_1, ..., k_n\right) O\left(p_1, p_2; k_1, ..., k_n\right) \sum_{spins} \left|\mathcal{A}_n\right|^2, \tag{2.1}$$

where $1/8/(p_1+p_2)^2$ is a normalisation factor taking into account the incoming flux and averages over the spins of the incoming particles. The phase space measure is given by

$$d\phi_n\left(p_1, p_2; k_1, ..., k_n\right) = \frac{1}{\prod N_j!}\prod_{i=1}^{n}\frac{d^3k_i}{(2\pi)^3 2E_i}\left(2\pi\right)^4\delta^4\left(p_1+p_2-\sum_{i=1}^{n}k_i\right). \tag{2.2}$$

There is a symmetry factor $1/N_j!$ for each type of particle, which occurs more than once in the final state. E_i is the energy of particle i:

$$E_i = \sqrt{\vec{k}_i^2 + m_i^2} \tag{2.3}$$

As the integrand can be a rather complicated function, the phase space integral is usually performed numerically by Monte Carlo integration.

The most important ingredient of formula (2.1) is the matrix element $\mathcal{A}_n$ for the process under consideration. This quantity is also called the amplitude or the Green function of the process. The amplitude has the perturbative expansion

$$\mathcal{A}_n = g^n \left(\mathcal{A}_n^{(0)} + g^2 \mathcal{A}_n^{(1)} + g^4 \mathcal{A}_n^{(2)} + g^6 \mathcal{A}_n^{(3)} + ... \right) \tag{2.4}$$

To the coefficient $\mathcal{A}_n^{(l)}$ contribute Feynman graphs with l loops and $(n+2)$ external legs. The recipe for the computation of $\mathcal{A}_n^{(l)}$ is as follows: Draw first all Feynman diagrams with the given number of external particles and l loops. Then translate each graph into a mathematical formula with the help of the Feynman rules. $\mathcal{A}_n^{(l)}$ is then given as the sum of all these terms.

2.1 Feynman rules. Feynman rules allow us to translate a Feynman graph into a mathematical formula. These rules are derived from the fundamental Lagrange density of the theory, but for our purposes it is sufficient to accept them as a starting point. The most important ingredients are internal propagators, vertices and external lines. For example, the rules for the propagators of a fermion or a gauge boson read

Fermion:  $= i \dfrac{\not{p} + m}{p^2 - m^2 + i\delta},$

Gauge boson:  $= \dfrac{-i}{k^2 + i\delta} \left(g_{\mu\nu} - (1-\xi) \dfrac{k_\mu k_\nu}{k^2} \right).$ (2.5)

Here p and k are the momenta of the fermion and the boson, respectively. m is the mass of the fermion. $\not{p} = p_\mu \gamma^\mu$ is a short-hand notation for the contraction of the momentum with the Dirac matrices. The metric tensor is denoted by $g_{\mu\nu}$ and the convention adopted here is to take the metric tensor as $g_{\mu\nu} = \text{diag}(1,-1,-1,-1)$. The propagator would have a pole for $p^2 = m^2$, or phrased differently $E = \pm\sqrt{\vec{p}^2 + m^2}$. When integrating over E, the integration contour has to be deformed to avoid these two poles. Causality dictates into which directions the contour has to be deformed. The pole on the negative real axis is avoided by escaping into the lower complex half-plane, the pole at the positive real axis is avoided by a deformation into the upper complex half-plane. Feynman invented the trick to add a small imaginary part $i\delta$ to the denominator, which keeps track of the directions into which the contour has to be deformed. I will usually suppress the $i\delta$-term in order to keep the notation compact.

The gauge boson propagator depends on an additional variable ξ, called the gauge-fixing parameter. This parameter can be chosen arbitrarily, but one must use the same choice for all diagrams contributing to an gauge-invariant subset of the amplitude. It also shows that there is no point in speaking about the "value" of an individual graph, as this value depends in general on our choice of the gauge

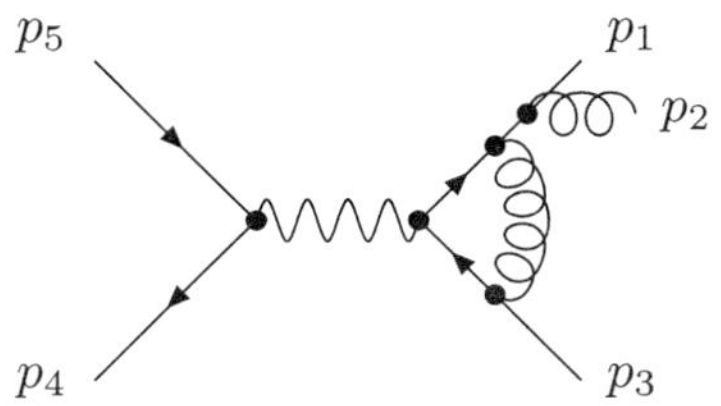

Figure 1 A one-loop Feynman diagram contributing to the process $e^+e^- \to qg\bar{q}$.

parameter. Only in the sum of graphs over a gauge-invariant set does this dependency on the gauge parameter cancel. The most economical choice is $\xi = 1$, which is called Feynman gauge.

As a typical example for an interaction vertex let us look at the vertex involving a fermion pair and a gauge boson:

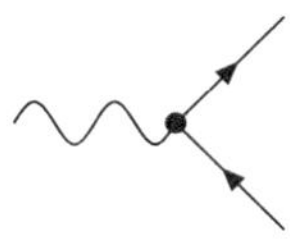

$$= \quad ig\gamma^\mu. \tag{2.6}$$

Here, g is the coupling constant and γ^μ denotes the Dirac matrices. At each vertex, we have momentum conservation: The sum of the incoming momenta equals the sum of the outgoing momenta.

To each external line we have to associate a factor, which describes the polarisation of the corresponding particle: There is a polarisation vector $\varepsilon^\mu(k)$ for each external gauge boson and a spinor $\bar{u}(p)$, $u(p)$, $v(p)$ or $\bar{v}(p)$ for each external fermion.

Furthermore there are a few additional rules: First of all, there is an integration

$$\int \frac{d^4k}{(2\pi)^4} \tag{2.7}$$

for each loop. Secondly, each closed fermion loop gets an extra factor of (-1). Finally, each diagram gets multiplied by a symmetry factor $1/S$, where S is the order of the permutation group of the internal lines and vertices leaving the diagram unchanged when the external lines are fixed.

Having stated the Feynman rules, let us look at an example: Fig. 1 shows a Feynman diagram contributing to the one-loop corrections for the process $e^+e^- \to qg\bar{q}$. At high energies we can ignore the masses of the electron and the light quarks. From the Feynman rules one obtains for this diagram (ignoring coupling and colour prefactors):

$$-\bar{v}(p_4)\gamma^\mu u(p_5)\frac{1}{p_{123}^2}\int \frac{d^4k_1}{(2\pi)^4}\frac{1}{k_2^2}\bar{u}(p_1)\not{\varepsilon}(p_2)\frac{\not{p}_{12}}{p_{12}^2}\gamma_\nu\frac{\not{k}_1}{k_1^2}\gamma_\mu\frac{\not{k}_3}{k_3^2}\gamma^\nu v(p_3). \tag{2.8}$$

Here, $p_{12} = p_1 + p_2$, $p_{123} = p_1 + p_2 + p_3$, $k_2 = k_1 - p_{12}$, $k_3 = k_2 - p_3$. Further $\not{\varepsilon}(p_2) = \gamma_\tau \varepsilon^\tau(p_2)$, where $\varepsilon^\tau(p_2)$ is the polarisation vector of the outgoing gluon. All external momenta are assumed to be massless: $p_i^2 = 0$ for $i = 1..5$. We can reorganise this formula into a part, which depends on the loop integration and a part, which does not. The loop integral to be calculated reads:

$$\int \frac{d^4k_1}{(2\pi)^4}\frac{k_1^\rho k_3^\sigma}{k_1^2 k_2^2 k_3^2}, \tag{2.9}$$

while the remainder, which is independent of the loop integration is given by

$$-\bar{v}(p_4)\gamma^{\mu}u(p_5)\frac{1}{p_{123}^2 p_{12}^2}\bar{u}(p_1)\not{\varepsilon}(p_2)\not{p}_{12}\gamma_{\nu}\gamma_{\rho}\gamma_{\mu}\gamma_{\sigma}\gamma^{\nu}v(p_3). \tag{2.10}$$

The loop integral in eq. (2.9) contains in the denominator three propagator factors and in the numerator two factors of the loop momentum. We call a loop integral, in which the loop momentum occurs also in the numerator a "tensor integral". A loop integral, in which the numerator is independent of the loop momentum is called a "scalar integral". The basic strategy, which will be discussed in detail in sect. 3, consists in reducing tensor integrals to scalar integrals. The scalar integral associated to eq. (2.9) reads

$$\int \frac{d^4k_1}{(2\pi)^4}\frac{1}{k_1^2 k_2^2 k_3^2}. \tag{2.11}$$

2.2 Regularisation. Before we start with the actual calculation of loop integrals, I should mention one complication: Loop integrals are often divergent ! Let us first look at the simple example of a scalar two-point one-loop integral with zero external momentum:

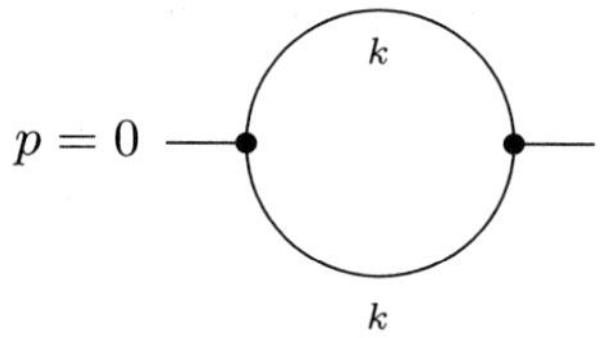

$$\begin{aligned} &= \int \frac{d^4k}{(2\pi)^4}\frac{1}{(k^2)^2} \\ &= \frac{1}{(4\pi)^2}\int\limits_0^\infty dk^2\frac{1}{k^2} = \frac{1}{(4\pi)^2}\int\limits_0^\infty \frac{dx}{x}. \end{aligned} \tag{2.12}$$

This integral diverges at $k^2 \to \infty$ as well as at $k^2 \to 0$. The former divergence is called ultraviolet divergence, the later is called infrared divergence. Any quantity, which is given by a divergent integral, is of course an ill-defined quantity. Therefore the first step is to make these integrals well-defined by introducing a regulator. There are several possibilities how this can be done, but the method of dimensional regularisation [22–24] has almost become a standard, as the calculations in this regularisation scheme turn out to be the simplest. Within dimensional regularisation one replaces the four-dimensional integral over the loop momentum by an D-dimensional integral, where D is now an additional parameter, which can be a non-integer or even a complex number. We consider the result of the integration as a function of D and we are interested in the behaviour of this function as D approaches 4. The D-dimensional integration still fulfils the standard laws for integration, like linearity, translation invariance and scaling behaviour [25, 26]. If f and g are two functions, and if a and b are two constants, linearity states that

$$\int d^Dk\,(af(k)+bg(k)) \;=\; a\int d^Dk f(k) + b\int d^Dk g(k). \tag{2.13}$$

Translation invariance requires that

$$\int d^Dk f(k+p) \;=\; \int d^Dk f(k). \tag{2.14}$$

for any vector p.
The scaling law states that

$$\int d^D k f(\lambda k) = \lambda^{-D} \int d^D k f(k). \tag{2.15}$$

The D-dimensional integral has also a rotation invariance:

$$\int d^D k f(\Lambda k) = \int d^D k f(k), \tag{2.16}$$

where Λ is an element of the Lorentz group $SO(1, D-1)$ of the D-dimensional vector-space. Here we assumed that the D-dimensional vector-space has the metric diag$(+1, -1, -1, -1, ...)$. The integral measure is normalised such that it agrees with the result for the integration of a Gaussian function for all integer values D:

$$\int d^D k \exp\left(\alpha k^2\right) = i\left(\frac{\pi}{\alpha}\right)^{\frac{D}{2}}. \tag{2.17}$$

We will further assume that we can always decompose any vector into a 4-dimensional part and a $(D-4)$-dimensional part

$$k^\mu_{(D)} = k^\mu_{(4)} + k^\mu_{(D-4)}, \tag{2.18}$$

and that the 4-dimensional and $(D-4)$-dimensional subspaces are orthogonal to each other:

$$k_{(4)} \cdot k_{(D-4)} = 0. \tag{2.19}$$

If D is an integer greater than 4, this is obvious. We postulate that these relations are true for any value of D. One can think of the underlying vector-space as a space of infinite dimension, where the integral measure mimics the one in D dimensions.

In practice we will always arrange things such that every function we integrate over D dimensions is rotational invariant, e.g. is a function of k^2. In this case the integration over the $(D-1)$ angles is trivial and can be expressed in a closed form as a function of D. Let us assume that we have an integral, which has a UV-divergence, but no IR-divergences. Let us further assume that this integral would diverge logarithmically, if we would use a cut-off regularisation instead of dimensional regularisation. It turns out that this integral will be convergent if the real part of D is smaller than 4. Therefore we may compute this integral under the assumption that $\text{Re}(D) < 4$ and we will obtain as a result a function of D. This function can be analytically continued to the whole complex plane. We are mainly interested in what happens close to the point $D = 4$. For an ultraviolet divergent one-loop integral we will find that the analytically continued result will exhibit a pole at $D = 4$. It should be mentioned that there are also integrals which are quadratically divergent, if a cut-off regulator is used. In this case we can repeat the argumentation above with the replacement $\text{Re}(D) < 2$.

Similarly, we can consider an IR-divergent integral, which has no UV-divergence. This integral will be convergent if $\text{Re}(D) > 4$. Again, we can compute the integral in this domain and continue the result to $D = 4$. Here we find that each IR-divergent loop integral can lead to a double pole at $D = 4$.

We will use dimensional regularisation to regulate both the ultraviolet and infrared divergences. The attentative reader may ask how this goes together, as we argued above that UV-divergences require $\text{Re}(D) < 4$ or even $\text{Re}(D) < 2$, whereas IR-divergences are regulated by $\text{Re}(D) > 4$. Suppose for the moment that we use dimensional regularisation just for the UV-divergences and that we use a second regulator for the IR-divergences. For the IR-divergences we could keep all external momenta off-shell, or introduce small masses for all massless particles or even raise the original propagators to some power ν. The exact implementation of this regulator is not important, as long as the IR-divergences are screened by this procedure. We then perform the loop integration in the domain where the integral is UV-convergent. We obtain a result, which we can analytically continue to the whole complex D-plane, in particular to $\text{Re}(D) > 4$. There we can remove the additional regulator and the IR-divergences are now regulated by dimensional regularisation. Then the infrared divergences will also show up as poles at $D = 4$.

In summary, within dimensional regularisation the initial divergences show up as poles in the complex D-plane at the point $D = 4$. What shall we do with these poles ? The answer has to come from physics and we distinguish again the case of UV-divergences and IR-divergences. The UV-divergences are removed through renormalisation. On the level of Feynman diagrams we can associate to any UV-divergent part a counter-term, which has exactly the same pole term at $D = 4$, but with an opposite sign, such that in the sum the two pole terms cancel.

As far as infrared-divergences are concerned we first note that any detector has a finite resolution. Therefore two particles which are sufficiently close to each other in phase space will be detected as one particle. Now let us look again at eqs. (2.1) and (2.4). The next-to-leading order term will receive contributions from the interference term of the one-loop amplitude $\mathcal{A}_n^{(1)}$ with the leading-order amplitude $\mathcal{A}_n^{(0)}$, both with n final state particles. This contribution is of order g^{2n+2}. Of the same order is the square of the leading-order amplitude $\mathcal{A}_{n+1}^{(0)}$ with $(n+1)$ final state particles. This contribution we have to take into account whenever our detector resolves only n particles. It turns out that the phase space integration over the regions where one or more particles become unresolved is also divergent, and, when performed in D dimensions, leads to poles with the opposite sign as the one encountered in the loop amplitudes. Therefore the sum of the two contributions is finite. The Kinoshita-Lee-Nauenberg theorem [27,28] guarantees that all infrared divergences cancel, when summed over all degenerate physical states. To make this cancellation happen in practice requires usually quite some work, as the different contributions live on phase spaces of different dimensions.

To summarise we are interested into loop integrals regulated by dimensional regularisation. As a result we seek the Laurent expansion around $D = 4$. It is common practice to parameterise the deviation of D from 4 by

$$D = 4 - 2\varepsilon. \tag{2.20}$$

Divergent loop integrals will therefore have poles in $1/\varepsilon$. In an l-loop integral ultraviolet divergences will lead to poles $1/\varepsilon^l$ at the worst, whereas infrared divergences can lead to poles up to $1/\varepsilon^{2l}$. We will also encounter integrals, where the dimension is shifted by units of two. In these cases we often write

$$D = 2m - 2\varepsilon, \tag{2.21}$$

where m is an integer, and we are again interested in the Laurent series in ε.

2.3 Dimensional regularisation schemes. Let us look again at the example in eqs. (2.8) to (2.10). We separated the loop integral from the remainder in eq. (2.10), which is independent of the loop integration. In this remainder the following string of Dirac matrices occurs:

$$\gamma_\nu \gamma_\rho \gamma_\mu \gamma_\sigma \gamma^\nu. \tag{2.22}$$

If we anti-commute the first Dirac matrix, we can achieve that the two Dirac matrices with index ν are next to each other:

$$\gamma_\nu \gamma^\nu. \tag{2.23}$$

In four dimensions this equals 4 times the unit matrix. What is the value within dimensional regularisation ? The answer depends on how we treat the Dirac algebra. Does the Dirac algebra remain in four dimensions or do we also continue the Dirac algebra to D dimensions ? There are several schemes on the market which treat this issue differently. To discuss these schemes it is best to look how they treat the momenta and the polarisation vectors of observed and unobserved particles. Unobserved particles are particles circulating inside loops or emitted particles not resolved within a given detector resolution. The most commonly used schemes are the conventional dimensional regularisation scheme (CDR) [26], where all momenta and all polarisation vectors are taken to be in D dimensions, the 't Hooft-Veltman scheme (HV) [22, 29], where the momenta and the helicities of the unobserved particles are D dimensional, whereas the momenta and the helicities of the observed particles are 4 dimensional, and the four-dimensional helicity scheme (FD) [30–32], where all polarisation vectors are kept in four dimensions, as well as the momenta of the observed particles. Only the momenta of the unobserved particles are continued to D dimensions.

The conventional scheme is mostly used for an analytical calculation of the interference of a one-loop amplitude with the Born amplitude by using polarisation sums corresponding to D dimensions. For the calculation of one-loop helicity amplitudes the 't Hooft-Veltman scheme and the four-dimensional helicity scheme are possible choices. All schemes have in common, that the propagators appearing in the denominator of the loop-integrals are continued to D dimensions. They differ how they treat the algebraic part in the numerator. In the 't Hooft-Veltman scheme the algebraic part is treated in D dimensions, whereas in the FD scheme the algebraic part is treated in four dimensions. It is possible to relate results obtained in one scheme to another scheme, using simple and universal transition formulae [33–35].

Therefore, if we return to the example above, we have

$$\gamma_\nu \gamma^\nu = \begin{cases} D \cdot \mathbf{1}, & \text{in the CDR and HV scheme,} \\ 4 \cdot \mathbf{1}, & \text{in the FD scheme.} \end{cases} \tag{2.24}$$

2.4 Loop integration in D dimensions. In this section I will discuss how to perform the D-dimensional loop integrals. It would be more correct to say that we exchange them for some parameter integrals. Our starting point is a one-loop integral with n external legs:

$$e^{\varepsilon \gamma_E} \mu^{2\varepsilon} \int \frac{d^D k}{i \pi^{D/2}} \frac{1}{(-P_1)(-P_2)...(-P_n)}, \tag{2.25}$$

where the propagators are of the form

$$P_i = \left(k - \sum_{j=1}^{i} p_j \right)^2 - m_i^2 \tag{2.26}$$

and p_j are the external momenta. The small imaginary parts $i\delta$ are not written explicitly. In eq. (2.25) there are some overall factors, which I inserted for convenience: μ is an arbitrary mass scale and the factor $\mu^{2\varepsilon}$ ensures that the mass dimension of eq. (2.25) is an integer. The factor $e^{\varepsilon\gamma_E}$ avoids a proliferation of Euler's constant γ_E in the final result. The integral measure is now $d^Dk/(i\pi^{D/2})$ instead of $d^Dk/(2\pi)^D$, and each propagator is multiplied by (-1). The reason for doing this is that the final result will be simpler.

In order to perform the momentum integration we proceed by the following steps:

1. Feynman or Schwinger parametrisation.
2. Shift of the loop momentum to complete the square, such that the integrand depends only on k^2.
3. Wick rotation.
4. Introduction of generalised spherical coordinates.
5. The angular integration is trivial. Using the definitions of Euler's Gamma and Beta functions, the radial integration can be performed.
6. This leaves only the non-trivial integration over the Feynman parameters.

Although I discuss here only one-loop integrals, the methods presented in this section are rather general and can be applied iteratively to l-loop integrals.

2.4.1 *Feynman and Schwinger parameterisation.* As already discussed above, the only functions we really want to integrate over D dimensions are the ones which depend on the loop momentum only through k^2. The integrand in eq. (2.25) is not yet in such a form. To bring the integrand into this form, we first convert the product of propagators into a sum. To do this, there are two techniques, one due to Feynman, the other one due to Schwinger. Let me start with the Feynman parameter technique. In its full generality it is also applicable to cases, where each factor in the denominator is raised to some power ν. The formula reads:

$$\begin{aligned}\prod_{i=1}^{n} \frac{1}{(-P_i)^{\nu_i}} &= \frac{\Gamma(\nu)}{\prod\limits_{i=1}^{n}\Gamma(\nu_i)} \int_0^1 \left(\prod_{i=1}^{n} dx_i\, x_i^{\nu_i-1}\right) \frac{\delta\left(1-\sum\limits_{i=1}^{n} x_i\right)}{\left(-\sum\limits_{i=1}^{n} x_i P_i\right)^{\nu}},\\ \nu &= \sum_{i=1}^{n} \nu_i. \end{aligned} \tag{2.27}$$

The proof of this formula can be found in many text books and is not repeated here. The price we have to pay for converting the product into a sum are $(n-1)$ additional integrations. Let us look at the example from eq. (2.11):

$$\begin{aligned}\frac{1}{(-k_1^2)(-k_2^2)(-k_3^2)} &= 2\int_0^1 dx_1 \int_0^1 dx_2 \int_0^1 dx_3 \frac{\delta(1-x_1-x_2-x_3)}{(-x_1k_1^2-x_2k_2^2-x_3k_3^2)^3} \\ &= 2\int_0^1 dx_1 \int_0^{1-x_1} dx_2 \frac{1}{(-x_1k_1^2-x_2k_2^2-(1-x_1-x_2)k_3^2)^3}. \end{aligned} \tag{2.28}$$

An alternative to Feynman parameters are Schwinger parameters. Here each propagator is rewritten as

$$\frac{1}{(-P)^{\nu}} = \frac{1}{\Gamma(\nu)} \int\limits_0^{\infty} dx \; x^{\nu-1} \exp(xP). \tag{2.29}$$

Therefore we obtain for our example

$$\frac{1}{(-k_1^2)(-k_2^2)(-k_3^2)} = \int\limits_0^{\infty} dx_1 \int\limits_0^{\infty} dx_2 \int\limits_0^{\infty} dx_3 \; \exp\left(x_1 k_1^2 + x_2 k_2^2 + x_3 k_3^2\right). \tag{2.30}$$

2.4.2 *Shift of the integration variable.* We can now complete the square and shift the loop momentum, such that the integrand becomes a function of k^2. This is best discussed by an example. We consider again eq. (2.28). With $k_2 = k_1 - p_{12}$ and $k_3 = k_2 - p_3$ we have

$$-x_1 k_1^2 - x_2 k_2^2 - x_3 k_3^2 = -\left(k_1 - x_2 p_{12} - x_3 p_{123}\right)^2 - x_1 x_2 s_{12} - x_1 x_3 s_{123}, \tag{2.31}$$

where $s_{12} = (p_1 + p_2)^2$ and $s_{123} = (p_1 + p_2 + p_3)^2$. We can now define

$$k_1' = k_1 - x_2 p_{12} - x_3 p_{123} \tag{2.32}$$

and using translational invariance our loop integral becomes

$$\int \frac{d^D k_1}{i\pi^{D/2}} \frac{1}{(-k_1^2)(-k_2^2)(-k_3^2)} =$$
$$2 \int \frac{d^D k_1'}{i\pi^{D/2}} \int\limits_0^1 dx_1 \int\limits_0^1 dx_2 \int\limits_0^1 dx_3 \frac{\delta(1 - x_1 - x_2 - x_3)}{\left(-{k_1'}^2 - x_1 x_2 s_{12} - x_1 x_3 s_{123}\right)^3}. \tag{2.33}$$

The integrand is now a function of ${k_1'}^2$.

2.4.3 *Wick rotation.* Having succeeded to rewrite the integrand as a function of k^2, we then perform a Wick rotation, which transforms Minkowski space into an Euclidean space. Remember, that k^2 written out in components in D- dimensional Minkowski space reads

$$k^2 = k_0^2 - k_1^2 - k_2^2 - k_3^2 - ... \tag{2.34}$$

(Here k_j denotes the j-th component of the vector k, in contrast to the previous section, where we used the subscript to label different vectors k_j. It should be clear from the context what is meant.) Furthermore, when integrating over k_0, we encounter poles which are avoided by Feynman's $i\delta$-prescription. In the complex k_0-plane we consider the integration contour shown in fig. 2. Since the contour does not enclose any poles, the integral along the complete contour is zero:

$$\oint dk_0 f(k_0) = 0. \tag{2.35}$$

If the quarter-circles at infinity give a vanishing contribution (it can be shown that this is the case) we obtain

$$\int\limits_{-\infty}^{\infty} dk_0 f(k_0) = -\int\limits_{i\infty}^{-i\infty} dk_0 f(k_0). \tag{2.36}$$

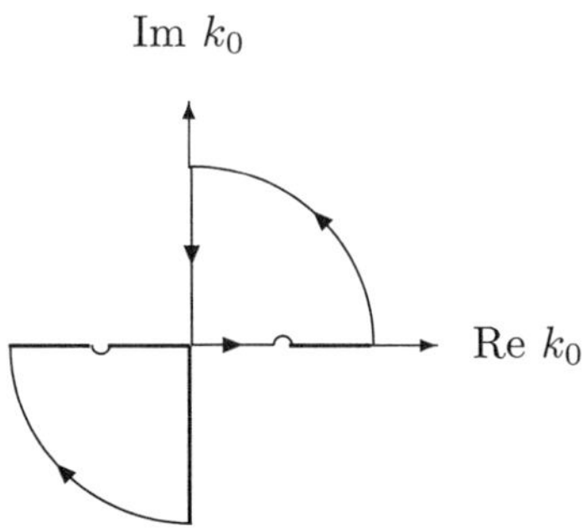

Figure 2 Integration contour for the Wick rotation. The little circles along the real axis exclude the poles.

We now make the following change of variables:

$$\begin{aligned} k_0 &= iK_0, \\ k_j &= K_j, \quad \text{for } 1 \leq j \leq D-1. \end{aligned} \tag{2.37}$$

As a consequence we have

$$\begin{aligned} k^2 &= -K^2, \\ d^Dk &= id^DK, \end{aligned} \tag{2.38}$$

where K^2 is now given with Euclidean signature:

$$K^2 = K_0^2 + K_1^2 + K_2^2 + K_3^2 + ... \tag{2.39}$$

Combining eq. (2.36) with eq. (2.37) we obtain for the integration of a function $f(k^2)$ over D dimensions

$$\int \frac{d^Dk}{i\pi^{D/2}} f(-k^2) = \int \frac{d^DK}{\pi^{D/2}} f(K^2), \tag{2.40}$$

whenever there are no poles inside the contour of fig. 2 and the arcs at infinity give a vanishing contribution. The integral on the r.h.s. is now over D-dimensional Euclidean space. Eq. (2.40) justifies our conventions, to introduce a factor i in the denominator and a minus sign for each propagator in eq. (2.25). These conventions are just such that after Wick rotation we have simple formulae.

2.4.4 *Generalised spherical coordinates.* We now have an integral over D-dimensional Euclidean space, where the integrand depends only on K^2. It is therefore natural to introduce spherical coordinates. In D dimensions they are given by

$$\begin{aligned} K_0 &= K \cos\theta_1, \\ K_1 &= K \sin\theta_1 \cos\theta_2, \\ &\dots \\ K_{D-2} &= K \sin\theta_1 ... \sin\theta_{D-2} \cos\theta_{D-1}, \\ K_{D-1} &= K \sin\theta_1 ... \sin\theta_{D-2} \sin\theta_{D-1}. \end{aligned} \tag{2.41}$$

In D dimensions we have one radial variable K, $D-2$ polar angles θ_j (with $1 \leq j \leq D-2$) and one azimuthal angle θ_{D-1}. The measure becomes

$$d^DK = K^{D-1} dK d\Omega_D, \tag{2.42}$$

where

$$d\Omega_D = \prod_{i=1}^{D-1} \sin^{D-1-i}\theta_i \, d\theta_i. \tag{2.43}$$

Integration over the angles yields

$$\int d\Omega_D = \int_0^{\pi} d\theta_1 \sin^{D-2}\theta_1 ... \int_0^{\pi} d\theta_{D-2} \sin\theta_{D-2} \int_0^{2\pi} d\theta_{D-1} = \frac{2\pi^{D/2}}{\Gamma\left(\frac{D}{2}\right)}, \tag{2.44}$$

where $\Gamma(x)$ is Euler's Gamma function. Note that the integration on the l.h.s of eq. (2.44) is defined for any natural number D, whereas the result on the r.h.s is an analytic function of D, which can be continued to any complex value.

2.4.5 *Euler's Gamma and Beta function.* It is now the appropriate place to introduce two special functions, Euler's Gamma and Beta function, which are used within dimensional regularisation to continue the results from integer D towards non-integer values. The Gamma function is defined for $\text{Re}(x) > 0$ by

$$\Gamma(x) = \int_0^\infty e^{-t} t^{x-1} dt. \tag{2.45}$$

It fulfils the functional equation

$$\Gamma(x+1) = x\,\Gamma(x). \tag{2.46}$$

For positive integers n it takes the values

$$\Gamma(n+1) = n! = 1 \cdot 2 \cdot 3 \cdot ... \cdot n. \tag{2.47}$$

At $x = 1/2$ it has the value

$$\Gamma\left(\frac{1}{2}\right) = \sqrt{\pi}, \tag{2.48}$$

which can also be inferred from the relation

$$\Gamma(x)\Gamma(1-x) = \frac{\pi}{\sin \pi x}. \tag{2.49}$$

For integers n we have the reflection identity

$$\frac{\Gamma(x-n)}{\Gamma(x)} = (-1)^n \frac{\Gamma(1-x)}{\Gamma(1-x+n)}. \tag{2.50}$$

The Gamma function $\Gamma(x)$ has poles located on the negative real axis at $x = 0, -1, -2, ...$ Quite often we will need the expansion around these poles. This can be obtained from the expansion around $x = 1$ and the functional equation. The expansion around $\varepsilon = 1$ reads

$$\Gamma(1+\varepsilon) = \exp\left(-\gamma_E \varepsilon + \sum_{n=2}^{\infty} \frac{(-1)^n}{n} \zeta_n \varepsilon^n\right), \tag{2.51}$$

where γ_E is Euler's constant

$$\gamma_E = \lim_{n\to\infty}\left(\sum_{j=1}^{n} \frac{1}{j} - \ln n\right) = 0.5772156649... \tag{2.52}$$

and ζ_n is given by

$$\zeta_n = \sum_{j=1}^{\infty} \frac{1}{j^n}. \tag{2.53}$$

For example we obtain for the Laurent expansion around $\varepsilon = 0$

$$\Gamma(\varepsilon) = \frac{1}{\varepsilon} - \gamma_E + O(\varepsilon). \tag{2.54}$$

Euler's Beta function is defined for $\mathrm{Re}(x) > 0$ and $\mathrm{Re}(y) > 0$ by

$$B(x,y) = \int_0^1 t^{x-1}(1-t)^{y-1} dt, \tag{2.55}$$

or equivalently by

$$B(x,y) = \int_0^\infty \frac{t^{x-1}}{(1+t)^{x+y}} dt. \tag{2.56}$$

The Beta function can be expressed in terms of Gamma functions:

$$B(x,y) = \frac{\Gamma(x)\Gamma(y)}{\Gamma(x+y)}. \tag{2.57}$$

2.4.6 *Result for the momentum integration.* We are now in a position to perform the integration over the loop momentum. Let us discuss again the example from eq. (2.33). After Wick rotation we have

$$I = \int \frac{d^Dk_1}{i\pi^{D/2}} \frac{1}{(-k_1^2)(-k_2^2)(-k_3^2)} = 2 \int \frac{d^DK}{\pi^{D/2}} \int d^3x \frac{\delta(1-x_1-x_2-x_3)}{\left(K^2 - x_1x_2s_{12} - x_1x_3s_{123}\right)^3}. \tag{2.58}$$

Introducing spherical coordinates and performing the angular integration this becomes

$$I = \frac{2}{\Gamma\left(\frac{D}{2}\right)} \int_0^\infty dK^2 \int d^3x \frac{\delta(1-x_1-x_2-x_3)\left(K^2\right)^{\frac{D-2}{2}}}{\left(K^2 - x_1x_2s_{12} - x_1x_3s_{123}\right)^3}. \tag{2.59}$$

For the radial integration we have after the substitution $t = K^2/(-x_1x_2s_{12} - x_1x_3s_{123})$

$$\int_0^\infty dK^2 \frac{\left(K^2\right)^{\frac{D-2}{2}}}{\left(K^2 - x_1x_2s_{12} - x_1x_3s_{123}\right)^3} = \left(-x_1x_2s_{12} - x_1x_3s_{123}\right)^{\frac{D}{2}-3} \int_0^\infty dt \frac{t^{\frac{D-2}{2}}}{(1+t)^3}. \tag{2.60}$$

The remaining integral is just the second definition of Euler's Beta function

$$\int_0^\infty dt \frac{t^{\frac{D-2}{2}}}{(1+t)^3} = \frac{\Gamma\left(\frac{D}{2}\right)\Gamma\left(3-\frac{D}{2}\right)}{\Gamma(3)}. \tag{2.61}$$

Putting everything together and setting $D = 4 - 2\varepsilon$ we obtain

$$\int \frac{d^D k_1}{i\pi^{D/2}} \frac{1}{(-k_1^2)(-k_2^2)(-k_3^2)} = \Gamma(1+\varepsilon) \int d^3x \, \delta(1 - x_1 - x_2 - x_3) \, x_1^{-1-\varepsilon} \left(-x_2 s_{12} - x_3 s_{123}\right)^{-1-\varepsilon}. \tag{2.62}$$

Therefore we succeeded in performing the integration over the loop momentum k at the expense of introducing a two-fold integral over the Feynman parameters.

As the steps discussed above always occur in any loop integration we can combine them into a master formula. If $\mathcal{U}$ and $\mathcal{F}$ are functions, which are independent of the loop momentum, we have for the integration over Minkowski space with dimension $D = 2m - 2\varepsilon$:

$$\int \frac{d^{2m-2\varepsilon}k}{i\pi^{m-\varepsilon}} \frac{(-k^2)^a}{\left[-\mathcal{U}k^2 + \mathcal{F}\right]^\nu} = \frac{\Gamma(m+a-\varepsilon)}{\Gamma(m-\varepsilon)} \frac{\Gamma(\nu - m - a + \varepsilon)}{\Gamma(\nu)} \frac{\mathcal{U}^{-m-a+\varepsilon}}{\mathcal{F}^{\nu-m-a+\varepsilon}}. \tag{2.63}$$

The functions $\mathcal{U}$ and $\mathcal{F}$ depend usually on the Feynman parameters and the external momenta and are obtained after Feynman parametrisation from completing the square. In eq. (2.63) we allowed additional powers $(-k^2)^a$ of the loop momentum in the numerator. This is a slight generalisation and will be useful later. Here we observe that the dependency of the result on a, apart from a factor $\Gamma(m+a-\varepsilon)/\Gamma(m-\varepsilon)$, occurs only in the combination $m+a-\varepsilon = D/2+a$. Therefore adding a power of $(-k^2)$ to the numerator is almost equivalent to consider the integral without this power in dimensions $D+2$.

There is one more generalisation: Sometimes it is convenient to decompose k^2 into a $(2m)$-dimensional piece and a remainder:

$$k_{(D)}^2 = k_{(2m)}^2 + k_{(-2\varepsilon)}^2. \tag{2.64}$$

If D is an integer greater than $2m$ we have

$$\begin{aligned} k_{(2m)}^2 &= k_0^2 - k_1^2 - \ldots - k_{2m-1}^2, \\ k_{(-2\varepsilon)} &= -k_{2m}^2 - \ldots - k_{D-1}^2. \end{aligned} \tag{2.65}$$

We also need loop integrals where additional powers of $(-k_{(-2\varepsilon)}^2)$ appear in the numerator. These are related to integrals in higher dimensions as follows:

$$\int \frac{d^{2m-2\varepsilon}k}{i\pi^{m-\varepsilon}} (-k_{(-2\varepsilon)}^2)^r f(k_{(2m)}^\mu, k_{(-2\varepsilon)}^2) = \frac{\Gamma(r-\varepsilon)}{\Gamma(-\varepsilon)} \int \frac{d^{2m+2r-2\varepsilon}k}{i\pi^{m+r-\varepsilon}} f(k_{2m}^\mu, k_{-2\varepsilon}^2). \tag{2.66}$$

Here, $f(k_{(2m)}^\mu, k_{(-2\varepsilon)}^2)$ is a function which depends on k_{2m}, k_{2m+1}, ..., k_{D-1} only through $k_{(-2\varepsilon)}^2$. The dependency on k_0, k_1, ..., k_{2m-1} is not constrained.
Finally it is worth noting that

$$\int \frac{d^{2m-2\varepsilon}k}{i\pi^{m-\varepsilon}} \left(-k^2\right)^a = \begin{cases} (-1)^a \, \Gamma(a+1), & \text{if } m + a - \varepsilon = 0, \\ 0, & \text{otherwise.} \end{cases} \tag{2.67}$$

The non-zero value for $m + a - \varepsilon = 0$ can be verified by expanding eq. (2.17) into a power series in α.

3 Tensor integrals

In the previous section we considered scalar integrals and integrals where the functional dependence on the loop momentum of the numerator of the integrand is particular simple, like for example through factors k^2 or $k^2_{(-2\varepsilon)}$. If we recall the example discussed in eqs. (2.8) and (2.9) we see that more general tensor structures occur. In this section I discuss systematic algorithms for the reduction of tensor integrals to scalar integrals. In sect. 3.1 I discuss the Passarino-Veltman reduction technique, which historically was the first systematic procedure and can be applied to one-loop integrals. In sect. 3.2 I present a more general reduction method, which applies to arbitrary l-loop integrals.

3.1 Passarino - Veltman reduction. For one-loop integrals a systematic algorithm has been first worked out by Passarino and Veltman [36]. Let us first introduce a notation for one-loop tensor integrals. For integrals with one, two or three external legs we write

$$\begin{aligned} A_0(m) &= e^{\varepsilon\gamma_E}\mu^{2\varepsilon}\int\frac{d^Dk}{i\pi^{D/2}}\frac{1}{(-k^2+m^2)}, \qquad (3.1)\\ B_{0,\mu,\mu\nu}(p,m_1,m_2) &= e^{\varepsilon\gamma_E}\mu^{2\varepsilon}\int\frac{d^Dk}{i\pi^{D/2}}\frac{1,k_\mu,k_\mu k_\nu}{(-k^2+m_1^2)(-(k-p)^2+m_2^2)},\\ C_{0,\mu,\mu\nu}(p_1,p_2,m_1,m_2,m_3) &= \\ e^{\varepsilon\gamma_E}\mu^{2\varepsilon}\int\frac{d^Dk}{i\pi^{D/2}}&\frac{1,k_\mu,k_\mu k_\nu}{(-k^2+m_1^2)(-(k-p_1)^2+m_2^2)(-(k-p_1-p_2)^2+m_3^2)}, \end{aligned}$$

with an obvious generalisation towards more external legs and higher rank tensor integrals. The notation implies that the numerator for B_0 equals 1, for B_μ the numerator equals k_μ and for $B_{\mu\nu}$ the numerator equals $k_\mu k_\nu$. The reduction technique according to Passarino and Veltman uses the fact that due to Lorentz symmetry the result can only depend on tensor structures which can be build from the external momenta p_j^μ and the metric tensor $g^{\mu\nu}$. We therefore write the tensor integrals in the most general form in terms of form factors times external momenta and/or the metric tensor. For example

$$\begin{aligned} B^\mu &= p^\mu B_1,\\ B^{\mu\nu} &= p^\mu p^\nu B_{21} + g^{\mu\nu}B_{22},\\ \\ C^\mu &= p_1^\mu C_{11} + p_2^\mu C_{12},\\ C^{\mu\nu} &= p_1^\mu p_1^\nu C_{21} + p_2^\mu p_2^\nu C_{22} + (p_1^\mu p_2^\nu + p_1^\nu p_2^\mu)\,C_{23} + g^{\mu\nu}C_{24}. \qquad (3.2) \end{aligned}$$

One then solves for the form factors B_1, B_{21}, B_{22}, C_{11}, etc. by first contracting both sides with the external momenta and the metric tensor $g^{\mu\nu}$. On the left-hand side the resulting scalar products between the loop momentum k^μ and the external momenta are rewritten in terms of inverse propagators, as for example

$$2p\cdot k = \left[-(k-p)^2+m_2^2\right] - \left[-k^2+m_1^2\right] + \left(p^2+m_1^2-m_2^2\right). \qquad (3.3)$$

The first two terms of the right-hand side above cancel propagators, whereas the last term does not involve the loop momentum anymore. The remaining step is to solve for the form-factors by inverting the matrix which one obtains on the right-hand side of equation (3.2).

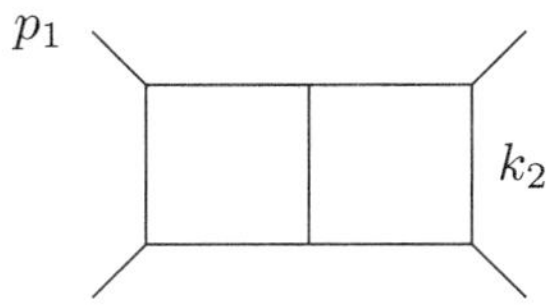

Figure 3 An example for an irreducible scalar product in the numerator: The scalar product $2p_1k_2$ cannot be expressed in terms of inverse propagators.

As an example we consider the two-point function: Contraction with p_μ or $p_\mu p_\nu$ and $g_{\mu\nu}$ yields

$$p^2 B_1 = -\frac{1}{2}\left(\left(m_2^2 - m_1^2 - p^2\right) B_0 - A_0(m_1) + A_0(m_2)\right),$$

$$\begin{pmatrix} p^2 & 1 \\ p^2 & D \end{pmatrix} \begin{pmatrix} B_{21} \\ B_{22} \end{pmatrix} = \begin{pmatrix} -\frac{1}{2}(m_2^2 - m_1^2 - p^2)B_1 - \frac{1}{2}A_0(m_2) \\ m_1^2 B_0 - A_0(m_2) \end{pmatrix}. \tag{3.4}$$

Solving for the form factors we obtain

$$\begin{aligned} B_1 &= -\frac{1}{2p^2}\left(\left(m_2^2 - m_1^2 - p^2\right) B_0 - A_0(m_1) + A_0(m_2)\right), \\ B_{21} &= \frac{1}{(D-1)p^2}\left(-\frac{D}{2}(m_2^2 - m_1^2 - p^2)B_1 - m_1^2 B_0 - \frac{D-2}{2}A_0(m_2)\right), \\ B_{22} &= \frac{1}{2(D-1)}\left((m_2^2 - m_1^2 - p^2)B_1 + 2m_1^2 B_0 - A_0(m_2)\right). \end{aligned} \tag{3.5}$$

Due to the matrix inversion in the last step determinants usually appear in the denominator of the final expression. For a three-point function we would encounter the Gram determinant

$$\Delta_3 = 4 \begin{vmatrix} p_1^2 & p_1 \cdot p_2 \\ p_1 \cdot p_2 & p_2^2 \end{vmatrix}. \tag{3.6}$$

One drawback of this algorithm is closely related to these determinants : In a phase space region where p_1 becomes collinear to p_2, the Gram determinant will tend to zero, and the form factors will take large values, with possible large cancellations among them. This makes it difficult to set up a stable numerical program for automated evaluation of tensor loop integrals.

The Passarino-Veltman algorithm is based on the observation, that for one-loop integrals a scalar product of the loop momentum with an external momentum can be expressed as a combination of inverse propagators. This property does no longer hold if one goes to two or more loops. Fig. (3) shows a two-loop diagram, for which the scalar product of a loop momentum with an external momentum cannot be expressed in terms of inverse propagators.

3.2 General reduction method. Let us now consider a general multi-loop tensor integral and assume that we follow the steps in sect. 2.4 as we did for scalar integrals. After the change of variables for the diagonalisation of the quadratic form, we have a polynomial in the Feynman or Schwinger parameters and the loop momentum k in the numerator. Integrals with an odd power of the loop momentum in the numerator vanish by symmetry, while integrals with an even power of the loop momentum can be related by Lorentz invariance to scalar integrals:

$$\begin{aligned}\int \frac{d^D k}{i\pi^{D/2}} k^\mu k^\nu f(k^2) &= -\frac{1}{D} g^{\mu\nu} \int \frac{d^D k}{i\pi^{D/2}} (-k^2) f(k^2), \quad (3.7)\\ \int \frac{d^D k}{i\pi^{D/2}} k^\mu k^\nu k^\rho k^\sigma f(k^2) &= \frac{1}{D(D+2)} \left(g^{\mu\nu} g^{\rho\sigma} + g^{\mu\rho} g^{\nu\sigma} + g^{\mu\sigma} g^{\nu\rho} \right) \\ & \quad \int \frac{d^D k}{i\pi^{D/2}} (-k^2)^2 f(k^2).\end{aligned}$$

The generalisation to arbitrary higher tensor structures is obvious. In the master formula eq. (2.63) we already observed that a factor $(-k^2)$ in the numerator is equivalent (apart from prefactors) to a shift in the dimension $D \to D+2$. Let us introduce an operator $\mathbf{D}^+$, which shifts the dimension:

$$\mathbf{D}^+ \int \frac{d^D k}{i\pi^{D/2}} f\left(k^2\right) = \int \frac{d^{(D+2)} k}{i\pi^{(D+2)/2}} f\left(k^2\right). \quad (3.8)$$

Then we have for example

$$\int \frac{d^D k}{i\pi^{D/2}} k^\mu k^\nu f(k^2) = -\frac{1}{2} g^{\mu\nu} \, \mathbf{D}^+ \int \frac{d^D k}{i\pi^{D/2}} f(k^2). \quad (3.9)$$

In addition, shifting the loop momentum like in $k' = k - xp$ introduces for tensor integrals the (Feynman or Schwinger) parameters x_j in the numerator. For the tensor reduction it is convenient to work temporarily with Schwinger parameters. Let us recall the formula (2.29) for Schwinger parameters:

$$\frac{1}{(-P)^\nu} = \frac{1}{\Gamma(\nu)} \int\limits_0^\infty dx \, x^{\nu-1} \exp(xP). \quad (3.10)$$

A Schwinger parameter x in the numerator is equivalent to raising the power of the original propagator by one unit: $\nu \to \nu+1$. It is convenient to denote by $\mathbf{i}^+$ the operator, which raises the power of propagator i by one.

$$\nu_i \mathbf{i}^+ \frac{1}{(-P_i)^{\nu_i}} = \nu_i \frac{1}{(-P_i)^{\nu_i+1}} = \frac{1}{\Gamma(\nu_i)} \int\limits_0^\infty dx_i \, x_i^{\nu_i-1} x_i \exp(x_i P_i). \quad (3.11)$$

Therefore we can consider an integral, where a Schwinger parameter occurs in the numerator as a scalar integral, where the corresponding propagator is raised to a higher power. As a consequence, by using an intermediate Schwinger parametrisation, we can express all tensor integrals in terms of scalar integrals [37, 38]. The price we have to pay is that these scalar integrals involve higher powers of the propagators and/or have shifted dimensions. Each integral can be specified by its topology, its value for the dimension D and a set of indices, denoting the powers of the propagators. In general the number of different integrals is quite large.

4 Scalar multi-loop integrals

In the previous section we saw that all tensor integrals can be reduced to scalar integrals, where the propagators are raised to some power and the dimension is shifted. In sect. 2.4 we discussed these scalar integrals and gave in eq. (2.63) a master formula which is applicable, once we have completed the square and shifted the loop momentum, such that the integrand depends only on k^2. This raises the question if the result of completing the square and shifting the loop momentum, which has to be done iteratively for each loop, can be read off directly from the underlying Feynman graph. This is indeed the case and in this section we will give the appropriate formulae.

4.1 Graph polynomials. We consider a scalar l-loop integral I_G in $D = 2m - 2\varepsilon$ dimensions with n propagators, corresponding to the graph G:

$$I_G = \left(e^{\varepsilon\gamma_E}\mu^{2\varepsilon}\right)^l \int \prod_{r=1}^{l} \frac{d^D k_r}{i\pi^{\frac{D}{2}}} \prod_{j=1}^{n} \frac{1}{(-q_j^2 + m_j^2)^{\nu_j}}, \tag{4.1}$$

The momenta q_j of the propagators are linear combinations of the external and the loop momenta. The propagators can we raised to a power ν_j. Introducing Feynman parameters and performing the momentum integrations as in sect. 2.4 one arrives at the following Feynman parameter integral [39]:

$$I_G = \left(e^{\varepsilon\gamma_E}\mu^{2\varepsilon}\right)^l \frac{\Gamma(\nu - lD/2)}{\prod\limits_{j=1}^{n}\Gamma(\nu_j)} \int\limits_0^1 \left(\prod_{j=1}^{n} dx_j\, x_j^{\nu_j - 1} \right) \delta(1 - \sum_{i=1}^{n} x_i) \frac{\mathcal{U}^{\nu-(l+1)D/2}}{\mathcal{F}^{\nu-lD/2}}. \tag{4.2}$$

Here, $\nu = \sum_{j=1}^{n} \nu_j$. The functions $\mathcal{U}$ and $\mathcal{F}$ can be straightforwardly derived from the topology of the corresponding Feynman graph G. Cutting l lines of a given connected l-loop graph such that it becomes a connected tree graph T defines a chord $\mathcal{C}(T, G)$ as being the set of lines not belonging to this tree. The Feynman parameters associated with each chord define a monomial of degree l. The set of all such trees (or 1-trees) is denoted by $\mathcal{T}_1$. The 1-trees $T \in \mathcal{T}_1$ define $\mathcal{U}$ as being the sum over all monomials corresponding to the chords $\mathcal{C}(T, G)$. Cutting one more line of a 1-tree leads to two disconnected trees (T_1, T_2), or a 2-tree. $\mathcal{T}_2$ is the set of all such pairs. The corresponding chords define monomials of degree $l+1$. Each 2-tree of a graph corresponds to a cut defined by cutting the lines which connected the 2 now disconnected trees in the original graph. The square of the sum of momenta through the cut lines of one of the two disconnected trees T_1 or T_2 defines a Lorentz invariant

$$s_T = \left(\sum_{j \in \mathcal{C}(T,G)} p_j \right)^2. \tag{4.3}$$

The function $\mathcal{F}_0$ is the sum over all such monomials times minus the corresponding invariant. The function $\mathcal{F}$ is then given by $\mathcal{F}_0$ plus an additional piece involving the internal masses m_j. In summary, the functions $\mathcal{U}$ and $\mathcal{F}$ are obtained from the graph as follows:

$$\begin{aligned}
\mathcal{U} &= \sum_{T \in \mathcal{T}_1} \Big[\prod_{j \in \mathcal{C}(T,G)} x_j \Big], \\
\mathcal{F}_0 &= \sum_{(T_1,T_2) \in \mathcal{T}_2} \Big[\prod_{j \in \mathcal{C}(T_1,G)} x_j \Big] (-s_{T_1}), \\
\mathcal{F} &= \mathcal{F}_0 + \mathcal{U} \sum_{j=1}^{n} x_j m_j^2 .
\end{aligned} \tag{4.4}$$

Let us consider as an example the following two-loop graph:

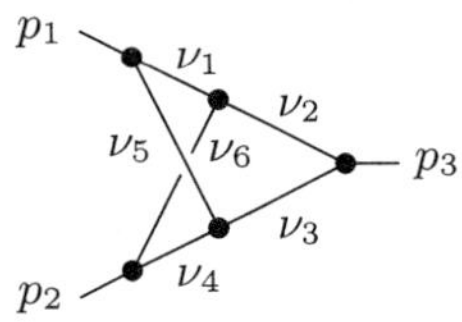

$$= \left(e^{\varepsilon\gamma_E}\mu^{2\varepsilon}\right)^2 \int \frac{d^Dk_1}{i\pi^{D/2}} \int \frac{d^Dk_2}{i\pi^{D/2}} \times \frac{1}{(-k_1^2)^{\nu_1}(-k_2^2)^{\nu_2}(-k_3^2)^{\nu_3}(-k_4^2)^{\nu_4}(-k_5^2)^{\nu_5}(-k_6^2)^{\nu_6}}. \tag{4.5}$$

For simplicity we assume that all internal propagators are massless. Then the functions $\mathcal{U}$ and $\mathcal{F}$ read:

$$\begin{aligned} \mathcal{U} &= x_{15}x_{23} + x_{15}x_{46} + x_{23}x_{46}, \\ \mathcal{F} &= \left(x_1x_3x_4 + x_5x_2x_6 + x_1x_5x_{2346}\right)\left(-p_1^2\right) \\ &\quad + \left(x_6x_3x_5 + x_4x_1x_2 + x_4x_6x_{1235}\right)\left(-p_2^2\right) \\ &\quad + \left(x_2x_4x_5 + x_3x_1x_6 + x_2x_3x_{1456}\right)\left(-p_3^2\right). \end{aligned} \tag{4.6}$$

Here we used the notation that $x_{ij...} = x_i + x_j + ...$ In general, $\mathcal{U}$ is a positive semi-definite function. Its vanishing is related to the UV sub-divergences of the graph. Overall UV divergences, if present, will always be contained in the prefactor $\Gamma(\nu - lD/2)$. In the Euclidean region, $\mathcal{F}$ is also a positive semi-definite function of the Feynman parameters x_j.

4.2 Feynman integrals and differential forms. Note that $\mathcal{U}$, $\mathcal{F}_0$ and $\mathcal{F}$ are homogeneous functions in the Feynman parameters x_j with degrees

$$\deg \mathcal{U} = l, \qquad \deg \mathcal{F} = \deg \mathcal{F}_0 = l + 1. \tag{4.7}$$

The function

$$f(x_1, ..., x_n) = \left(\prod_{j=1}^{n} x_j^{\nu_j - 1}\right) \frac{\mathcal{U}^{\nu-(l+1)D/2}}{\mathcal{F}^{\nu - lD/2}} \tag{4.8}$$

is then a homogeneous function of degree

$$\deg f = -n. \tag{4.9}$$

It follows that

$$\left(n + \sum_{j=1}^{n} x_j \frac{\partial}{\partial x_j}\right) f = 0. \tag{4.10}$$

If we now define the differential form

$$\omega = \sum_{j=1}^{n} (-1)^{j-1}\, x_j\, dx_1 \wedge ... \wedge \hat{dx_j} \wedge ... \wedge dx_n, \tag{4.11}$$

where the hat indicates that the corresponding term is omitted, then $f\omega$ is closed:

$$d\,(f\omega) = 0. \tag{4.12}$$

If $f\omega$ vanishes on any hyper-surface $x_j = 0$, then we can write the Feynman integral as [40]

$$I_G = \left(e^{\varepsilon\gamma_E}\mu^{2\varepsilon}\right)^l \frac{\Gamma(\nu - lD/2)}{\prod\limits_{j=1}^{n}\Gamma(\nu_j)} \int\limits_M f\omega, \tag{4.13}$$

where M is any hyper-surface covering the solid angle $x_j > 0$.

5 One-loop integrals

The simplest, but most important loop integrals are the one-loop integrals. We have a good understanding of these integrals and I will present the main results in this section. For one-loop tensor integrals we can use the Passarino-Veltman method, which reduces any tensor integral to a combination of scalar integrals. Note that this method does not shift the dimension, nor does it raise the powers of the propagators. An important result is that any scalar integral with more than four external legs can be reduced to scalar integrals with no more than 4 external legs. Therefore the set of basic one-loop integrals is rather limited. I will discuss this reduction in sect. 5.1. A second important result is that all one-loop integrals can be expressed through the logarithm and the dilogarithm. No other transcendental functions occur. I will discuss the appearance of the dilogarithm in sect. 5.2 and 5.3. Although we discussed reduction methods for tensor integrals already in sect. 3, the methods presented there are rather general and need not be the most efficient ones. Any tensor integral $I_{\mu\nu...}$ is usually contracted into a tensor $J^{\mu\nu...}$ independent of the loop integration. If we have additional information on $J^{\mu\nu...}$ more efficient algorithms for the tensor reduction of $I_{\mu\nu...}$ can be derived. An example for one-loop integrals is discussed in sect. 5.4.

5.1 Reduction to integrals with no more than four external legs. In this section we discuss the reduction of scalar integrals with more than four external legs to a basic set of scalar one-, two-, three- and four-point functions. It is a long known fact, that higher point scalar integrals can be expressed in terms of this basic set [41,42], however the practical implementation within dimensional regularisation was only worked out recently [43–47]. The one-loop n-point functions with $n \geq 5$ are always UV-finite, but they may have IR-divergences. Let us first assume that there are no IR-divergences. Then the integral is finite and can be performed in four dimensions. In a space of four dimensions we can have no more than four linearly independent vectors, therefore it comes to no surprise that in an one-loop integral with five or more propagators, one propagator can be expressed through the remaining ones. This is the basic idea for the reduction of the higher point scalar integrals. With slight modifications it can be generalised to dimensional regularisation. I will discuss the method for massless one-loop integrals

$$I_n = e^{\varepsilon \gamma_E} \mu^{2\varepsilon} \int \frac{d^D k}{i\pi^{\frac{D}{2}}} \frac{1}{(-k^2)(-(k-p_1)^2)...(-(k-p_1-...p_{n-1})^2)}. \quad (5.1)$$

With the notation

$$q_i = \sum_{j=1}^{i} p_j \quad (5.2)$$

one can associate two matrices S and G to the integral in eq. (5.1). The entries of the $n \times n$ kinematical matrix S are given by

$$S_{ij} = (q_i - q_j)^2, \quad (5.3)$$

and the entries of the $(n-1) \times (n-1)$ Gram matrix are defined by

$$G_{ij} = 2 q_i q_j. \quad (5.4)$$

For the reduction one distinguishes three different cases: Scalar pentagons (i.e. scalar five-point functions), scalar hexagons (scalar six-point functions) and scalar integrals with more than six propagators.

Let us start with the pentagon. A five-point function in $D = 4-2\varepsilon$ dimensions can be expressed as a sum of four-point functions, where one propagator is removed, plus a five-point function in $6-2\varepsilon$ dimensions [43]. Since the $(6-2\varepsilon)$-dimensional pentagon is finite and comes with an extra factor of ε in front, it does not contribute at $\mathcal{O}(\varepsilon^0)$. In detail we have

$$I_5 = -2\varepsilon B I_5^{6-2\varepsilon} - \sum_{i=1}^{5} b_i I_4^{(i)} = -\sum_{i=1}^{5} b_i I_4^{(i)} + \mathcal{O}(\varepsilon), \tag{5.5}$$

where $I_5^{6-2\varepsilon}$ denotes the $(6-2\varepsilon)$-dimensional pentagon and $I_4^{(i)}$ denotes the four-point function, which is obtained from the pentagon by removing propagator i. The coefficients B and b_i are obtained from the kinematical matrix S_{ij} as follows:

$$b_i = \sum_j \left(S^{-1}\right)_{ij}, \qquad B = \sum_i b_i. \tag{5.6}$$

The six-point function can be expressed as a sum of five-point functions [44] without any correction of $\mathcal{O}(\varepsilon)$

$$I_6 = -\sum_{i=1}^{6} b_i I_5^{(i)}. \tag{5.7}$$

The coefficients b_i are again related to the kinematical matrix S_{ij}:

$$b_i = \sum_j \left(S^{-1}\right)_{ij}. \tag{5.8}$$

For the seven-point function and beyond we can again express the n-point function as a sum over $(n-1)$-point functions [47]:

$$I_n = -\sum_{i=1}^{n} r_i I_{n-1}^{(i)}. \tag{5.9}$$

In contrast to eq. (5.7), the decomposition in eq. (5.9) is no longer unique. A possible set of coefficients r_i can be obtained from the singular value decomposition of the Gram matrix

$$G_{ij} = \sum_{k=1}^{4} U_{ik} w_k \left(V^T\right)_{kj}. \tag{5.10}$$

as follows [48]

$$r_i = \frac{V_{i5}}{W_5}, \quad 1 \le i \le n-1, \qquad r_n = -\sum_{j=1}^{n-1} r_j, \qquad W_5 = \frac{1}{2}\sum_{j=1}^{n-1} G_{jj} V_{j5}. \tag{5.11}$$

5.2 The three-point function. As an example for the appearance of the dilogarithm let us discuss the one-loop three-point function with no internal masses and the kinematical configuration $p_1^2 \neq 0$, $p_2^2 \neq 0$ and $p_3^2 = (p_1+p_2)^2 \neq 0$.

$$C_0 = e^{\varepsilon\gamma_E}\mu^{2\varepsilon} \int \frac{d^D k}{i\pi^{D/2}} \frac{1}{(-k^2)(-(k-p_1)^2)(-(k-p_1-p_2)^2)}. \tag{5.12}$$

The integral is finite and can be evaluated in four dimensions.

$$C_0 = \int_0^1 dx \int_0^x dy \frac{1}{-x^2p_3^2 - y^2p_2^2 + xy(p_1^2 - p_2^2 - p_3^2) - xp_3^2 + y(p_3^2 - p_1^2)} + \mathcal{O}(\varepsilon). \tag{5.13}$$

I follow here closely the original work of 't Hooft and Veltman [49]. We make the change of variables $y' = y - \alpha x$ and choose α as a root of the equation

$$-\alpha^2 p_2^2 + \alpha\left(p_1^2 - p_2^2 - p_3^2\right) - p_3^2 = 0. \tag{5.14}$$

With this choice we eliminate the quadratic term in x. We then perform the x-integration and we end up with three integrals of the form

$$\int_0^1 \frac{dy}{y - y_0}\left[\ln\left(ay^2 + by + c\right) - \ln\left(ay_0^2 + by_0 + c\right)\right] \tag{5.15}$$

Factorising the arguments of the logarithms, these integrals are reduced to the type

$$R = \int_0^1 \frac{dy}{y - y_0}\left[\ln(y - y_1) - \ln(y_0 - y_1)\right]. \tag{5.16}$$

This integral is expressed in terms of a new function, the dilogarithm, as follows:

$$R = \mathrm{Li}_2\left(\frac{y_0}{y_1 - y_0}\right) - \mathrm{Li}_2\left(\frac{y_0 - 1}{y_1 - y_0}\right), \tag{5.17}$$

provided $-y_1$ and $1/(y_0 - y_1)$ have imaginary part of opposite sign, otherwise additional logarithms occur.

5.3 The dilogarithm. The dilogarithm is defined by

$$\mathrm{Li}_2(x) = -\int_0^1 dt \frac{\ln(1 - xt)}{t} = -\int_0^x dt \frac{\ln(1 - t)}{t}. \tag{5.18}$$

If we take the main branch of the logarithm with a cut along the negative real axis, then the dilogarithm has a cut along the positive real axis, starting at the point $x = 1$. For $|x| \leq 1$ the dilogarithm has the power series expansion

$$\mathrm{Li}_2(x) = \sum_{n=1}^{\infty} \frac{x^n}{n^2}. \tag{5.19}$$

Some important numerical values are

$$\mathrm{Li}_2(0) = 0, \quad \mathrm{Li}_2(1) = \frac{\pi^2}{6}, \quad \mathrm{Li}_2(-1) = -\frac{\pi^2}{12}, \quad \mathrm{Li}_2\left(\frac{1}{2}\right) = \frac{\pi^2}{12} - \frac{1}{2}(\ln 2)^2.$$

The dilogarithm with argument x can be related to the dilogarithms with argument $(1 - x)$ or $1/x$:

$$\begin{aligned} \mathrm{Li}_2(x) &= -\mathrm{Li}_2(1 - x) + \frac{1}{6}\pi^2 - \ln(x)\ln(1 - x), \\ \mathrm{Li}_2(x) &= -\mathrm{Li}_2\left(\frac{1}{x}\right) - \frac{1}{6}\pi^2 - \frac{1}{2}\left(\ln(-x)\right)^2. \end{aligned} \tag{5.20}$$

Another important relation is the five-term relation:

$$\mathrm{Li}_2(xy) = \mathrm{Li}_2(x) + \mathrm{Li}_2(y) + \mathrm{Li}_2\left(\frac{xy-x}{1-x}\right) + \mathrm{Li}_2\left(\frac{xy-y}{1-y}\right) + \frac{1}{2}\ln^2\left(\frac{1-x}{1-y}\right). \tag{5.21}$$

5.4 Spinor techniques. The reduction methods for tensor integrals discussed in sect. 3 are rather general and independent of the tensor structure into which the tensor integral is contracted. By taking into account information from this external tensor structure, more efficient reduction algorithms can be derived [50–55]. I will discuss as an example a method for one-loop integrals with massless propagators. The method is most conveniently explained within the FD-scheme of dimensional regularisation. In this scheme we can assume without loss of generality that the tensor structure $J_{\mu_1...\mu_r}$ is given by

$$J_{\mu_1...\mu_r} = \langle a_1 - |\gamma_{\mu_1}| b_1 - \rangle \dots \langle a_r - |\gamma_{\mu_r}| b_r - \rangle, \tag{5.22}$$

where $\langle a_i - |$ and $|b_j - \rangle$ are Weyl spinors of definite helicity. Therefore we consider tensor integrals of the form

$$I_n^r = e^{\varepsilon\gamma_E}\mu^{2\varepsilon} \langle a_1 - |\gamma_{\mu_1}| b_1 - \rangle \dots \langle a_r - |\gamma_{\mu_r}| b_r - \rangle \int \frac{d^D k}{i\pi^{\frac{D}{2}}} \frac{k_{(4)}^{\mu_1} \dots k_{(4)}^{\mu_r}}{k^2(k-p_1)^2 \dots (k-p_1-\dots p_{n-1})^2}, \tag{5.23}$$

where $k_{(4)}^\mu$ denotes the projection of the D dimensional vector k^μ onto the four-dimensional subspace. The quantity $\langle a - |\gamma_\mu| b - \rangle$ is a vector in a complex vector-space of dimension 4 and can therefore be expressed as a linear combination of four basis vectors.

The first step for the construction of the reduction algorithm based on spinor methods is to associate to each n-point loop integral a pair of two light-like momenta l_1 and l_2, which are linear combinations of two external momenta p_i and p_j of the loop integral under consideration [53]. Obviously, this construction only makes sense for three-point integrals and beyond, as for two-point integrals there is only one independent external momentum. We write

$$l_1 = \frac{1}{1-\alpha_1\alpha_2}(p_i - \alpha_1 p_j), \qquad l_2 = \frac{1}{1-\alpha_1\alpha_2}(-\alpha_2 p_i + p_j), \tag{5.24}$$

where α_1 and α_2 are two constants, which can be determined from p_i and p_j.

In the second step we use l_1 and l_2 to write $\langle a - |\gamma_\mu| b - \rangle$ as a linear combination of the basis vectors

$$\langle l_1 - |\gamma_\mu| l_1 - \rangle, \quad \langle l_2 - |\gamma_\mu| l_2 - \rangle, \quad \langle l_1 - |\gamma_\mu| l_2 - \rangle, \quad \langle l_2 - |\gamma_\mu| l_1 - \rangle. \tag{5.25}$$

The contraction of $k_{(4)}^\mu$ with the first or second basis vector leads to

$$2k_l l_1 = \frac{1}{1-\alpha_1\alpha_2}\left[2p_i k_l - \alpha_1 2p_j k_l\right], \qquad 2k_l l_2 = \frac{1}{1-\alpha_1\alpha_2}\left[-\alpha_2 2p_i k_l + 2p_j k_l\right], \tag{5.26}$$

and therefore reduces immediately the rank of the tensor integral. Repeating this procedure we end up with integrals, where the numerator is given by products of

$$\langle l_1 - |\not{k}_{(4)}| l_2 - \rangle \quad \text{and} \quad \langle l_2 - |\not{k}_{(4)}| l_1 - \rangle, \tag{5.27}$$

plus additional reduced integrals. Therefore the tensor integral is now in a standard form. In the next step one shows that any product of factors as in eq. (5.27) can

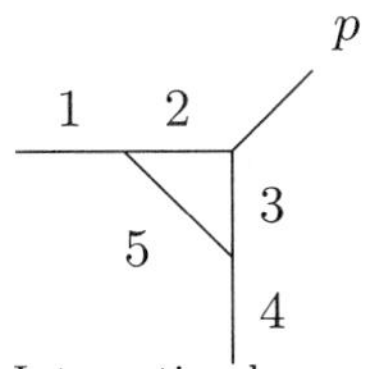

Figure 4 The triangle trick: Integration by parts eliminates the propagators 1, 2, 3 or 4.

be reduced. This reduces all tensor integrals to rank 1 integrals. Finally, the rank 1 integrals are reduced to scalar integrals.

6 Advanced methods

In sects. 2 to 4 we discussed general l-loop integrals and showed that all tensor integrals can be reduced to scalar integrals and that the integration over the l independent loop momenta can be performed at the expense of introducing additional parameter integrals. The integrand for the Feynman parameter integral can be read off directly from the underlying Feynman graph. In this section I will present methods to compute these Feynman parameter integrals. As the tensor reduction usually introduces a huge number scalar integrals, which differ by the powers of the propagators or the by the dimension, I start in sect. 6.1 and 6.2 with techniques, which reduce the unknown integrals to a smaller set, called "master integrals". In sects. 6.3 to 6.5 I will present techniques for the analytic calculation of Feynman parameter integrals. Finally I review in sect. 6.6 a method for the numerical computation of the coefficients of the Laurent expansion.

6.1 Integration by parts. Integration-by-part identities [56] are based on the fact, that the integral of a total derivative is zero:

$$\int \frac{d^D k}{i\pi^{D/2}} \frac{\partial}{\partial k^\mu} v^\mu f(k, p_i) = 0. \tag{6.1}$$

Here, k is the loop momentum, the p_i's are external momenta with respect to this loop integration and v can either be a loop momentum or an external momentum. Working out the derivative yields a relation among several scalar integrals.

As an example we look at the triangle trick [57]: Assume that we have inside a multi-loop integral the building block shown in fig. 4 with massless propagators and $p^2 = 0$:

$$\begin{aligned} T(\nu_1, \nu_2, \nu_3, \nu_4, \nu_5) = & \frac{1}{(-k_1^2)^{\nu_1} \left(-(k_1-p)^2\right)^{\nu_4}} \\ & \times \int \frac{d^D k_2}{i\pi^{D/2}} \frac{1}{(-k_2^2)^{\nu_2} \left(-(k_2-p)^2\right)^{\nu_3} \left(-(k_2-k_1)^2\right)^{\nu_5}}. \end{aligned} \tag{6.2}$$

Then by choosing $k = k_2$ and $v = k_2$ in eq. (6.1) we obtain

$$\left[(D - 2\nu_2 - \nu_3 - \nu_5) - \nu_3 \mathbf{3}^+ \mathbf{2}^- - \nu_5 \mathbf{5}^+ \left(\mathbf{2}^- - \mathbf{1}^-\right)\right] T(\nu_1, \nu_2, \nu_3, \nu_4, \nu_5) = 0. \tag{6.3}$$

The choice $k = k_2 - p$ and $v = k_2 - p$ yields a second relation:

$$\left[(D - \nu_2 - 2\nu_3 - \nu_5) - \nu_2 \mathbf{2}^+ \mathbf{3}^- - \nu_5 \mathbf{5}^+ \left(\mathbf{3}^- - \mathbf{4}^-\right)\right] T(\nu_1, \nu_2, \nu_3, \nu_4, \nu_5) = 0. \tag{6.4}$$

If we assume that ν_1 and ν_2 are positive integers we can eliminate by repeated use of eq. (6.3) either propagator 1 or 2. Note that if propagator 2 is eliminated, propagators 3 and 5 form a rather trivial one-loop insertion. Similar considerations apply to propagators 3 and 4 by using eq. (6.4).

6.2 Reduction to master integrals. In the example discussed above we could use the triangle trick to simplify the integrand. In general, integration-by-part identities lead to relations among various scalar integrals. Additional relations among scalar integrals are obtained from the invariance of scalar integrals under Lorentz transformations [58]. A scalar integral is evidently invariant under an infinitesimal Lorentz transformation, parametrised as

$$p^\mu \to p^\mu + \delta p^\mu = p^\mu + \delta\epsilon^\mu_\nu p^\nu \qquad \text{with} \qquad \delta\epsilon^\mu_\nu = -\delta\epsilon^\nu_\mu \,. \tag{6.5}$$

This implies that

$$\left(p_1^\nu \frac{\partial}{\partial p_{1\mu}} - p_1^\mu \frac{\partial}{\partial p_{1\nu}} + \ldots + p_n^\nu \frac{\partial}{\partial p_{n\mu}} - p_n^\mu \frac{\partial}{\partial p_{n\nu}} \right) I(p_1,\ldots,p_n) = 0 \,, \tag{6.6}$$

where $I(p_1,...,p_n)$ is a scalar integral with the external momenta p_i. Again, working out the derivatives yields a relation among several scalar integrals.
Each relation is linear in the scalar integrals and in principle one could use Gauss elimination to reduce the set of scalar integrals to a small set of master integrals. In practice this approach is inefficient. An efficient algorithm has been given by Laporta [59]. The starting point is to introduce an order relation for scalar integrals. This can be done in several ways, a possible choice is to order the topologies first: A scalar integral corresponding to a topology T_1 is considered to be "smaller" than an integral with topology T_2, if T_1 can be obtained from T_2 by pinching of some propagators. Within each topology, the scalar integrals can be ordered according to the powers of the propagators and the dimension of space-time. Laporta's algorithm is based on the fact that starting from a specific topology, integration-by-part and Lorentz-invariance relations only generate relations involving this topology and "smaller" ones. To avoid to substitute a specific identity into a large number of other identities, one starts from the "smallest" topology and generates all relevant relations for this topology. Inside this class, integrals with higher powers of the propagators or higher dimensions are then expressed in terms of a few master integrals. These manipulations involve only a small subset of the complete system of integrals and relations and can therefore be done efficiently. Once this topology is completed, one moves on to the next topology, until all topologies have been considered. All integrals, which cannot be eliminated by this procedure are called master integrals. These master integrals have to be evaluated explicitly.

6.3 Mellin-Barnes representation. In sect. 4 we saw that the Feynman parameter integrals depend on two graph polynomials $\mathcal{U}$ and $\mathcal{F}$, which are homogeneous functions of the Feynman parameters. In this section we will continue the discussion how these integrals can be performed and exchanged against a (multiple) sum over residues. The case, where the two polynomials are absent is particular simple:

$$\int\limits_0^1 \left(\prod_{j=1}^n dx_j \, x_j^{\nu_j - 1} \right) \delta(1 - \sum_{i=1}^n x_i) \quad = \quad \frac{\prod\limits_{j=1}^n \Gamma(\nu_j)}{\Gamma(\nu_1 + ... + \nu_n)}. \tag{6.7}$$

With the help of the Mellin-Barnes transformation we now reduce the general case to eq. (6.7). The Mellin-Barnes transformation reads

$$\left(A_1 + A_2 + ... + A_n\right)^{-c} = \frac{1}{\Gamma(c)} \frac{1}{\left(2\pi i\right)^{n-1}} \int\limits_{-i\infty}^{i\infty} d\sigma_1 ... \int\limits_{-i\infty}^{i\infty} d\sigma_{n-1} \tag{6.8}$$
$$\times \Gamma(-\sigma_1)...\Gamma(-\sigma_{n-1})\Gamma(\sigma_1 + ... + \sigma_{n-1} + c)\ A_1^{\sigma_1}...A_{n-1}^{\sigma_{n-1}} A_n^{-\sigma_1 - ... - \sigma_{n-1} - c}$$

Each contour is such that the poles of $\Gamma(-\sigma)$ are to the right and the poles of $\Gamma(\sigma+c)$ are to the left. This transformation can be used to convert the sum of monomials of the polynomials $\mathcal{U}$ and $\mathcal{F}$ into a product, such that all Feynman parameter integrals are of the form of eq. (6.7). Therefore we exchange the Feynman parameter integrals against multiple complex contour integrals. As this transformation converts sums into products it is the "inverse" of Feynman parametrisation. The contour integrals are then performed by closing the contour at infinity and summing up all residues which lie inside the contour. Here it is useful to know the residues of the Gamma function:

$$\text{res}\ \left(\Gamma(\sigma + a), \sigma = -a - n\right) = \frac{(-1)^n}{n!}, \qquad \text{res}\ \left(\Gamma(-\sigma + a), \sigma = a + n\right) = -\frac{(-1)^n}{n!}. \tag{6.9}$$

Therefore we obtain (multiple) sum over residues. Techniques to manipulate these sums are discussed in the next section. In particular simple cases the contour integrals can be performed in closed form with the help of two lemmas of Barnes. Barnes first lemma states that

$$\frac{1}{2\pi i} \int\limits_{-i\infty}^{i\infty} d\sigma \Gamma(a+\sigma)\Gamma(b+\sigma)\Gamma(c-\sigma)\Gamma(d-\sigma) = \frac{\Gamma(a+c)\Gamma(a+d)\Gamma(b+c)\Gamma(b+d)}{\Gamma(a+b+c+d)}, \tag{6.10}$$

if none of the poles of $\Gamma(a+\sigma)\Gamma(b+\sigma)$ coincides with the ones from $\Gamma(c-\sigma)\Gamma(d-\sigma)$. Barnes second lemma reads

$$\frac{1}{2\pi i} \int\limits_{-i\infty}^{i\infty} d\sigma \frac{\Gamma(a+\sigma)\Gamma(b+\sigma)\Gamma(c+\sigma)\Gamma(d-\sigma)\Gamma(e-\sigma)}{\Gamma(a+b+c+d+e+\sigma)}$$
$$= \frac{\Gamma(a+d)\Gamma(b+d)\Gamma(c+d)\Gamma(a+e)\Gamma(b+e)\Gamma(c+e)}{\Gamma(a+b+d+e)\Gamma(a+c+d+e)\Gamma(b+c+d+e)}. \tag{6.11}$$

6.4 Expansion in a small parameter for specific classes of transcendental functions. From the Mellin-Barnes representation we obtain multiple sums involving Gamma functions. The small parameter ε appears in the arguments of the Gamma functions. We are interested in the Laurent expansion in ε. In this section we discuss techniques to manipulate multiple sums. As an example we consider again the scalar integral already encountered in eq. (2.62), but this time in dimension $D = 2m - 2\varepsilon$ and each propagator P_j is raised to the power ν_j. A short calculation leads to

$$\int \frac{d^{2m-2\varepsilon}k_1}{i\pi^{m-\varepsilon}} \frac{1}{(-k_1^2)^{\nu_1}} \frac{1}{(-k_2^2)^{\nu_2}} \frac{1}{(-k_3^2)^{\nu_3}} = (-s_{123})^{m-\varepsilon-\nu_{123}} \frac{1}{\Gamma(\nu_1)\Gamma(\nu_2)} \tag{6.12}$$
$$\times \frac{\Gamma(m-\varepsilon-\nu_1)\Gamma(m-\varepsilon-\nu_{23})}{\Gamma(2m-2\varepsilon-\nu_{123})} \sum_{n=0}^{\infty} \frac{\Gamma(n+\nu_2)\Gamma(n-m+\varepsilon+\nu_{123})}{\Gamma(n+1)\Gamma(n+\nu_{23})} (1-x)^n,$$

where $x = s_{12}/s_{123}$, $\nu_{23} = \nu_2 + \nu_3$ and $\nu_{123} = \nu_1 + \nu_2 + \nu_3$. The infinite sum in the last line of (6.12) is a hyper-geometric function, where the small parameter ε occurs in the Gamma functions. More complicated loop integrals yield additional classes of infinite sums. One often encounters the following types of infinite sums:
Type A:

$$\sum_{i=0}^{\infty} \frac{\Gamma(i+a_1)}{\Gamma(i+a_1')} \cdots \frac{\Gamma(i+a_k)}{\Gamma(i+a_k')} x^i \tag{6.13}$$

Up to prefactors the hyper-geometric functions ${}_{J+1}F_J$ fall into this class. The example discussed above is also contained in this class.
Type B:

$$\sum_{i=0}^{\infty}\sum_{j=0}^{\infty} \frac{\Gamma(i+a_1)}{\Gamma(i+a_1')} \cdots \frac{\Gamma(i+a_k)}{\Gamma(i+a_k')} \frac{\Gamma(j+b_1)}{\Gamma(j+b_1')} \cdots \frac{\Gamma(j+b_l)}{\Gamma(j+b_l')}$$
$$\times \frac{\Gamma(i+j+c_1)}{\Gamma(i+j+c_1')} \cdots \frac{\Gamma(i+j+c_m)}{\Gamma(i+j+c_m')} x^i y^j \tag{6.14}$$

An example for a function of this type is given by the first Appell function F_1.
Type C:

$$\sum_{i=0}^{\infty}\sum_{j=0}^{\infty} \binom{i+j}{j} \frac{\Gamma(i+a_1)}{\Gamma(i+a_1')} \cdots \frac{\Gamma(i+a_k)}{\Gamma(i+a_k')} \frac{\Gamma(i+j+c_1)}{\Gamma(i+j+c_1')} \cdots \frac{\Gamma(i+j+c_m)}{\Gamma(i+j+c_m')} x^i y^j \tag{6.15}$$

Here, an example is given by the Kampé de Fériet function S_1.
Type D:

$$\sum_{i=0}^{\infty}\sum_{j=0}^{\infty} \binom{i+j}{j} \frac{\Gamma(i+a_1)}{\Gamma(i+a_1')} \cdots \frac{\Gamma(i+a_k)}{\Gamma(i+a_k')} \frac{\Gamma(j+b_1)}{\Gamma(j+b_1')} \cdots \frac{\Gamma(j+b_l)}{\Gamma(j+b_l')}$$
$$\times \frac{\Gamma(i+j+c_1)}{\Gamma(i+j+c_1')} \cdots \frac{\Gamma(i+j+c_m)}{\Gamma(i+j+c_m')} x^i y^j \tag{6.16}$$

An example for a function of this type is the second Appell function F_2.
Note that in these examples there are always as many Gamma functions in the numerator as in the denominator. We assume that all a_n, a_n', b_n, b_n', c_n and c_n' are of the form "integer + const $\cdot\, \varepsilon$". The generalisation towards the form "rational number + const $\cdot\, \varepsilon$" is briefly discussed in sect. 6.4.3. The task is now to expand these functions systematically into a Laurent series in ε.

6.4.1 *Nested sums.* In this section I review the underlying mathematical structure for the systematic expansion of the functions in (6.13)-(6.16). I discuss properties of particular forms of nested sums, which are called Z-sums and show that they form a Hopf algebra [60–62]. This Hopf algebra admits as additional structures a conjugation and a convolution product. Z-sums are defined by

$$Z(n;m_1,...,m_k;x_1,...,x_k) = \sum_{n\geq i_1>i_2>\ldots>i_k>0} \frac{x_1^{i_1}}{i_1{}^{m_1}} \ldots \frac{x_k^{i_k}}{i_k{}^{m_k}}. \quad (6.17)$$

k is called the depth of the Z-sum and $w = m_1 + ... + m_k$ is called the weight. If the sums go to Infinity ($n = \infty$) the Z-sums are multiple polylogarithms [63]:

$$Z(\infty;m_1,...,m_k;x_1,...,x_k) = \text{Li}_{m_1,...,m_k}(x_1,...,x_k). \quad (6.18)$$

For $x_1 = ... = x_k = 1$ the definition reduces to the Euler-Zagier sums [64, 65]:

$$Z(n;m_1,...,m_k;1,...,1) = Z_{m_1,...,m_k}(n). \quad (6.19)$$

For $n = \infty$ and $x_1 = ... = x_k = 1$ the sum is a multiple ζ-value [66]:

$$Z(\infty;m_1,...,m_k;1,...,1) = \zeta_{m_1,...,m_k}. \quad (6.20)$$

The usefulness of the Z-sums lies in the fact, that they interpolate between multiple polylogarithms and Euler-Zagier sums.

In addition to Z-sums, it is sometimes useful to introduce as well S-sums. S-sums are defined by

$$S(n;m_1,...,m_k;x_1,...,x_k) = \sum_{n\geq i_1\geq i_2\geq\ldots\geq i_k\geq 1} \frac{x_1^{i_1}}{i_1{}^{m_1}} \ldots \frac{x_k^{i_k}}{i_k{}^{m_k}}. \quad (6.21)$$

The S-sums reduce for $x_1 = ... = x_k = 1$ (and positive m_i) to harmonic sums [67–69]:

$$S(n;m_1,...,m_k;1,...,1) = S_{m_1,...,m_k}(n). \quad (6.22)$$

The S-sums are closely related to the Z-sums, the difference being the upper summation boundary for the nested sums: $(i-1)$ for Z-sums, i for S-sums. The introduction of S-sums is redundant, since S-sums can be expressed in terms of Z-sums and vice versa. It is however convenient to introduce both Z-sums and S-sums, since some properties are more naturally expressed in terms of Z-sums while others are more naturally expressed in terms of S-sums. An algorithm for the conversion from Z-sums to S-sums and vice versa can be found in [60].

The Z-sums form an algebra. The unit element in the algebra is given by the empty sum

$$e = Z(n). \quad (6.23)$$

The empty sum $Z(n)$ equals 1 for non-negative integer n. Before I discuss the multiplication rule, let me note that the basic building blocks of Z-sums are expressions of the form

$$\frac{x_j^n}{n^{m_j}}, \quad (6.24)$$

which will be called "letters". For fixed n, one can multiply two letters with the same n:

$$\frac{x_1^n}{n^{m_1}} \cdot \frac{x_2^n}{n^{m_2}} = \frac{(x_1x_2)^n}{n^{m_1+m_2}}, \quad (6.25)$$

e.g. the x_j's are multiplied and the degrees are added. Let us call the set of all letters the alphabet A. As a short-hand notation I will in the following denote a letter just by $X_j = x_j^n/n^{m_j}$. A word is an ordered sequence of letters, e.g.

$$W = X_1, X_2, ..., X_k. \quad (6.26)$$

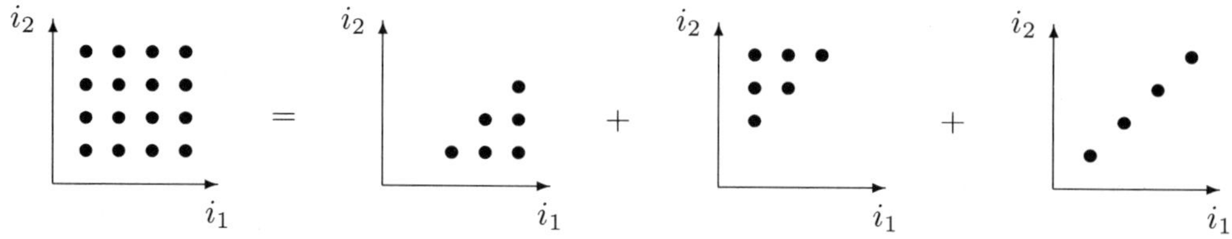

Figure 5 Sketch of the proof for the multiplication of Z-sums. The sum over the square is replaced by the sum over the three regions on the r.h.s.

The word of length zero is denoted by e. The Z-sums defined in (6.17) are therefore completely specified by the upper summation limit n and a word W. A quasi-shuffle algebra $\mathcal{A}$ on the vector-space of words is defined by [70]

$$\begin{aligned} e \circ W &= W \circ e = W, \\ (X_1, W_1) \circ (X_2, W_2) &= X_1, (W_1 \circ (X_2, W_2)) + X_2, ((X_1, W_1) \circ W_2) \\ &\quad +(X_1 \cdot X_2), (W_1 \circ W_2). \end{aligned} \tag{6.27}$$

Note that "$\cdot$" denotes multiplication of letters as defined in eq. (6.25), whereas "$\circ$" denotes the product in the algebra $\mathcal{A}$, recursively defined in eq. (6.27). This defines a quasi-shuffle product for Z-sums. It is also called a mixable shuffle product [71] or stuffle product [66]. The recursive definition in (6.27) translates for Z-sums into

$$\begin{aligned} & Z_{m_1,\ldots,m_k}(n) \times Z_{m_1',\ldots,m_l'}(n) \\ &= \sum_{i_1=1}^{n} \frac{1}{i_1^{m_1}} Z_{m_2,\ldots,m_k}(i_1 - 1) Z_{m_1',\ldots,m_l'}(i_1 - 1) \\ &\quad + \sum_{i_2=1}^{n} \frac{1}{i_2^{m_1'}} Z_{m_1,\ldots,m_k}(i_2 - 1) Z_{m_2',\ldots,m_l'}(i_2 - 1) \\ &\quad + \sum_{i=1}^{n} \frac{1}{i^{m_1+m_1'}} Z_{m_2,\ldots,m_k}(i - 1) Z_{m_2',\ldots,m_l'}(i - 1). \end{aligned} \tag{6.28}$$

The proof that Z-sums obey the quasi-shuffle algebra is sketched in Fig. 5. The outermost sums of the Z-sums on the l.h.s of (6.28) are split into the three regions indicated in Fig. 5. A simple example for the multiplication of two Z-sums is

$$\begin{aligned} & Z(n; m_1; x_1) Z(n; m_1; x_2) = \\ & \quad Z(n; m_1, m_2; x_1, x_2) + Z(n; m_2, m_1; x_2, x_1) + Z(n; m_1 + m_2; x_1 x_2). \end{aligned} \tag{6.29}$$

The quasi-shuffle algebra $\mathcal{A}$ is isomorphic to the free polynomial algebra on the Lyndon words. If one introduces a lexicographic ordering on the letters of the alphabet A, a Lyndon word is defined by the property

$$W < V \tag{6.30}$$

for any sub-words U and V such that $W = U, V$. Here U, V means just concatenation of U and V.

The Z-sums form actual a Hopf algebra. It is convenient to phrase the coalgebra structure in terms of rooted trees. Z-sums can be represented as rooted trees without any side-branchings. As a concrete example the pictorial representation of a sum of depth three reads:

$$Z(n; m_1, m_2, m_3; x_1, x_2, x_3) = \sum_{i_1=1}^{n} \sum_{i_2=1}^{i_1-1} \sum_{i_3=1}^{i_2-1} \frac{x_1^{i_1}}{{i_1}^{m_1}} \frac{x_2^{i_2}}{{i_2}^{m_2}} \frac{x_3^{i_3}}{{i_3}^{m_3}} = \quad (6.31)$$

The outermost sum corresponds to the root. By convention, the root is always drawn on the top. Trees with side-branchings are given by nested sums with more than one sub-sum, for example:

$$\sum_{i=1}^{n} \frac{x_1^i}{i^{m_1}} Z(i-1; m_2, x_2) Z(i-1; m_3; x_3) = \quad (6.32)$$

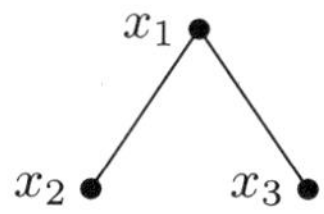

Of course, due to the multiplication formula, trees with side-branchings can always be reduced to trees without any side-branchings. The coalgebra structure is now formulated in terms of rooted trees. I first introduce some notation how to manipulate rooted trees, following the notation of Kreimer and Connes [72, 73]. An elementary cut of a rooted tree is a cut at a single chosen edge. An admissible cut is any assignment of elementary cuts to a rooted tree such that any path from any vertex of the tree to the root has at most one elementary cut. An admissible cut maps a tree t to a monomial in trees $t_1 \circ ... \circ t_{k+1}$. Note that precisely one of these subtrees t_j will contain the root of t. Denote this distinguished tree by $R^C(t)$, and the monomial delivered by the k other factors by $P^C(t)$. The counit $\bar{e}$ is given by

$$\begin{aligned} \bar{e}(e) &= 1, \\ \bar{e}(t) &= 0, \quad t \neq e. \end{aligned} \quad (6.33)$$

The coproduct Δ is defined by the equations

$$\begin{aligned} \Delta(e) &= e \otimes e, \\ \Delta(t) &= e \otimes t + t \otimes e + \sum_{\text{adm. cuts } C \text{ of } t} P^C(t) \otimes R^C(t), \\ \Delta(t_1 \circ ... \circ t_k) &= \Delta(t_1)(\circ \otimes \circ)...(\circ \otimes \circ)\Delta(t_k). \end{aligned} \quad (6.34)$$

The antipode $\mathcal{S}$ is given by

$$\begin{aligned} \mathcal{S}(e) &= e, \\ \mathcal{S}(t) &= -t - \sum_{\text{adm. cuts } C \text{ of } t} \mathcal{S}\left(P^C(t)\right) \circ R^C(t), \\ \mathcal{S}(t_1 \circ ... \circ t_k) &= \mathcal{S}(t_1) \circ ... \circ \mathcal{S}(t_k). \end{aligned} \quad (6.35)$$

Since the multiplication in the algebra is commutative the antipode satisfies

$$\mathcal{S}^2 = \text{id}. \quad (6.36)$$

Let me give some examples for the coproduct and the antipode for Z-sums:

$$\begin{aligned} \Delta Z(n; m_1; x_1) &= e \otimes Z(n; m_1; x_1) + Z(n; m_1; x_1) \otimes e, \\ \Delta Z(n; m_1, m_2; x_1, x_2) &= e \otimes Z(n; m_1, m_2; x_1, x_2) + Z(n; m_1, m_2; x_1, x_2) \otimes e \\ &\quad + Z(n; m_2; x_2) \otimes Z(n; m_1; x_1), \end{aligned} \quad (6.37)$$

$$\begin{aligned} \mathcal{S}Z(n; m_1; x_1) &= -Z(n; m_1; x_1), \\ \mathcal{S}Z(n; m_1, m_2; x_1, x_2) &= Z(n; m_2, m_1; x_2, x_1) + Z(n; m_1 + m_2; x_1 x_2). \end{aligned}$$

The Hopf algebra of nested sums has additional structures if we allow expressions of the form

$$\frac{x_0^n}{n^{m_0}} Z(n; m_1, ..., m_k; x_1, ..., x_k), \tag{6.38}$$

e.g. Z-sums multiplied by a letter. Then the following convolution product

$$\sum_{i=1}^{n-1} \frac{x^i}{i^m} Z(i-1; ...) \frac{y^{n-i}}{(n-i)^{m'}} Z(n-i-1; ...) \tag{6.39}$$

can again be expressed in terms of expressions of the form (6.38). An example is

$$\begin{aligned} &\sum_{i=1}^{n-1} \frac{x^i}{i} Z_1(i-1) \frac{y^{n-i}}{(n-i)} Z_1(n-i-1) = \\ &\quad \frac{x^n}{n} \left[Z\left(n-1; 1, 1, 1; \frac{y}{x}, \frac{x}{y}, \frac{y}{x}\right) + Z\left(n-1; 1, 1, 1; \frac{y}{x}, 1, \frac{x}{y}\right) \right. \\ &\quad \left. + Z\left(n-1; 1, 1, 1; 1, \frac{y}{x}, 1\right) \right] + (x \leftrightarrow y). \end{aligned} \tag{6.40}$$

In addition there is a conjugation, e.g. sums of the form

$$-\sum_{i=1}^{n} \binom{n}{i} (-1)^i \frac{x^i}{i^m} S(i; ...) \tag{6.41}$$

can also be reduced to terms of the form (6.38). Although one can easily convert between the notations for S-sums and Z-sums, expressions involving a conjugation tend to be shorter when expressed in terms of S-sums. The name conjugation stems from the following fact: To any function $f(n)$ of an integer variable n one can define a conjugated function $C \circ f(n)$ as the following sum

$$C \circ f(n) = \sum_{i=1}^{n} \binom{n}{i} (-1)^i f(i). \tag{6.42}$$

Then conjugation satisfies the following two properties:

$$\begin{aligned} C \circ 1 &= 1, \\ C \circ C \circ f(n) &= f(n). \end{aligned} \tag{6.43}$$

An example for a sum involving a conjugation is

$$\begin{aligned} &-\sum_{i=1}^{n} \binom{n}{i} (-1)^i \frac{x^i}{i} S_1(i) = \\ &\quad S\left(n; 1, 1; 1-x, \frac{1}{1-x}\right) - S(n; 1, 1; 1-x, 1). \end{aligned} \tag{6.44}$$

Finally there is the combination of conjugation and convolution, e.g. sums of the form

$$-\sum_{i=1}^{n-1} \binom{n}{i} (-1)^i \frac{x^i}{i^m} S(i; ...) \frac{y^{n-i}}{(n-i)^{m'}} S(n-i; ...) \tag{6.45}$$

can also be reduced to terms of the form (6.38). An example is given by

$$-\sum_{i=1}^{n-1} \binom{n}{i} (-1)^i \; S(i;1;x)\, S(n-i;1;y) = \tag{6.46}$$
$$\frac{1}{n}\left\{S(n;1;y) + (1-x)^n \left[S\left(n;1;\frac{1}{1-\frac{1}{x}}\right) - S\left(n;1;\frac{1-\frac{y}{x}}{1-\frac{1}{x}}\right)\right]\right\}$$
$$+\frac{(-1)^n}{n}\left\{S(n;1;x) + (1-y)^n \left[S\left(n;1;\frac{1}{1-\frac{1}{y}}\right) - S\left(n;1;\frac{1-\frac{x}{y}}{1-\frac{1}{y}}\right)\right]\right\}.$$

6.4.2 *Expansion of hyper-geometric functions.* In this section I discuss how the algebraic tools introduced in the previous section can be used for the Laurent expansion of the transcendental functions (6.13)-(6.16). First I give some motivation for the introduction of Z-sums: The essential point is that Z-sums interpolate between multiple polylogarithms and Euler-Zagier-sums, such that the interpolation is compatible with the algebra structure. On the one hand, we expect multiple polylogarithm to appear in the Laurent expansion of the transcendental functions (6.13)-(6.16), a fact which is confirmed a posteriori. Therefore it is important that multiple polylogarithms are contained in the class of Z-sums. On the other the expansion parameter ε occurs in the functions (6.13)-(6.16) inside the arguments of Gamma-functions. The basic formula for the expansion of Gamma functions reads

$$\Gamma(n+\varepsilon) = \Gamma(1+\varepsilon)\Gamma(n)\left[1 + \varepsilon Z_1(n-1) + \varepsilon^2 Z_{11}(n-1) \right.$$
$$\left. + \varepsilon^3 Z_{111}(n-1) + ... + \varepsilon^{n-1} Z_{11...1}(n-1)\right], \tag{6.47}$$

containing Euler-Zagier sums for finite n. As a simple example I discuss the expansion of

$$\sum_{i=0}^{\infty} \frac{\Gamma(i+a_1+t_1\varepsilon)\Gamma(i+a_2+t_2\varepsilon)}{\Gamma(i+1)\Gamma(i+a_3+t_3\varepsilon)} x^i \tag{6.48}$$

into a Laurent series in ε. Here a_1, a_2 and a_3 are assumed to be integers. Up to prefactors the expression in (6.48) is a hyper-geometric function ${}_2F_1$. Using $\Gamma(x+1) = x\Gamma(x)$, partial fractioning and an adjustment of the summation index one can transform (6.48) into terms of the form

$$\sum_{i=1}^{\infty} \frac{\Gamma(i+t_1\varepsilon)\Gamma(i+t_2\varepsilon)}{\Gamma(i)\Gamma(i+t_3\varepsilon)} \frac{x^i}{i^m}, \tag{6.49}$$

where m is an integer. Now using (6.47) one obtains

$$\Gamma(1+\varepsilon) \sum_{i=1}^{\infty} \frac{(1+\varepsilon t_1 Z_1(i-1)+...)\,(1+\varepsilon t_2 Z_1(i-1)+...)}{(1+\varepsilon t_3 Z_1(i-1)+...)} \frac{x^i}{i^m}. \tag{6.50}$$

Inverting the power series in the denominator and truncating in ε one obtains in each order in ε terms of the form

$$\sum_{i=1}^{\infty} \frac{x^i}{i^m} Z_{m_1...m_k}(i-1) Z_{m_1'...m_l'}(i-1) Z_{m_1''...m_n''}(i-1) \tag{6.51}$$

Using the quasi-shuffle product for Z-sums the three Euler-Zagier sums can be reduced to single Euler-Zagier sums and one finally arrives at terms of the form

$$\sum_{i=1}^{\infty} \frac{x^i}{i^m} Z_{m_1 \dots m_k}(i-1), \tag{6.52}$$

which are special cases of multiple polylogarithms, called harmonic polylogarithms $H_{m,m_1,\dots,m_k}(x)$. This completes the algorithm for the expansion in ε for sums of the form (6.13). Since the one-loop integral discussed in (6.12) is a special case of (6.13), this algorithm also applies to the integral (6.12). In addition, this algorithm shows that in the expansion of hyper-geometric functions ${}_{J+1}F_J(a_1, \dots, a_{J+1}; b_1, \dots, b_J; x)$ around integer values of the parameters a_k and b_l only harmonic polylogarithms appear in the result.

The algorithm for the expansion of sums of type (6.13) used the multiplication formula for Z-sums to pass from (6.51) to (6.52). To expand double sums of type (6.14) one needs in addition the convolution product (6.39). To expand sums of type (6.15) the conjugation (6.41) is needed. Finally, for sums of type (6.16) the combination of conjugation and convolution as in (6.45) is required. More details can be found in [60].

Let me come back to the example of the one-loop Feynman integral discussed in eq. (2.62). For $\nu_1 = \nu_2 = \nu_3 = 1$ and $m = 2$ in (6.12) one obtains:

$$\begin{aligned} &\int \frac{d^{4-2\varepsilon}k_1}{i\pi^{2-\varepsilon}} \frac{1}{(-k_1^2)} \frac{1}{(-k_2^2)} \frac{1}{(-k_3^2)} \\ &= \frac{\Gamma(-\varepsilon)\Gamma(1-\varepsilon)\Gamma(1+\varepsilon)}{\Gamma(1-2\varepsilon)} \frac{\left(-p_{123}^2\right)^{-1-\varepsilon}}{1-x} \sum_{n=1}^{\infty} \varepsilon^{n-1} H_{\underbrace{1,\dots,1}_{n}}(1-x). \end{aligned} \tag{6.53}$$

In this special case all harmonic polylogarithms can be expressed in terms of powers of the standard logarithm:

$$H_{\underbrace{1,\dots,1}_{n}}(1-x) = \frac{(-1)^n}{n!} (\ln x)^n. \tag{6.54}$$

This particular example is very simple and one recovers the well-known all-order result

$$\frac{\Gamma(1-\varepsilon)^2\Gamma(1+\varepsilon)}{\Gamma(1-2\varepsilon)} \frac{\left(-p_{123}^2\right)^{-1-\varepsilon}}{\varepsilon^2} \frac{1-x^{-\varepsilon}}{1-x}, \tag{6.55}$$

which (for this simple example) can also be obtained by direct integration.

6.4.3 *Expansion around half-integer values.* Up to now we only discussed the case where the Gamma functions are expanded around an integer value. The extension to rational numbers is straightforward, if the Gamma functions always occur in ratios of the form

$$\frac{\Gamma(n+a-\frac{p}{q}+b\varepsilon)}{\Gamma(n+c-\frac{p}{q}+d\varepsilon)}, \tag{6.56}$$

where the same rational number p/q occurs in the numerator and in the denominator [74]. The generalisation of eq. (6.47) reads

$$\Gamma\left(n+1-\frac{p}{q}+\varepsilon\right) = \frac{\Gamma\left(1-\frac{p}{q}+\varepsilon\right)\Gamma\left(n+1-\frac{p}{q}\right)}{\Gamma\left(1-\frac{p}{q}\right)} \times \exp\left(-\frac{1}{q}\sum_{l=0}^{q-1}\left(r_q^l\right)^p \sum_{k=1}^{\infty}\varepsilon^k \frac{(-q)^k}{k} Z(q\cdot n;k;r_q^l)\right) \tag{6.57}$$

and introduces the q-th roots of unity

$$r_q^p = \exp\left(\frac{2\pi i p}{q}\right). \tag{6.58}$$

With the help of the q-th roots of unity we may express any Z-sum $Z(n;...)$ as a combination of Z-sums $Z(q\cdot n;...)$, where the summation goes now up to $q\cdot n$. If the occurrence of rational numbers is not balanced as in eq. (6.56), the sums can be performed only in special cases. These include the important cases of binomial sums

$$\sum_{n=1}^{\infty}\binom{2n}{n}\frac{z^n}{n^{m_0}}S(n;m_1,...,m_k;x_1,...,x_k), \tag{6.59}$$

and inverse binomial sums

$$\sum_{n=1}^{\infty}\frac{1}{\binom{2n}{n}}\frac{z^n}{n^{m_1}}. \tag{6.60}$$

6.5 Differential equations. An alternative approach to the computation of Feynman parameter integrals is based on differential equations [75–77]. To evaluate these integrals the procedure used in [58, 78, 79] consists in finding first for each master integral a differential equation, which this master integral has to satisfy. The derivative is taken with respect to an external scale, or a ratio of two scales. It turns out that the resulting differential equations are linear, inhomogeneous first order equations of the form

$$\frac{\partial}{\partial y}T(y)+f(y)T(y)=g(y), \tag{6.61}$$

where y is usually a ratio of two kinematical invariants and $T(y)$ is the master integral under consideration. The inhomogeneous term $g(y)$ is usually a combination of simpler master integrals. In general, a first order linear inhomogeneous differential equation is solved by first considering the corresponding homogeneous equation. One introduces an integrating factor

$$M(y)=e^{\int f(y)dy}, \tag{6.62}$$

such that $T_0(y)=1/M(y)$ solves the homogeneous differential equation ($g(y)=0$). This yields the general solution of the inhomogeneous equation as

$$T(y)=\frac{1}{M(y)}\left(\int g(y)M(y)dy+C\right), \tag{6.63}$$

where the integration constant C can be adjusted to match the boundary conditions. This method yields a master integral in the form of transcendental functions (e.g. for example hyper-geometric functions), which still have to be expanded in the small parameter ε of dimensional regularisation. Although this can be done systematically and was discussed above, the authors of [58, 78, 79] followed a different road: Factoring out a trivial dimension-full normalisation factor, one writes down an ansatz for the solution of the differential equation as a Laurent expression in ε. Each term in this Laurent series is a sum of terms, consisting of basis functions times some unknown (and to be determined) coefficients. This ansatz is inserted into the differential equation and the unknown coefficients are determined order by order from the differential equation. This approach will succeed if we know in advance the right set of basis functions. The basis functions are taken as a subset of multiple polylogarithms.

6.6 Sector decomposition. In sect. 4 we presented the general formula for a scalar l-loop integral as a Feynman parameter integral. We are interested in the Laurent expansion in ε of this integral. In this section I will discuss an algorithm, which allows to compute the coefficients of the Laurent expansion numerically for a given kinematical configuration of external momenta. The major challenge such an algorithm has to face is the disentanglement of overlapping singularities. An example for an overlapping singularity is given by

$$\int d^3x \, \delta\left(1 - \sum_{i=1}^{3} x_i\right) \frac{x_1^{-\varepsilon} x_2^{-\varepsilon}}{x_1(x_1 + x_2)}. \tag{6.64}$$

The term $1/(x_1 + x_2)$ is an overlapping singularity. Sector decomposition [80–82] is a convenient tool to disentangle overlapping singularities. For the example in eq. (6.64) one splits the integration region into two sectors $x_1 > x_2$ and $x_1 < x_2$. In the first sector one rescales x_2 as $x_2' = x_2/x_1$, while in the second sector one rescales $x_1' = x_1/x_2$. Binoth and Heinrich [82] gave a systematic algorithm for integrals of the form of eq. (4.2). First the integration domain is mapped from the $(n-1)$-dimensional simplex to the $(n-1)$-dimensional hypercube. This is done as follows: First the integration domain is split into n parts, using the identity

$$\int_0^1 d^n x \quad = \quad \sum_{l=1}^{n} \int_0^1 d^n x \prod_{\substack{j=1 \\ j \neq l}}^{n} \theta(x_l - x_j) \,, \tag{6.65}$$

such that the integral I_G becomes a sum over n integrals $I_G^{(l)}$, where in each "primary sector" l the variable x_l is the largest one. The variables are transformed in each primary sector l as follows:

$$x_j \quad = \quad \begin{cases} x_l t_j, & j < l, \\ x_l, & j = l, \\ x_l t_{j-1}, & j > l. \end{cases} \tag{6.66}$$

By construction, x_l factorises from $\mathcal{U}$ and $\mathcal{F}$. x_l is eliminated in each G_l using

$$\int dx_l/x_l \; \delta(1 - x_l(1 + \sum_{k=1}^{n-1} t_k)) = 1 \,. \tag{6.67}$$

This ensures that the singular behaviour leading to poles in ε still comes from the region of small t's. By applying the sector decomposition iteratively, one finally arrives at a form where all singularities are factorised explicitly in terms of factors of Feynman parameters like $t_j^{-1-\kappa\varepsilon}$. Subtractions of the form

$$\int_0^1 dt_j\, t_j^{-1-\kappa\varepsilon} f(t_j) = -\frac{1}{\kappa\varepsilon} f(0) + \int_0^1 dt_j\, t_j^{-1-\kappa\varepsilon} \left[f(t_j) - f(0)\right] \tag{6.68}$$

for each j, where $\lim_{t_j \to 0} f(t_j)$ is finite by construction, allow to extract all poles and lead to integrals which are finite and can be integrated numerically.

7 Multiple polylogarithms

In this section I discuss multiple polylogarithms. The multiple polylogarithms are defined by

$$\mathrm{Li}_{m_1,...,m_k}(x_1,...,x_k) = \sum_{i_1>i_2>...>i_k>0} \frac{x_1^{i_1}}{{i_1}^{m_1}} \cdots \frac{x_k^{i_k}}{{i_k}^{m_k}}. \tag{7.1}$$

They are special cases of Z-sums:

$$\mathrm{Li}_{m_1,...,m_k}(x_1,...,x_k) = Z(\infty; m_1,...,m_k; x_1,...,x_k). \tag{7.2}$$

The multiple polylogarithms contain as the notation already suggests as subsets the classical polylogarithms $\mathrm{Li}_n(x)$ [83], as well as Nielsen's generalised polylogarithms [84]

$$S_{n,p}(x) = \mathrm{Li}_{n+1,1,...,1}(x,\underbrace{1,...,1}_{p-1}), \tag{7.3}$$

and the harmonic polylogarithms [85]

$$H_{m_1,...,m_k}(x) = \mathrm{Li}_{m_1,...,m_k}(x,\underbrace{1,...,1}_{k-1}). \tag{7.4}$$

Multiple polylogarithms have been studied extensively in the literature by physicists [19, 67, 78, 85–88] and mathematicians [63, 66, 89–98]. Here I summarise the most important properties. Being special cases of Z-sums they obey the quasi-shuffle Hopf algebra for Z-sums. In addition there is a second algebra structure, derived from the iterated integral representation.

7.1 Definition through iterated integrals. Multiple polylogarithms have been defined in this article via the sum representation (7.1). In addition, they admit an integral representation. From this integral representation a second algebra structure arises, which turns out to be a shuffle Hopf algebra. To discuss this second Hopf algebra it is convenient to introduce for $z_k \neq 0$ the following functions

$$G(z_1,...,z_k;y) = \int_0^y \frac{dt_1}{t_1 - z_1} \int_0^{t_1} \frac{dt_2}{t_2 - z_2} \cdots \int_0^{t_{k-1}} \frac{dt_k}{t_k - z_k}. \tag{7.5}$$

In this definition one variable is redundant due to the following scaling relation:

$$G(z_1,...,z_k;y) = G(xz_1,...,xz_k;xy) \tag{7.6}$$

If one further defines

$$g(z;y) = \frac{1}{y-z}, \tag{7.7}$$

then one has

$$\frac{d}{dy}G(z_1, ..., z_k; y) = g(z_1; y)G(z_2, ..., z_k; y) \tag{7.8}$$

and

$$G(z_1, z_2, ..., z_k; y) = \int_0^y dt\, g(z_1; t)G(z_2, ..., z_k; t). \tag{7.9}$$

One can slightly enlarge the set and define $G(0, ..., 0; y)$ with k zeros for z_1 to z_k to be

$$G(0, ..., 0; y) = \frac{1}{k!}(\ln y)^k. \tag{7.10}$$

This permits us to allow trailing zeros in the sequence $(z_1, ..., z_k)$ by defining the function G with trailing zeros via (7.9) and (7.10). To relate the multiple polylogarithms to the functions G it is convenient to introduce the following short-hand notation:

$$G_{m_1,...,m_k}(z_1, ..., z_k; y) = G(\underbrace{0, ..., 0}_{m_1-1}, z_1, ..., z_{k-1}, \underbrace{0..., 0}_{m_k-1}, z_k; y) \tag{7.11}$$

Here, all z_j for $j = 1, ..., k$ are assumed to be non-zero. One then finds

$$\text{Li}_{m_1,...,m_k}(x_1, ..., x_k) = (-1)^k G_{m_1,...,m_k}\left(\frac{1}{x_1}, \frac{1}{x_1 x_2}, ..., \frac{1}{x_1 ... x_k}; 1\right). \tag{7.12}$$

The inverse formula reads

$$G_{m_1,...,m_k}(z_1, ..., z_k; y) = (-1)^k \,\text{Li}_{m_1,...,m_k}\left(\frac{y}{z_1}, \frac{z_1}{z_2}, ..., \frac{z_{k-1}}{z_k}\right). \tag{7.13}$$

Eq. (7.12) together with (7.11) and (7.5) defines an integral representation for the multiple polylogarithms. To make this more explicit I first introduce some notation for iterated integrals

$$\int_0^\Lambda \frac{dt}{t-a_n} \circ ... \circ \frac{dt}{t-a_1} = \int_0^\Lambda \frac{dt_n}{t_n - a_n} \int_0^{t_n} \frac{dt_{n-1}}{t_{n-1} - a_{n-1}} \times ... \times \int_0^{t_2} \frac{dt_1}{t_1 - a_1} \tag{7.14}$$

and the short hand notation:

$$\int_0^\Lambda \left(\frac{dt}{t}\circ\right)^m \frac{dt}{t-a} = \int_0^\Lambda \underbrace{\frac{dt}{t} \circ ... \frac{dt}{t}}_{m \text{ times}} \circ \frac{dt}{t-a}. \tag{7.15}$$

The integral representation for $\text{Li}_{m_1,...,m_k}(x_1, ..., x_k)$ reads then

$$\text{Li}_{m_1,...,m_k}(x_1, ..., x_k) = (-1)^k \int_0^1 \left(\frac{dt}{t}\circ\right)^{m_1-1} \frac{dt}{t-b_1}$$

$$\circ \left(\frac{dt}{t}\circ\right)^{m_2-1} \frac{dt}{t-b_2} \circ ... \circ \left(\frac{dt}{t}\circ\right)^{m_k-1} \frac{dt}{t-b_k}, \tag{7.16}$$

where the b_j's are related to the x_j's

$$b_j = \frac{1}{x_1 x_2 ... x_j}. \tag{7.17}$$

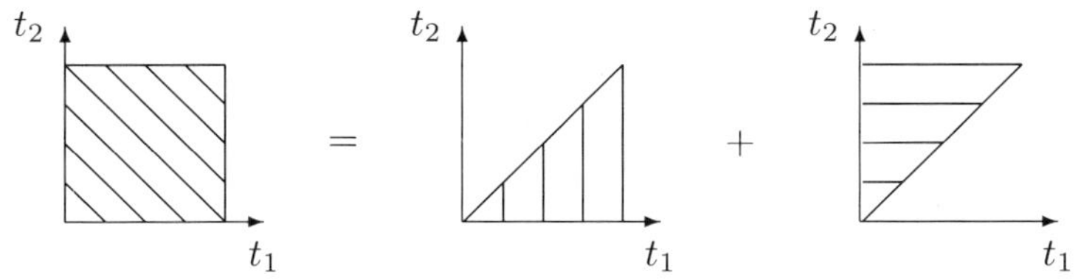

Figure 6 Sketch of the proof for the multiplication of two G-functions. The integral over the square is replaced by two integrals over the upper and lower triangle.

From the iterated integral representation (7.5) a second algebra structure for the functions $G(z_1, ..., z_k; y)$ (and through (7.12) also for the multiple polylogarithms) is obtained as follows: We take the z_j's as letters and call a sequence of ordered letters $w = z_1, ..., z_k$ a word. Then the function $G(z_1, ..., z_k; y)$ is uniquely specified by the word $w = z_1, ..., z_k$ and the variable y. The neutral element e is given by the empty word, equivalent to

$$G(;y) \quad = \quad 1. \tag{7.18}$$

A shuffle algebra on the vector space of words is defined by

$$\begin{aligned} e \circ w &= w \circ e = w, \\ (z_1, w_1) \circ (z_2, w_2) &= z_1, (w_1 \circ (z_2, w_2)) + z_2, ((z_1, w_1) \circ w_2). \end{aligned} \tag{7.19}$$

Note that this definition is very similar to the definition of the quasi-shuffle algebra (6.27), except that the third term in (6.27) is missing. In fact, a shuffle algebra is a special case of a quasi-shuffle algebra, where the product of two letters is degenerate: $X_1 \cdot X_2 = 0$ for all letters X_1 and X_2 in the notation of Sect. 6.4.1. The definition of the shuffle product (7.19) translates into the following recursive definition of the product of two G-functions:

$$\begin{aligned} & G(z_1, ..., z_k; y) \times G(z_{k+1}, ..., z_n; y) = \\ & \quad \int\limits_0^y \frac{dt}{t - z_1} G(z_2, ..., z_k; t) G(z_{k+1}, ..., z_n; t) \\ & \quad + \int\limits_0^y \frac{dt}{t - z_{k+1}} G(z_1, ..., z_k; t) G(z_{k+2}, ..., z_n; t) \end{aligned} \tag{7.20}$$

The proof is sketched in fig. 6.

For the discussion of the coalgebra part for the functions $G(z_1, ..., z_k; y)$ we may proceed as in Sect. 6.4.1 and associate to any function $G(z_1, ..., z_k; y)$ a rooted tree without side-branchings as in the following example:

$$G(z_1, z_2, z_3; y) \quad = \tag{7.21}$$

The outermost integration (involving z_1) corresponds to the root. The formulae for the coproduct (6.34) and the antipode (6.35) apply then also to the functions $G(z_1, ..., z_k; y)$.

A shuffle algebra is simpler than a quasi-shuffle algebra and one finds for a shuffle algebra besides the recursive definitions of the product, the coproduct and the antipode also closed formulae for these operations. For the product one has

$$G(z_1, ..., z_k; y)\, G(z_{k+1}, ..., z_{k+l}; y) \quad = \quad \sum_{shuffle} G\left(z_{\sigma(1)}, ..., z_{\sigma(k+l)}; y\right), \tag{7.22}$$

where the sum is over all permutations which preserve the relative order of the strings $z_1, ..., z_k$ and $z_{k+1}, ..., z_{k+l}$. This explains the name "shuffle product". For the coproduct one has

$$\Delta G(z_1, ..., z_k; y) \quad = \quad \sum_{j=0}^{k} G(z_1, ..., z_j; y) \otimes G(z_{j+1}, ..., z_k; y) \tag{7.23}$$

and for the antipode one finds

$$\mathcal{S} G(z_1, ..., z_k; y) \quad = \quad (-1)^k G(z_k, ..., z_1; y). \tag{7.24}$$

The shuffle multiplication is commutative and the antipode satisfies therefore

$$\mathcal{S}^2 \quad = \quad \text{id}. \tag{7.25}$$

From (7.24) this is evident.

7.2 The antipode and integration-by-parts. Integration-by-parts has always been a powerful tool for calculations in particle physics. By using integration-by-parts one may obtain an identity between various G-functions. The starting point is as follows:

$$\begin{aligned} G(z_1, ..., z_k; y) &= \int_0^y dt \left(\frac{\partial}{\partial t} G(z_1; t) \right) G(z_2, ..., z_k; y) \\ &= G(z_1; y) G(z_2, ..., z_k; y) - \int_0^y dt\; G(z_1; t) g(z_2; t) G(z_3, ..., z_k; y) \\ &= G(z_1; y) G(z_2, ..., z_k; y) - \int_0^y dt \left(\frac{\partial}{\partial t} G(z_2, z_1; t) \right) G(z_3, ..., z_k; y). \end{aligned} \tag{7.26}$$

Repeating this procedure one arrives at the following integration-by-parts identity:

$$\begin{aligned} & G(z_1, ..., z_k; y) + (-1)^k G(z_k, ..., z_1; y) \\ & \quad = \quad G(z_1; y) G(z_2, ..., z_k; y) - G(z_2, z_1; y) G(z_3, ..., z_k; y) + ... \\ & \qquad -(-1)^{k-1} G(z_{k-1}, ... z_1; y) G(z_k; y), \end{aligned} \tag{7.27}$$

which relates the combination $G(z_1, ..., z_k; y) + (-1)^k G(z_k, ..., z_1; y)$ to G-functions of lower depth. This relation is useful in simplifying expressions. Eq. (7.27) can also be derived in a different way. In a Hopf algebra we have for any non-trivial element w the following relation involving the antipode:

$$\sum_{(w)} w^{(1)} \cdot \mathcal{S}(w^{(2)}) \quad = \quad 0. \tag{7.28}$$

Here Sweedler's notation has been used. Sweedler's notation writes the coproduct of an element w as

$$\Delta(w) = \sum_{(w)} w^{(1)} \otimes w^{(2)}. \tag{7.29}$$

Working out the relation (7.28) for the shuffle algebra of the functions $G(z_1, ..., z_k; y)$, we recover (7.27).

We may now proceed and check if (7.28) provides also a non-trivial relation for the quasi-shuffle algebra of Z-sums [99]. This requires first some notation: A composition of a positive integer k is a sequence $I = (i_1, ..., i_l)$ of positive integers such that $i_1 + ... i_l = k$. The set of all composition of k is denoted by $\mathcal{C}(k)$. Compositions act on Z-sums as

$$\begin{aligned}
&(i_1, ..., i_l) \circ Z(n; m_1, ..., m_k; x_1, ..., x_k) \\
&\quad = Z\left(n; m_1 + ... + m_{i_1}, m_{i_1+1} + ... + m_{i_1+i_2}, ..., m_{i_1+...+i_{l-1}+1} + ...\right. \\
&\qquad \left. + m_{i_1+...+i_l}; x_1 ... x_{i_1}, x_{i_1+1} ... x_{i_1+i_2}, ..., x_{i_1+...+i_{l-1}+1} ... x_{i_1+...+i_l}\right), \quad (7.30)
\end{aligned}$$

e.g. the first i_1 letters of the Z-sum are combined into one new letter, the next i_2 letters are combined into the second new letter, etc.. With this notation for compositions one obtains the following closed formula for the antipode in the quasi-shuffle algebra:

$$\mathcal{S} Z(n; m_1, ..., m_k; x_1, ..., x_k) = (-1)^k \sum_{I \in \mathcal{C}(k)} I \circ Z(n; m_k, ..., m_1; x_k, ..., x_1) \tag{7.31}$$

From (7.28) we then obtain

$$\begin{aligned}
&Z(n; m_1, ..., m_k; x_1, ..., x_k) + (-1)^k Z(n; m_k, ..., m_1; x_k, ..., x_1) \\
&\quad = - \sum_{adm.\ cuts} P^C \left(Z(n; m_1, ..., m_k; x_1, ..., x_k)\right) \\
&\qquad \cdot \mathcal{S}\left(R^C \left(Z(n; m_1, ..., m_k; x_1, ..., x_k)\right)\right) \\
&\qquad - (-1)^k \sum_{I \in \mathcal{C}(k) \setminus (1,1,...,1)} I \circ Z(n; m_k, ..., m_1; x_k, ..., x_1). \quad (7.32)
\end{aligned}$$

Again, the combination $Z(n; m_1, ..., m_k; x_1, ..., x_k) + (-1)^k Z(n; m_k, ..., m_1; x_k, ..., x_1)$ reduces to Z-sums of lower depth, similar to (7.27). We therefore obtained an "integration-by-parts" identity for objects, which don't have an integral representation. We first observed, that for the G-functions, which have an integral representation, the integration-by-parts identities are equal to the identities obtained from the antipode. After this abstraction towards an algebraic formulation, one can translate these relations to cases, which only have the appropriate algebra structure, but not necessarily a concrete integral representation. As an example we have

$$\begin{aligned}
&Z(n; m_1, m_2, m_3; x_1, x_2, x_3) - Z(n; m_3, m_2, m_1; x_3, x_2, x_1) = \\
&\quad Z(n; m_1; x_1) Z(n; m_2, m_3; x_2, x_3) - Z(n; m_2, m_1; x_2, x_1) Z(n; m_3; x_3) \\
&\quad - Z(n; m_1 + m_2; x_1 x_2) Z(n; m_3; x_3) + Z(n; m_2 + m_3, m_1; x_2 x_3, x_1) \\
&\quad + Z(n; m_3, m_1 + m_2; x_3, x_1 x_2) + Z(n; m_1 + m_2 + m_3; x_1 x_2 x_3), \quad (7.33)
\end{aligned}$$

which expresses the combination of the two Z-sums of depth 3 as Z-sums of lower depth. The analog example for the shuffle algebra of the G-function reads:

$$G(z_1, z_2, z_3; y) - G(z_3, z_2, z_1; y) = G(z_1; y)G(z_2, z_3; y) - G(z_2, z_1; y)G(z_3; y). \tag{7.34}$$

Multiple polylogarithms obey both the quasi-shuffle algebra and the shuffle algebra. Therefore we have for multiple polylogarithms two relations, which are in general independent.

7.3 Numerical evaluation. At the end of the day of an analytic calculation physicists would like to get a number. This requires a method for the numerical evaluation of multiple polylogarithms. As an example I first discuss the numerical evaluation of the dilogarithm [49]:

$$\text{Li}_2(x) = -\int\limits_0^x dt \frac{\ln(1-t)}{t} = \sum_{n=1}^{\infty} \frac{x^n}{n^2} \tag{7.35}$$

The power series expansion can be evaluated numerically, provided $|x| < 1$. Using the functional equations

$$\begin{aligned} \text{Li}_2(x) &= -\text{Li}_2\left(\frac{1}{x}\right) - \frac{\pi^2}{6} - \frac{1}{2}\left(\ln(-x)\right)^2, \\ \text{Li}_2(x) &= -\text{Li}_2(1-x) + \frac{\pi^2}{6} - \ln(x)\ln(1-x). \end{aligned} \tag{7.36}$$

any argument of the dilogarithm can be mapped into the region $|x| \leq 1$ and $-1 \leq \text{Re}(x) \leq 1/2$. The numerical computation can be accelerated by using an expansion in $[-\ln(1-x)]$ and the Bernoulli numbers B_i:

$$\text{Li}_2(x) = \sum_{i=0}^{\infty} \frac{B_i}{(i+1)!}\left(-\ln(1-x)\right)^{i+1}. \tag{7.37}$$

The generalisation to multiple polylogarithms proceeds along the same lines [100]: Using the integral representation

$$G_{m_1,...,m_k}(z_1, z_2, ..., z_k; y) = \tag{7.38}$$
$$\int\limits_0^y \left(\frac{dt}{t}\circ\right)^{m_1-1} \frac{dt}{t-z_1} \left(\frac{dt}{t}\circ\right)^{m_2-1} \frac{dt}{t-z_2} ... \left(\frac{dt}{t}\circ\right)^{m_k-1} \frac{dt}{t-z_k}$$

one transforms all arguments into a region, where one has a converging power series expansion:

$$\begin{aligned} G_{m_1,...,m_k}(z_1, ..., z_k; y) &= \sum_{j_1=1}^{\infty} ... \sum_{j_k=1}^{\infty} \frac{1}{(j_1 + ... + j_k)^{m_1}} \left(\frac{y}{z_1}\right)^{j_1} \\ &\times \frac{1}{(j_2 + ... + j_k)^{m_2}} \left(\frac{y}{z_2}\right)^{j_2} ... \frac{1}{(j_k)^{m_k}} \left(\frac{y}{z_k}\right)^{j_k}. \end{aligned} \tag{7.39}$$

The multiple polylogarithms satisfy the Hölder convolution [66]. For $z_1 \neq 1$ and $z_w \neq 0$ this identity reads

$$G(z_1, ..., z_w; 1) = \sum_{j=0}^{w} (-1)^j G\left(1 - z_j, 1 - z_{j-1}, ..., 1 - z_1; 1 - \frac{1}{p}\right) G\left(z_{j+1}, ..., z_w; \frac{1}{p}\right). \tag{7.40}$$

The Hölder convolution can be used to accelerate the convergence for the series representation of the multiple polylogarithms.

7.4 Related functions. In the literature one often encounters log-sine integrals. These are closely related to inverse binomial sums and can be expressed in terms of multiple polylogarithms. Here I briefly summarise the results from the literature. The following inverse binomial sum can be evaluated with elementary functions as follows:

$$\Gamma\left(\frac{1}{2}\right) \sum_{n=1}^{\infty} \frac{\Gamma(n+1)}{\Gamma\left(n+\frac{1}{2}\right)} \frac{x^n}{n} = \frac{2\sqrt{x}\arcsin\left(\sqrt{x}\right)}{\sqrt{1-x}}. \tag{7.41}$$

In the literature, evaluations of inverse binomial sums of higher weights are given in terms of log-sine functions [101–108] :

$$\Gamma\left(\frac{1}{2}\right) \sum_{n=1}^{\infty} \frac{\Gamma(n+1)}{\Gamma\left(n+\frac{1}{2}\right)} \frac{z^n}{n^m} = -\sum_{j=0}^{m-2} \frac{(-2)^j}{j!(m-2-j)!} \left(\ln 4z\right)^{m-2-j} \mathrm{Ls}_{j+2}^{(1)}(\theta), \tag{7.42}$$

where $\theta = 2\arcsin\sqrt{z}$ and the log-sine functions are defined by

$$\mathrm{Ls}_j^{(k)}(\theta) = -\int_0^\theta d\theta' \, (\theta')^k \ln^{j-k-1}\left|2\sin\frac{\theta'}{2}\right|. \tag{7.43}$$

By analytic continuation the log-sine functions are then related to polylogarithms [108]. A simple example is given by

$$\mathrm{Ls}_2^{(0)}(\theta) = \mathrm{Cl}_2(\theta), \tag{7.44}$$

involving the Clausen function Cl_2. The Clausen functions Cl_n are given in terms of polylogarithms by

$$\mathrm{Cl}_n(\theta) = \begin{cases} \frac{1}{2i}\left[\mathrm{Li}_n\left(e^{i\theta}\right) - \mathrm{Li}_n\left(e^{-i\theta}\right)\right], & n \text{ even}, \\ \frac{1}{2}\left[\mathrm{Li}_n\left(e^{i\theta}\right) + \mathrm{Li}_n\left(e^{-i\theta}\right)\right], & n \text{ odd}. \end{cases} \tag{7.45}$$

8 Outlook

In the previous section we discussed methods to perform the Feynman parameter integrals. Unfortunately not every Feynman integral can be solved with these methods. In this section I will try to give an outlook towards open questions and future directions. One fundamental question is related to the type of functions which occur in the results of Feynman integrals. In sect. 8.1 I briefly discuss the massless master two-loop two-point function. Here multiple zeta values are sufficient to express the result. However, the class of multiple polylogarithms (which contains the multiple zeta values) might not be sufficiently large to accommodate the results of all Feynman integrals. Indications come from the three-loop two-point function with equal internal masses, where complete elliptic integrals occur.

This is shortly discussed in sect. 8.2. In this article we considered mainly integrals associated to individual Feynman graphs. The physical scattering amplitude is the sum over all Feynman graphs. It is not unusual that the result for a scattering amplitude can be expressed more elegantly than the results for individual Feynman graphs. In this context I review in sect. 8.3 a method for the computation of loop amplitudes based on unitarity, which by-passes traditional Feynman graphs. I conclude this section with a review of a conjecture, which states that the l-loop amplitude in maximally supersymmetric Yang-Mills theory is basically determined by the one-loop amplitude.

8.1 The two-loop two-point function. In massless theories all two-loop two-point functions can be derived from the following master integral:

$$\hat{I}^{(2,5)}(m-\varepsilon,\nu_1,\nu_2,\nu_3,\nu_4,\nu_5) = c_\Gamma^{-2}\left(-p^2\right)^{\nu_{12345}-2m+2\varepsilon} \int \frac{d^D k_1}{i\pi^{D/2}} \int \frac{d^D k_2}{i\pi^{D/2}} \frac{1}{\left(-k_1^2\right)^{\nu_1}\left(-k_2^2\right)^{\nu_2}\left(-k_3^2\right)^{\nu_3}\left(-k_4^2\right)^{\nu_4}\left(-k_5^2\right)^{\nu_5}}, \tag{8.1}$$

where $k_3 = k_2 - p$, $k_4 = k_1 - p$, $k_5 = k_2 - k_1$, $D = 2m - 2\varepsilon$ and

$$c_\Gamma = \frac{\Gamma(1+\varepsilon)\Gamma(1-\varepsilon)^2}{\Gamma(1-2\varepsilon)}. \tag{8.2}$$

The trivial dependence of this integral on $(-p^2)$ is already factored out in eq. (8.1). Therefore for given values of m and ν_j, the Laurent series in ε contains only numbers. For a long time it has been an open question what type of numbers one encounters in this Laurent expansion. With the help of the methods discussed in sect. 6.4 it was possible to prove that only multiple zeta values occur in the Laurent expansion of the two-loop integral $\hat{I}^{(2,5)}(m-\varepsilon,\nu_1,\nu_2,\nu_3,\nu_4,\nu_5)$, if all powers of the propagators are of the form $\nu_j = n_j + a_j\varepsilon$, where the n_j are positive integers and the a_j are non-negative real numbers [109]. As an example one has

$$(1-2\varepsilon)\,\hat{I}^{(2,5)}(2-\varepsilon,1+\varepsilon,1+\varepsilon,1+\varepsilon,1+\varepsilon,1+\varepsilon) = 6\zeta_3 + 9\zeta_4\varepsilon + 372\zeta_5\varepsilon^2 + \left(915\zeta_6 - 864\zeta_3^2\right)\varepsilon^3 + \left(18450\zeta_7 - 2592\zeta_4\zeta_3\right)\varepsilon^4 + \left(50259\zeta_8 - 76680\zeta_5\zeta_3 - 2592\zeta_{6,2}\right)\varepsilon^5 + \left(905368\zeta_9 - 200340\zeta_6\zeta_3 - 130572\zeta_5\zeta_4 + 66384\zeta_3^3\right)\varepsilon^6 + \mathcal{O}(\varepsilon^7). \tag{8.3}$$

8.2 Elliptic integrals. Laporta considered a three-loop two-point function with equal internal masses [110]. This integral is given by

$$A = \int\limits_0^\infty \frac{dl}{\lambda(l,-1,-1)} \int\limits_0^\infty \frac{dm}{\lambda(m,l,-1)\lambda(m,-1,-1)}. \tag{8.4}$$

Based on strong numerical evidence, this integral can be written as

$$A = K(w_-)K(w_+), \quad w_\pm = \frac{z_\pm}{z_\pm - 1}, \quad z_\pm = -\left(2-\sqrt{3}\right)^4\left(4\pm\sqrt{15}\right)^2, \tag{8.5}$$

where $K(x)$ is the complete elliptic integral of the first kind:

$$K(x) = \int\limits_0^1 \frac{dt}{\sqrt{1-t^2}\sqrt{1-xt^2}}. \tag{8.6}$$

8.3 The unitarity based method. We conclude the discussion of techniques for loop calculations with a method based on unitarity [6, 111]. I will discuss this method for one-loop amplitudes. Within the unitarity based approach one chooses first a basis of integral functions $I_i \in \mathcal{S}$. For one-loop calculations in massless QCD a possible set consists of scalar boxes, triangles and two-point functions. The loop amplitude $A^{(1)}$ is written as a linear combination of these functions

$$A^{(1)} = \sum_i c_i I_i + r. \tag{8.7}$$

The unknown coefficients c_i are to be determined. r is a rational function in the invariants, not proportional to any element of the basis of integral functions. The integral functions themselves are combinations of rational functions, logarithms, logarithms squared and dilogarithms. The latter three can develop imaginary parts in certain regions of phase space, for example

$$\begin{aligned} \mathrm{Im} \ln \left(\frac{-s - i\delta}{-t - i\delta} \right) &= -\pi \left[\theta(s) - \theta(t) \right], \\ \mathrm{Im}\, \mathrm{Li}_2 \left(1 - \frac{(-s - i\delta)}{(-t - i\delta)} \right) &= -\ln \left(1 - \frac{s}{t} \right) \mathrm{Im} \ln \left(\frac{-s - i\delta}{-t - i\delta} \right). \end{aligned} \tag{8.8}$$

Knowing the imaginary parts, one can reconstruct uniquely the corresponding integral functions. In general there will be imaginary parts corresponding to different channels (e.g. to the different possibilities to cut a one-loop diagram into two parts). The imaginary part in one channel of a one-loop amplitude can be obtained via unitarity from a phase space integral over two tree-level amplitudes. With the help of the Cutkosky rules we have

$$\mathrm{Im}\, A^{(1)} = \mathrm{Im} \int \frac{d^D k}{(2\pi)^D} \frac{1}{k_1^2 + i\delta} \frac{1}{k_2^2 + i\delta} A_L^{(0)} A_R^{(0)}. \tag{8.9}$$

$A^{(1)}$ is the one-loop amplitude under consideration, $A_L^{(0)}$ and $A_R^{(0)}$ are tree-level amplitudes appearing on the left and right side of the cut in a given channel. Lifting eq. (8.9) one obtains

$$A^{(1)} = \int \frac{d^D k}{(2\pi)^D} \frac{1}{k_1^2 + i\delta} \frac{1}{k_2^2 + i\delta} A_L^{(0)} A_R^{(0)} + \text{ cut free pieces}, \tag{8.10}$$

where "cut free pieces" denote contributions which do not develop an imaginary part in this particular channel. By evaluating the cut, one determines the coefficients c_i of the integral functions, which have an imaginary part in this channel. Iterating over all possible cuts, one finds all coefficients c_i. One advantage of a cut-based calculation is that one starts with tree amplitudes on both sides of the cut, which are already sums of Feynman diagrams. Therefore cancellations and simplifications, which usually occur between various diagrams, can already be performed before we start the calculation of the loop amplitude.

In general, a cut-based calculation leaves as ambiguity the ration piece r in eq. (8.7), which can not be obtained with this technique. One example for such an ambiguity would be

$$\int \frac{d^D k}{(2\pi)^D} \frac{k^\mu k^\nu - \frac{1}{3} q^\mu q^\nu + \frac{1}{12} g^{\mu\nu} q^2}{k^2 (k-q)^2}. \tag{8.11}$$

This term does not have a cut and will therefore not be detected in a cut-based calculation. However, Bern, Dixon, Dunbar and Kosower [111] have proven the following power counting criterion: If a one-loop amplitude has in some gauge a representation, in which all n-point loop integrals have at most $n-2$ powers of the loop momentum in the numerator (with the exception of two-point integrals, which are allowed to have one power of the loop momentum in the numerator), then the loop amplitude is uniquely determined by its cuts. This does not mean that the amplitude has no cut-free pieces, but rather that all cut-free pieces are associated with some integral functions. In particular $\mathcal{N} = 4$ supersymmetric amplitudes satisfy the power-counting criterion above and are therefore cut-constructible.

QCD does in general not satisfy the power-counting theorem and leaves as an ambiguity the rational function r. In principle one can obtain the rational piece r by calculating higher order terms in ε within the cut-based method. At one-loop order an arbitrary scale $\mu^{2\varepsilon}$ is introduced in order to keep the coupling dimensionless. In a massless theory the factor $\mu^{2\varepsilon}$ is always accompanied by some kinematical invariant $s^{-\varepsilon}$ for dimensional reasons. If we write symbolically

$$A^{(1)} = \frac{1}{\varepsilon^2} c_2 \left(\frac{s_2}{\mu^2}\right)^{-\varepsilon} + \frac{1}{\varepsilon} c_1 \left(\frac{s_1}{\mu^2}\right)^{-\varepsilon} + c_0 \left(\frac{s_0}{\mu^2}\right)^{-\varepsilon}, \tag{8.12}$$

the cut-free pieces $c_0(s_0/\mu^2)^{-\varepsilon}$ can be detected at order ε:

$$c_0 \left(\frac{s_0}{\mu^2}\right)^{-\varepsilon} = c_0 - \varepsilon c_0 \ln\left(\frac{s_0}{\mu^2}\right) + O(\varepsilon^2). \tag{8.13}$$

8.4 Iterative structures of loop amplitudes. I would like to conclude this section with an outlook towards current research related to iterative structures of loop amplitudes. The loop amplitudes in maximally supersymmetric $\mathcal{N} = 4$ Yang-Mills theory (MSYM) are a popular play-ground, as calculations in this theory tend to be simpler compared to the corresponding ones in QCD. In addition, one focusses on the leading-colour contributions to the loop amplitudes. Here only planar diagrams contribute. Let us denote by $M_n^{(l)} = \mathcal{A}_n^{(l)}/\mathcal{A}_n^{(0)}$ the ratio of the l-loop amplitude to the corresponding tree amplitude. It is conjectured that the two-loop n-point function in MSYM is related to the one-loop n-point function by [112]

$$\begin{aligned} M_n^{(2)}(\varepsilon) &= \frac{1}{2}\left(M_n^{(1)}(\varepsilon)\right)^2 + f^{(2)}(\varepsilon) M_n^{(1)}(2\varepsilon) - \frac{5}{4}\zeta_4 + \mathcal{O}(\varepsilon), \\ f^{(2)}(\varepsilon) &= -\zeta_2 - \zeta_3\varepsilon - \zeta_4\varepsilon^2. \end{aligned} \tag{8.14}$$

This conjecture has been verified for $n = 4$. For $n = 5$ some partial results are available, which support the conjecture [113]. In addition it is believed that this iterative structure generalises to higher loops. For three loops it is conjectured [114] that

$$\begin{aligned} M_n^{(3)}(\varepsilon) &= -\frac{1}{3}\left(M_n^{(1)}(\varepsilon)\right)^3 + M_n^{(1)}(\varepsilon) M_n^{(2)}(\varepsilon) + f^{(3)}(\varepsilon) M_n^{(1)}(3\varepsilon) \\ &\quad + \left(\frac{341}{216} + \frac{2}{9}c_1\right)\zeta_6 + \left(-\frac{17}{9} + \frac{2}{9}c_2\right)\zeta_3^2 + \mathcal{O}(\varepsilon), \\ f^{(3)}(\varepsilon) &= \frac{11}{2}\zeta_4 + (6\zeta_5 + 5\zeta_2\zeta_3)\,\varepsilon + \left(c_1\zeta_6 + c_2\zeta_3^2\right)\varepsilon^2 \end{aligned} \tag{8.15}$$

Again, this conjecture has been verified for $n = 4$. The two constants c_1 and c_2 cannot be determined from the $n = 4$ calculation.

9 Summary

In this article I discussed loop integrals which occur in a perturbative approach to quantum field theory. I reviewed standard techniques, which allow us to exchange the integration over the loop momenta against Feynman parameter integrals as wells as methods, which reduce tensor integrals to scalar integrals. The important sub-class of one-loop integrals was discussed in detail. The main part of this lecture was devoted to the computation of Feynman parameter integrals, with an emphasis on the mathematical structures underlying these computations. One encounters iterated structures as nested sums or iterated integrals, which form a Hopf algebra with a shuffle or quasi-shuffle product. In the final results multiple polylogarithms appear, and their properties have been discussed at length. The last section addressed open questions and conjectures.

References

[1] T. van Ritbergen, J. A. M. Vermaseren, and S. A. Larin, *The four-loop beta function in quantum chromodynamics*, Phys. Lett. **B400** (1997) 379–384, [hep-ph/9701390].

[2] S. Laporta and E. Remiddi, *The analytical value of the electron (g-2) at order alpha**3 in qed*, Phys. Lett. **B379** (1996) 283–291, [hep-ph/9602417].

[3] S. G. Gorishnii, A. L. Kataev, and S. A. Larin, *The o (alpha-s**3) corrections to sigma-tot (e+ e- → hadrons) and gamma (tau- → tau-neutrino + hadrons) in qcd*, Phys. Lett. **B259** (1991) 144–150.

[4] S. Moch, J. A. M. Vermaseren, and A. Vogt, *The three-loop splitting functions in qcd: The non-singlet case*, Nucl. Phys. **B688** (2004) 101–134, [hep-ph/0403192].

[5] A. Vogt, S. Moch, and J. A. M. Vermaseren, *The three-loop splitting functions in qcd: The singlet case*, Nucl. Phys. **B691** (2004) 129–181, [hep-ph/0404111].

[6] Z. Bern, L. Dixon, and D. A. Kosower, *A two-loop four-gluon helicity amplitude in qcd*, JHEP **01** (2000) 027, [hep-ph/0001001].

[7] Z. Bern, L. Dixon, and A. Ghinculov, *Two-loop correction to bhabha scattering*, Phys. Rev. **D63** (2001) 053007, [hep-ph/0010075].

[8] Z. Bern, A. De Freitas, and L. J. Dixon, *Two-loop amplitudes for gluon fusion into two photons*, JHEP **09** (2001) 037, [hep-ph/0109078].

[9] Z. Bern, A. De Freitas, L. J. Dixon, A. Ghinculov, and H. L. Wong, *Qcd and qed corrections to light-by-light scattering*, JHEP **11** (2001) 031, [hep-ph/0109079].

[10] Z. Bern, A. De Freitas, and L. Dixon, *Two-loop helicity amplitudes for gluon gluon scattering in qcd and supersymmetric yang-mills theory*, *JHEP* **03** (2002) 018, [hep-ph/0201161].

[11] C. Anastasiou, E. W. N. Glover, C. Oleari, and M. E. Tejeda-Yeomans, *Two-loop QCD corrections to the scattering of massless distinct quarks*, Nucl. Phys. **B601** (2001) 318–340, [hep-ph/0010212].

[12] C. Anastasiou, E. W. N. Glover, C. Oleari, and M. E. Tejeda-Yeomans, *Two-loop QCD corrections to massless identical quark scattering*, Nucl. Phys. **B601** (2001) 341–360, [hep-ph/0011094].

[13] C. Anastasiou, E. W. N. Glover, C. Oleari, and M. E. Tejeda-Yeomans, *One-loop qcd corrections to quark scattering at nnlo*, Phys. Lett. **B506** (2001) 59–67, [hep-ph/0012007].

[14] C. Anastasiou, E. W. N. Glover, C. Oleari, and M. E. Tejeda-Yeomans, *Two-loop qcd corrections to massless quark gluon scattering*, Nucl. Phys. **B605** (2001) 486–516, [hep-ph/0101304].

[15] E. W. N. Glover, C. Oleari, and M. E. Tejeda-Yeomans, *Two-loop qcd corrections to gluon gluon scattering*, Nucl. Phys. **B605** (2001) 467–485, [hep-ph/0102201].

[16] T. Binoth, E. W. N. Glover, P. Marquard, and J. J. van der Bij, *Two-loop corrections to light-by-light scattering in supersymmetric qed*, JHEP **05** (2002) 060, [hep-ph/0202266].

[17] L. W. Garland, T. Gehrmann, E. W. N. Glover, A. Koukoutsakis, and E. Remiddi, *The two-loop qcd matrix element for e+ e- → 3jets*, Nucl. Phys. **B627** (2002) 107–188, [hep-ph/0112081].

[18] L. W. Garland, T. Gehrmann, E. W. N. Glover, A. Koukoutsakis, and E. Remiddi, *Two-loop qcd helicity amplitudes for e+ e- → 3jets*, Nucl. Phys. **B642** (2002) 227–262, [hep-ph/0206067].
[19] S. Moch, P. Uwer, and S. Weinzierl, *Two-loop amplitudes with nested sums: Fermionic contributions to e+ e- → q anti-q g*, Phys. Rev. **D66** (2002) 114001, [hep-ph/0207043].
[20] A. Denner, S. Dittmaier, M. Roth, and L. H. Wieders, *Complete electroweak o(alpha) corrections to charged- current e+ e- → 4fermion processes, Phys. Lett.* **B612** (2005) 223–232, [hep-ph/0502063].
[21] A. Denner, S. Dittmaier, M. Roth, and L. H. Wieders, *Electroweak corrections to charged-current e+ e- → 4 fermion processes: Technical details and further results*, Nucl. Phys. **B724** (2005) 247–294, [hep-ph/0505042].
[22] G. 't Hooft and M. J. G. Veltman, *Regularization and renormalization of gauge fields*, Nucl. Phys. **B44** (1972) 189–213.
[23] C. G. Bollini and J. J. Giambiagi, *Dimensional renormalization: The number of dimensions as a regularizing parameter*, Nuovo Cim. **B12** (1972) 20–25.
[24] G. M. Cicuta and E. Montaldi, *Analytic renormalization via continuous space dimension*, Nuovo Cim. Lett. **4** (1972) 329–332.
[25] K. G. Wilson, *Quantum field theory models in less than four-dimensions*, Phys. Rev. **D7** (1973) 2911–2926.
[26] J. Collins, *Renormalization.* Cambridge University Press, 1984.
[27] T. Kinoshita, *Mass singularities of feynman amplitudes*, J. Math. Phys. **3** (1962) 650–677.
[28] T. D. Lee and M. Nauenberg, *Degenerate systems and mass singularities*, Phys. Rev. **133** (1964) B1549–B1562.
[29] P. Breitenlohner and D. Maison, *Dimensional renormalization and the action principle*, Commun. Math. Phys. **52** (1977) 11–38.
[30] Z. Bern and D. A. Kosower, *The computation of loop amplitudes in gauge theories*, Nucl. Phys. **B379** (1992) 451–561.
[31] S. Weinzierl, *Equivariant dimensional regularization*, hep-ph/9903380.
[32] Z. Bern, A. De Freitas, L. Dixon, and H. L. Wong, *Supersymmetric regularization, two-loop qcd amplitudes and coupling shifts*, Phys. Rev. **D66** (2002) 085002, [hep-ph/0202271].
[33] Z. Kunszt, A. Signer, and Z. Trocsanyi, *One loop helicity amplitudes for all 2 → 2 processes in qcd and n=1 supersymmetric yang-mills theory*, Nucl. Phys. **B411** (1994) 397–442, [hep-ph/9305239].
[34] A. Signer, *Helicity method for next-to-leading order corrections in qcd*, Ph.D. thesis, Diss. ETH Nr. 11143.
[35] S. Catani, M. H. Seymour, and Z. Trocsanyi, *Regularization scheme independence and unitarity in qcd cross sections*, Phys. Rev. **D55** (1997) 6819–6829, [hep-ph/9610553].
[36] G. Passarino and M. J. G. Veltman, *One loop corrections for e+ e- annihilation into mu+ mu- in the weinberg model*, Nucl. Phys. **B160** (1979) 151.
[37] O. V. Tarasov, *Connection between feynman integrals having different values of the space-time dimension*, Phys. Rev. **D54** (1996) 6479–6490, [hep-th/9606018].
[38] O. V. Tarasov, *Generalized recurrence relations for two-loop propagator integrals with arbitrary masses*, Nucl. Phys. **B502** (1997) 455–482, [hep-ph/9703319].
[39] C. Itzykson and J. B. Zuber, *Quantum field theory*, . New York, Usa: Mcgraw-hill (1980) 705 P.(International Series In Pure and Applied Physics).
[40] R. Scharf and J. B. Tausk, *Scalar two loop integrals for gauge boson selfenergy diagrams with a massless fermion loop*, Nucl. Phys. **B412** (1994) 523–552.
[41] D. B. Melrose, *Reduction of feynman diagrams*, Nuovo Cim. **40** (1965) 181–213.
[42] W. L. van Neerven and J. A. M. Vermaseren, *Large loop integrals, Phys. Lett.* **B137** (1984) 241.
[43] Z. Bern, L. J. Dixon, and D. A. Kosower, *Dimensionally regulated pentagon integrals*, Nucl. Phys. **B412** (1994) 751–816, [hep-ph/9306240].
[44] T. Binoth, J. P. Guillet, and G. Heinrich, *Reduction formalism for dimensionally regulated one-loop n- point integrals*, Nucl. Phys. **B572** (2000) 361–386, [hep-ph/9911342].
[45] J. Fleischer, F. Jegerlehner, and O. V. Tarasov, *Algebraic reduction of one-loop feynman graph amplitudes*, Nucl. Phys. **B566** (2000) 423–440, [hep-ph/9907327].
[46] A. Denner and S. Dittmaier, *Reduction of one-loop tensor 5-point integrals*, Nucl. Phys. **B658** (2003) 175–202, [hep-ph/0212259].

[47] G. Duplancic and B. Nizic, *Reduction method for dimensionally regulated one-loop n-point feynman integrals*, Eur. Phys. J. **C35** (2004) 105–118, [hep-ph/0303184].
[48] W. T. Giele and E. W. N. Glover, *A calculational formalism for one-loop integrals*, JHEP **04** (2004) 029, [hep-ph/0402152].
[49] G. 't Hooft and M. J. G. Veltman, *Scalar one loop integrals*, Nucl. Phys. **B153** (1979) 365–401.
[50] R. Pittau, *A simple method for multi-leg loop calculations*, Comput. Phys. Commun. **104** (1997) 23–36, [hep-ph/9607309].
[51] R. Pittau, *A simple method for multi-leg loop calculations. ii: A general algorithm*, Comput. Phys. Commun. **111** (1998) 48–52, [hep-ph/9712418].
[52] S. Weinzierl, *Reduction of multi-leg loop integrals*, Phys. Lett. **B450** (1999) 234, [hep-ph/9811365].
[53] F. del Aguila and R. Pittau, *Recursive numerical calculus of one-loop tensor integrals*, JHEP **07** (2004) 017, [hep-ph/0404120].
[54] R. Pittau, *Formulae for a numerical computation of one-loop tensor integrals*, hep-ph/0406105.
[55] A. van Hameren, J. Vollinga, and S. Weinzierl, *Automated computation of one-loop integrals in massless theories*, Eur. Phys. J. **C41** (2005) 361–375, [hep-ph/0502165].
[56] K. G. Chetyrkin and F. V. Tkachov, *Integration by parts: The algorithm to calculate beta functions in 4 loops*, Nucl. Phys. **B192** (1981) 159–204.
[57] T. van Ritbergen and R. G. Stuart, *On the precise determination of the fermi coupling constant from the muon lifetime*, Nucl. Phys. **B564** (2000) 343–390, [hep-ph/9904240].
[58] T. Gehrmann and E. Remiddi, *Differential equations for two-loop four-point functions*, Nucl. Phys. **B580** (2000) 485–518, [hep-ph/9912329].
[59] S. Laporta, *High-precision calculation of multi-loop feynman integrals by difference equations*, Int. J. Mod. Phys. **A15** (2000) 5087–5159, [hep-ph/0102033].
[60] S. Moch, P. Uwer, and S. Weinzierl, *Nested sums, expansion of transcendental functions and multi-scale multi-loop integrals*, J. Math. Phys. **43** (2002) 3363–3386, [hep-ph/0110083].
[61] S. Weinzierl, *Symbolic expansion of transcendental functions, Comput. Phys. Commun.* **145** (2002) 357–370, [math-ph/0201011].
[62] S. Moch and P. Uwer, *Xsummer: Transcendental functions and symbolic summation in form*, Comput. Phys. Commun. **174** (2006) 759–770, [math-ph/0508008].
[63] A. B. Goncharov, *Multiple polylogarithms, cyclotomy and modular complexes*, Math. Res. Lett. **5** (1998) 497, [(available at http://www.math.uiuc.edu/K-theory/0297)].
[64] L. Euler, *Meditationes circa singulare serierum genus*, Novi Comm. Acad. Sci. Petropol. **20** (1775) 140.
[65] D. Zagier, *Values of zeta functions and their applications*, First European Congress of Mathematics, Vol. II, Birkhauser, Boston (1994) 497.
[66] J. M. Borwein, D. M. Bradley, D. J. Broadhurst, and P. Lisonek, *Special values of multiple polylogarithms*, Trans. Amer. Math. Soc. **353:3** (2001) 907, [math.CA/9910045].
[67] J. A. M. Vermaseren, *Harmonic sums, mellin transforms and integrals*, Int. J. Mod. Phys. **A14** (1999) 2037, [hep-ph/9806280].
[68] J. Blumlein and S. Kurth, *Harmonic sums and mellin transforms up to two-loop order*, Phys. Rev. **D60** (1999) 014018, [hep-ph/9810241].
[69] J. Blumlein, *Algebraic relations between harmonic sums and associated quantities*, hep-ph/0311046.
[70] M. E. Hoffman, *Quasi-shuffle products*, J. Algebraic Combin. **11** (2000) 49, [math.QA/9907173].
[71] L. Guo and W. Keigher, *Baxter algebras and shuffle products*, Adv. in Math. **150** (2000) 117, [math.RA/0407155].
[72] D. Kreimer, *On the hopf algebra structure of perturbative quantum field theories*, Adv. Theor. Math. Phys. **2** (1998) 303–334, [q-alg/9707029].
[73] A. Connes and D. Kreimer, *Hopf algebras, renormalization and noncommutative geometry*, Commun. Math. Phys. **199** (1998) 203–242, [hep-th/9808042].
[74] S. Weinzierl, *Expansion around half-integer values, binomial sums and inverse binomial sums*, J. Math. Phys. **45** (2004) 2656–2673, [hep-ph/0402131].

[75] A. V. Kotikov, *Differential equations method: New technique for massive feynman diagrams calculation*, Phys. Lett. **B254** (1991) 158–164.
[76] A. V. Kotikov, *Differential equation method: The calculation of n point feynman diagrams*, Phys. Lett. **B267** (1991) 123–127.
[77] E. Remiddi, *Differential equations for feynman graph amplitudes*, *Nuovo Cim.* **A110** (1997) 1435–1452, [hep-th/9711188].
[78] T. Gehrmann and E. Remiddi, *Two-loop master integrals for gamma* → 3jets: The planar topologies*, Nucl. Phys. **B601** (2001) 248–286, [hep-ph/0008287].
[79] T. Gehrmann and E. Remiddi, *Two-loop master integrals for gamma* → 3jets: The non-planar topologies*, Nucl. Phys. **B601** (2001) 287–317, [hep-ph/0101124].
[80] K. Hepp, *Proof of the bogolyubov-parasiuk theorem on renormalization*, Commun. Math. Phys. **2** (1966) 301–326.
[81] M. Roth and A. Denner, *High-energy approximation of one-loop feynman integrals*, Nucl. Phys. **B479** (1996) 495–514, [hep-ph/9605420].
[82] T. Binoth and G. Heinrich, *An automatized algorithm to compute infrared divergent multi-loop integrals*, Nucl. Phys. **B585** (2000) 741–759, [hep-ph/0004013].
[83] L. Lewin, *Polylogarithms and associated functions*. North Holland, Amsterdam, 1981.
[84] N. Nielsen, *Der Eulersche Dilogarithmus und seine Verallgemeinerungen*, Nova Acta Leopoldina (Halle) **90** (1909) 123.
[85] E. Remiddi and J. A. M. Vermaseren, *Harmonic polylogarithms*, Int. J. Mod. Phys. **A15** (2000) 725, [hep-ph/9905237].
[86] T. Gehrmann and E. Remiddi, *Numerical evaluation of harmonic polylogarithms*, Comput. Phys. Commun. **141** (2001) 296–312, [hep-ph/0107173].
[87] T. Gehrmann and E. Remiddi, *Numerical evaluation of two-dimensional harmonic polylogarithms*, Comput. Phys. Commun. **144** (2002) 200–223, [hep-ph/0111255].
[88] T. Gehrmann and E. Remiddi, *Analytic continuation of massless two-loop four-point functions*, Nucl. Phys. **B640** (2002) 379–411, [hep-ph/0207020].
[89] R. M. Hain, *Classical polylogarithms*, alg-geom/9202022.
[90] A. B. Goncharov, *Multiple polylogarithms and mixed tate motives*, math.AG/0103059.
[91] A. B. Goncharov, *Polylogarithms, regulators and arakelov motivic complexes*, math.AG/0207036.
[92] A. B. Goncharov, *Galois symmetries of fundamental groupoids and noncommutative geometry*, math.AG/0208144.
[93] P. Elbaz-Vincent and H. Gangl, *On poly(ana)logs i*, Comp. Math. **130** (2002) 161, [math.KT/0008089].
[94] H. Gangl, *Functional equations of higher logarithms*, math.KT/0207222.
[95] M. P. H. M. Minh and J. van der Hoeven, *Shuffle algebra and polylogarithms*, Discrete Math. **225:1-3** (2000) 217.
[96] P. Cartier, *Fonctions polylogarithmes, nombres polyzeta et groupes pro-unipotents*, Séminaire Bourbaki (2001) 885.
[97] J. Ecalle, *Ari/gari, la dimorphie et l'arithmétique des multizêtas: un premier bilan*, (available at http://www.math.u-psud.fr/ biblio/ppo/2002/ppo2002-23.html).
[98] G. Racinet, *Doubles mélanges des polylogarithmes multiples aux racines de l'unité*, math.QA/0202142.
[99] S. Weinzierl, *Algebraic algorithms in perturbative calculations*, hep-th/0305260.
[100] J. Vollinga and S. Weinzierl, *Numerical evaluation of multiple polylogarithms*, Comput. Phys. Commun. **167** (2005) 177, [hep-ph/0410259].
[101] O. M. Ogreid and P. Osland, *Summing one- and two-dimensional series related to the euler series*, J. Comput. Appl. Math. **98** (1998) 245–271, [hep-th/9801168].
[102] J. Fleischer and M. Y. Kalmykov, *Single mass scale diagrams: Construction of a basis for the epsilon-expansion*, Phys. Lett. **B470** (1999) 168–176, [hep-ph/9910223].
[103] A. I. Davydychev, *Explicit results for all orders of the epsilon-expansion of certain massive and massless diagrams*, Phys. Rev. **D61** (2000) 087701, [hep-ph/9910224].
[104] J. Fleischer, A. V. Kotikov, and O. L. Veretin, *Analytic two-loop results for selfenergy- and vertex-type diagrams with one non-zero mass*, Nucl. Phys. **B547** (1999) 343–374, [hep-ph/9808242].
[105] J. M. Borwein, D. J. Broadhurst, and J. Kamnitzer, *Central binomial sums, multiple clausen values and zeta values*, Exper. Math. **10** (2001) 25–34, [hep-th/0004153].

[106] A. I. Davydychev and M. Y. Kalmykov, *New results for the epsilon-expansion of certain one-, two- and three-loop feynman diagrams*, Nucl. Phys. **B605** (2001) 266–318, [hep-th/0012189].

[107] M. Y. Kalmykov and O. Veretin, *Single-scale diagrams and multiple binomial sums*, Phys. Lett. **B483** (2000) 315–323, [hep-th/0004010].

[108] A. I. Davydychev and M. Y. Kalmykov, *Massive feynman diagrams and inverse binomial sums*, hep-th/0303162.

[109] I. Bierenbaum and S. Weinzierl, *The massless two-loop two-point function*, Eur. Phys. J. **C32** (2003) 67–78, [hep-ph/0308311].

[110] S. Laporta, *High-precision epsilon expansions of massive four-loop vacuum bubbles*, Phys. Lett. **B549** (2002) 115–122, [hep-ph/0210336].

[111] Z. Bern, L. Dixon, D. C. Dunbar, and D. A. Kosower, *Fusing gauge theory tree amplitudes into loop amplitudes*, Nucl. Phys. **B435** (1995) 59–101, [hep-ph/9409265].

[112] C. Anastasiou, Z. Bern, L. J. Dixon, and D. A. Kosower, *Planar amplitudes in maximally supersymmetric yang-mills theory*, Phys. Rev. Lett. **91** (2003) 251602, [hep-th/0309040].

[113] F. Cachazo, M. Spradlin, and A. Volovich, *Iterative structure within the five-particle two-loop amplitude*, hep-th/0602228.

[114] Z. Bern, L. J. Dixon, and V. A. Smirnov, *Iteration of planar amplitudes in maximally supersymmetric yang-mills theory at three loops and beyond*, *Phys. Rev.* **D72** (2005) 085001, [hep-th/0505205].

Fields Institute Communications
Volume **50**, 2007

The Transition Temperature of the Weakly Interacting Bose Gas

J. Zinn-Justin
DSM/DAPNIA, CEA/Saclay
F91191 Gif-sur-Yvette cédex, France
and
Institut de Mathmatiques de Jussieu–Chevaleret
Université de Paris VII
Paris, France
zinn@dsm-mail.saclay.cea.fr

Abstract. The effect of a weak repulsive two-body interaction on the transition temperature of a dilute Bose gas at fixed density has been controversial for a long time. We show here that the *effect is non-perturbative* in nature.

The *Renormalization group* then enables one to prove that the critical temperature T_c increases linearly with the strength of the interaction, parametrized in terms of the s-wave scattering length. However, the coefficient cannot be obtained from perturbative calculations.

Recognizing that the hamiltonian of the system, which also describes the *Helium superfluid transition*, is the $N = 2$ example of the general N vector model, one generalizes the problem to arbitrary N. The coefficient of $\Delta T_c/T_c$ can then be expanded in powers of $1/N$.

The leading order result, which is *independent of N for non-trivial reasons*, is in reasonable agreement with estimates coming from numerical simulations.

1 Introduction

The effect of a weak repulsive two-body interaction on the transition temperature of a dilute Bose gas at fixed density has been controversial for a long time [1, 2, 3, 4, 5, 6]. It has been slowly realized that the effect was non-perturbative in nature. It has recently been argued theoretically [7] that T_c increases linearly with the strength of the interaction parametrized in terms of the scattering length a. Since the coefficient cannot be obtained from perturbation theory, in Ref. [7] a simple self-consistent approximation was used to derive an explicit estimate.

In [8], first the linear behaviour was derived more directly, using general renormalization group arguments. Then, recognizing that the hamiltonian of the system

---

2000 *Mathematics Subject Classification.* Primary 81T16; Secondary 76F30.

under study, which also describes the helium superfluid transition, is a particular example of the general N vector model for $N = 2$, the problem was generalized to arbitrary N. This generalization makes new tools available [9, 10, 11]; in particular, the coefficient of $\Delta T_c/T_c$ can be calculated by carrying out an expansion in $1/N$. In [8], the leading order in $1/N$ was calculated explicitly. The result happens to be independent of N, for non-trivial reasons. The calculation involves subtle technical points which are most easily dealt with by dimensional regularization [12]. Surprisingly, the result is in reasonable agreement with the most recent numerical simulations [13].

This paper is organized as follows: In section 2, we recall the basis of Bose–Einstein condensation. In section 3, we describe the field integral representation of the partition function of the dilute Bose gas (see e.g., ref. [14, 15]) and lay out the basics of the problem. Then, in section 4, we derive the linear behaviour of the shift of the critical temperature for weak two-body interactions, using renormalization group arguments. Finally, in section 5, we present the general N-vector model and we calculate the leading order $1/N$ contribution to the temperature shift. After the article [8], the $1/N$ correction to the shift has been calculated [16] and other methods employed to calculate the amplitude of the shift. For a recent review see ref. [17].

2 Independent bosons: Bose–Einstein condensation

Equation of state. The equation of state of a gas relates the average number of particles $\langle \mathbf{N} \rangle$, the chemical potential μ and the inverse temperature $T = 1/\beta$. In the case of independent bosons, the equation of state takes the simple form

$$\langle \mathbf{N} \rangle = \sum_i \langle n_i \rangle , \quad \langle n_i \rangle = \frac{1}{\mathrm{e}^{\beta(\omega_i - \mu)} - 1} , \tag{2.1}$$

where $\langle n_i \rangle$ is the average occupation number of the level of energy ω_i.

The equation is defined only for $\mu < \inf_i \omega_i$.

This expression can also be expressed in terms of the one-particle hamiltonian operator $H^{(1)}$, with eigenvalues ω_i, as

$$\langle \mathbf{N} \rangle = \mathrm{tr} \, \frac{1}{\mathrm{e}^{\beta(H^{(1)} - \mu)} - 1} . \tag{2.2}$$

Bose–Einstein (BE) condensation. For a quantum hamiltonian operator of the form $H^{(1)}(\hat{p}, \hat{q})$ with a discrete spectrum, the equation of state, in the high temperature limit $\beta \delta\omega \ll 1$, reduces to its semi-classical approximation. In d dimensions,

$$\langle \mathbf{N} \rangle \sim \frac{1}{(2\pi\hbar)^d} \int \frac{\mathrm{d}^d p \, \mathrm{d}^d q}{\mathrm{e}^{\beta(H^{(1)}(\mathbf{p},\mathbf{q}) - \mu)} - 1} . \tag{2.3}$$

The average number of particles $\langle \mathbf{N} \rangle$ is an increasing function of $\mu \le \mu_c = \inf H^{(1)}(\mathbf{p}, \mathbf{q})$. If the integral converges for $\mu = \mu_c$, one faces the apparent paradox that $\langle \mathbf{N} \rangle$ is bounded at fixed temperature.

The solution to the paradox is that the *semi-classical approximation is no longer valid for the lowest energy level*, which yields a divergent contribution for $\mu = E_{\text{ground state}}$. Returning to the exact expression, one infers that *the additional particles accumulate into the ground state*: this is the essence of BE condensation.

In the rather theoretical example of particles of mass m in a box of linear size L (a harmonic potential is closer to the experimental conditions), high temperature is equivalent to $L \to \infty$. One finds, for $d > 2$, the critical density

$$n_c = \langle \mathbf{N} \rangle L^{-d} \underset{L\to\infty}{\sim} \frac{1}{(2\pi\hbar)^d} \int \frac{\mathrm{d}^d p}{\mathrm{e}^{\beta \mathbf{p}^2/2m} - 1} = \zeta(d/2)/\lambda^d, \tag{2.4}$$

where $\zeta(z)$ is the Riemann function, and λ the thermal wave length:

$$\lambda = \hbar\sqrt{2\pi/mT}\,.$$

Alternatively, at density n fixed, the equation of state has no solution for temperatures $T < T_c(n) \propto (\hbar^2/m)n^{2/d}$.

In what follows, we will only consider this situation, that is, particles in a box in the infinite volume limit.

3 Statistical field theory of the weakly interacting Bose gas

We want now to investigate the effect of weak interactions between bosons.

The partition function. The partition function of a *dilute gas* of identical bosons of mass m, at temperature $T = 1/\beta$, can be expressed as an integral

$$\mathcal{Z} = \int [\mathrm{d}\psi(t,x)\mathrm{d}\psi^*(t,x)]\mathrm{e}^{-\mathcal{S}(\psi^*,\psi)/\hbar},$$

over fields ψ^*, ψ periodic in euclidean time,

$$\psi(0,x) = \psi(\beta,x), \quad \psi^*(0,x) = \psi^*(\beta,x),$$

associated with boson creation and annihilation.

We assume here that the gas is *sufficiently dilute* for the two-body interactions to be weak and three-body or higher interactions to be totally negligible.

Since one is interested only in long wavelength phenomena, *the two-body potential can be replaced by a delta-function* and parametrized in terms of the s-wave scattering length a (positive because the interaction is assumed repulsive).

For $d = 3$, the effective euclidean action of the system may then be written as

$$\begin{aligned}\mathcal{S}(\psi^*,\psi) \;&=\; -\int_0^\beta \mathrm{d}t \int \mathrm{d}^3x \left[\psi^*(t,x)\left(\hbar\frac{\partial}{\partial t} + \frac{\hbar^2}{2m}\nabla_x^2 + \mu\right)\psi(t,x)\right.\\ &\left.+\frac{2\pi\hbar^2 a}{m}\left(\psi^*(t,x)\psi(t,x)\right)^2\right],\end{aligned}$$

where μ is the chemical potential.

The condition that the interaction is weak implies that $a \ll \lambda$, where λ is the thermal wavelength

$$\lambda = \hbar\sqrt{2\pi/mT}\,.$$

Equation of state and two-point function. Quite generally, the equation of state can be expressed in terms of the $\langle\psi^*\psi\rangle$ correlation function G. In the Fourier representation,

$$n = T\frac{\partial \ln \mathcal{Z}}{\partial \mu} = T\int \frac{\mathrm{d}^d k}{(2\pi)^d} \sum_\nu \tilde{G}(k,\omega_\nu;\mu), \tag{3.1}$$

where $\omega_\nu = 2\pi\nu T$ are Matsubara frequencies, the quantization of ω being a consequence of the periodic boundary conditions in euclidean time.

A $U(1)$ phase transition of superfluid Helium type occurs at a critical chemical potential μ_c given by

$$\tilde{G}^{-1}(k=0,\omega=0;\mu_c)=0\,,$$

where the correlation length diverges.

In the limit of free bosons

$$\tilde{G}^{-1}(k,\omega;\mu)=k^2/2m-i\omega-\mu\,.$$

Then, in a box of linear size L and $d>2$, one recovers $\mu_c=0$ and the critical density (2.4). The general equation of state (3.1) implies that, in the limit of vanishing repulsive interactions, the phase transition of the interacting model reduces to the BE condensation of the free Bose gas. In $d=3$ dimensions, the relation between condensation temperature and density n reads

$$T_c^0(n)\underset{d=3}{\propto}(\hbar^2/m)n^{2/3}.$$

Thus, at leading order in the interaction, the shift in the transition temperature at fixed density, $\Delta T_c=T_c-T_c^0$, can be related, to the change Δn in the density at fixed T_c by

$$\frac{\Delta T_c}{T_c}=-\frac{2}{3}\frac{\Delta n}{n}\,.$$

The theory of critical phenomena tells us that the variation of the critical temperature in systems with dimension $d<4$ depends primarily on contributions from the small momenta or large distance (or IR) region. This property simplifies the problem, since *to leading order* the IR properties are only sensitive to the $\omega_\nu=0$ Matsubara frequency. The entire calculation can thus be cast in terms of a *classical statistical field theory*. The integration over the other non-zero Matsubara frequencies yields corrections to the effective action. They do not affect the temperature shift at leading order.

4 Effective classical statistical field theory and renormalization group

After a rescaling the field ψ in order to introduce more conventional field theory normalizations, and a parametrization in terms of two real fields ϕ_1,ϕ_2:

$$\psi=\sqrt{mT}(\phi_1+i\phi_2),$$

the partition function takes the form of the field integral

$$\mathcal{Z}=\int[\mathrm{d}\phi(x)]\exp\left[-\mathcal{S}(\phi)\right]$$

with

$$\mathcal{S}(\phi)=\int\left\{\frac{1}{2}\left[\partial_\mu\phi(x)\right]^2+\frac{1}{2}r\phi^2(x)+\frac{u}{4!}\left[\phi^2(x)\right]^2\right\}\mathrm{d}^dx\,, \tag{4.1}$$

where $r=-2mT\mu$ and, for $d=3$, $u=96\pi^2a/\lambda^2$.

The euclidean action reduces to the ordinary $O(2)$ symmetric $(\phi^2)^2$ field theory, which also describes the universal properties of the superfluid Helium transition.

A UV large momentum cutoff $\Lambda\sim 1/\lambda\propto\sqrt{mT}$ is provided by the higher frequency modes. In the action, we have kept the dimension d arbitrary in order to use dimensional regularization later.

The goal is to obtain the leading order non-trivial contribution at criticality (in the massless theory) to

$$n = 4mT\rho\,, \quad \rho = \int^{\Lambda} \frac{\mathrm{d}^d k}{(2\pi)^d} \frac{1}{\tilde{\Gamma}^{(2)}(k)},$$

where $\tilde{\Gamma}^{(2)}(k)$ is the vertex two-point function, inverse of the two-point correlation function.

Because the interactions are weak, one may imagine calculating the change in the transition temperature by perturbation theory. However, the perturbative expansion for a critical theory does not exist for any fixed dimension $d < 4$. A discussion of this problem requires *RG arguments.*

Renormalization group equations. We introduce the dimensionless coupling constant (the large momentum cutoff Λ is proportional to $1/\lambda$)

$$g = \Lambda^{d-4} u \propto (a/\lambda)^{d-2} \ll 1\,.$$

At T_c, $\tilde{\Gamma}^{(2)}(k)$ satisfies the RG equation

$$\left(\Lambda \frac{\partial}{\partial \Lambda} + \beta(g) \frac{\partial}{\partial g} - \eta(g)\right) \tilde{\Gamma}^{(2)}(k, \Lambda, g) = 0\,. \tag{4.2}$$

For the $O(N)$ symmetric $(\phi^2)^2$ field theory, perturbative calculations yield

$$\beta(g) = -(4-d)g + \frac{N+8}{48\pi^2} g^2 + \mathcal{O}(g^3), \quad \eta(g) = \frac{N+2}{18} \frac{1}{(4\pi)^4} g^2 + \mathcal{O}(g^3).$$

We recall here that, in the framework of the $\varepsilon = 4-d$ expansion, the β-function has, below four dimensions, a non-trivial zero [19]

$$g^* = 48\pi^2 \varepsilon/(N+8) + \mathcal{O}(\varepsilon^2),$$

which governs the large distance behaviour of connected correlation functions. Precise estimates of the fixed point values and corresponding critical exponents can be found in [20].

Solution. Equation (4.2), together with dimensional analysis, implies that the vertex two-point function has the general form

$$\tilde{\Gamma}^{(2)}(p, \Lambda, g) = p^2 Z(g) F\big(p/\Lambda(g)\big) \tag{4.3}$$

with

$$\beta(g) \frac{\partial \ln Z(g)}{\partial g} = \eta(g), \quad \beta(g) \frac{\partial \ln \Lambda(g)}{\partial g} = -1\,.$$

Therefore,

$$Z(g) = \exp \int_0^g \frac{\eta(g')}{\beta(g')} \mathrm{d}g' = 1 + \mathcal{O}(g^2)$$

and

$$\delta\rho \underset{g\to 0}{\sim} \int^{\Lambda} \frac{\mathrm{d}^3 p}{(2\pi)^3} \frac{1}{p^2} \left(\frac{1}{F(p/\Lambda(g))} - 1\right).$$

On dimensional grounds $\Lambda(g)$ is proportional to Λ. The function $\Lambda(g)$ is then obtained by integration:

$$\Lambda(g) = g^{1/(4-d)} \Lambda \exp\left[-\int_0^g \mathrm{d}g' \left(\frac{1}{\beta(g')} + \frac{1}{(4-d)g'}\right)\right].$$

Crossover scale. The quantity $\Lambda(g)$ is a crossover scale separating a universal long-distance regime governed by the non-trivial zero g^* of the β-function, from a universal short distance regime governed by the gaussian fixed point $g = 0$. Such a regime exists only if $\Lambda(g) \ll \Lambda$, that is, if there is an intermediate scale between the IR and the microscopic scales; otherwise only the IR behaviour can be observed. In a generic situation g is of order unity and, thus, $\Lambda(g)$ is of order Λ: the universal large momentum region is absent.

Instead, $\Lambda(g) \ll \Lambda$ implies

$$g^{1/(4-d)} \exp\left[-\int_0^g \mathrm{d}g' \left(\frac{1}{\beta(g')} + \frac{1}{(4-d)g'}\right)\right] \ll 1 .$$

Since g (equal to a/λ for $d = 3$), is $\ll 1$, this condition is satisfied in the present situation.

Then,

$$\tilde{\Gamma}^{(2)}(p) \propto p^{2-\eta} \text{ for } p \ll \Lambda(g), \quad \tilde{\Gamma}^{(2)}(p) \propto p^2 \text{ for } \Lambda(g) \ll p \ll \Lambda,$$

with $\eta \equiv \eta(g^*) \approx 0.035$ for $N = 2$. *Linear behaviour.* We now show that the condition $\Lambda(g) \ll \Lambda$ implies $\Delta T_c \propto \Lambda(g)$.

First, from perturbation theory one infers that for $d = 3$, the function $F(p)$ (equation (4.3)) behaves for large p as

$$F(p) = \tilde{\Gamma}^{(2)}(p)/p^2 = 1 + \mathcal{O}(\ln p/p^2).$$

Therefore, the first correction to the density is convergent at large momentum and independent of the cutoff procedure:

$$\delta\rho = \int \frac{\mathrm{d}^3 p}{(2\pi)^3} \frac{1}{p^2} \left(\frac{1}{F(p/\Lambda(g))} - 1\right).$$

Similarly, the IR behaviour implies that this integral is IR convergent.

Setting $p = \Lambda(g)k$, one then finds the general form

$$\delta\rho = \Lambda(g) \int \frac{\mathrm{d}^3 k}{(2\pi)^3} \frac{1}{k^2} \left(\frac{1}{F(k)} - 1\right).$$

The g dependence is entirely contained in $\Lambda(g)$. For g small, one concludes

$$\frac{\delta\rho}{\rho} \propto g \sim (-3/2c_0) a n^{1/3} ,$$

a linear behaviour that, however, is non-perturbative! Moreover, the amplitude $-3c_0/2$ is universal.

5 The N-vector model. The large N expansion at order $1/N$

Since the function $F(p)$ cannot be obtained from a perturbative calculation, we consider the $O(N)$ symmetric generalization of the model corresponding to the euclidean action (4.1): the field $\phi(x)$ then has N real components ϕ_i, $i = 1, \ldots, N$.

The advantage of such a generalization is that it provides us with a tool, the large N expansion, which allows calculating at the critical point. The large N limit is taken at Nu fixed.

The first non-trivial correction to $\tilde{\Gamma}^{(2)}(p)$ appears only at order $1/N$ [18, 14, 11]:

$$\tilde{\Gamma}^{(2)}(p) = p^2 + \frac{2}{N} \int \frac{\mathrm{d}^d q}{(2\pi)^d} \frac{1}{(6/Nu) + B(q)} \left(\frac{1}{(p+q)^2} - \frac{1}{q^2}\right) + \mathcal{O}\left(\frac{1}{N^2}\right),$$

where $B(q)$ is the one-loop contribution (the bubble diagram) to the perturbative four-point function

$$B(q) = \int \frac{\mathrm{d}^d k}{(2\pi)^d} \frac{1}{k^2(k+q)^2} \underset{q\to 0}{\sim} b(\varepsilon) q^{-\varepsilon},$$

($\varepsilon = 4 - d > 0$) which is UV finite for $d < 4$. For $d = 3$, $b(1) = 1/8$.

One evaluates

$$\delta\rho = -\frac{2}{N} \int \frac{\mathrm{d}^d p}{(2\pi)^{2d}} \frac{1}{p^4} \frac{\mathrm{d}^d q}{(6/Nu) + b(\varepsilon) q^{-\varepsilon}} \left(\frac{1}{(p+q)^2} - \frac{1}{q^2} \right)$$

by keeping the dimension d generic and using dimensional regularization. This allows exchanging the order between the p and q integrations, something which is not allowed directly in dimension 3. In the $d = 3$ limit, the two integrations yield $(1/32\pi^2)(Nu/6)$.

As expected, $\delta\rho \propto u$:

$$\delta\rho = -u/96\pi^2 = -\frac{a}{\lambda^2}.$$

One finally obtains the change in the transition temperature [8]:

$$\begin{aligned} \frac{\Delta T_c}{T_c} &= \frac{8\pi}{3\zeta(3/2)} \frac{a}{\lambda} = c_0 a n^{1/3} \text{ with} \\ c_0 &= \frac{8\pi}{3\zeta(3/2)^{4/3}} = 2.33\ldots. \end{aligned}$$

For non-trivial reasons, the leading order result does not depend on N. The calculation involves a $1/N$ correction, but the result is proportional to u, which appears only in the combination uN.

Nevertheless, the result is valid only for N large. Taking into account, the $1/N$ correction [16], one finds $c_0 = 1.71\ldots$. A number of other methods, lattice calculations [13], summation of perturbative expansions, suggest $c_0 \approx 1.3$. These $1/N$ estimates have a precision typical of other $1/N$ calculations.

6 Conclusions

Using RG arguments, we have shown that the properties of the dilute, weakly interacting Bose gas remain dominated by the UV fixed point up to large length scales; this is why one can still refer to BE condensation when discussing the phase transition of the interacting Bose gas.

RG arguments enabled us to confirm that the relative shift of the transition temperature at fixed density is proportional to the dimensionless combination $an^{1/3}$ for weak interactions. This result is non-perturbative and the proportionality coefficient, which is *universal*, cannot be obtained from perturbation theory.

Therefore, a *non-perturbative method*, the large N expansion, has been introduced that allows a systematic, analytic, calculation of this coefficient as a power series in $1/N$, where eventually one has to set $N = 2$.

The leading order contribution is formally of order $1/N$ multiplied by a function of aN, which is kept fixed in the large N limit. Because for $d = 3$ the result is linear in a, the $1/N$ factor somewhat surprisingly cancels and the result is independent of N. Adding the 30% $1/N$ correction, one finds a value in reasonable agreement with the most recent numerical estimates.

References

[1] T.D. Lee and C.N. Yang, *Phys. Rev.* **112**, 1419 (1957).

[2] K. Huang, in *Stud. Stat. Mech*, **II**, J. de Boer and G.E. Uhlenbeck, eds. (North Holland Publ., Amsterdam, 1964), 1.

[3] T. Toyoda, *Ann. Phys.* (NY) **141**, 154 (1982).

[4] H.T.C. Stoof, *Phys. Rev.* **A45**, 8398 (1992); M. Bijlsma and H.T.C. Stoof, *Phys. Rev.* **A54**, 5085, 1996.

[5] P. Grüter, D. Ceperley, and F. Laloë, *Phys. Rev.* Lett.**79**, 3549, 1997.

[6] M. Holzmann, P. Grüter, and F. Laloë, cond-mat/9809356, *Euro. Phys. J.* B (in press).

[7] G. Baym, J.-P. Blaizot, M. Holzmann, F. Laloë, and D. Vautherin, *Phys. Rev. Lett.* 83 (1999) 1703 [cond-mat/9905430].

[8] G. Baym, J.-P. Blaizot and J. Zinn-Justin, *Euro. Phys. Lett.* 49 (2000) 150, [cond-mat/9907241];

[9] H.E. Stanley, *Phys. Rev.* 176 (1968) 718; R. Abe, Prog. Theor. Phys. 48 (1972) 1414; *ibidem* 49 (1973) 113; S.K. Ma, *Phys. Rev. Lett.* 29 (1972) 1311, *Phys. Rev.* A7 (1973) 2172; M. Suzuki, *Phys. Lett.* 42A (1972) 5, *Prog. Theor. Phys.* 49 (1973) 424; K.G. Wilson, *Phys. Rev.* D7 (1973) 2911.

[10] See also the contributions of S.K. Ma and E. Brézin, J.C. Le Guillou and J. Zinn-Justin to *Phase Transitions and Critical Phenomena* vol. 6, C. Domb and M.S. Green eds. (Academic Press, London 1976).

[11] For a review see M. Moshe, J. Zinn-Justin, *Quantum field theory in the large N limit: a review*, *Phys. Rept.* 385 (2003) 69, [hep-th/0306133];

[12] J. Ashmore, *Lett. Nuovo Cimento* 4 (1972) 289; G. 't Hooft and M. Veltman, *Nucl. Phys.* B44 (1972) 189; C.G. Bollini and J.J. Giambiaggi, *Phys. Lett.* 40B (1972) 566, *Nuovo Cimento* 12B (1972).

[13] N. Prokof'ev, B. Svistunov, V.A. Kashurnikov, [cond-mat/0103149]; N. Prokof'ev, B. Svistunov, [cond-mat/0103146]; P. Arnold, G.D. Moore, *Phys.Rev.* E64 (2001) 066113; *Erratum-ibid.* E68 (2003) 049902 [cond-mat/0103227]; P. Arnold, G.D. Moore, *Phys.Rev.Lett.* 87 (2001) 120401 cond-mat/0103228; see also M. Holzmann, G. Baym, J.-P. Blaizot F. Laloë, *Phys. Rev. Lett.* 87, 120403 (2001) [cond-mat/0103595].

[14] J. Zinn-Justin, *Quantum field theory and critical phenomena*, Oxford Univ. Press (Oxford 1989, fourth ed. 2002).

[15] J. Zinn-Justin, *Path integrals in quantum mechanics*, Oxford Univ. Press (Oxford 2004).

[16] P. Arnold and B. Tomasik, *Phys. Rev.* A 62 (2000) 063604 [cond-mat/0005197]; see also P. Arnold, G.D. Moore and B. Tomasik, *Phys. Rev.* A65 (2002) 013606 [cond-mat/0107124].

[17] J. O. Andersen, *Theory of the weakly interacting Bose gas, Reviews of Modern Physics* 76 (2004) 599, [cond-mat/0305138] and enclosed references.

[18] R.A. Ferrel and D.J. Scalapino, *Phys. Rev. Lett.* 29 (1972) 413; A. Aharony, *Phys. Rev.* B10 (1974) 2834.

[19] The idea of the ε-expansion is due to K.G. Wilson and M.E. Fisher, *Phys. Rev. Lett.* 28 (1972) 240.

[20] Precise estimates of critical exponents have been reported in J.C. Le Guillou and J. Zinn-Justin, *Phys. Rev. Lett.* 39 (1977) 95; *Phys. Rev.* B21 (1980) 3976; more recent improved estimates are found in R. Guida and J. Zinn-Justin, *J. Phys. A* 31 (1998) 8103 [cond-mat/9803240].

[21] J.-L. Kneur, M.B. Pinto, and R.O. Ramos, *Phys. Rev. Lett.* 89, 210403 (2002) [cond-mat/0207089]; B. Kastening, *Phys. Rev.* A 69, 043613 (2004) [cond-mat/0309060].

Titles in This Series

24 **Edward Bierstone, Boris Khesin, Askold Khovanskii, and Jerrold E. Marsden, Editors,** The Arnoldfest: Proceedings of a conference in honour of V. I. Arnold for his sixtieth birthday, 1999

23 **Myrna H. Wooders, Editor,** Topics in mathematical economics and game theory. Essays in honor of Robert J. Aumann, 1999

22 **Graciela Chichilnisky, Editor,** Topology and markets, 1999

21 **Shigui Ruan, Gail S. K. Wolkowicz, and Jianhong Wu, Editors,** Differential equations with applications to biology, 1998

20 **Peter A. Fillmore and James A. Mingo, Editors,** Operator algebras and their applications II, 1998

19 **William G. Dwyer, Stephen Halperin, Richard Kane, Stanley O. Kochman, Mark E. Mahowald, and Paul S. Selick (Editor-in-Chief), Editors,** Stable and unstable homotopy, 1998

18 **Panos M. Pardalos and Henry Wolkowicz, Editors,** Topics in semidefinite and interior-point methods, 1998

17 **Joachim J. R. Cuntz and Masoud Khalkhali, Editors,** Cyclic cohomology and noncommutative geometry, 1997

16 **Victor P. Snaith, Editor,** Algebraic K-theory, 1997

15 **Stephen P. Braham, Jack D. Gegenberg, and Robert J. McKellar, Editors,** Sixth Canadian conference on general relativity and relativistic astrophysics, 1997

14 **Mourad E. H. Ismail, David R. Masson, and Mizan Rahman, Editors,** Special functions, q-series and related topics, 1997

13 **Peter A. Fillmore and James A. Mingo, Editors,** Operator algebras and their applications, 1997

12 **Dan-Virgil Voiculescu, Editor,** Free probability theory, 1997

11 **Colleen D. Cutler and Daniel T. Kaplan, Editors,** Nonlinear dynamics and time series: Building a bridge between the natural and statistical sciences, 1997

10 **Jerrold E. Marsden, George W. Patrick, and William F. Shadwick, Editors,** Integration algorithms and classical mechanics, 1996

9 **W. H. Kliemann, W. F. Langford, and N. S. Namachchivaya, Editors,** Nonlinear dynamics and stochastic mechanics, 1996

8 **Larry M. Bates and David L. Rod, Editors,** Conservative systems and quantum chaos, 1996

7 **William F. Shadwick, Perinkulam Sambamurthy Krishnaprasad, and Tudor Stefan Ratiu, Editors,** Mechanics day, 1996

6 **Anna T. Lawniczak and Raymond Kapral, Editors,** Pattern formation and lattice gas automata, 1996

5 **John Chadam, Martin Golubitsky, William Langford, and Brian Wetton, Editors,** Pattern formation: Symmetry methods and applications, 1996

4 **William F. Langford and Wayne Nagata, Editors,** Normal forms and homoclinic chaos, 1995

3 **Anthony Bloch, Editor,** Hamiltonian and gradient flows, algorithms and control, 1994

2 **K. A. Morris, Editor,** Control of flexible structures, 1993

1 **Michael J. Enos, Editor,** Dynamics and control of mechanical systems: The falling cat and related problems, 1993